Assessment of Groundwater Resources and Management

Assessment of Groundwater Resources and Management

Editors

AL. RAMANATHAN
PROSUN BHATTACHARYA
A.K. KESHARI
JOCHEN BUNDSCHUH
D. CHANDRASEKHARAM
S.K. SINGH

TECHSAR™
NEW DELHI

Published and Published by
Techsar Pvt. Ltd.
4435-36/7, Ansari Road, Daryaganj
New Delhi-110 002 (India)
E-mail: tech_info@techsarworld.com
Website: www.techsarworld.com

ISBN: 978-81-90675-72-7

Reprint 2025

Preface

This volume is addressing the issue to facilitate the joint analysis of groundwater management studies and problems by scientist, engineers, managers and other scholars from natural and applied sciences. The geological aspect of groundwater science is universal and applied in nature to have a sustainable water resource development with social, economic, ecological, cultural and aesthetic background. Since 99% of the world's fresh available water is groundwater; yet, the vast majority of financial resources are directed to surface water found in rivers and lakes. This serious imbalance requires urgent redress. Significant financial support is required for basic groundwater research if sustainable development is to be a realistic goal. As a fresh water resource, groundwater has major advantages over surface water. This is a basic idea that explicitly appears in almost all paper of this book. The authors have tried to focus their task on those topics that seemed to us more urgent and relevant and have paid much attention to questions related to management of aquifers or to the chemical pollution groundwater resources. The long-term problem and the key issue in developing countries, where majority of world population live where at present enormous groundwater abstraction occurs.

Groundwater problem is of several folds, starting from inorganic pollution to organic pollution of available ground water and drinking water along with the problem of available quantity of ground water particularly in the arid and semiarid countries. The intensive over use of groundwater in the recent past in most developing countries in large part of the world is a well known fact. Therefore, adequate knowledge and governmental initiatives is urgently needed for its quantity and quality management high-level water decision-makers of respective countries making best use of the available scientific and technological advances in hydrogeology and related disciplines in this changing social-economic dimension of the fast moving world.

The different authors have put forth different approach and views on the issues and discussed here and we (editors) here acted as facilitators to the dissipate proper information in a systematic scientific manner confirmed by authentic ground truth/facts to audience and leave it to the best interpretation and use of this data by scientific community and planers working on hydrogeology and make their choice. However, we try to make it understandable to every one the concept of groundwater management and sustainability through this book.

The International Conference "IGC-2006" on Groundwater held in New Delhi in February 2006 was conceived in order to provide a platform to bring together earth scientists, professionals from chemical and engineering science disciplines, public health professionals and social scientists involved with the management and development of groundwater resources. The merit of the contributions will be judged through a standard peer-review process in order to ensure high scientific quality of the publication. The book is expected to reflect the state of art of current understanding of all these issues related to management of groundwater resources and its sustainable use. This book is brought out as a Special Publication of the INternational Soceity of Groundwater for Sustainable Development (ISGSD).

The organizers of the IGC-2006 would like to thank the Swedish International Development Cooperation Agency (Sida) in New Delhi for the final support to the organization of the conference. We deeply appreciate the support we have received from the Vice Chancellor of the Jawaharlal Nehru University, New Delhi as a co-organiser of the conference. Further, we would also like to thank the UNICEF, New Delhi, India especially Dr. Paul Devrill and Ross Nickson, the Department of Science and Technology, Ministry of Environment and Forest, Government of India and Dr. Bhanu Neupune from the UNESCO, New Delhi, India for their financial support to organize the conference. We would also like to thank Prof. Dr. Anders Flodström, the former President of the Royal Institute of Technology (KTH) and the present President Prof. Dr. Peter Gudmundson for their keen interest in this important conference and the financial support. Last but not the least, we do express our deep sense of gratitude to Mr. Krishan Makhijani and the entire editorial team of the I.K. International Publishing House Pvt. Limited during the production of this book.

AL. Ramanathan (India)
Prosun Bhattacharya (Sweden)
A.K. Keshari (India)
Jochen Bundschuh (Costa Rica)
D. Chandrashekharam (India)
S.K. Singh (India)

The Editors

AL. RAMANATHAN is an Professor in the School of Environmental Sciences, Jawaharlal Nehru University, and New Delhi, India. His research area is in the field of Hydrogeochemistry of the ground water in urban, hard rock and coastal groundwater with application of computer graphics and statistical models and hydro geochemical models to ground water resource management. He has been teaching the ground water courses to PG students and research students for the past two decades. He has taught in numerous universities in India and abroad. He has guided few research students for their research degrees on ground water modeling aspects and guiding few more at present on these aspects for their research degrees from India and abroad. He has widely travelled all over the world for the advanced research work in ground water. He is also a recipient of various international and national scholarship including STA Japan, Linnaeus-Palme Exchange Programme, International Programme Office, Sida, Sweden, IFS, Sweden, Indo-Australian, Indo-Russian, ILTP (DST) Russia, SASNET and Sida, Sweden to have interactions and collaborations with other eminent scientists working in this field. He has authored numerous articles in reputed refereed journals and authored five books in these aspects.

PROSUN BHATTACHARYA received his Ph.D. degree in Petrology and Sedimentary Geochemistry from the University of Delhi, India in 1990. He joined Kungliga Tekniska Högskolan in 1993 and started his carrier as a Research Scientist and worked on the genesis of high fluoride groundwater in Western India. He developed expertise on solid phase characterization through detailed soil chemical studies and their bearing on the high fluoride groundwater in the region. Since 1995, he has been working on heavy metal contaminated soils which have resulted from different industrial activities, particularly the wood preservations sites contaminated by CCA, ammunition industries, and shooting in Sweden along with the studies on chemical remediation. Since 1995, he is actively engaged with the studies on the genesis of high arsenic in groundwaters of the sedimentary aquifers in the Bengal Delta Plain, West Bengal and Bangladesh and devising sustainable techniques for safeguarding groundwater resources. At present his primary research interest is directed towards the understanding the mechanisms of arsenic mobilization in groundwaters of the sedimentary aquifers based on hydrogeochemical, sediment geochemical and stable isotopic characteristics in Bengal Delta Plain (BDP), Huhhot Alluvial Basin (HAB), Mekong Delta (MD) and the Chaco-Pampean Plain of Argentina. He established the Groundwater Arsenic Research Group (GARG) at the Department of Land and Water Resources Engineering, KTH in 1999, a group which is dedicated on research on the groundwater arsenic occurrences in different parts of the world. At present GARG is collaborating with a number of universities and research organizations in India, Bangladesh, China, Australia, Argentina, Costa Rica, Bolivia, Pakistan and Ghana. He has received the title as a Docent (Associate Professor) in Groundwater Chemistry at the KTH in April 2002. Prosun has authored nearly 170 international publications in peer reviewed journals, books and conference proceedings and has organized a number of international workshops on Arsenic in Groundwaters of Sedimentary Aquifers.

ASHOK K. KESHARI is an Associate Professor of Water Resources Engineering and Geoinformatics at Department of Civil Engineering, Indian Institute of Technology Delhi. He has earlier worked as Scientist at National Institute of Hydrology, Roorkee and in North Eastern Electric Power Coorporation, Shillong and at IIT Kanpur for short periods. He has been Postdoctoral Fellow and Visiting Researcher at Kongju National University, Korea in years 2001-2002, and has also visited Japan, USA and Denmark. He specializes in hydrogeology, groundwater flow and pollution modeling, remote sensing & GIS, hydrologic simulation & design, FEM & optimization, environmental & river hydraulics, hazard & risk assessment and project economics. Dr. Keshari has published more than 120 papers and has carried out many real life projects for UP Bhumi Sudhar Nigam, Tata Infotech Limited, CGWB, WAPCOS, SASE, DRDO, ISRO, DST, CES, MCD, Pondicherry PWD, ORG Informatics, Haryana Distillery, etc., some of them are World Bank funded projects. He is the member of Indian Association of Hydrologic Sciences (IAHS), Indian Association of Hydrologists (IAH) and is also member of Bureau of Indian Standards (BIS), Indo French Unit for Water and Waste Technologies (IFUWWT), etc. He has been the reviewer of many international and national journals.

JOCHEN BUNDSCHUH completed his PhD on numerical modelling of heat transport in aquifers in Tübingen in 1990. For over 14 years he is working in international academic and technical co-operation in different fields of sustainable use of groundwater systems and connected disciplines, including the water-related economic, social, health, and political aspects. The last 12 years he spent in different Latin-American countries, where his main tasks comprised capacity building for institutional strengthening and human resources development, applied research, and project management. The countries comprise: Paraguay, Argentina, Brazil, Uruguay, Brazil, Mexico, Bolivia, Costa Rica, Honduras, Guatemala, Panama, Pakistan, India, Bangladesh, Middle East, Tunisia and South Africa. From 1993 to 99 he served as expert of the German Agency of Technical Cooperation (GTZ) and as long-term professor for the DAAD (German Academic Exchange Service) in Argentine to install a centre for water and environment offering training, consulting and research. Currently (since 2001) he is integrated expert of the GTZ/CIM programme for technical cooperation between Germany and Costa Rica and deals with the sustainable use of surface and groundwater resources for power generation. This mission comprises to develop and apply capacity building programmes, research, and to identify and formulate project proposals. Additionally to these principal activities, he is involved in many consultancies, capacity building programmes and projects in Latin-America, but also in Africa and Asia, which are comprising-in between others-research of groundwater contamination by arsenic originating from geogenic resources.

D. CHANDRASEKHARAM is a Professor of Geology in the Department of Earth Sciences, and Head, Centre of Studies in Resources Engineering, Indian Institute of Technology Bombay. He is an active researcher involved in research related to arsenic and fluoride pollution in groundwater in India. Over the last 20 years his research interests include geochemistry of natural and thermal waters and gases. He has been nominated as a "Senior Associate" to The Abdus Salam International Centre for Theoretical Physics, Trieste, Italy for a period of five years from January 2002.

PROF. S.K. SINGH is currently the Head of Civil and Environment Engineering of Delhi College of Engineering. He has guided, several Ph.D scholars, 39 M.E. thesis and more than hundred UG projects and has undertaken R&D projects sponsored by UGC and DST, GOI. He is member of several high level committees of Ministry of Environmental and Forests, Ministry of Rural Development (Govt. of India) CAPART, AICTE, BIS, etc. He is life member and office bearer of professional bodies such as Institute of Engineers (I), Indian Water Works Association, New York Academy of Sciences, New York American Society of Civil Engineers Indian Section, Indian Association of Environmental Management, Nagpur, Indian Science Congress Association, Calcutta Council for National Health Development, New Indian Society for Technical Education, New Delhi.

Contents

Groundwater Scenario and its Management in Shillong Urban Agglomerate, Meghalaya

1

A.K. Biswal[1] and P. Kalita[2]

[1]Central Groundwater Board (SER), Bhubaneswar, Orissa, India

[2]Central Groundwater Board (SUO), Shillong, Meghalaya, India

ABSTRACT

Greater Shillong Water Supply Scheme (GSWSS) is the only major water supply scheme which supplies treated surface water to Shillong urban agglomerate. There is no major groundwater supply scheme. About 80% of the total households are dependent on supplied surface water for their domestic purpose. Recent valley fills (in Polo area) and quartzites constitute the major aquifer system in the study area. Two sets of fractures are observed within 100 m depth. The average seasonal water level fluctuation is around 1.19 m whereas the long-term water level does not show any significant trend. Most of the bore wells drilled in the area yield 5 m^3/hr to 10 m^3/hr. The depth at which water struck lies between 2.43 m to 100 m. The high yielding bore wells are generally associated with NE-SW trending lineaments and the intersection of NE-SW and NW-SE lineaments. Most of the springs in the study area are of gravitative type with discharge varying from 8100 lpd to 187488 lpd*. The springs at Umkdait, Lengkarding, Pynther, and Madanrting indicate a very close correspondence to rainfall. The low yielding springs are least affected by rainfall. The stage of groundwater development is 69.25% and qualitatively the groundwater is potable.

MLD* = Million liters per day, *lpd = Liters per day, msl = mean sea level, m bgl = metre below ground level

1 INTRODUCTION

Shillong, the state capital and district headquarters of East Khasi Hills district, Meghalaya is situated at the height of about 1500 m above mean sea level (msl). It is famous for its soothing and healing environment and is considered one of the best health resort and is better known as "Scotland of the east".

The geographic confines of the Shillong urban agglomerate stretches from Umshorshar in the north, Madanrting in the south and Umpling in the east to Upper Shillong in west. The area lies between north latitudes 25° 33′15″ to 25° 36′ 02″ and east longitude 91° 50′50″ to 91° 56′17″ and falls in the survey of India toposheet no. 78 O/14.

Shillong urban agglomerate is divided in 7 administrative divisions. The population of the area was 267881 (2001 census) spread over an area of 25.4 sq. km. The following table (Table 1) gives the detail of the population in different divisions of Shillong urban area. Table 2 represents the estimated population of different administrative divisions from 2004 to 2010.

Table 1 Population in Shillong urban area (town-wise and 2001 census).

S.N.	*Units/Towns*	*Male*	*Female*	*Total*
1.	Mawlai	18411	19830	38241
2.	Nongthymmai	17015	17194	34209
3.	Madanrting	8478	8222	16700
4.	Shillong Municipality	66129	6674	132876
5.	Shillong Cantonment	7069	5316	12385
6.	Pynthormukhrah	11387	10721	22108
7.	Nongmensung	5927	5435	11362
	Total in Shillong U.A.	34416	133465	267881

Source: District level statistics, 2002, Directorate of Economics and Statistics.

Table 2 Estimated population in Shillong U.A. (town wise) from 2004 to 2010.

Sl. No.	*Units/Towns*	*2004*	*2005*	*2006*	*2007*	*2008*	*2009*	*2010*
1.	Mawlai	40701	41556	42429	43320	44230	45159	46107
2.	Nongthymmai	36732	37614	38517	39441	40388	41357	42350
3.	Madanrting	20059	21323	22666	24094	25612	27226	28941
4.	Shillong Municipality	133194	133301	133408	133515	133662	133729	133836
5.	Shillong Cantonment	12798	12939	13081	13225	13370	13517	13666
6.	Pynthermukhrah	25519	26769	28081	29457	30900	32414	34004
7.	Nongmensung	11915	12114	12316	12522	12731	12944	13160
	Total in Shillong U.A.	280918	285616	290498	295574	300893	306346	312064

Source: Statistical handbook, Meghalaya, 2003, Directorate of Economics and Statistics.

Shillong owes its importance due to its strategic location. About 55% of the total workers of the town are engaged in non-agricultural activities. The percentage of workers engaged in industry is 10.5, in trade is 17.7 and in transport communication etc. is 8. So far morphology of the city is concerned it is found that the residential areas account for 74% of the total area of the city and defense 14%. The commercial areas occupies 3.5% of the city.

Shillong U.A. is not well developed from industrial point of view.

2 HYDROMETEROLOGY

The area experiences tropical cool climate which is directly, influenced by the south-west monsoon originating from the Bay of Bengal and the Arabian Sea. Though it rains intermittently for the whole year, but the wettest period of the year is mid-May to mid-October and about three-fourths of the total rainfall is received during this period only. The average annual rainfall (1991 – 2002) in Shillong U.A. is 2241.85 mm and ranges from 1676.7 mm in 1994 to 2587.7 mm in 1991. The minimum and maximum ambient temperature is 5° to 25°C. During south-west monsoon period the relative humidity is 85%.

3 GEOMORPHOLOGY

The area can be described as a part of central upland zone which falls in the central Shillong plateau region with an average altitude of 5000 ft (1524 m) (msl). The hill ranges with intervening narrow valleys constitute the main physiographic features in the area. The rivers Umkhrah, Umshirpi and Umpling with numerous other minor streams flowing towards east, north and north-north-east respectively mainly drain the area.

4 PRESENT WATER SUPPLY AND DEMAND

With the increased urbanization during the last two decades there has been a lot of thrust on water resources in the urban area. Consequently, there is a big gap between water supply and demand. During the dry period the situation becomes worse. The demand for water may further increase in future since the limits of city are expanding along with the increase in population.

The principal source of water supply in Shillong U.A. is surface water by Greater Shillong Water Supply Scheme (GSWSS). There is no major groundwater scheme in the area to supply water for drinking and other domestic purposes. PHED and Shillong Municipality Board also play a role to some extent in this regard. Table 3(A) explains the total water supply in Shillong urban agglomerate and Table 3(B) town-wise.

Table 3 (A) Total water supply in Shillong U.A.

Sl.No.	*Supplier*	*Total water supply (MLD*)*
1.	GSWSS	22.46
2.	PHED	5.272
3.	Shillong Municipality Board	1.35
	Total in Shillong U.A.	29.082

Source: EE (Monitoring cell), PHED, Govt. of Meghalaya, Shillong.

Table 3 (B) Town-wise supply in Shillong U.A. in 2004.

Sl.No.	*Town/Units*	*Total water suply(MLD)*
1.	Mawlai	2.89
2.	Nongthymmai	0.9
3.	Madanrting	0.2
4.	Shillong Municipality	23.32
5.	Shillong Cantonment	0.23
6.	Pynthormukhrah	1.1
7.	Nongmensung	0.442

Source: EE (Monitoring cell), PHED, Govt. of Meghalaya, Shillong.

From the above tables it is observed that groundwater forms only a minor part of the requirement of the city. Table 4 explains the present demand, supply and shortfall.

Table 4 Present demand, supply and shortfall.

Town/Unit	*Population in 2001*	*Projected population in 2004*	*Per capita supply (LPCD)*	*Present demand (MLD)*	*Present supply (in2004-05 (MLD)*	*Scarcity (MLD)*	*Present per capita supply* (LPCD)*
Mawlai	16700	40701	135	5.50	2.89	2.61	71
Nongthymmai	34209	36732	-do-	4.85	0.9	3.95	24.5
Madanrting	22108	20059	-do-	2.71	0.2	2.51	10
Shillong Municipality	132876	133194	-do-	17.98	23.32	Surplus (5.34)	175
Shillong Cantonment	12385	12798	-do-	1.73	0.23	1.5	18
Pynthermukhrah	38241	26769	-do-	3.61	1.1	2.51	41
Nongmensung	11371	11915	-do-	1.60	0.44	1.15	37
Total in Shillong U.A.	267881	282168	-do-	38.1	29.08	9.02	103

Source: Statistical handbook, 2003 & EE (Monitoring cell) Deptt.of PHED, Govt. of Meghalaya.

The present supply through surface water source is only 29.082 as against the total requirement of 38.1 MLD leaving a deficit of 9.02 MLD. To bridge this gap mostly spring water and in Polo area groundwater from the dug wells is used to augment the water supply. The average per capita water supply is 103 LPCD. However, in Shillong municipality area there is surplus of water supply of 5.34 MLD.

It is observed that (2001 census) about 80% of the total households use surface water for drinking and other domestic purposes. Only 10.11% are dependent on spring and 7.26% depend on groundwater. As per the projected estimates, the demand of water supply in 2010 will be approximately 41 MLD.

5 HYDROGEOLOGY

The area forms a part of the ancient plateau of pre-Cambrian Indian peninsular shield. The main rock type is quartzite, phyllite, conglomerate with acidic and basic intrusive belonging to Shillong group of rocks of pre-Cambrian age. The quartzite rocks are fine to coarse grained and massive in nature. The strike of these quartzite rocks is in the NE-SW with a dip angle varying from 25° to 35° towards SE. They are jointed mainly in two directions, namely, E-W and N-S. The rocks are highly weathered being grayish white to yellowish white in colour. Recent valley fill deposits occur mainly along the channels of major streams of the area.

In general, there are two trends of lineaments i.e. NE-SW, which is mostly prominent, and NW-SE (Fig. 1).

Quartzites and recent valley fills (in Polo area) constitute the major aquifer system in the area. Groundwater occurs under unconfined condition in the weathered rock and residuum. The movement and occurrence of groundwater is controlled by secondary porosity created by weathering and fractures. Weathering of 1m to 25 m is commonly noticed in the area. Further groundwater occurs under semi-confined condition especially in the fractured rocks, which is seen below the weathered rock. Groundwater development in the urban agglomeration is both by dug wells generally confining to the weathered zone and bore wells, which mainly tap, fractured zone in the hard rocks. These fractures sometime extend very deep occurring even beyond 60 m bgl, but otherwise generally close before 60 m depth. Geophysical resistivity investigations carried out at various locations in Shillong U.A. with the help of ABTM

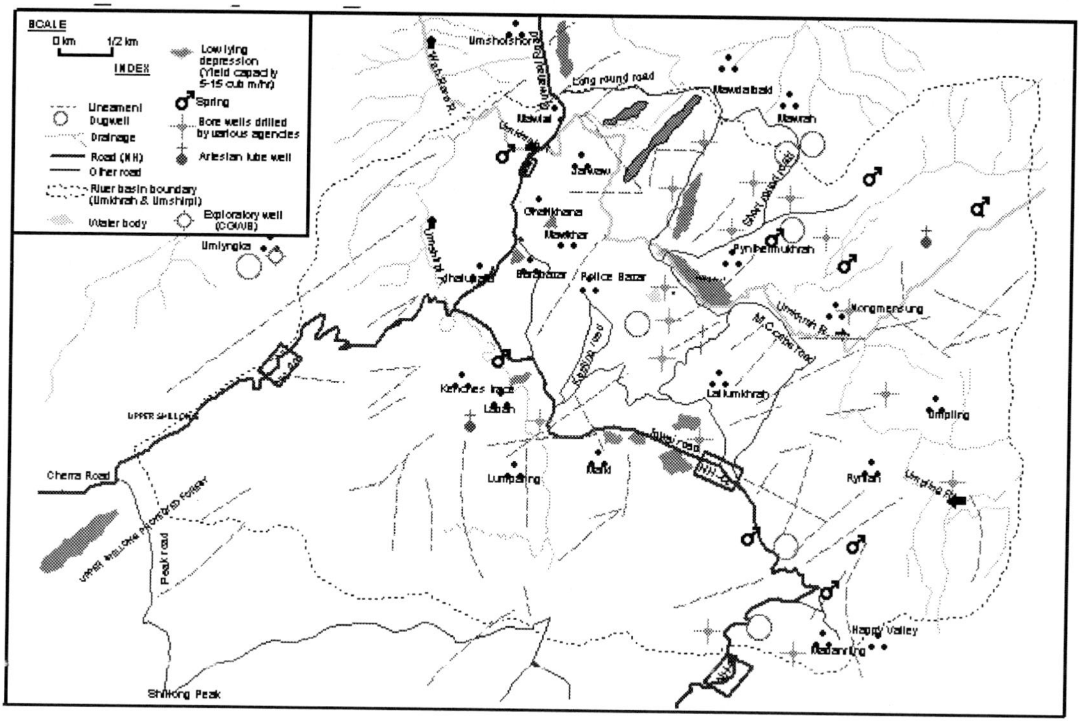

Fig. 1 Hydrogeological map of Shillong urban agglomerate.

Terrameter reveals that the thickness of top soil cover ranges from 1 m – 8.5 m and the thickness of weathering zone ranges from 6.16 m bgl to 41.2 m bgl. There are generally two sets of fractures within 100 m bgl. The 1st set of fracture is present within 30 m bgl and 2nd set is within 100 m bgl. Sometimes 3rd set is also present which is within 120 m bgl.

The mode of groundwater abstraction by dug wells is only observed in polo valley area and is mostly for domestic purposes. The depth to water level in dry period varies from 0.5 to 5.56 m bgl and in post-monsoon period, it is from 0.15 to 5.1 m bgl. The average seasonal water level fluctuation is of the order of 0.35 m to 2.95 m whereas the long-term water levels do not show any significant trend.

The piezometric head in dry period varies from 2.63 m bgl to 17.20 m bgl. In post-monsoon period it is from 1.97 m bgl to 14.32 m bgl at the same locations respectively. Two numbers of tube wells yield free flow with a discharge ranging from 0.13 to 3.6 m^3per hour respectively.

From the bore wells drilled by the PHED and other private agencies in the area, it has been observed that with depth ranges from 31.08 to 126 m bgl with the maximum number of tube wells located in Pynthermukhrah. In these wells weathered thickness varies from 3.04 to 41.69 m while discharge varies from 0.9 to 50 m^3/hr with maximum wells yielding 5 to 10 m^3/hr. The depth at which water struck lies is between 2.43 m to 100 m.

It has been observed that the high-yielding bore wells are generally associated with NE-SW trending lineaments and the intersection of NE-SW and NW-SE lineaments and mostly located in polo valley area. Fractured quartzite is the only aquifer in the area.

Spring plays a major role for the water requirement of the people. Hydrogeological studies suggest that the lineaments produced by joints, fractures, and faults play a significant role on the hydrogeological regime of a catchment. Spring discharge is controlled by rainfall, land use, vegetation, grazing incidence and geomorphology of the recharge zone. Geologically, the springs in the study area are of gravitative type. The discharge of the spring in dry period varies from 8100 lpd to 51,840 lpd and in wet period, it varies from 11,293 lpd to 1,87,488 lpd. With the above observation the hydrogeological map of the Shillong urban agglomerate has been prepared and presented in Fig. 1.

It has been observed from the discharge pattern (Fig. 2) in relation to rainfall that the discharge of the spring declined sharply at Umkdait, Lengkarding, Pynther, and Madanrting indicating a very close correspondence to rainfall. A low to moderate decline in discharge was recorded in other locations. These springs were least affected by rainfall during post-rainy season. It has also been observed that the low yielding springs are least affected by rainfall in the study area.

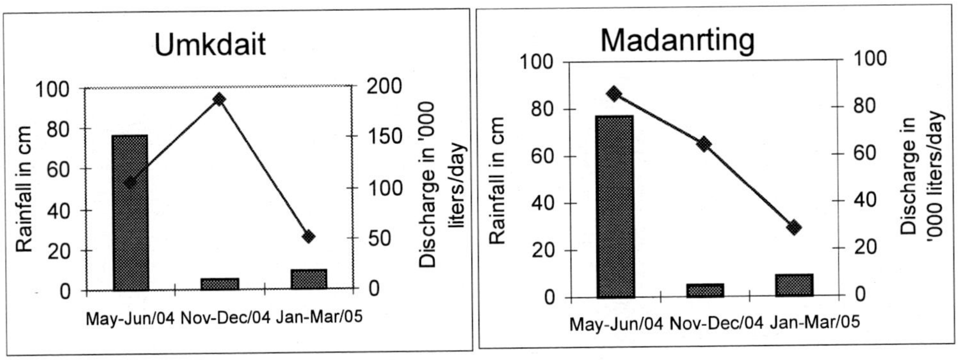

Fig. 2 Discharge pattern of springs.

6 GROUNDWATER RESOURCE

A preliminary assessment of groundwater resource indicates that the total groundwater resource is 41.8 HaM and annual groundwater draft is 20.95 HaM thus calculating the stages of groundwater development to 69.25%.

7 GROUNDWATER QUALITY

The evaluation of groundwater quality is very important as it is usually determined by its chemical characteristics. The quality of groundwater depends upon the nature of rock formation, physiography, soil environment, recharge and discharge conditions of the area. The chemical quality is studied through sampling nine water samples from various sources including dug wells tube wells and springs.

It has been observed that the pH value of groundwater varies from 4.8 to 7.8 indicating slightly acidic to alkaline nature. EC varies from 50 to 361 μs/cm at 25°C indicating very low to moderate mineralisation. The carbonate is totally absent while bicarbonate is found to vary from 12 to 159 mg/l. The chloride content varies from 7.1 to 35 mg/l and fluoride content from 0.09 to 0.66 mg/l. Total hardness (TH) varies from 10 to 125 mg/l while Fe varies from 0.03 to 8.4 mg/l. Therefore, all the constituents except Fe at Nongmensung and pH at Pynther are well within the permissible limits fixed by BIS (IS-10500-91). The groundwater at Pynther is slightly acidic and at Nongmensung, which is a free flow tube well, Fe content is more than the prescribed limit.

In general, there is no chemical hazard for drinking water and qualitatively the water is fresh and potable.

8 CONCLUSIONS AND RECOMMENDATIONS

Shillong, better known, as "Scotland of the east" is the state capital of Meghalaya and is situated at the height of about 1500 m above msl. The principal source of water supply in Shillong urban agglomerate is surface water by GSWSS. Quartzites and recent valley fills (in Polo area) constitute the major aquifer system in the area. Most of the bore wells yield low to moderate discharge with some exceptional high yielding bore wells, which are generally associated with NE-SW and the intersection of NE-SW and NW-SE trending lineaments. There are generally two sets of fractures within 30 and 100 m bgl. Geologically, the springs in the study area are of gravitative type. The discharge of the spring is controlled by rainfall, land use, vegetation, grazing incidence and geomorphology of the recharge zone. The low yielding springs are least affected by rainfall in the study area. Each and every spring in the area should be developed fully to mitigate the water scarcity in the study area. The Fe content in groundwater at Nongmensung is more than the prescribed limit. However, in overall, the groundwater of the study area is potable. As the stages of groundwater development reaches 69.25%, caution may be exercised in planning future development, with regard to quantum of additional groundwater withdrawal. Only PHED should be allowed for further groundwater exploitation. Proper identification of the recharge area is necessary for recharge to deeper aquifers. Monitoring of piezometric head of deeper aquifer to assess the impact of heavy withdrawal of groundwater need to be taken up. Dug wells with diameter (2 – 3 m) and depth (10 – 12 m) in low-lying depression area is recommended for groundwater extraction. Even when the area receives good rainfall throughout the year, there is very low recharge to the groundwater. It is due to the fact that the availability of the recharge area is very less. Therefore, construction of artificial recharge structures such as check dams, ponds, channels, contour ditches, bunds, and sub-surface dykes at suitable sites where plenty of open spaces are available across small streams which drain into big rivers, is needed to be taken up to augment the groundwater.

Acknowledgements

The authors are very much grateful to Dr. Saleem Romani, Chairman, Central Groundwater Board for his kind permission for preparation and presentation of this paper. The authors are also thankful to Mr. A.R. Bakshi, the Regional Director, C.G.W.B., NER, Guwahati for his general guidance and supervision and to Mr. S. Chakladar, Regional Director, C.G.W.B., SER,

Bhubaneswar for his kind support during the paper preparation work. The authors are also very thankful to the various state govt. departments for their support in providing various data during the survey. The authors are also indebted to Mr. P.K. Mohapatra, Sc.D, Mr. N.C. Nayak, Sc.B and Mr. P.K. Nayak, Sc.B, CGWB, SER, Bhubaneswar for their help extended during the preparation of the paper.

References

Census of India final population Totals, 2001.

District level statistics, 2002, Directorate of Economics and Statistics, Govt. of Meghalaya.

Groundwater Resource Estimation Methodology, 1997, Central Groundwater Board, Ministry of Water Resources, New Delhi.

Negi, G.C.S. and Joshi, Varun, Geohydrology of springs in a mountain watershed: The need for problem solving research., Current Science, Vol. 71, No. 10, 25 November 1996.

Short-term water supply investigation at Shillong Municipality Board, Riat Khawan Forest, Shillong, September 2001, Central Groundwater Board, NER, Guwahati.

Short-term water supply investigation at NEEPCO complex, Polo, Shillong, September 2001, Central Groundwater Board, NER, Guwahati.

Short-term water supply investigation at Chief Minister's residence, Shillong, September 2001, Central Groundwater Board, NER, Guwahati.

Short-term water supply investigation at Power Gridco complex, Lapalang, Shillong, October 2001, Central Groundwater Board, NER, Guwahati.

Short-term water supply investigation at DGAR, Assam Rifles, Laitumkhrah, Shillong, February 2003, Central Groundwater Board, NER, Guwahati.

Short term water supply investigation at SE falls complex, Shillong, March 2003, Central Groundwater Board, NER, Guwahati.

Short-term water supply investigation at Garrison Engineer, Kench Trace, Shillong, March 2003, Central Groundwater Board, NER, Guwahati.

Short-term water supply investigation at MES pump house for Garrison Engineer, Happy valley, Shillong, March 2003, Central Groundwater Board, NER, Guwahati.

Short-term water supply investigation at Holiday Home, Shillong, March 2003, Central Groundwater Board, NER, Guwahati.

Statistical handbook, 2003, Directorate of Economics and Statistics, Govt. of Meghalaya.

Problems and Perspectives used Groundwater Sources in Case of Emergency and Projecting the Drainage in Arid Zone of Central Asia

2

I.Sh. Normatov and A.E. Kireev

Institute of Water Problems, Hydropower and Ecology Academy of Sciences Republic of Tajikistan, Dushanbe, Tajikistan

ABSTRACT

By the end of 19th century there were some 7–8 million people in the region. Irrigated land amounted to about 3.4 million hectares and was equipped by an irrigation network. Today the region's population has increased seven fold, and irrigated areas have broadened twice—up to 7.5–7.7 million hectares. Arid climate and irrigated nature contribute to land and pasture land degradation, which leads to significant decrease of agricultural productivity. Significant parts of irrigated areas are subject to salinization, which is 16% and above in Tajikistan, up to 30% in Kazakhstan and about 70% in Turkmenistan. Therefore, the major tasks in the field of melioration and obtaining the supplementary sources of irrigation water has great importance.

1 INTRODUCTION

In Central Asian Region (CAR) water is one of the most important factors defining the possibility for life and development in an arid zone.

New independent states of CAR acquired independence in 1991, namely, the Republic of Kazakhstan, Kyrgyz Republic, Republic of Tajikistan, Turkmenistan and Republic of Uzbekistan. General area of region territory is about 3882 thousand Km^2 with the population of more than 53 million people.

In each and every individual case the possible complex of measures and their mutual combination must be defined in view of natural conditions, technical economic purpose orientation. Surface and underground water-carrying horizons, vertical drainage, open and closed horizontal drainage, ant-filtering covers, machine irrigation and biotechnical drainage should be used widely in the complex of engineering constructions and measures in irrigation-melioration building.

Often while conducting meliorative hydrogeological works, the proper attention is not paid either to water balance and the balance of territorial groundwater or to possible changes in position of level and mineralization of groundwater. For irrigated lands the main components of subterranean water balance are filtration loss of irrigation waters, total evaporation from the soil and artificial drainage throw of water.

2 MEASURES TAKEN FOR PREPARING UNDERGROUND WATER SOURCES

Underground water get their input from the outside atmospheric precipitation and also probably from the condensed waters and generally precipitate on irrigated grounds superficially and in underground waters. But in most cases dominating, powerful and constantly acting source of feed for underground and pressure underground waters of quaternary deposits are filtered loss of superficial waters going to ground from channel, an irrigate network, watering cards and downturn with throwing waters. In numerous cases the alternative water supply sources comes from the ground water resources. At existing water use for such feed of underground waters is spent up to 30–50% and more from common water - gate of superficial waters in irrigated system.

The above marked picture of a feed and expenditure of underground waters is characteristic as a whole for irrigated grounds of republic. But between their separate geomorphological elements the precisely expressed distinction is in this respect observed. The ground water occurs in a depth of 3-5m below the ground and it serves as basic filtered water supply which compensate the loss of surfaces waters. On these grounds the drain of underground waters from an external mountain frame is directed also. In most cases size last within the limits of tens and hundreds liters per second. An objective large parameter in this respect is the small area of cross-section irrigated alluvium caught of terrace, on which is carried out under channel a drain of earth waters from mountains. The capacity of alluvium here is usually no more than 5–10 m, and width of caught terrace – within the limits of several tens and, at the best, of hundreds meters. Earth waters of a considered part of foothill plain in the party of the located below files and in separate rare cases in a local hydrographic network are spent, and mainly without essential change of a degree them mineralization and chemical structure (as the drain occurs mainly on well washed out more thickly pebbles). But such drain already has essential meaning in an additional feed of underground earth and pressure head waters on the grounds, located below on a flow, a plain part of foothill plain (bottom part of cones carrying out), described weak natural draining and superficial deposit of earth waters. The horizontal movements of filtered irrigated water contribute to ground water as well as the drainage water dump and evaporation loss from soil occurs simultaneously. On such grounds these factors dominate the account of balance of underground waters. Their smaller part is spent usually considerably for a underground drain in a local hydrographic network on the normally advanced cones carrying out.

On old-irrigated grounds, basic mass of underground waters, formed on them, makes horizontal moving on well washed out pebbles without essential change of a degree of mineralization and chemical compound.

The significant area of spreading powerful pebble water-bearing horizons provides an opportunity of creation in them practically of unlimited adjusting capacity. Preliminary in them operating a level of underground waters on 5–10 m and more can ensure adjusting capacity, in most cases quite sufficient for use water-bearing pebbles of horizons as underground reservoirs for seasonal and of many-years regulation of a drain of yet not used part of superficial waters (plenty of flood waters and not vegetative the period).

Water-bearing horizons in pebbles, described raised water-permeability, large square spreading and significant total powerfulness, practically provide an opportunity of development of a wide network of high-efficiency chinks of a vertical drainage and centralized economical drinking water supply. Debit of them can make from tens up to 100-200 L/sec and more, and effect drainage many tens and hundreds ha.

On the basic objects of irrigation of Northern Tajikistan and Kizilsu-Yakhsu valleys, the network of such chinks can supply the most economic joint decision (in comparison with machine irrigation in a combination with horizontal drainage) tasks on melioration of irrigated grounds and covering of deficiency in irrigated to water. On the water-supplied objects of irrigation, requiring in land improvement of the salted grounds the chinks in pebbles can ensure more economical decision of meliorative task with passing use of basic weight pumped out fresh and concerning fresh underground waters for irrigation and washings, and also for centralized economical drinking water supply having here and important meliorative meaning (at the expense of the termination winter water serving for this purpose on irrigating network). Marked debit in most cases is possible at downturn of a dynamic level on 10–15 m (specific debit from 35 up to 10 L/sec and more) and depth of chinks about 50–100 m.

Deposit of chinks of a considered type, in the volume number by a diameter up to 800–1200 mm, is provided already with available self-propelled machine tools high-speed rotor drilling with direct and return washing. The large diameters of drilling simplifying a task of creation of the powerful grave-sandy filter around stranger of a chink sharply reduce at the end specific cost of pumped out underground water.

The rather small productivity existing horizontal drain in cover small-ground is conditioned sometimes not only lowered filtered ability the last, but also small dip of drain is direct under a level of underground waters. In particular, it is caused also by that in already wetted and strongly becoming swollen small-ground practically was not presented possible to put in pawn deeper drainage.

At pumping out is superficial (0-1-2 m) deposit of underground waters from a single chink or their small group, when they do not cover completely file subject land-improvement, in an external contour already on small distance from a chink the depth up to a water mirror can be of less design norm of drainage. Here, alongside with outflow to a chink, earth waters will be spent and for total evaporation from a soil cover. Naturally, at the expense of such joint action, chink and the evaporations the general area drainage and radius of action are increased, but the area with design norm of drainage can make only small part from potentially possible (proceeding from volume pumping out and size of the drainage module). So, at joint work of five chinks the area of grounds with design norm of drainage from potentially possible can make only about 30–40%, and at work of one chink even 5–7%. Unfortunately, this situation is not yet clear to a wide circle of the experts – designers and hydrogeologists.

As a result the overestimation of the drainage module and underestimated of hydraulic interrelation between pressure head waters, that provided the overestimated quantity of a vertical drainage. The stocks of underground waters are expedient for considering together with basic irrigate - land-improve by parameters of irrigated grounds of republic.

3 THE STATE OF ARTIFICIAL DRAINAGE IN TAJIKISTAN

In Tajikistan the underground waters for irrigated lands are spent mainly on natural and artificial drainage flow and the total evaporation from the ground surface. The very fact of evaporation from the ground surface is the cause of that there are still great areas of salinized lands. At the first stage of developing the virgin lands with the shallow soil the main part of filtration losses is to fill up the volume (static) store of underground waters, thus causing the corresponding rise of their level.

The surface and underground waters in Tajikistan are characterized by the considerable variety in mineralization degree and chemical composition. The least mineralized groundwater is observed near the filtration river sites and irrigation channels. The most marked difference in chemical composition of republican irrigation waters is observed in chloral ion contents, on the base of which the evaluation of ground salinization degree is given for the areas with sulfate and sulfate chloride salinization.

It is possible to establish some laws in changing of mineralization and chemical composition of underground waters in quarter deposits on irrigated lands, taking into account the chemical composition of irrigation waters of Vakhsh River and river waters in Hissar Valley. Irrigation waters of Vakhsh River contain up to 0.1–0.15 g/L of chloral ion. As to river waters of Hissar Valley, its content decreases to 0.002–0.004 g/L. Compared to Vakhsh Valley the precipitation in Hissar Valley is 2-3 times higher, the total evaporation from the ground surface is some times lower, and groundwater running is higher.

It should be mentioned that in Tajikistan the land salinization does not result in falling of free soluble salts into solid sediment. This is explained by the fact that lands in Tajikistan, which are subject to salinizing, are characterized by the running of groundwater. At the same time sulfate and other more hardly soluble salts fall into solid sediment, right up to forming the independent layers in soil grounds.

Lands, irrigated for the long time, are characterized by comparatively broad development of ground salinization. The part of these lands in the republic is over 116,000 ha. This situation is caused by the scanty development of drainage or its neglected condition, and as well by the insufficient depth of horizontal drains. As to amelioration, the only thing is poor in conditions of two-tier geological structure of these lands the main part of salts is focused in the most upper layer of soil grounds. At the same time, limited salt store and broadly developed bedding pebbles, which easily absorb water, with fresh subterranean and forceful waters are potentially favorable factors for relatively fast ameliorative improvement of these lands, if vertical drainage is possible to apply.

In the republic the average part of horizontal drainage is 20 m/ha. According to data, obtained from large-scale topography, the average part of lands with the depth of groundwater to 4 metre, when they my influence actively the ground surface, is about 37% of the total topographed area. About 60% out of them are salinized to a different degree, including 20% of greatly salinized lands. The total area of salinized lands correlates with the areas, where the groundwater (mineralization is over 2 g/L) lies in the depth of 2 m. The area of greatly salinized lands correlates with that, where mineralization of groundwater is over 5 g/L.

On lands, irrigated for a long time the maximum level is observed almost everywhere in August-September, irrespective of their geomorphologic geological structure. Minimum level is observed in February-March. This course in level regime of groundwater is caused by the corresponding consumption regime of river waters and water supply for irrigation. It disguises the role of the rest factors, on which the level regime depends, including such a great one as the total evaporation from the ground surface in the hot season of the year.

But the considerable change in level regime in future may occur on the land massifs, where vertical drainage will be applied. As a result of pumping out on vast territories the level may be on the depth of not over 3–5 m, when its regime fluctuations have no ameliorative meaning.

Mineralization and chemical composition of subterranean and forceful waters on irrigated lands in Tajikistan are also subject to regime changes. According to year seasons the most marked changes in this respect are observed on greatly salinized lands. There in the hot season of the year the groundwater of the surface shallow grounds with sharply increased mineralization is spent on the total evaporation from the ground surface, thus increasing temporarily the salt store in the soil grounds of aeration zone and on the land surface. The waters with lesser mineralization come to their place from the bedding pebbles. By the next vegetation period the adverse effect takes place – salts dissolve in the groundwater due to precipitation.

Large-scale application of vertical drainage on salinized lands will cause the gradual freshening of groundwater in the surface shallow grounds. But in the upper horizon of bedding pebbles, through which primarily stored salts will be pumped out, the gradual increase in waters mineralization will be observed for a number of years until at least to a time where change of initial water storage takes place within the whole range of rocks, during water pumping. In consequence, a gradual decrease mineralization of pumped out water may be observed. Finally, after corresponding stabilization under new conditions, mineralization degree of groundwater will be defined by the difference between for the total evaporation from the ground surface and the total depth filtration water throw from the irrigation network into exploited water-carrying horizon.

Mineralization and chemical composition of forceful waters of the second, third and deeper horizons (100–150 m) under the influence of vertical drainage in the most upper one (the first) will not be subject to considerable changes. Water from these horizons, as the least subject to pollution, is worthwhile being used for centralized water supply.

The essential change in mineralization and chemical composition of groundwater should be expected on the developed virgin lands with thick surface shallow grounds. At the first stage of developing these lands the descending waters currents will wash the salts out of the upper layers and transmit them into deeper horizons. When at the next stage the level approaches the day surface, the groundwater may have already somewhat less mineralization than at the initial state. At this moment the artificial drainage must begin functioning, gradually throwing the water and salts out of irrigated lands. The drainage flowing may be up to 30–40% of the total water-intake into irrigation systems. Due to this correlation between water-intake and the surface flow of drainage and waters, the groundwater, already suitable for irrigation, may gradually being formed in the zone of active influence of drainage.

Filtration coefficient of water-carrying pebbles changes from the tenth parts of meter up to many tens and more meters per day along the separate irrigation sites. According to data from the main part, on the main irrigation massifs the value of filtration coefficient varies from 10–20 up to 30–40 meters per day.

In the upper layer of the surface shallow grounds, which is 3–5 meters thick and functions more actively under horizontal drainage, the filtration coefficient in most cases is 0.1–0.5 m/d.

The considerable area, where the strong pebbly water-carrying horizons are spread, gives an opportunity to create almost unlimited regulation capacity there. The preliminary cutting off the level of underground waters by 5–10 m and more may ensure the regulation capacity, which is sufficient in most cases for using the water-carrying pebbly horizons as underground reservoirs to regulate seasonally and for many years the flowing of that part of surface waters, which is not used yet.

Water-carrying horizons in pebbles are characterized by the high ability to absorb water, great area spreading and considerable total capacity (power). They give an opportunity to develop the broad network of highly productive chinks of vertical drainage and centralized water supply. Their debit may vary from tens up to 100–200 l/s and higher and draining effect may reach many tens and hundreds ha.

In irrigation sites, supplied with water and requiring ameliorative improvement of salinized lands, chinks in pebbles may ensure more economical decision of ameliorative task. The above-mentioned debit is possible in most cases, when the dynamic level decreases by 10-15 m and the depth of chinks is approximately 50–100 m.

Large diameters of drilling, simplifying the task of creating the powerful gravel-sand filter around chink strainer, finally will reduce considerably the specific cost of pumped out underground water.

4 CONCLUSION

As a result of our experience in agricultural use of non regulated surface water flow the irrigated land uses more than 30% of the total water in take into irrigation system in our country.

Therefore, it is necessary, first of all, to regulate the water use. The filtration coefficient (K_0) is often used for calculation of horizontal drains. In many cases the filtration properties of watered soils are changed over the short distances horizontally and vertically from the thousandth parts of meter up to several meters per day. The capacity of water-carrying horizons reaches many tens and hundreds meters, and the flakiness of the last horizon impedes the vertical water exchange. It is almost impossible to give the filtration coefficient value with proper accuracy for these conditions with relatively limited number of initial parameters.

Usually only the initial mineralization of groundwater in the upper part of water-carrying horizon is taken as a principle in project studies, aimed to change the existing meliorative and hydrogeological situation. The dynamic process of groundwater mineralization and, as the main thing, the possibility of managing them is not taken into account. By this approach it is possible to cause the land salinization in the near future, having at first fresh water and applying non-washing irrigation regime.

Critical correlations between bed depth and mineralization of groundwater for project conditions should be defined in view of other factors. First of all, the project depth down to groundwater should be defined, which will define in its turn both the permissible degree of mineralization and necessary rate for prophylactic washing. Finally, the economical technical equipment for decreasing the level of groundwater should be taken into consideration, as it is limited for horizontal drainage.

Development of salinized lands with groundwater bedded in not deeply is usually carried out in complex with general washing and artificial drainage. Project washing water is defined usually by simple multiplication of total saline store, which is to be removed out of root stratum, to specific water consumption. This rate is sufficient only for regular water distribution over the area. If the water distribution over the area is not regular, the main water mass is filtered near the drains. As a result of it the main part of the land remains unwashed. In this case the potential of the permanent network, which is to support the level of groundwater at the depth of 1–2 m, and is enough to reach to the groundwater derived from the surface washing waters. This leads both to decreasing manifold the productivity of drains and using the regulating capacity of soils in aeration zone and the time factors.

Reference

Pankratov P.A. 1979. *K probleme hydrogeological – meliorativnogo obosnovaniya meropriyatii po ispolzovaniyu Vodnich resursov I meloraciya zemel Tajikistana.*

Sustainable Water Management and Reuse by Natural Treatment Methods

3

R. Arthur James[1©], K.V. Emmanuel[2] , V.R. Saras Chand[3], S. Nagarajan[3] and A. Rajendran[1]

[1]Department of Marine Science, Bharathidasan University, Tiruchirapalli – 620 024. Tamil Nadu, India.
[2]Indian Leather Industries Foundation (ILIFO), Periamet, Chennai – 600 003. India.
[3]Centre for Environmental Studies, Anna University, Chennai – 600 025.
[©]Email: james_fse@yahoo.com

ABSTRACT

In the present environmental scene, water scarcity is becoming increasingly alarming. This problem can be tackled in many ways like optimized use, waste reduction and reuse of wastewater, etc. This paper focuses on various natural treatment methods and reuse of domestic and certain industrial wastewaters and their working procedures. The performance of combined anaerobic and aerobic treatment system with different natural filter media (gravel, slag and pebbles) and their efficiencies were examined for domestic wastewater. The gravitational flow rate to the system including Anaerobic Baffled Reactor (ABR), Anaerobic Filter (AnF) and Aerobic Filter (AF) was kept constant giving hydraulic retention time of 7.6 h, 7.2 h and 11.4 h respectively. In anaerobic filter, gravel media show higher efficiency than slag media. Small pebble (8–10 mm) with reed in the aerobic filter show higher efficiency (94%). Combination of these three systems gives excellent alternatives (90 to 95% efficiency) to conventional treatment system.

1 INTRODUCTION

Water is an essential basis for life and has utmost importance for health and dignity and without continuous access to clean water, social and economic development is simply not feasible. Production of food and industrial goods as well as human well-being in social communities are highly dependent on this resource, which became a key issue of international development strategies in the last decades. The challenge is obvious. The pressure on water resources has increased tremendously despite the fact that significant progress in the improvement of water supply and water sanitation could be observed over the last decades (James *et al.*, 2006). The dynamic growth of both the global population as well as the economic development has simply absorbed these achievements. At the beginning of the new millennium 1.1 billion people are without access to clean water and 2.4 billion people have no access to sanitation (UNEP, 2000). The overexploitation of water resources has caused a direct threat to agricultural production in a number of regions and polluted water impacts industrial production negatively. More than half of the world's major rivers are seriously depleted and polluted, degrading and poisoning the surrounding ecosystems and threatening the health and livelihood of the people who depend on them (Subramanian, 2000). One of the most significant effects of poor water supply and water sanitation are severe health hazards. For instance, approximately 4 billion cases of diarrhoea annually cause 2.2 million deaths, mostly among children under the age of five for instance. Water, sanitation, and hygiene interventions would be able to reduce diarrhoea on average by between one-quarter and one-third. Asian rivers also have a biological oxygen demand (BOD) some 1.4 times the global average, as well as three times as many bacteria from human waste as the global average (UNEP, 2000). On Global water budget, only 0.69% is available as freshwater in the form of rivers, lakes and groundwater. In the present-day environment, good quality of water is becoming increasingly scarce. This problem can be tackled in many ways i.e., optimized use, waste reduction and reuse of waste-water, etc. In India, most of the sewage is untreated and allowed to drain through rivers, streams, canals, lakes and finally to ocean. Considering this reported fact, less than 8% of population is provided with proper sewage and only less than 4% of the total sewage is treated (UNEP 2000). These scenarios urge us to understand and respond with immediate necessary steps to follow wastewater treatment and reuse.

1.1 Surface and Groundwater Pollution

Over the last 200 years the world has seen a tremendous increase of its population and economic activities. Along with a dynamic economic growth and a deep social transformation, the increasing demand for natural resources has been a main feature of this developing process especially in industrialized countries. But latest since the 1950s, a dynamic growth of population, rapidly changing consumption and production patterns, increasing urbanization and the tremendously expanding pressure on natural resources have become a common trend. Water supply for the population and availability of water for the agricultural and industrial production were two of the main features of developing strategies. Latest since the second half of the 19th century, comprehensive water sanitation became a key element of urban and rural water management strategies (Sasse 1998). Today, the challenges for water strategies worldwide are obvious: About one-third of the world's population lives in countries suffering from moderate-to-high water stress (water stress is defined as a situation where water consumption is more than 10 per cent of renewable freshwater resources in a given area.) Increasing water stress will be a common feature in both urban and rural areas. In rural areas, the introduction of irrigation agriculture respectively overexploitation of water resources and poor management of irrigation schemes have severely contributed to water stress. Projections suggest that by the year 2025, around 58% of the world population will live in urban and semi-urban areas. Estimates show that the urban population of the less developed world is expected to nearly double in size between 2000 and 2030 from a little under 2 billion to nearly

4 billion people with the greatest. At the same time, the current deterioration of the urban infrastructure is obvious with per capita water supplies decreasing by a third between 1970 and 1990. Water shortages, pollution and unsanitary water conditions contribute to an urban water and health crisis. Mainly the poorest households are facing poor water supply and sanitation services. While gaining access to water was a main feature of public attention for the longest time, the issue of water quality problems has become more and more obvious. Untreated sewage, chemical discharges, petroleum leaks and spills, dumping in old mines and pits, and agricultural chemicals that are washed off or seep downward from farm fields have caused a negative impact on ecosystems, people and business.

1.2 Genesis of Natural Water Treatment

Problems with treatment and disposal of municipal sewage is becoming increasingly prominent with the population explosion and increasing urbanization. While industrial wastewater treatment is realized to a greater extent, providing adequate sewerage and treatment facilities are still a far cry in most developing countries. Although collection of storm or rainwaters and drainage into natural water bodies were present from ancient times, collection of wastewaters for the purpose of treatment began in the late 19th century and in India in the mid-20th century. As populations centralised over an area, the availability of large areas of land for disposal were becoming scarce and severe nuisance was caused by odour, clogging and declining health conditions. Thereby the need for disposal and treatment of wastewaters arose. The common method for treatment of sewage was the collection and retention of wastewaters in septic tanks. The treated effluent was then discharged into the open/underground through soak pits. This method, however, was soon found inadequate to remove organics (septic tanks removes, besides the solids, only around 30% soluble BOD) and resulted in groundwater contamination, prompting the development of collection and removal of sewage through sewerage networks (Emmanuel, 2004).

The earlier forms of anaerobic filters were the septic tank type reactors modified with a filter at the mouth of the outlets to prevent the draining of larger particles. It was then seen that these filters added on some sort of treatment and degradation of organic solids which otherwise would have clogged the filter in a short time. This observation led to the development of the presently seen anaerobic filters. Other forms of anaerobic reactors followed including anaerobic lagoon, baffled reactors, etc. However, anaerobic systems soon lost the attraction, as their organic removal rate was not adequate to meet regulatory standards and generally consumed large area besides creating odour nuisance. Systems that consume less space, more efficient and aesthetically acceptable were the need of the hour, which led to the adoption of aerobic systems for sewage treatment. All common forms of aerobic systems viz., aerated lagoon, activated sludge plants; aerobic filters (trickling filters, roughing filters, etc.) were found extremely efficient in removing the organic content in the wastewaters even when supplied with large quantities of wastewaters. However, disadvantages associated with the aerobic systems too have been noted such as: high power requirement, high precision of operation, monitoring and maintenance and vulnerability of the system to situations such as prolonged power failure, etc.

In this background search for systems which do not consume much power or chemicals, require lower amount of operation/management control and do not need much space have gained momentum and of late, filters, both aerobic and anaerobic have emerged as a likely solution.

1.3 Study Area

Many works had been conducted in domestic wastewaters in the lab and pilot scale study showing a high efficiency in the low strength wastewater. In the present study, 15000 litre capacity field station was used to study the reduction rate, treatment efficiencies for multiple natural treatment system. To facilitate the above mentioned system, a series of interconnected Anaerobic Baffled Reactor (ABR), Anaerobic Filter (AnF) and Aerobic Filter (AF) was operated for a period of 6 months (Fig. 1a and 1b). The details of design and size of the units are given in Table 1.

Fig. 1a Anaerobic Baffled Reactor (ABR).

Fig.1b Gravel and slag media in Anaerobic Filter (AnF).

Table 1 Specifications of combined treatment system.

S. No.	Characteristic of the units	Unit	ABR	AnF	AF
1.	Length of each compartment	Metre	0.6	0.4	1.5
2.	Width of each compartment	Metre	2.3	0.5	1.0
3.	Depth of each compartment	Metre	1.63	1.2	0.5
4.	Number of compartments	—	7	6	5
5.	Flow rate in each section	Litre per hour	2000	200	66
6.	Size of the filter media	Millimetre	—	80–90	0.5–10
7.	Uniformity coefficient of the media	%	—	8–10	20–30

1.4 Methodology

The anaerobic baffled reactor and anaerobic filter units have been covered with mass concreting with small cover slabs of width 20 cm to facilitate sampling from each compartment separately. A ruttner sampler was used to collect the samples from the bottom of the chamber. Initially the ABR was seeded with sludge from the anaerobically digested sewage. Regular feeding from the ABR into each section of the anaerobic filter was maintained at 200 L/h for 20 h per day. Similarly, each compartment of the aerobic filter was maintained at 66 L/h for 20 h/d for six months.

Samples of the raw sewage admitted into the collection-cum-settling tank were sampled regularly. The duplicate effluent from six compartments of the anaerobic baffled reactor, twelve compartments from the anaerobic filter and five outlets of the aerobic filters were collected at twice a month intervals. For all the samples, one litre of the same was collected and preserved in the refrigerator at 4°C. Quality control tools such as x and s control chart were executed during the time of analysis to confirm the accuracy. The individual parameters and detailed method of analysis are given in Table 2.

Table 2 Analytical method adopted for the study.

S. No.	Parameters	Methodology / Instrument
1.	pH	pH meter - Model pH 197, WTW, Germany
2.	Electrical Conductivity	Conductivity meter – Model Conductivity 197, WTW, Germany
3.	Total Suspended Solids	Gravimetric method
4.	BOD	Winkler's method
5.	COD	COD Analyzer (Open Reflux)
6.	Sulphates	Turbidimetric method
7.	Phosphates	Stannous Chloride method
8.	Total Alkalinity	Titration method
9.	Total Iron	Phenanthroline method
10.	Chloride	Argentometric method

2 RESULTS AND DISCUSSION

2.1 Anaerobic Baffled Reactor

The main cleaning agents in wastewater treatment are a variety of specialized bacteria that convert, mostly through fermentation, the organic materials of sewage into carbon dioxide, methane and hydrogen gases. To derive energy from organic compounds such as carbohydrates, bacteria use either fermentation or respiration processes. The process of fermentation, used by most bacteria, takes place in an anaerobic environment and releases energy to the cell for growth, movement, temperature regulation, and a host of other survival processes. Respiration is a more complex process that requires oxygen to yield energy (Metcalf and Eddy, 1996). The ABR has a series of baffles to force the wastewater containing organic pollutants to flow under and over the baffles as it passes from the inlet to the outlet (Table 3). Bacteria within the reactor gently rise and settle due to flow characteristics and gas production, but move down the reactor at a slow rate. The main driving force behind reactor design has been to enhance the solids retention capacity. This can be observed from the present study, TSS reduces to 90% when compared with all other parameters. At the first chamber, the elevated concentration of 930 mg/L was observed and it reduces to 210 mg/L in the sixth compartment. Hence, ABR treats the organics more or less 50% and it can be observed during the six months of study period. The inlet COD was observed as 350 mg/L in the first chamber and got reduced to 170 mg/L in sixth chamber. Similarly, BOD is from 210 to 110 mg/L.

2.2 Anaerobic Filter

The anaerobic filter process is emerging as an efficient and economic method for the treatment of wastewater. Inexpensive organic stabilization coupled with efficient pathogen control makes the anaerobic filter process viable and hence broadens its applications. This process is a column filled with various types of solid media used for the treatment of the carbonaceous organic matter in wastewater. It is packed with material on which microorganisms can attach and which facilitates the development of a dense bio film and the filter material should have a large surface area per unit of volume, provide rough texture for bacterial adhesion, have high durability, and not easily clog (Bodik *et al.*, 2002). In the present system, the bio film is retained within the reactor on filters made from materials such as gravel and slag. In the case of gravel media, it shows slightly better efficiency than slag media and can be seen in Fig. 2a and 2b with gradual increase in trend towards time upto 4th month. In both cases, it is clearly evident that PO_4^{3-} is not removed considerably as it was expected especially in slag media since it has iron in it. The BOD at the inlet of the gravel media varied from 100 to 110 mg/L and the outlet it reduces to 40 to 50 mg/L. Similarly, at slag media the reduction ranges were from 100 to 60 mg/L. Sulphate concentration was not shown much variation between input and output as it ranges from 45 mg/L to 38 mg/L (Table 4).

2.3 Aerobic Filter (Reed Bed)

The operational pattern involves a self regulating dynamics of a specially designed soil eco-system. It encompasses the life interaction of various species of bacteria, fungi, micro-organisms, roots of wetland plant (halophytes), filter bed media (sieved sand in the range of 2–5 mm and pebbles of 4–10 mm) and water. The halophytes transfer atmospheric oxygen through

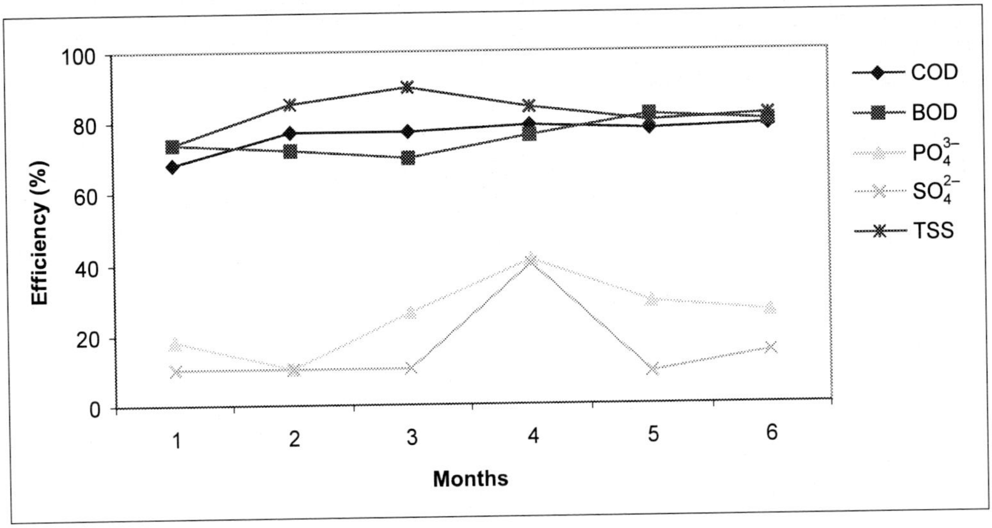

Fig. 2a Removal efficiency for gravel media in AnF.

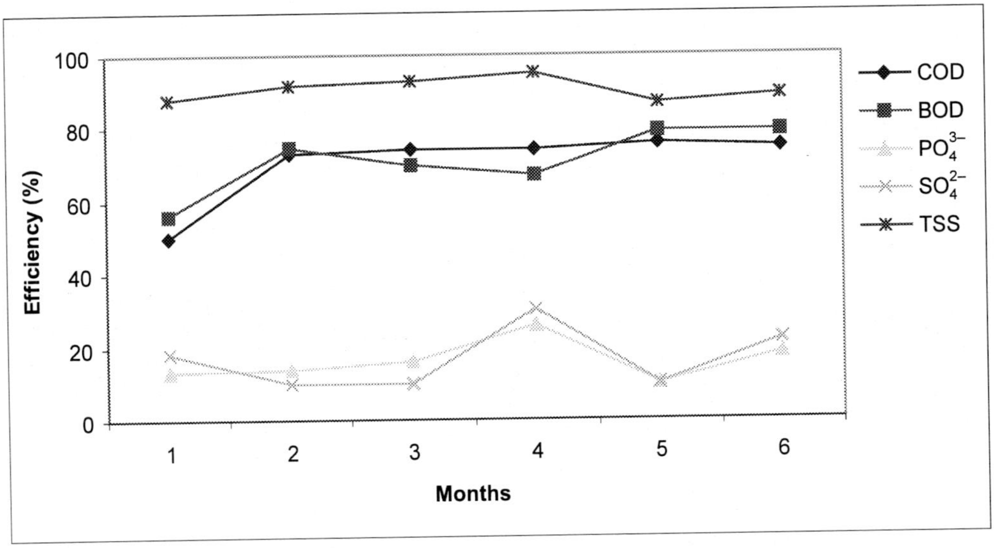

Fig. 2b Removal efficiency for slag media in AnF.

their root system and create suitable conditions for the growth of aerobic bacteria. Since the process occurs in a complex filter bed, aerobic and anaerobic zones exist side by side including different types of chemical reactions balancing bacterial growth. Once the wastewater enters the root zone horizontally the organic pollutants are decomposed biochemically by micro-organisms present in the rhizosphere of the halophytes. The filter bed media provide favorable conditions for both bacterial growth as well as sufficient hydraulic loading. Further the aerobic bacteria remove the nitrogen by nitrification as nitrates and anaerobic bacteria convert the nitrates into nitrogen gas by denitrification. The application of reed bed extends to municipal sewage, food processing, textiles, distilleries, oil refineries, tanneries and pharmaceuticals.

From this study the planted gravel filter and pebble (8–10 mm) got maximum treatment capacity and smooth operation. To confirm the consecutive overall treatment efficiency for the combined system, COD concentration shows steady performance during six months time (from 100 mg/L to 40 mg/L) and the reduction rates were observed by reduction in concentration curve by 90% (from 380 mg/L in ABR to 40 mg/L in AnF).

Table 3 Chemical composition of raw wastewater.

S.No	*Parameters*	*Value*		*Average*
		Min.	*Max.*	
1.	pH	6.78	7.43	—
2.	Total Dissolved Solids (mg/L)	1110	1390	1240
3.	Total Suspended Solids (mg/L)	140	340	220
4.	Chemical Oxygen Demand (mg/L)	280	460	380
5.	Biochemical Oxygen Demand (mg/L)	140	290	210
6.	Chloride (mg/L)	472	568	540
7.	Total Nitrogen (mg/L)	60	78	74
8.	Phosphates (mg/L)	10.5	28.3	18.6
9.	Sulphate (mg/L)	74	110	90

Table 4 Comparison of removal efficiency with different types of wastewater in anaerobic filter.

Type of wastewater	*Loading rate*	*HRT (hr)*	*COD reduction (%)*	*BOD reduction (%)*	*Reference*
Domestic	0.25 g COD l^{-1} d^{-1}	7.2	89	94.5	James *et al.*, 2006
Municipal Wastewater	0.1-0.25 g COD l^{-1} d^{-1}	6 – 20	46 – 92	41 – 96	Bodik *et al.*, (2002)
Industrial Dairy	8 kg COD $m^{-3}d^{-1}$	–	50 – 85	–	Garrido *et al.*, (2001)
Synthetic Wastewater	–	10 – 20	46 – 90	41 – 95	Bodik *et al.*, (2000)
Dairy	21 kg COD m^{-3} d^{-1}	12	80	–	Ince *et al.*, (2000)
Sewage Wastewater	–	5.5	–	96 – 97	Kim *et al.*, (1997)
Synthetic Wastewater	4.4 g COD $l^{-1}d^{-1}$	23	77	–	Akunna *et al.*, (1994)
Tuna fish Processing	11-13 kg COD m^{-3} d^{-1}	1 – 4 days	90		Veiga *et al.*, (1994)
Dairy (PVC media)	6 kg COD $m^{-3}d^{-1}$	0.9 – 2.8 days	80	–	Anderson *et al.*, (1994)
Pharmaceutical	1.1 to 16.66 kg COD m^{-3} d^{-1}	1.25 – 5 days	82.5 – 84.2	–	Roy *et al.*, (1993)
Vegetable-Tanning	7.1 g COD l^{-1} d^{-1}	23.5	84.6	87.5	Routh and Dhaneswar (1988)
Domestic Wastewater	0.02 lb COD ft^{-3} d^{-1}	24	78	79	Kobayashi *et al.*, (1983)

3 CONCLUSION

This study indicates a very good performance on combination of anaerobic baffled reactor followed by anaerobic filter and aerobic filter (reed bed). The system could achieve very low BOD and COD which even conventional treatment units fail to achieve. In the anaerobic filter, gravel media proved better performance than slag media. Hence, by designing these systems in such a way that gravitational flow pattern can be employed to run the unit without any care as well as to reduce operational cost. The outflow treated water can be utilised for gardening, toilet flushing and also for groundwater recharging to enhance the groundwater potential in sustainable way. Since the chemical quality of treated water meets well within the permissible limit, the reuse of water for secondary purposes can be highly encouraged.

Acknowledgements

The authors thank Mr. Pedro Kraemer and FEDINA-BORDA-DEWATS coordination programme for financial support to carry out this work. They also thank the Director, Centre for Environmental Studies, Anna University to provide logistical support to run the programme successfully.

References

Akunna, J., Claude Bizeau, Rene Moletta, Nicloas Bernet and Alain Heduit (1994), "Combined organic carbon and complete nitrogen removal using Anaerobic and Aerobic Upflow filters", *Water Science and Technology*, 30, 297–306.

Anderson G.K., Kasapgil and Ince O. (1994), "Comparison of Porous and non-porous media in upflow anaerobic filters when treating Dairy Wastewater", *Water Research*, 28, 1619–1624.

APHA - AWWA - WPCE (1998), "Standards Methods for the Examination of Water and Wastewater" Washington.

Bodik I., Kratochvíl B., Herdová B., Tapia G. and Gašpariková E. (2002), "Municipal wastewater treatment in the anaerobic-aerobic baffled filter reactor at ambient temperature", *Water Science & Technology* Vol.46, No.8, pp. 127–135.

Bodík, I, Herdová B. and Kratochvíl K. (2000), "The application of Anaerobic filter for Municipal Wastewater treatment" *Chemical Papers* 54 (3) 159–164.

Emmanuel K.V. (2004) "Decentralised wastewater treatment for domestic purpose", Technical report, *ILIFO*, pp. 130.

Garrido J.M., Omil F., Arrojo B., Mendez R., Lema J.M. (2001), "Carbon and nitrogen removal from a waste water of an industrial dairy laboratory with a coupled anaerobic filter-sequencing batch reactor system", *Water Science and Technology*, 43, 249–256.

James, R.A., Emmanuel, K.V., Rose Scaria and Thanasekaran K (2006). "Evaluation of domestic wastewater treatment using various natural filter media", *Asian Journal of Water, Pollution and Environment*, Vol.3, No.1, pp.103–110.

Kim Y., Kazuhiro Mikawa., Toshiaki Saito., Kazuhiro Tanaka and Hiroyoshi Emori (1997) "Development of novel anaerobic/aerobic filter process for nitrogen removal using immobilized nitrifier pellets", *Water Science and Technology*, Vol.36, No.12, pp. 151–158.

Kobayashi, H. A. Stenstrom K. and Mah A. (1983), "Treatment of Low Strength Domestic Wastewater using the Anaerobic filter", *Water Research*, Vol. 17, No. 8, pp. 903–909.

Metcalf and Eddy (1996), "Wastewater Engineering - Treatment, Disposal and Reuse", Tata McGraw-Hill, New Delhi.

Roy, M.D., Nandy T and Kaul S.N. (1993), "Anaerobic Upflow Bioreactor for Pharmaceutical Waste Treatment", *Indian Journal of Environmental Protection*, 13, 41–46.

Routh, T. and Dhaneswar R.S. (1988), "Comparative Performance of Anaerobic Filter, Upflow Sludge Blanket and Baffled Filter for Treating Vegetable Tanning Wastewater", IAWPC TECH, ANNUAL 15, 1–9.

Sasse L. (1998). "DEWATS - Decentralised Wastewater Treatment in Developing Countries", BORDA, Bremen, Germany.

Subramanian, V (2000). "Water", Kingston International Publishers, Surrey, UK, pp. 256.

UNEP (2000). Global Environment Outlook.

Veiga M.C., Mendez R. and Lema J. M. (1994), "Anaerobic filter and DSFF reactors in anaerobic treatment of tuna processing wastewater", Water Science and Technology, 30, 425–432.

Enhancing the Recharge Capabilities of Aravalli Ridge Around the National Capital Territory, Delhi, India

4

S.K. Mohiddin, Shashank Shekhar, Uma Kapoor and P.N. Singh
Central Ground Water Board, New Delhi, India

ABSTRACT

Delhi quartzite ridge, the northernmost tip of Aravalli range, acts as the most potential recharge zone for the aquifers underlying not only the ridge area but also underlying the surrounding alluvial areas. Groundwater regime in quartzitic terrain is in hydrological continuum with that of adjacent alluvium as inferred based on the detailed Groundwater flow patterns and depth to water levels in the study area. Moreover, because of the continuous recharge from the ridge, the fresh/saline water interface is pushed to deeper levels near the ridge which rises to shallow depths away from the ridge. The aquifer in quartzite ridge is abetting against the aquifer systems in alluvium. Thus, quartzitic terrain, highly fractured/jointed in nature with substantial thickness of overburden material, provides good recharge environment for aquifers underlying the surrounding areas. Detailed watershed pattern in the area has been studied. The area can be divided into 26 watersheds of different dimensions. Out of these, 17 watersheds fall in NCT, Delhi and 9 watersheds in Gurgaon and Faridabad districts of Haryana. It is estimated that about 6.875 MCM of rainfall runoff is available in a normal rainfall year from the 17 watersheds of NCT, Delhi. Though the runoff available in these watersheds is being completely harnessed through check dams and bundharas constructed during the historical period which created open ponding for recharge purpose, the desired impact on groundwater regime is not being observed as the sites for construction of check dams and bundharas were not selected based on the proper scientific studies and infiltration characteristics of an area. Check dams constructed recently by Central Groundwater Board and Delhi Development Authority in JNU-IIT-Sanjay Van-Vasant Kunj-II and Kushaknala watersheds which are having the recharge efficiency of about 90 to 95%. 21 new sites have been identified for construction of nalabunds and check dams which can have high efficient recharge rates. Apart from these nalabunds and check dams, recent urbanization of the ridge area has resulted into generation of lot of runoff from the urbanized environments which can also be utilized for recharging the depleted aquifers through cost efficient structures like recharge shafts and trenches associated with recharge tubewells. The designing characters of recharge structures in urbanized environments of Pushp Vihar Sector-I are based on *in-situ* hydrogeological and hydrological environments. It is estimated that the annual water harvesting capacity of Pushp Vihar Sector-I is about 36275 m^3 in and average rainfall year.

1 INTRODUCTION

The National Capital Territory (NCT) of Delhi lies between 28° 24′ 17″ to 28° 53′ 00″ N and 76° 50′ 24″ to 77° 20′ 37″ E (Fig. 1) and covers an area of 1485 km^2 and a current population of approximately 15 million. The ridge is the principal watershed of the area and acts as a groundwater divide between western and eastern parts of Delhi area. Delhi ridge which is northern part of the Harchandpur ridge controls the occurrence and movement of groundwater including the recharging characters of the aquifers in the area. The major rock types present the area around the NCT, include metasediments, represented by the quartzites interbedded with mica schists, and garnet staurolite schists of the Neoproterozoic Delhi Supergroup. These are intruded by pegmatites and quartz veins. These rocks had earlier grouped into Alwar formation of the Delhi Group. Now the status of the Delhi Group has changed to Delhi Super - group leading to reassignment of the various earlier formations within the Delhi Supergroup. Among the various geological units of the Delhi Supergroup, the principal lithology is dominated by quartzite. The geological succession exposed in NCT region is presented in Table 1.

Table 1 Geological formations exposed in the NCT region around Delhi.

Period	*Formation*	*Description*
	Newer Alluvium (Younger alluvium)	Unconsolidated, inter-bedded lenses of sand, silt gravel and clay confined to flood plains of Yamuna river.
Quaternary	Older Alluvium	Unconsolidated inter-bedded, inter-fingering deposit sand, clay and kankar, moderately sorted, thickness variable, at places more than 300 metres.
Precambrian	Pegmatite and Quartz Veins	Well stratified, thick-bedded brown to buff colour, hard and compact, intruded locally by pegmatite and quartz veins inter-bedded with mica schists.
	Quartzites and Minor Schist Bands	

2 RAINFALL PATTERN IN DELHI

Delhi experiences annual average rainfall of 61 cm out of which 50 cm is during monsoon months, i.e. July, August and September. The spatial variation also occurs from lower annual rainfall in Nangloi and Shahdara to highest in Chandrawal and Delhi University area although extremes of rainfall could occur in any areas (Fig. 2). Frequency analysis of one-day storm on the basis of available 70 years data for Lodi Road observatory from 1901–1970 was done by IMD and one day rainfall values likely for various periods were obtained as under:

Table 2 One day rainfall in different return periods.

Return Period	*Estimated one day rainfall (mm)*
2 yrs	99.3
5 yrs	147.3
10 yrs	179.1
25 yrs	219.2
50 yrs	249.0
100 yrs	278.8

Precipitation is lost to evaporation unless it occurs in bulk. This is because rainfall upto 2.5 mm a day serves merely to wet the surface of the earth, evaporating subsequently to simply add to the humidity in the atmosphere. Water recharge thus occurs only on days when rainfall exceeds ~2.5 mm. Precipitation in excess of 2.5 mm a day occurs in Delhi mainly during the monsoon i.e. July, August and September. The normal monthly rainfall in Delhi is 19 cm in July, 19 cm in August and 12 cm in September. The rest of the year accounts for the balance of roughly 11 cm but does not figure in the groundwater budget.

The number of rainy days more than 2.5 mm is 8, 8 and 4 for July, August and September respectively. The Pan Evaporation rate is 6.9, 5.0 and 6.0 mm/day for July, August and September respectively. Evaporation from a pan is different from that of the soil. Typically monsoon wet soil is known to have an evaporation rate of 70% of the pan rate. Since soil retains moisture, a similar evaporation rate obtains the following days. During the monsoon, the soil is wet so even when there is no rainfall evaporation occurs. The 70% of the total monsoon rainfall (19 + 19 + 12 = 50 cm) is 35 cm, which is evaporation loss and, therefore, we have 15 cm of balance usable rainfall for the year which can be used for natural recharge. To recharge the runoff generated from the 50 cm rainfall, artificial recharge to groundwater is essential and shall be promoted as a committed source of water supply in Delhi.

3 NATURAL RECHARGE ZONES IN NCT, DELHI

The hydrogeological investigations and ground water flow pattern in NCT Delhi, clearly depicts that Delhi quartzite which is highly fractured and weathered in character is good recharge zone (Fig. 1). Because of weathered and fractured nature of quartz rocks, infiltration in this ridge is very high resulting in limited runoff. Based on CGWB exploratory drilling, four to five prominent fracture zones are interconnected, this ridge area is acting as single aquifer system. 85% of rainfall is being percolated into this aquifer system through joints and fractures in hard rocks. The sub-surface cross section prepared along JNU-Palam Road and Shekawati lines shows that quartzite ridge is abutting against the alluvium of surrounding areas. Highly weathered and fractured zones in the ridge area are present against the aquifers in alluvium as inferred from the cross-section. This shows that ground water situation in fracture zone of hard rock is in continuity with the ground water situation in alluvium area. Depth to ground water levels and ground water flows are in continuum from hard rock to alluvium formation in all over the ridge area. Thus, whole ridge in NCT, Delhi is good recharge zone.

The second prominent recharge zone is Najafgarh drain extending from Dhansa to Kakrola. Downstream of Kakrola, Najafgarh drain is mainly receiving ground water and thus is a discharge zone.

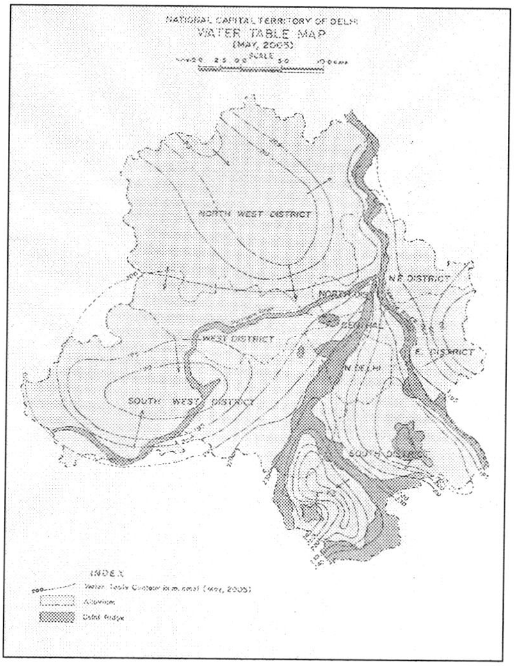

Fig. 1 Groundwater flow in NCT, Delhi.

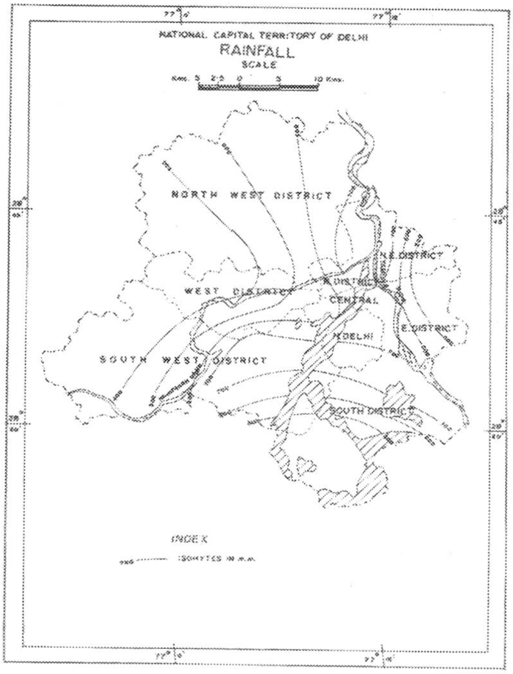

Fig. 2 Rainfall pattern in NCT, Delhi.

The third recharge zone is Western Yamuna Canal which is recharging the ground water all along its course in NCT, Delhi. This has been corroborated by the ground water flow pattern as well as the behaviour of fresh/saline water interface. A detailed study has been done on chemical quality of ground water from the varying depths along the Western Yamuna Canal. A thick column of fresh ground water aquifer system exists down to a depth of 50-70 m in the vicinity of Western Yamuna Canal and thickness decreases away from the canal (Mohiddin, 2001). A linear relationship exists between electrical conductivity of ground water and distance of sampling point from Western Yamuna Canal clearly depicts the contribution of seepage and infiltration component of fresh water into the aquifer in this area.

4 ARTIFICIAL RECHARGE ZONES OF NCT, DELHI

Artificial recharge to ground water schemes in an area are designed based on hydrogeological characters like soil cover, nature of aquifer system at shallow depths, depth to water level, Rainfall pattern, stage of groundwater development and rate of decline in ground water levels. Design of recharge techniques of an area need to be taken into consideration of all above factors. Based on hydrogeological characters of Delhi and surrounding NCR areas (Noida, Greater Noida, Ghaziabad, Meerut, Faridabad, Gurgaon and Bahadurgarh districts) are divided into six zones (C.G.W.B., 2003). They are:

1. Area underlain by Newer Alluvium with water levels more than 8 m: Areas falling in this category are Vasundhara Enclave, Preet Vihar, Vivek Vihar areas of East Delhi, Narela Singhola areas of North Delhi, Noida and parts of Greater Noida, Ghaziabad Municipal Corporation Area and Industrial Areas of Ghaziabad district, Meerut district, Industrial Areas of Delhi-Mathura Road of Faridabad district.
2. Area occupied by Older Alluvium with water levels 8 to 20 m: Areas falling in this category are Connaught place, Lodhi Road and Lodhi Colony, Parts of Lutyen Delhi, Karol Bagh, IARI Pusa, Parts of South-west district-Dwarka, Najafgarh, Parts of Bahadurgarh, Southern part of Gurgaon, Southern part of Faridabad district.
3. Area occupied by Older Alluvium with water levels 20 to 40 m: Areas falling in this zones are: Central parts of Chattarpur Basin, Gadaipur , Ghittorni and Surrounding areas, DLF-I, II and III, Udyog Vihar, Parts of Gurgaon Adjoining Delhi, Parts of South district-AIIMS, Green Park, Hauzkhas, R.K. Puram, Malviya Nagar.
4. Area occupied by Older Alluvium with water levels 40 to 70 m: Areas along Mehrauli-Badarpur Road, Tughlakabad-Pushp Vihar, Lado Sarai, Saket, Parts of Gurgaon development area.
5. Area occupied by Older Alluvium of depth 10 to 20 m followed by weathered and fractured quartzite with water levels varying between 20 and 40 m. (Fringe areas of ridge): Parts of C.R. Park, Kalkaji, Parts of Connaught place, Rajendra Nagar, Parts of Karol Bagh, Parts of Vasant Vihar.
6. Area occupied by weathered and fractured quartzite with water levels varying between 20 and 60 m: Delhi ridge extending from Kamla Nehru ridge to South Delhi, Vasant Kunj, Vasant Vihar, Okhla Industrial Area.

5 DESIGN OF MODEL ARTIFICIAL RECHARGE SCHEME IN URBAN HOUSING AREA AT PUSHP VIHAR

Pushp Vihar colony is underlain by alluvium of 70 to 100 m thick alluvium comprising of predominantly silty clay mixed with mixed with kankar in varying proportions. The litholog of the piezometer constructed in Pushp Vihar Sector-5 shows that alluvium is of silty clay in nature mixed with kankar upto the depth of 60 m followed by sticky clay. Minor intercalations of clay is present in the depth range of 42 to 45 m. Hard rock is encountered at the depth of 93 m. Depth to water levels in the Pushp Vihar colony is 55.0 m below ground level. Before designing the scheme, preliminary information regarding the hydrological and hydrogeological details of an area are collected and analyzed. The detailed information collected include layout plan of the area, details of existing storm water drains, details of existing ground water abstraction structures like tubewells, dug wells, etc., existing Land Use Pattern, other details like flooding and ponding and quality of ponded water.

Apart from the above information, the following hydrogeological information is also analysed for preparation of Artificial Recharge Scheme are:

1. Hydrogeology: Soil cover, nature of aquifer system, depth to water levels and chemical quality of ground water.
2. Area contributing runoff: How much is the area, land use pattern.
3. Hydrometeorological characters: How much is the rainfall, pattern of rainfall.

The Recharge scheme has been designed to catch both roof top rainwater and runoff from roads, greens and bare grounds (Storm runoff) for recharge purpose. The total area of Sector-1, Pushp Vihar is about 362750 m^2. The annual normal rainfall in this part of Delhi is 500 mm out of which about 90% occurs during monsoon season (based on long term rainfall data of Mehrauli Meteorological station). It is estimated that about 36275 m^3 of rainfall runoff is available in the area (Table 3).

Table 3 The Annual Rainwater harvesting capacity of each structure in Pushp Vihar Sector-1, New Delhi.

Structure No.	*Area (m^2)*	*Runoff Coefficients*	*Rainfall (m)*	*Runoff (m^3)*
S-1	36000	0.20	0.5	3600
S-2	39500	0.20	0.5	3950
S-3	37500	0.20	0.5	3750
S-4	Overflow of S-1, S-2 &S-3			
S-5	56250	0.20	0.5	5625
S-6	40500	0.20	0.5	4050
S-7	45000	0.20	0.5	4500
S-8	54000	0.20	0.5	5400
S-9	24750	0.20	0.5	2475
S-10	29250	0.20	0.5	2925
S-11	Areas where water-logging is taking place			
Total	**362750**			**36275**

Based on the hydrogeological and hydrological parameters, technical designs of recharge structures were prepared. To utilize the available runoff and as per the pattern of storm water drainage and flooding taking place in this sector, 11 recharge structures were recommended. The recharge trenches/shafts are designed based on the intensity of rainfall as 20 mm/hr because the rainfall in this part of Delhi is less than the normal rainfall of NCT, Delhi which is 611.5 mm per annum (Fig. 3). The dewatered aquifer present up to a depth of 60 m has been utilized for recharge purposes. The depth to water levels are in the range of 50 to 55 m below ground level, recharge tubewells of depth 50 to 55 m are recommended to be constructed in these recharge trenches. About 18 m of slotted pipe against the dewatered aquifer extending from 36 m to 54 m is recommended to recharge the dewatered aquifer system. As the discharge of tubewells in this colony varies between 300 and 400 LPM, the design recharge rate of the tubewells has been taken as 300 LPM for 153 mm diameter recharge tubewell. Based on the all these parameters design dimensions and storage capacity of the trenches are worked out and shown in Table 4.

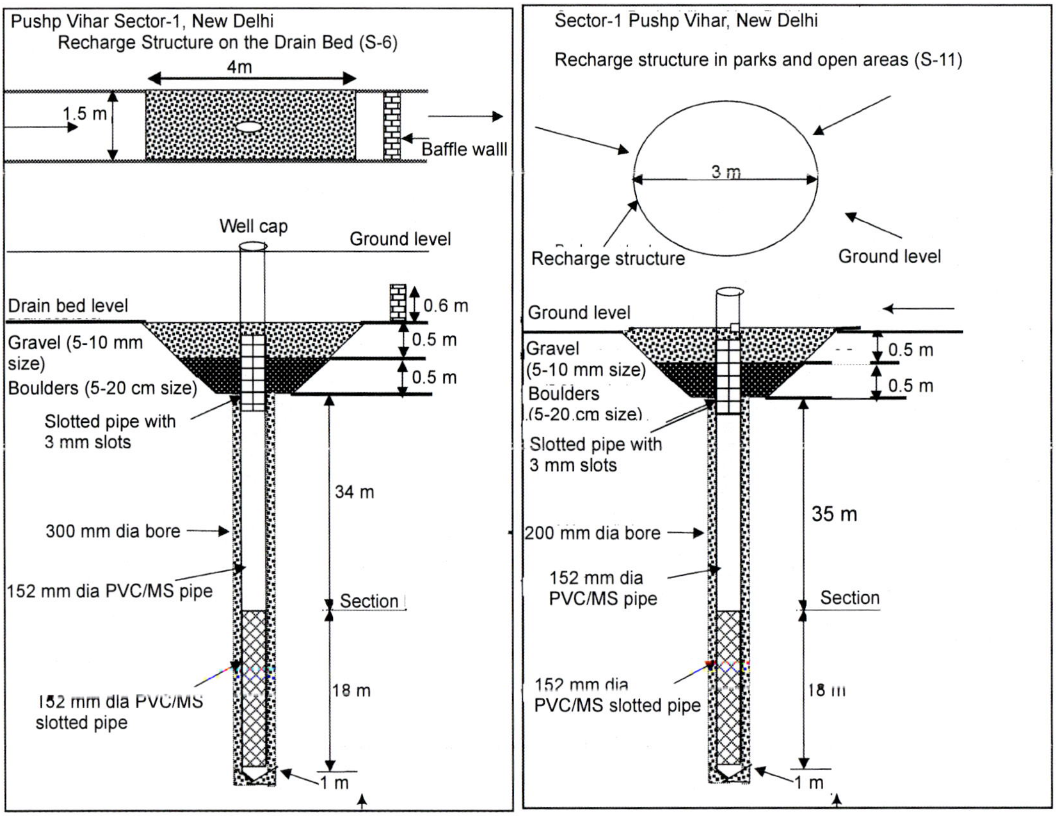

Fig. 3 Technical designs of recharge trenches/shafts in Pushp Vihar Sector-1, New Delhi.

Fig. Contd.

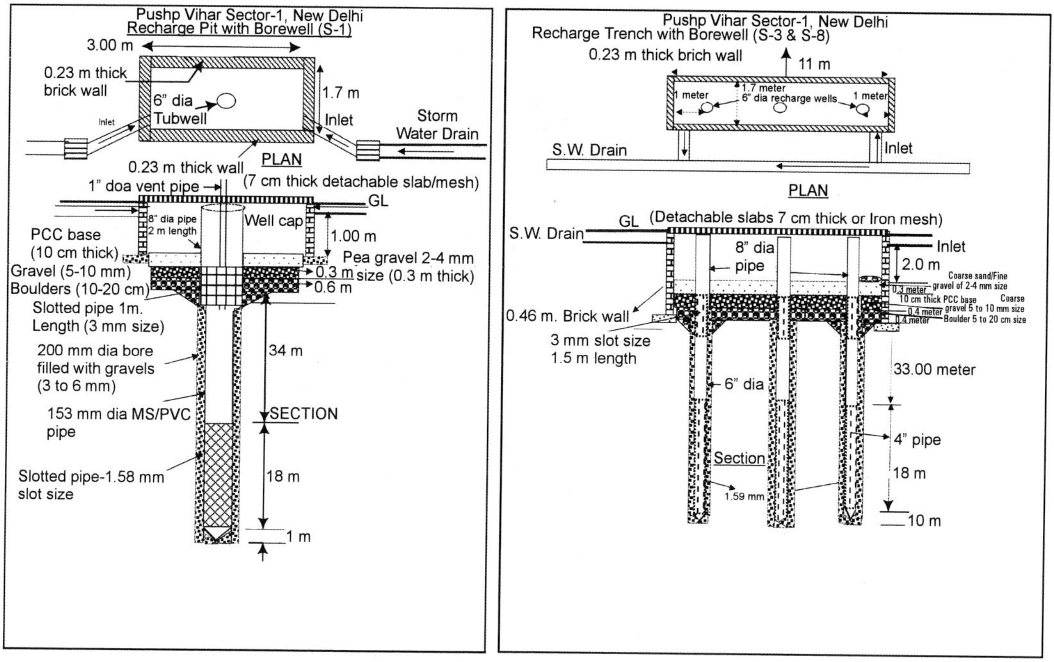

Fig. 3 Technical designs of recharge trenches/shafts in Pushp Vihar Sector-1, New Delhi.

Table 4 Runoff Available in the Sector-I, Pushp Vihar of hourly intensity rainfall.

	Area (m^2)	*Runoff Coefficient*	*Rainfall in m (intensity 20 mm/hr) (m)*	*Runoff availability per storm (m^3)*	*Dimensions of Recharge Structures*
S-1	36000	0.2	0.02	144	3*1.7*2.0 m
S-2	39500	0.2	0.02	158	4*1.7*2.0 m
S-3	37500	0.2	0.02	150	9*1.7*3.2 m
S-4	Overflow				7*1.7*3.0 m
S-5	56250	0.2	0.02	225	6*2.5*2.5 m
S-6	40500	0.2	0.02	162	4*1.7*3.0 m (Structure on drain bed)
S-7	45000	0.2	0.02	180	8*1.7*3.0 m
S-8	54000	0.2	0.02	216	9*1.7*3.2 m
S-9	24750	0.2	0.02	99	7*1.7*3.0 m
S-10	29250	0.2	0.02	117	7*1.7*3.0 m

6 DESIGNING OF RAIN WATER HARVESTING STRUCTURES IN URBAN STORM WATER DRAINS

There are 17 small drains directly outfalling into the Yamuna. Some of the important drains are: Magazine road drain, Metcalf House drain, Maharani Bagh drain, Delhi Gate drain, Kalkaji drain, Tughlakabad drain. All these drains originate in Delhi ridge and flow through the highly urbanized environments collecting domestic waste water along their course and discharging the same into the Yamuna. The upstream stretches of all these drains are dry and during monsoon season only storm water flows from these drains. Small nalabunds can be constructed in these drains to recharge the unconfined aquifers present in the ridge area. Based on the detailed hydrogeological investigations, designing of nalabunds in Kushak Nala which originates near Birla Mandir and flows in south-westerly direction has been carried out. The effective catchment of the Kushak Nala is 3.5 km^2. It is estimated that 0.142 MCM rainfall runoff generates from this catchment. Recharge to ground water is done through two gabion bunds and 2 nalabunds. The nala bed is underlain by 6 to 7 m thick overburden followed by weathered and fractured quartzites. Expected recharge from these structures are 0.11 mcm/yr and expected rise in ground water levels is 0.21 m in 3.5 km^2 area.

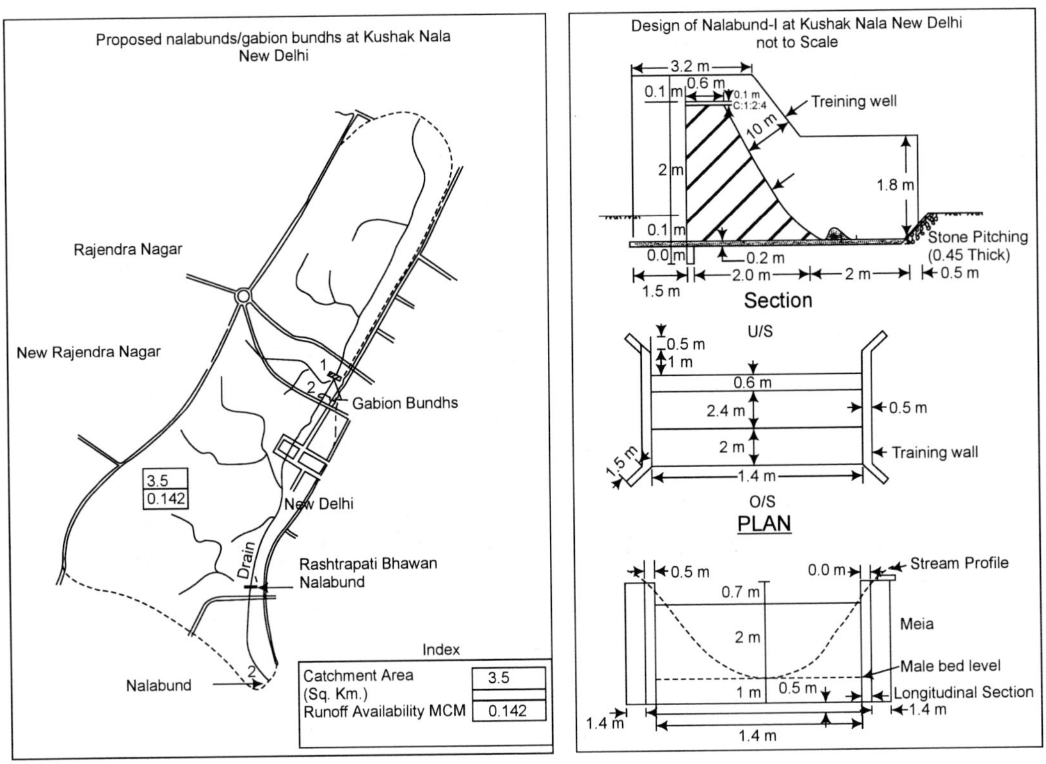

Fig. 4 Recharge structures in Kushak Nala, New Delhi.

7 FEASIBILITY OF CHECK-DAMS IN MICRO-WATERSHEDS IN URBANISED ENVIRONMENTS

Delhi ridge occupies an area of 145 km^2 and predominantly occupied by Alwar quartzites of Delhi Supergroup. The chemical quality of ground water in this area is fresh throughout the depth. About 25 mini and micro-watersheds are present in the ridge and surrounding forested area (Fig. 5). The runoff from these micro-watersheds can be used for recharging the depleted

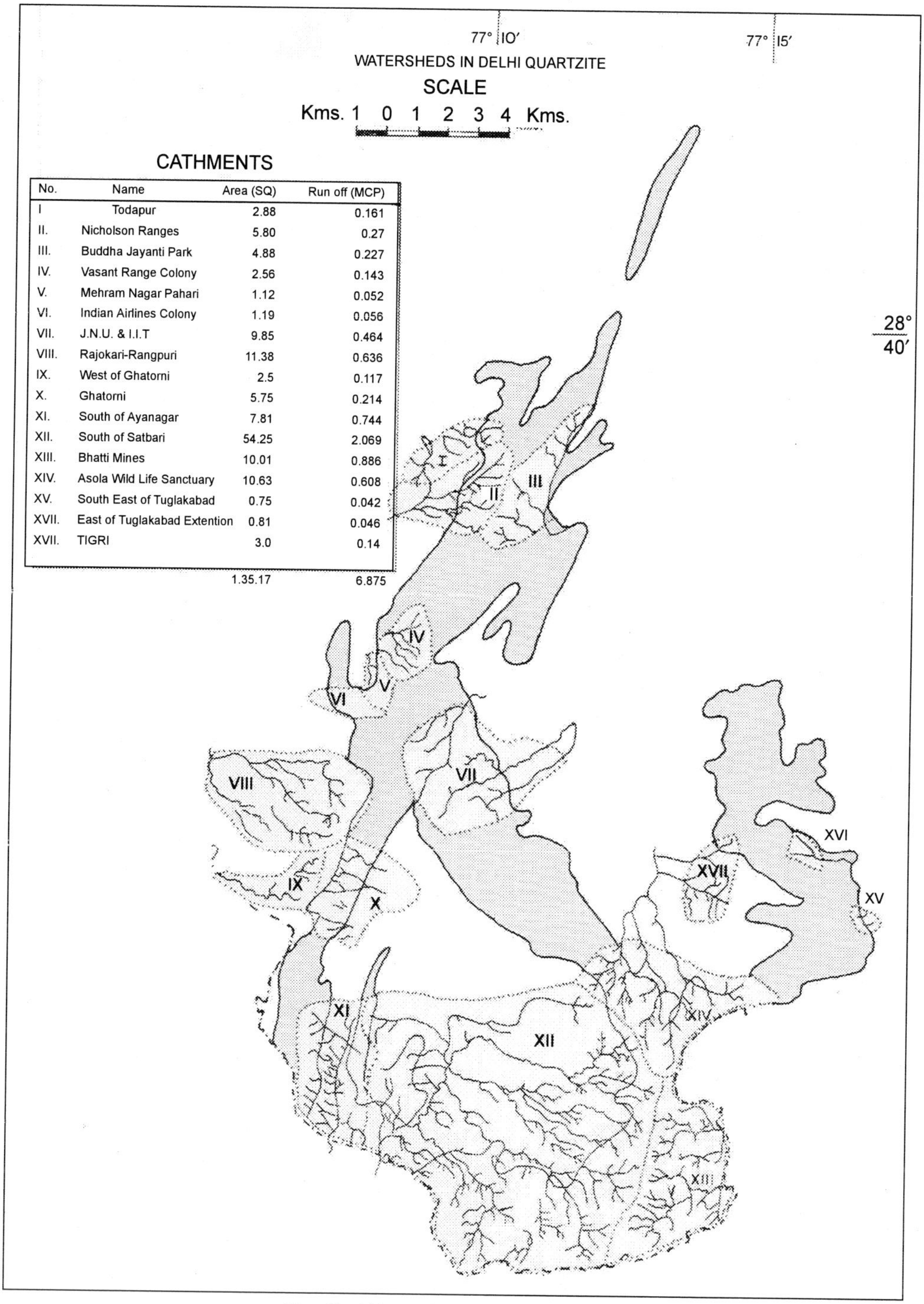

CATHMENTS

No.	Name	Area (SQ)	Run off (MCP)
I	Todapur	2.88	0.161
II.	Nicholson Ranges	5.80	0.27
III.	Buddha Jayanti Park	4.88	0.227
IV.	Vasant Range Colony	2.56	0.143
V.	Mehram Nagar Pahari	1.12	0.052
VI.	Indian Airlines Colony	1.19	0.056
VII.	J.N.U. & I.I.T	9.85	0.464
VIII.	Rajokari-Rangpuri	11.38	0.636
IX.	West of Ghatorni	2.5	0.117
X.	Ghatorni	5.75	0.214
XI.	South of Ayanagar	7.81	0.744
XII.	South of Satbari	54.25	2.069
XIII.	Bhatti Mines	10.01	0.886
XIV.	Asola Wild Life Sanctuary	10.63	0.608
XV.	South East of Tuglakabad	0.75	0.042
XVII.	East of Tuglakabad Extention	0.81	0.046
XVII.	TIGRI	3.0	0.14
		1.35.17	6.875

Fig. 5 Watersheds in Delhi ridge.

aquifers which is otherwise going waste. Central Ground Water Board has constructed four check dams in JNU-IIT watershed. Apart from these, Delhi Development Authority has constructed four more check dams to arrest the complete runoff in the same watershed. Highly fractured and weathered sections with thick zones of de-watered aquifers have been selected based on hydrogeological and geophysical investigations. Because of the location of these check dams in highly porous environments, the recharging efficiency of these reservoirs created are about 98%.

Selection of sites and Designing of Check dams in Vasant Kunj Phase-II, New Delhi: Delhi Development Authority is developing the 315 ha of land present adjacent to Nelson Mandela Marg as happening place with institutions and shopping malls. Because of high urbanization of surrounding areas and dependency of water supply on ground water source, the water levels in the area are declining at a faster rate. As this area is underlain by two distinct geological formations viz: Delhi quartzite and Alluvium. Delhi quartzite exposures are present in the southern and western parts of the area and form part of Delhi ridge. Alluvium is exposed mainly in the dissected valleys along the streamlets in eastern and north-eastern parts near Munrika Marg and Nelson Mandela Marg. The thickness of alluvium is about 20 m. Alluvium primarily consists of clay and silt followed by sand intermixed with kankar. Groundwater mainly occurs as semi-confined to confined conditions in fractured and jointed quartzitic formation. Four to five prominent fractured zones are deciphered in the area at a depth varying between 40 – 50 m, 55 – 65 m, 70 – 75 m, 80 – 95 m and 110 – 115 m below ground levels. Depth to water levels (April, 2000) is 26 to 30 m bgl. Shallow water levels are observed in southern and central parts of the area. Seasonal fluctuation of 5 to 6 m is observed in this area. Two check dam sites were selected considering the thick zone of alluvium followed by highly weathered and fractured quartzites. These two check dams have been constructed on one sub-watershed. The drainage characters of micro-watershed has been worked out and tabulated below.

Table 5 Drainage characters of sub-watershed SW-1.

Sub-watershed	*Catchment area in sq.km*	*Order of the main stream*	*Axial length in km*	*Average width in km*	*Form factor*	*General slope in %age*
SW-1	2.822	4	2.95	0.957	0.324	1.0

In the present sub-watershed, because of presence of lot of depressions and quarries and vegetation cover, the runoff coefficient is worked out as 14.5%. Because of highly undulating nature of this terrain and presence of number of ponds and quarries, most of the runoff generated is being accumulated in abandoned quarries and ponds. Part of runoff generated because of constructional activities can be stored by making the small check dams. Assuming that 80% of the monsoon and 50% of the no-monsoon rainfall will generate runoff and on the basis of computed runoff coefficient, the expected storm water availability is about 0.1906 MCM. Out of total runoff available in this micro-watershed, 0.03751 MCM of runoff water is being imponded at the two check dams constructed in JNU campus and 0.1631 MCM is imponded in two check dams of Vasant Kunj Ph-II. Detailed designing parameters of one check dam in Vasant Kunj Ph-II are as follows:

Storage Volume: Storage volume of check dam reservoir between different contour intervals (Fig. 6) were computed by applying the cone formulae $V = h/3$ ($A1 + A2 +$ Sq.rt of $A1$.

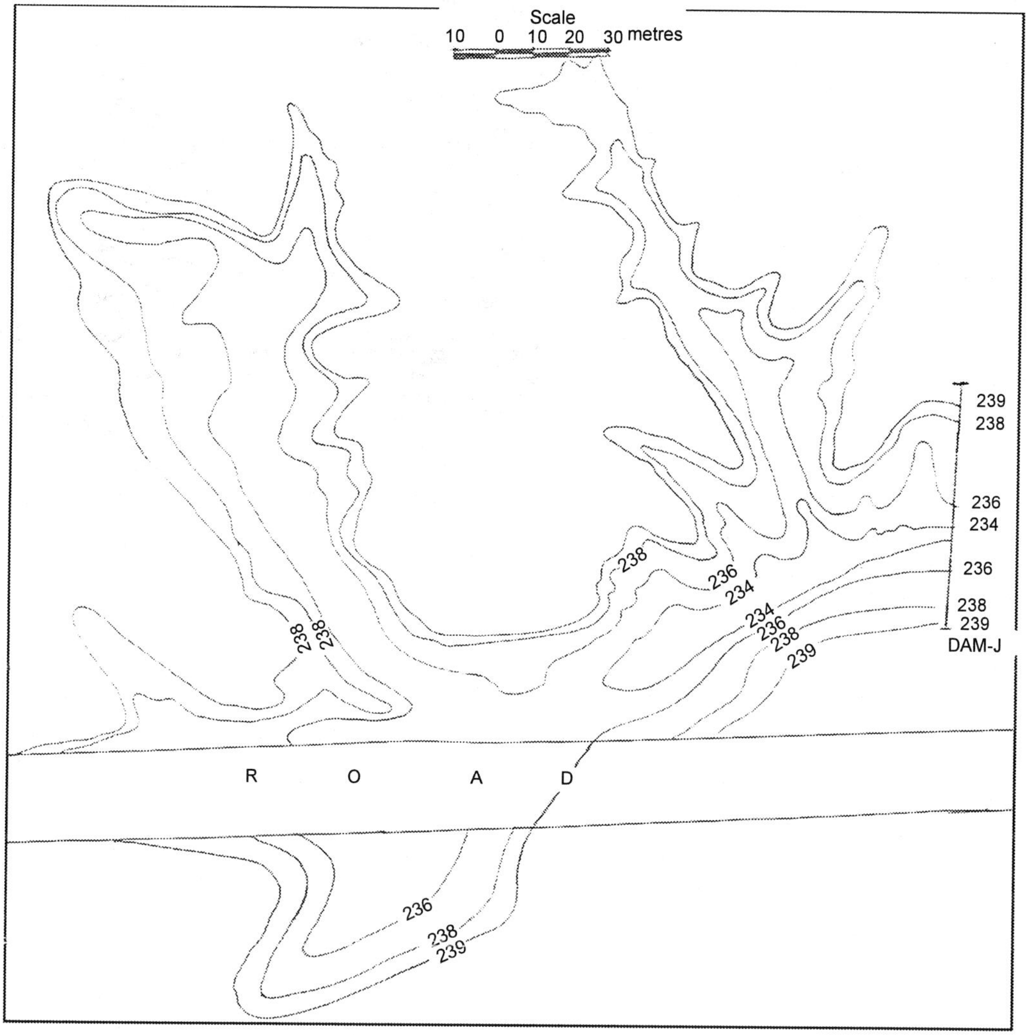

Fig. 6 Topographical contour map of reservoir site of check dam-1.

$A2$) where V = Total Storage volume, h = Contour interval and $A1$ & $A2$ are the areas occupied by preceding and succeeding contours. Elevation-Area-capacity curves for check dam-1 were prepared and the storage volume is worked out as follows:

Table 6 Water spread area and storage capacities of Check dam-1, Vasant Kunj Ph-II, Delhi

Contour Intervals (m)	*Water Spread Surface area (Sq. m)*	*Storage (Cu. m)*	
		Between contours	*Cumulative*
234-236	8800	5924	5924
236-238	18500	17448	23372

Based on available data, the components of water resources of the reservoir has been computed in the Table 7. Considering the soil strength in the dam site, it is proposed to construct check dam of 4 m height which can have a maximum storage capacity of 0.01407 MCM.

Table 7 Water resource balance in Check dam-1, Vasant Kunj Ph-II, Delhi

Month	*Normal Rainfall*	*Previous Balance*	*Runoff generated*	*Total storage*	*Maximum storage*	*Overflow*	*Rate of Evaporation and Infiltration*	*Total loss of water due to evaporation and infiltration*	*Balance*
	mm	MCM	MCM	MCM	MCM	MCM	Mm	MCM	MCM
June	38.8		0.0073	0.0073	0.0234	Nil	69.3	0.0019	0.0054
July	191.6	0.0054	0.0358	0.0412	0.0234	0.01783	60.5	0.0112	0.0122
Aug	197.4	0.0122	0.0368	0.049	0.0234	0.0256	50.5	0.0093	0.0141
Sep	105.3	0.01407	0.0196	0.0337	0.0234	0.0103	50.7	0.0094	0.014

Peak discharges and designing of Spillway Length of Check dam-1, Vasant Kunj Ph-II, Delhi

For designing the spillway, the highest flood discharge is required which were computed by applying the following empirical and rational formula:

Dicken (1965) formula: $Qp = CA^{3/4}$

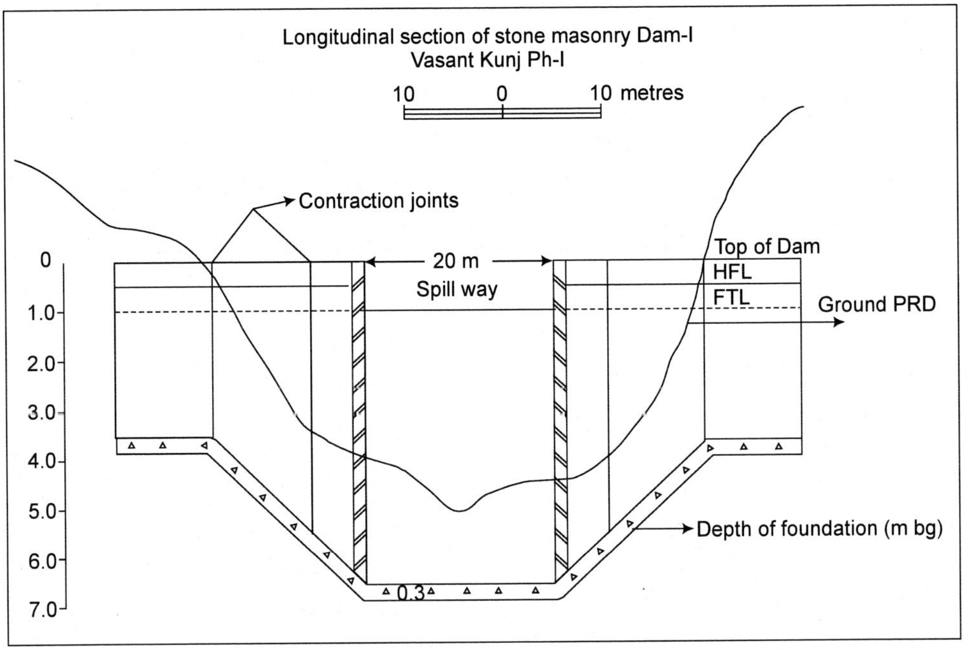

Fig. 7 Cross-section of check dam-1, Vasant Kunj Ph-II, New Delhi.

Where, Qp = Peak discharge (cumec)

A = Catchment area = 1.61 sq. km

C = Constant i.e. 11.5 for rugged topography like Vasant Kunj

The spillway being provided is rectangular weir and therefore, the discharge is given by

$$Q = CLH^{1.5}$$

Where, Q = Discharge (Cusec)

C = Coefficient of discharge

L = Length (feet)

H = Head (feet)

The spillway length for the check dam is worked out as 20 m (Fig. 7). Considering the fact these check dams are present in urbanized environments, 1.5 m foundation depth is recommended.

8 CONCLUSION

Delhi receives normal rainfall of 611.8 mm in 27 rainy days, which generates about 193 MCM of surface runoff. This can be tapped and stored in both surface and sub-surface reservoirs for which space is plenty in Delhi. Delhi ridge is one of the most potential zone in Delhi because of not only the highly undulating nature of the terrain, but also it is underlain by highly fractured and jointed nature of rocks exposed in Delhi ridge. The 26 watersheds draining the quartzitic ridge has the potentiality of recharging about 6.875 MCM of water. The sites selected based on remote sensing, geophysical and surface infiltration studies have shown very encouraging results of infiltration efficiency upto 95%. Implementation of urban rainwater harvesting projects in residential urbanized areas can also contribute substantially for recharging the depleted aquifers.

References

Central Ground Water Board, New Delhi, 1999. Report on Artificial recharge project in JNU, IIT and Sanjay Van Area of NCT, Delhi

Central Ground Water Board, New Delhi, 2001. Activities and Achievements of CGWB on rainwater harvesting and artificial recharge. pp. 31.

Central Ground Water Board, New Delhi, 2003. Ground Water in Delhi-Improving the sustainability through rain water harvesting.

Mohiddin S.K. 2000. Artificial recharge to ground water in Vasant Kunj Ph-II area. Unpublished CGWB technical report.

Mohiddin S.K. 2001. Assessment of vulnerability of fresh ground water aquifers for industrial pollution in Bawana industrial estate. Unpublished CGWB technical series report.

Sikka V.M, Singh S.B, Chakraborty D, Mohiddin S.K., Dalel Singh (1998); NCT of Delhi-"A test field area for artificial recharge studies" proceedings of seminar on artificial recahrge of ground water organized by CGWB held on 15th and 16th December at New Delhi. pp. I-31 to I-36.

Bangladesh Government Policy in Context of Arsenic Hazards and the Effectiveness of Activated Alumina in Arsenic Mitigation

5

Rezaul Kabir Chowdhury[1] and Muhammad Azizul Hoque[2]

[1]Department of Civil and Environmental Engineering, Shah Jalal University of Science and Technology Sylhet-3114, Bangladesh, E-mail: rezaul-cee@sust.edu

[2]Assistant Professor, Department of Civil and Environmental Engineering, Shah Jalal University of Science and Technology, Sylhet-3114, Bangladesh, E-mail: aziz-cee@yahoo.com

ABSTRACT

80% of the total population in Bangladesh drink water from shallow tube wells. 53 districts out of 64 are exposed to arsenic concentration greater than 50 µg/L. In this paper extent of arsenic contamination in groundwater has been discussed. Activities of various governmental and non-governmental organizations have also been delineated. Supply of safe water through alternate options, rehabilitation/treatment of arsenocosis patients and research and development activities on environmental impact of arsenic are the three major policy statements of the Govt. of Bangladesh in context of arsenic hazards. Adsorption of arsenic contaminated water on activated alumina can be used as an effective small-scale option of arsenic treatment.

1 INTRODUCTION

Although the first water supply in Bangladesh was surface water based, groundwater received priority in the subsequent development of water supply in the country. Withdrawal of groundwater through handpump tube wells in the rural Bangladesh has been taken up as early as 1928. In the context of very high prevalence of diarrhoeal diseases, groundwater received priority as a source of safe water as it is free from diseases producing microorganisms. Since 1928, an estimated 6-8 million hand tube wells have been sunk in Bangladesh to provide drinking water to about 97% of the population. The two main aquifers systems with intermediate impermeable layers and three main modes of tube wells based water supplies in Bangladesh are illustrated in Fig. 1 (DPHE, 2000).

The numbers of manually operated shallow and deep tube wells sunk by Department of Public Health Engineering (DPHE) are 894,941 and 81,384 respectively (DPHE, 2000). In Bangladesh the presence of arsenic in groundwater was first detected in 1993 at Barogharia union of Chapai Nawabganj district. Considering the weight of the problem, water sample testing activities were started in 1995 by various governmental (GO) and non-governmental (NGO) organizations. At present about 35 million people are at risk of arsenic toxicity in Bangladesh. A total of 53 districts out of 64 are arsenic affected having concentration beyond Bangladesh Standard. Nearly 10,560 cases of Arsenicosis patients have been identified. In 2004, Bangladesh Government prepared and enacted National Policy and Action Plan for Arsenic Mitigation. In this paper an attempt has been undertaken to describe the government policies with respect to arsenic problem mitigation in Bangladesh. Experimental results of arsenic polluted water treatment by adsorption on activated alumina have also been delineated in this paper.

2 LEVEL OF ARSENIC CONTAMINATION

In two phases, British Geological Survey (BGS), DPHE and Mott MacDonald Limited (MML) examined 3534 distributed water samples from 61 districts of Bangladesh (except 3 hilly districts) in an approximate grid of 6 km × 6 km (DPHE, BGS and MML, 1999; BGS and DPHE, 2001). On an average 58 samples per district and 8 samples per Upazila were analyzed. This study provides a reasonable distribution of arsenic contamination in Bangladesh. 42% and 25% of all tube well samples were found arsenic contaminated in excess of 10 μg/L (10 ppb) and 50 μg/L (50 ppb) respectively. In case of shallow tube wells, 46% and 27% samples exceeded 10 μg/L and 50 μg/L respectively whereas for deep tube wells (>150 m), arsenic content of only 5% samples exceeded 10 μg/L and 1% exceeded 50 μg/L. The maximum acceptable level of arsenic in Bangladesh is 50 μg/L (DoE, 1991) and the provisional WHO guideline value is 0.01 mg/L.

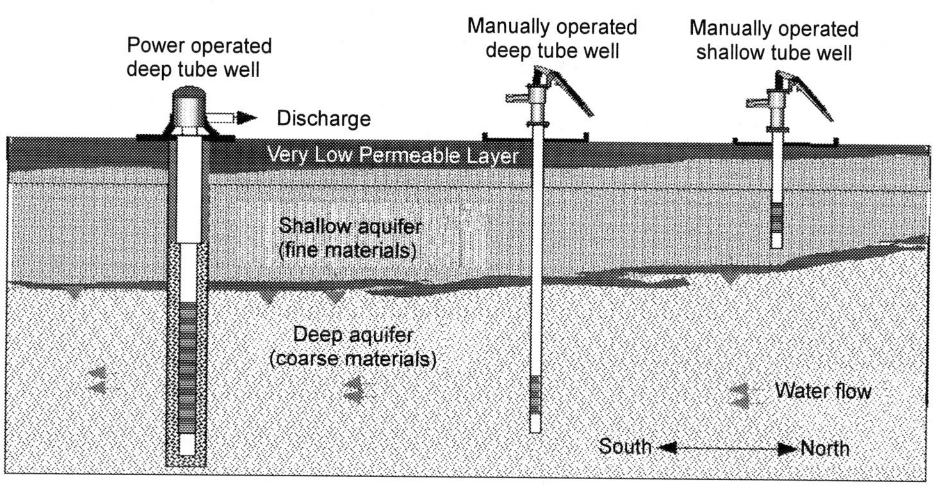

Fig. 1 Main aquifers and tube wells for water supply in Bangladesh (DPHE, 2000).

According to the Bangladesh Arsenic Mitigation Water Supply Project (BAMWSP), arsenic contaminated upazilas have been defined as those upazilas having at least one tubewell producing arsenic contaminated water (>50 μg/L). The numbers of administrative areas with at least one sample exceeding Bangladesh Drinking Water Quality Standard are shown in Table 1. On the basis of BGS/DPHE test results, percentages of shallow tubewells (STW) yielding water in excess of corresponding arsenic concentrations has been shown in Fig. 2. Figure 3 shows the map of the intensity of arsenic contamination of groundwater in different parts of Bangladesh (BGS/DPHE, 2001).

Table 1 Administrative areas with at least one tube well exceeding Bangladesh standard and WHO guideline value (BGS / DPHE, 2001).

Type of Administrative Areas	*Total Number of Areas*	*Number of Administrative Areas*			
		Bangladesh Standard (50 μg/L)		*WHO Guideline Value (10 μg/L)*	
		< 50 μg/L	*≥ 50 μg/L*	*< 10 μg/L*	*≥ 10 μg/L*
Division	6	0	6	0	6
District	61	8	53	1	60
Upazila	433	184	249	39	394

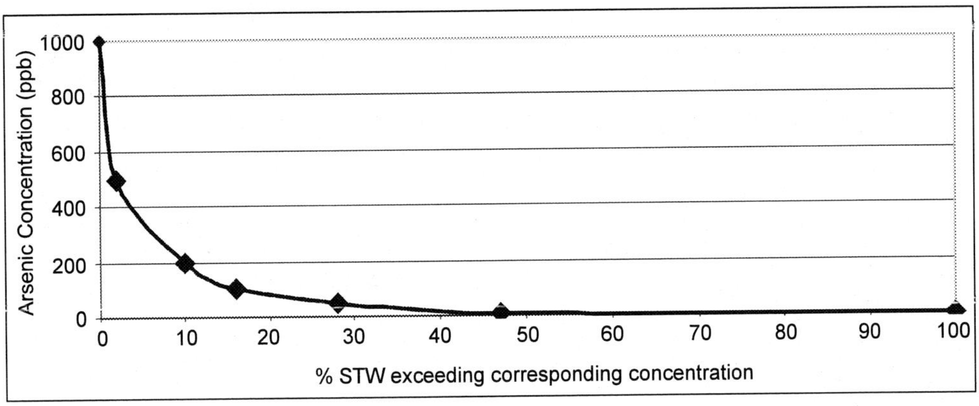

Fig. 2 Percentage of STW exceeding corresponding arsenic concentration.

2.1 Population Coverage by Different Options

The population coverage by different modes of water supply in Bangladesh has been presented in Table 2. The figures have been computed on the basis of data available from DPHE, 2000; BBS, 1998; BBS, 2001 and WHO, 1999. It is apparent from Table 2 that about 103 million people, which is about 80% of the total population of Bangladesh, depend on shallow tubewells for drinking water. Therefore, a shift from shallow tube well to alternative water supply options is very essential in acute arsenic problem areas. Prospective options are deep tube well and development of surface water based water supply system like Slow Sand Filters (SSFs) commonly known as Pond Sand Filters (PSFs).

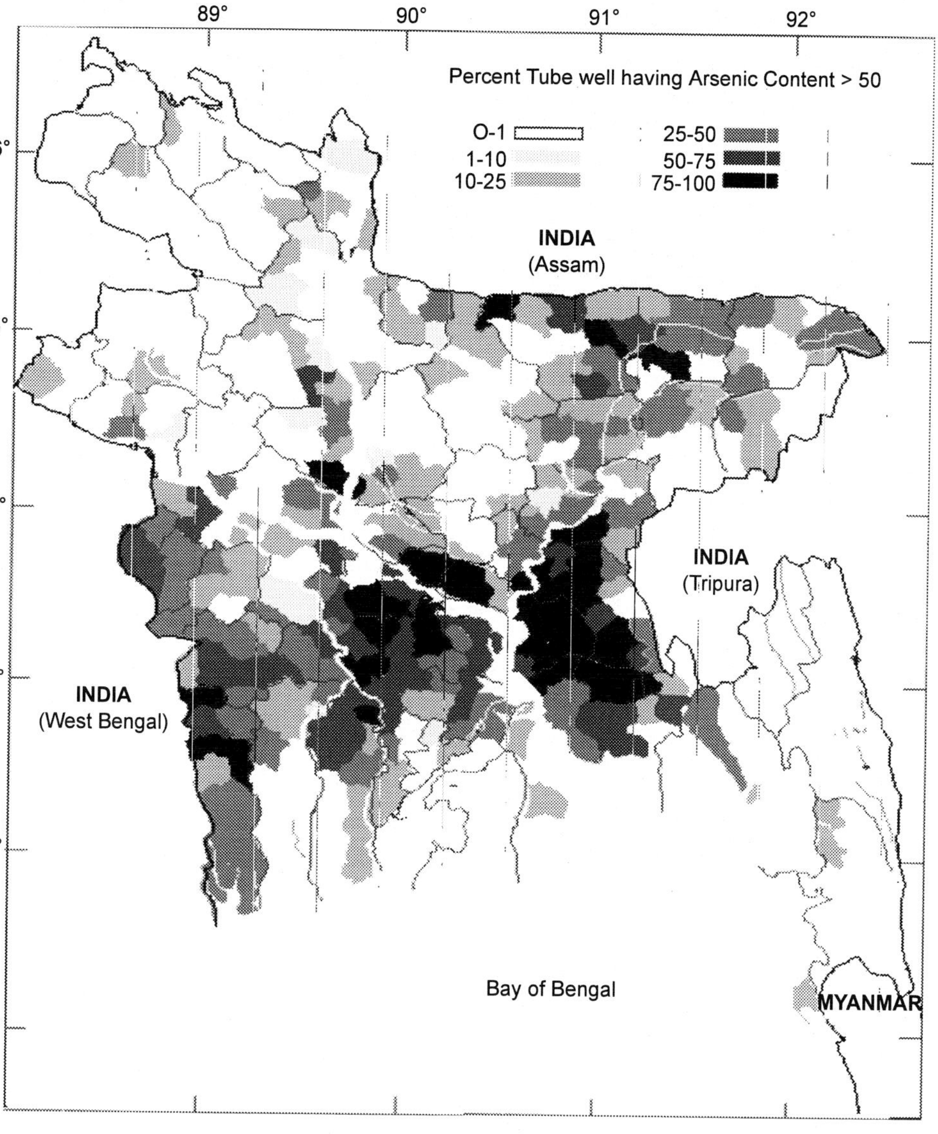

Fig. 3 Intensity of arsenic contaminated tube wells (> 50 ppb) in Bangladesh (DPHE, 2001).

2.2 Population Exposed to Arsenic

The population of Bangladesh exposed to arsenic contamination exceeding Bangladesh Standard (50 μg/L) has been presented in Table 3. The computation is based on DPHE, 2000; BGS and DPHE, 2001; BBS, 2001. Arsenic contamination has been found in excess of 50 μg/L in 15 production wells supplying piped water to about 0.75 million in urban areas, 1% deep tube wells, and 27.4% shallow tube wells. The population exposed to arsenic contamination in

Table 2 Population coverage by different modes of water supply (DPHE, 2000).

Modes of Water Supply	*Population Coverage (Million)*			*%*
	Urban	*Rural*	*Total*	*Coverage*
Piped water supply	13.10	–	13.10	10
Manually Operated Deep Tube wells	–	8.20	8.20	6
Manually Operated Shallow Tube wells	16.86	86.14	103.00	80
Dug/Ring Well	0.28	1.02	1.30	1
*PSF, VSST, SST, RWH	–	1.50	1.50	1
Others	–	2.15	2.15	2
Total	**30.24**	**99.01**	**129.25**	**100**

*PSF: Pond Sand Filter; VSST: Very Shallow Shrouded Tube well, SST: Shallow Shrouded Tube well, RWH: Rain Water Harvesting.

excess of 50 μg/L by those tube wells has been estimated as 29.05 million. Population exposed to different levels of arsenic from drinking water has been computed and presented in Fig. 4. The arsenic safe modes of water supplies are piped water in urban areas, deep tube wells (DTWs), uncontaminated shallow tube wells (STWs) and alternative water supply options such as pond sand filters, dug well, very shallow shrouded tube wells, rainwater harvesting, etc.

Table 3 Population exposed to arsenic from drinking water (> 50 ppb)(DPHE, 2000).

Modes of Water Supply	*Population Coverage, Million*	*Percentage of Tube well Contaminated with As (> 50 μg/L)*	*Population Exposed to As (> 50 μg/L), Million*
Piped water supply	13.10	–	0.75
Manually Operated Deep Tube wells	8.20	1	0.08
Manually Operated Shallow Tube wells	103.00	27.4	28.22
Dug well	1.30	0	0
PSF, VSST, SST, RWH, etc.	1.50	0	0
Others	2.15	0	0
Total	**129.25**		**29.05**

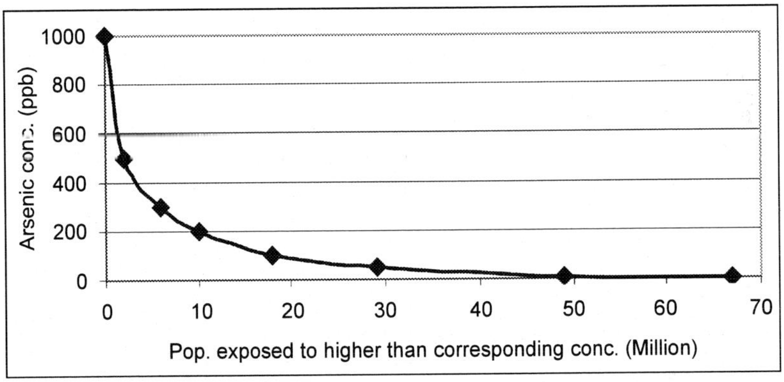

Fig. 4 Population exposed to different levels of arsenic concentration in drinking water (DPHE, 2000).

3 ACTIVITIES OF GO & NGOS

Many governmental (GO) and non-governmental (NGO) organizations are working in Bangladesh in order to investigate the extent of arsenic in groundwater and to provide alternate water supply modes to the community. Many projects have been completed and many are going on for arsenic mitigation. Summary of major activities of various GOs and NGOs are given below.

3.1 Department of Public Health Engineering (DPHE)

Under the assistance of UNICEF, DPHE has been testing arsenic in samples of tubewell water with field kit from July 1997 all over the country. Over 50 thousand tests were conducted. A total of 170 exploratory drillings were done in arsenic contaminated areas, where 95% of the tubewells having depth > 200 m shows no arsenic contamination. Awareness building activities were also done under this project. The cost of the project was Tk 199.59 lakh. The project started on October 1996 and ended in June 2000.

Under the assistance of British DFID and British Geological Survey (BGS), DPHE conducted a systematic and comprehensive survey in 61 districts (except in 3 hilly districts) of Bangladesh. About 3500 water samples were collected from field and tested in DPHE laboratories in Bangladesh and BGS laboratory in England. The cost of the project was Tk 609.29 lakh with PA Tk 578.35 lakh. The project started in December 1997 and ended in March 2000.

DPHE under the assistance of JICA conducted a study to investigate the deep groundwater and possibility to use it for the mitigation of arsenic problem in three western districts of Bangladesh (Jessore, Jhenaidah and Chuadanga). The study started May 2000 and continued up to March 2002. The cost of the project was Tk 2599.04 lakh with PA of Tk 2460.61 lakh.

DPHE-DANIDA Arsenic Mitigation Pilot Project has been taken up in the south-eastern part of Bangladesh. The estimated cost of the project is about Tk 7070.57 lakh. The main component of the project was to establish deep tube wells, mini piped water schemes and household treatment units.

3.2 Bangladesh Arsenic Mitigation Water Supply Project (BAMWSP)

BAMWSP was conceptualized jointly by the Government of Bangladesh, the World Bank and the Swiss Agency for Development and Cooperation. The project was launched in September 1998. The overall objective of the project is to reduce mortality and morbidity in rural and urban population caused by arsenic contamination of groundwater within sustainable management strategies. Project activities include understanding of the arsenic problem through national survey; strengthening implementation capacity of the local government entities and community based organizations; and onsite arsenic mitigation through sub-project development and implementation by the community based organizations (CBO). BAMWSP initiated a nationwide screening, community development and arsenic mitigation programme.

3.3 NGO Forum for Drinking Water Supply and Sanitation

NGO Forum is a national apex networking and service delivery organization in the water supply and sanitation sector dedicated to contribute in the improvement of the public health status of the poor and disadvantaged women, children and men of Bangladesh. NGO Forum comprises over 600 partner NGOs, CBOs and private sector actors who implement water supply and sanitation programmes at the un-served and under-served rural and urban communities. NGO Forum has so far undertaken seven projects and programmes on arsenic mitigation ranging from installing alternative arsenic safe water sources to establishing an information support unit for NGOs. The organization is providing training in 147 upazillas and 100 municipalities. NGO Forum has introduced a number of mitigation options including pond sand filters, arsenic removal plants, rainwater harvesters, dug wells, and piped water system at various regions of Bangladesh.

3.4 Bangladesh Rural Advancement Committee (BRAC)

BRAC, a national private development organization, in collaboration with DPHE and UNICEF, has been conducting an action research programme on community-based arsenic mitigation in two upazillas (Jhikorgacha and Sonargaon) since June 1999. The objectives of this project are to assess the technical viability and effectiveness of various alternative safe water options as well as their social acceptance. BRAC also screens tubewells for arsenic contamination, builds awareness among communities, identifies arsenic patients and provides medical care. They have recently started low-cost piped water supply schemes in several villages in collaboration with the Rural Development Academy and local communities. Under these schemes, local communities contribute 20% of the capital costs and 100% operations and maintenance costs of the projects.

3.5 Dhaka Community Hospital (DCH)

DCH is a trust owned organization and was established in 1988. It is the pioneer organization in the field of arsenic mitigation in Bangladesh. DCH was one of the first institutions to detect arsenic patients in the country. Since then DCH has been conducting arsenic detection and patient identification programmes in many parts of Bangladesh. DCH also manages arsenic patients and builds public awareness through its various programmes.

3.6 Grameen Bank

In October 1997, Grameen Bank started an arsenic mitigation pilot project with assistance of UNICEF and DPHE in Kochua upazilla of Chandpur district. The first phase of the project was started on 20th August 1999 in three unions. In December 1999, the project area was extended to 12 unions. The main objective of the project includes screening of tubewells, awareness and capacity building of the community, and community-based solutions for arsenic contaminated areas.

4 NATIONAL POLICY AND ACTION PLAN FOR ARSENIC MITIGATION

Ministry of Local Government and Rural Development (LGRD) of the Government of Bangladesh has prepared and enacted the National Policy 2004 and Action Plan for Arsenic Problem Mitigation in March 2004. Main features of the policy are given below:

Policy statement

- Safe drinking and cooking water will be ensured through implementation of alternate water supply system in all arsenic prone areas of the country.
- All arsenocosis patients will be identified and kept in an effective management system.
- Impact of arsenic on agriculture and environment will be identified through research and necessary actions will be taken.

Mitigation of arsenic problem

- Public awareness: Raising public awareness regarding the impact of arsenic pollution in groundwater, sources of alternate water supply options and regarding the treatment of arsenocosis patients.
- Alternate options: Selection and implementation of alternate arsenic free technological options that follow drinking water quality standards of Environment Conservation Act, 1995. Use of surface water will be given priority instead of groundwater. At least one source of safe water will have to be ensured for 50 families. Piped water supply system will be developed if possible.
- Arsenocosis patients: Identification and efficient management of arsenocosis patients, trained health care related persons, treatment facilities of arsenocosis patients have to be incorporated in the hospitals and health centres.
- Capacity building: Development of modern laboratory and test kit for analyzing water samples, human resources development for monitoring and surveillance of water supply systems, technical assistance for government organizations, appointment of specialist in health centres.

Institutional aspect

- A high-level committee will be formed to monitor all works as per the national policy.
- A technical committee will be formed to implement arsenic mitigation programmes.
- A strategy plan will be formed to control withdrawal and management of groundwater. Groundwater would be considered a non-renewable natural resource.

Research and development (R & D)

- Research and development works for field kit innovation.
- Research on alternate water supply options like improvement of PSF, rainwater harvesting, development of combined rougfing filter and SSF, development of piped water supply system, installation of small-scale treatment plants and research on existing plants.
- Research on deep aquifer.
- Research on arsenic removal technology development.

5 EFFECTIVENESS OF ACTIVATED ALUMINA FOR ARSENIC REMOVAL

In order to remove arsenic from groundwater, adsorption column study with activated alumina was carried out by Hoque, 2003 under different conditions using synthetic groundwater samples and distilled water in order to investigate the effects of arsenic concentration and oxidation state of arsenic.

5.1 Experimental Design

To conduct experimental run, down flow columns were designed to carry out the study. Graduated glass column of 1 cm^2 in cross-sectional area was used as laboratory test column. The experimental set-up is shown in Fig. 5. A special gravel filter was placed at the bottom of the activated alumina bed to avoid flow restriction. The inlet of the alumina column was connected to a feed tank (25 L plastic bucket) with a flexible rubber tube. To obtain sufficient head, the feed tank was placed on a high platform. To maintain constant head, an overflow tube was connected to the column at a higher position than the inflow connection. In order to minimize the overflow rate, a float system was introduced into the feed tank keeping in mind the constant head of flow.

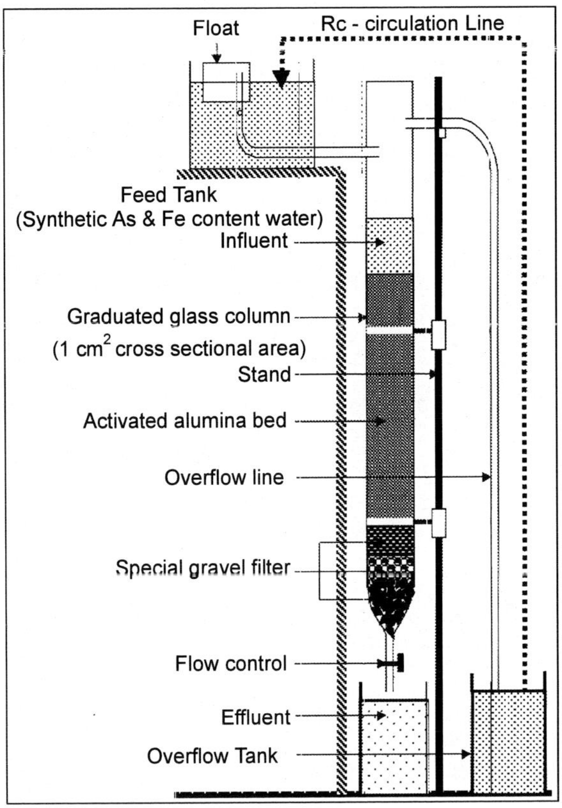

Fig. 5 Experimental design for test bed (Hoque, 2003).

Results of the experiments were expressed in terms of (i) Bed Volume (BV) upto the Maximum Contaminant Level (MCL) of arsenic which is the ratio of the quantity of water treated (ΣQ) upto arsenic MCL in the effluent and the volume (V) of activated alumina packed in the column (ii) quantity of adsorbed arsenic by activated alumina (mg/g), and (iii) overall arsenic removal efficiency of the activated alumina bed upto 50 ppb arsenic level (%).

5.2 Result and Discussion

- Removal of pentavalent arsenic, As (V)

Low arsenic concentration in the influent results lower value of residual arsenic in effluent after the adsorption of arsenic by activated alumina bed, whereas, for higher values of arsenic concentration in influent result high residual arsenic. Influence of As (V) concentration on treated bed volume and quantity of adsorbed arsenic by activated alumina has been shown in Table 4.

Table 4 Influence of As (V) concentration on bed volume and adsorption on activated alumina (Hoque, 2003).

Influent iron = 1.0 μg/L and pH = 7.0

Influent As(V) Conc. (μg/L)	*Bed Volume upto 50 μg/L As level*	*Adsorbed As (V) by Activated Alumina (mg/g)*
100	2115	0.15
300	1383	0.35
500	375	0.16

Arsenic removal efficiency with respect to As (V) concentration is illustrated in Fig. 6. Arsenic removal efficiency was found to be 78.7%, 93.7% and 96.5% for As (V) concentration of 100 μg/L, 300 μg/L and 500 μg/L respectively. Arsenic removal efficiency by activated alumina is directly proportional to arsenic concentration gradient between activated alumina surface and the surrounding.

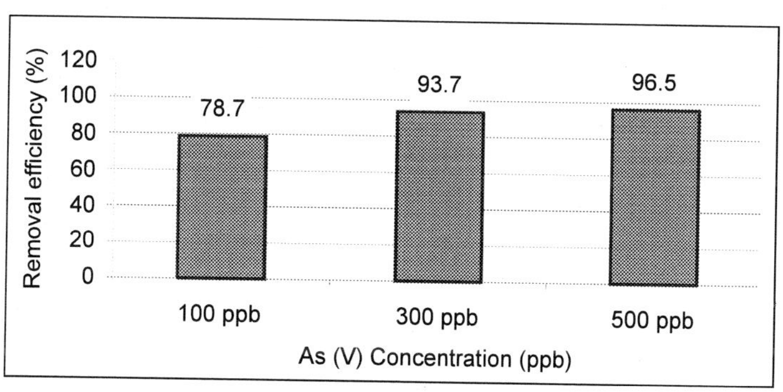

Fig. 6 Arsenic removal efficiency with respect to As (V) concentration (Hoque, 2003).

- Removal of trivalent arsenic, As (III)

Influence of As (III) concentration on treated bed volume and quantity of adsorbed arsenic by activated alumina is shown in Table 5.

Table 5 Influence of As (III) conc. on bed volume and adsorption on activated alumina (Hoque, 2003) Influent iron = 1.0 mg/l and pH =7.0.

Influent As (III) concentration (µg/L)	*Bed Volume upto 50 µg/L As level*	*Adsorbed As (III) by Activated Alumina (µg/L)*
100	372	0.03
300	176	0.04
500	100	0.04

Arsenic removal efficiency was found to be 84.5%, 93.4% and 95.9% for 100 µg/L, 300 µg/L and 500 µg/L arsenic concentrations respectively as shown in Fig. 7. Actually higher arsenic concentration gradient between activated alumina surface and its surrounding cause increase of arsenic adsorption rate on activated alumina surface. As a result, arsenic removal efficiency of activated alumina bed increases when influent arsenic concentration increases.

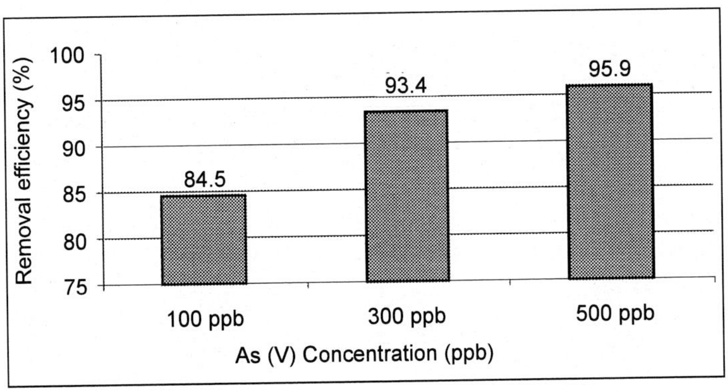

Fig. 7 Arsenic removal efficiency with respect to As (III) concentration (Hoque, 2003).

6 CONCLUSION

Presence of arsenic in groundwater is intensifying the water crisis of Bangladesh. Although more than a decade has passed after the first observation of arsenic in groundwater, however people are suffering for safe drinking and cooking water. Small-scale treatment plant for arsenic removal is feasible for the poor. Small/large-scale community based piped water supply system can be a good option.

References

BBS (1998) *Sample Vital Registration System*, Bangladesh Bureau of Statistics, Ministry of Planning, Govt. of People's Republic of Bangladesh.

BBS (1999) *Statistical Year Book of Bangladesh,* Bangladesh Bureau of Statistics, Ministry of Planning, Govt. of People's Republic of Bangladesh.

BBS (2001) *Population Census 2001: Preliminary Report,* Statistics Division, Bangladesh Bureau of Statistics, Ministry of Planning, Government of Bangladesh.

BGS and DPHE (2001) Arsenic *Contamination of Groundwater in Bangladesh,* Kinniburg, D.G. and Smedley. P.L. (Editors), Volume 2: Final Report. British Geological Survey Report WC/00/19.

DPHE, (2000) *Year Book (1998-99): Rural Water Supply,* Planning Circle, Department of Public Health Engineering, Bangladesh.

DPHE, BGS, and MML (1999) *Groundwater Studies for Arsenic Contamination in Bangladesh,* Final Report (Phase-1), Department of Public Health Engineering, British Geological Survey and Mott McDonald Ltd.

DPHE and UNICEF (1994) Study *to Forecast Declining Groundwater Level in Bangladesh,* Department of Public Health Engineering, Bangladesh.

Hoque, M.A. (2003) *A study of the effectiveness of activated alumina in removing arsenic from groundwater.* M.Sc. Engg. Thesis, Bangladesh University of Engineering and Technology, Dhaka-1000, Bangladesh.

Management of Groundwater Resources through Community Participation Approach in the Coastal Saline Region of Gujarat

6

Shital Lodhia
Centre for Development Alternatives, E-71 Akash, Bodak Dev, Ahmedabad, India

ABSTRACT

Overwithdrawal of groundwater in the coastal region of Gujarat has made water table lower. This lowering of water table has resulted in saline water intrusion due to reverse hydraulic gradient and capillary action in the coastal region. Implementing regulatory and economic reforms to control groundwater use is rather difficult for the state. In this situation, community based regulatory approach is the most appropriate approach. This paper presents a case study from the Gujarat coastal region in which community based regulatory mechanism has facilitated local people to maintain their livelihood by using groundwater in an efficient manner. Two major strategies, namely, use of drip irrigation and promoting horticultural crops, are studied in detail for their implications on groundwater resources. The study has shown that community based strategies have drastically reduced withdrawal of groundwater for those who have adopted above strategies.

1 INTRODUCTION

The coastal zone of Gujarat is widely affected by soil and water salinity. However, the nature and extent of salinity vary over the space and are directly related to the hydrology, coastal geomorphology and human interventions in the coastal environs. In addition to very low amount of precipitation, the coastal region is suffering from high salinity as a result of natural processes as well as manmade activities.

Agriculture intensity has increased the amount of water used per unit output. Change in cropping pattern along with free electricity has further raised the demand of groundwater for agricultural purpose in the coastal region of Gujarat. As a result, the rate of withdrawal of groundwater exceeded the rate of recharge. Overwithdrawal of groundwater has reduced groundwater level. The coastal region of Saurashtra is made up of deep porous geological formation. Reducing water table in such porous geological formation has resulted in saline water intrusion due to reverse hydraulic gradient and capillary action in this area. The level of groundwater development in the coastal talukas of Saurashtra have remained more than 85 per cent since 1991 indicating the alarming situation of declining water table in the coastal region. This decline in groundwater table has further accelerated the process of salinity ingression.

Implementing regulatory and economic reforms to control groundwater use is rather difficult for the state. In this situation, community based regulatory approach is the most appropriate approach. This paper presents a case study from the Gujarat coastal region. Community based regulatory approach here has facilitated local people to maintain their livelihood by using groundwater in an efficient manner. The case study has looked into the community-based mechanism adopted by village level institutions namely; Pani Samiti, Kharash Atkao Juth (KAJ) and Kharash Atkao Sankalan Samiti (KASS) supported by AKRSP (I). An attempt has been made here to understand the level of awareness and knowledge of local people regarding the activities carried out by village level institutions. Two major strategies, namely, use of micro irrigation devices and promoting horticultural crops, are studied in detail for their implications on groundwater resources. The study has shown that community based strategies have drastically reduced withdrawal of groundwater of farmers who have adopted these two strategies. It has also improved the livelihood of local people, who were otherwise vulnerable and exposed to frequent droughts.

The study includes the survey of three salinity-affected coastal villages, namely, Kukaswada, Virol and Farangta. All the three villages are selected from Mangrol taluka located in the costal belt of Gujarat. In order to assess the role of level institutions and their impact on villagers, a representative sample from village households is selected. A stratified random sampling method is used for selecting sample households. A detailed house listing was carried out in all the three villages for selecting sample. Households were classified as per landholding categories for all the villages. From each landholding category, around 20% of households were selected on a random base. A total of 210 farmers were selected according to different landholding categories.

2 GROUNDWATER RESOURCES OF THE REGION

Hydrogeologically, Mangrol taluka of the coastal area is formed by younger (quaternary) sedimentary. The miliolite aquifers occupy around 10–15 km wide coastal belts with 50 m of thickness. This forms phreatic aquifers, which contains highly variable permeability. It is assumed that salinity intrusion in the coastal region of Gujarat is accelerating at an average annual rate of 5.8 km towards land from the coastline.

Gross draft in Mangrol taluka has increased from 39 mcm in 1991 to 50 mcm in 1997. The level of groundwater development has also increased from 70% in 1991 to 90% in 1997. In fact, during the year of 2001 the level of development is observed to be more than 100 per cent putting the taluka in overexploited category. This indicates the alarming situation of over-withdrawal of groundwater that has resulted in decline and deterioration of groundwater.

Water balance analysis of water in the coastal belt of Mangrol taluka indicates that much of the rainwater (69%) falling in the coastal belt of Mangrol taluka flows to the sea. High run-off

Table 1 Groundwater potential in Mangrol taluka of coastal Gujarat.

Year	*Total groundwater resources in mcm*	*Gross draft in mcm*	*Level of groundwater development in %*	*Category*
1991	46.03	38.96	69.7	Grey
1997	68.93	49.57	90.07	Dark
2001	Not Published	Not Published	>100	Overexploited

Source: GWRDC, Gandhinagar

Table 2 Accumulation of rainwater, flow to sea and interface intrusion in the coastal belt.

Indicators of salinity ingression	*Mangrol Taluka*
Length reach (km)	28
Gaj width	6
Rainfall mm/yr	864
Accumulated rainwater in mcm/year	14.69
Flow to sea mcm/yr	10.00
% of rainfall flowing to sea	69
Intrusion interface (m)	1400
GWL at toe in metre	0.75

SIPC, Rajkot

along with less upstream recharge makes the balance of groundwater disturbed and finally lowers the water table. On the other hand, the withdrawal of groundwater continues to be more and more. As a result of this, groundwater level declines and the process of marine intrusion accelerates due to the reverse hydraulic process.

Intrusion interface data shows that seawater intrusion has reached more than one km in Mangrol taluka. Moreover, data on GWL at Toe shows that the lowest surface of water table is less than one metre above the mean sea level. This connotes the danger of disappearing of transition zone of interface that makes the groundwater totally saline, which may not become sweet even though much of the freshwater is recharged. The rational use of water and proper management using new technology for checking the freshwater flow is of an immediate necessity for the taluka. It will not only solve the problem of drinking water for this region but also help in preventing salinity intrusion.

3 INSTITUTIONAL DEVELOPMENT AND COMMUNITY CAPACITY BUILDING

The close linkage between local institutions and natural resource management has increasingly recognized the importance of local institutions for natural resource management. It is well established and accepted that even in a less developed stage, local institutions can help in using resources in a sustainable manner and protecting natural resources from further deterioration. One of the major objectives of the institutional development is capacity building. The present study looks into the role of institutional development and capacity building in promoting people to adopt horticultural crops that use less water, as well as in implementing water efficient technology to protect overwithdrawal of groundwater for preventing salinity ingression.

Basically, three types of local institutions have been working to address the salinity issues in the coastal villages of Mangrol taluka for the last four years. The first effort under community organizing strategy was the formation of Pani Samiti. The major functions of Pani Samiti are to promote people's movement and to plan for physical interventions. Later on KAJ (Kharash Atkao Juth) was formed to address specific issues of salinity prevention. Expanding the community organizing strategy, KASS (Kharas Atkao Sankalan Samiti) was developed for coordinating KAJ of different villages and helping in exploring market linkages. All the three institutions are supported by AKRSP (I) in their technical and financial matter.

The major activities carried out by these institutions with the support of AKRSP (I) are listed below:

- Community awareness building: Awareness and capacity building measures have remained very effective tools for making people participate in salinity ingression prevention activities. The major activities of institutions include various mass awareness campaigns, frequent meetings, folk theatres, street play, foot march, *Kharas Atkao cricket tournament, Mahila Milan, Kharas Atkao Milan, Sarpanch Milan,* various kinds of training and workshops; and exposure trip for motivating people to participate actively in the efficient use of groundwater.
- Promoting micro irrigation devices like drip irrigation and sprinklers.
- Selecting households for the construction of Roof Rainwater Harvesting Structures (RRWHS)
- Selection and construction of well sealing. Well sealing is an effective tool to prevent mixing of the saline water with sweet well water. In this technique, bottom of the well is sealed up to two feet with cement and concrete.
- Watershed management and watershed plus activities.
- Promoting groundwater recharge techniques like creation of check dams, percolation tanks, the construction of stone embankments on fields and well recharging in the lower level of farms.
- Well water testing by using a kit of water testing for making the localities aware about the drinking water quality.
- Promoting horticultural crops by demonstrative farm practising, training and exposures.

4 CREATING AWARENESS TO PARTICIPATE IN DEVELOPMENT ACTIVITIES

It is important to see how these institutions have encouraged people to use groundwater efficiently in their villages. The villagers are very enthusiastic about the working with all the institutions. Above 80% of the villagers agreed that the work of these institutions is helpful in some way or the other. It has been found during the survey that the people are aware of the various development related programmes conducted by these institutions in their villages.

These institutions have created awareness among the villagers to know their rights and how their participation in development work can improve the situations. Women are encouraged not only to participate in these institutions but also to take part in decision-making activities. Knowing about the importance and benefits of creating their own groups, people have also started participating in other community development institutions like Mahila-mandal, youth association and other village level institutions.

Table 3 Number of male and female members in village institutions.

Village	*Pani Samiti*		*KASS*		*KAJ*		*Total*	
	Female	*Male*	*Female*	*Male*	*Female*	*Male*	*Female*	*Male*
Kukaswada	0	0	4	14	56	167	60	181
Virol	12	12	0	0	0	0	12	12
Farangta	0	0	0	0	25	37	25	37
Total	**12**	**12**	**4**	**14**	**81**	**204**	**97**	**230**

Source: AKRSP (I), 2004

Table 4 Percentage of sample households who are the members of various organizations.

Member of any association or mandali	*(%)*
Panchayat	5.8
Mahila mandal	12.32
Youth association	1.45
Other Village level Institutions	36.96
Pani Samiti	15.22
KAJ	28.26
KASS	3.5

Source: Fieldwork, 2004

Around 36% of the sampled households have one or the other member actively participating in institutional activities. Approximately 28% of sampled households have their family members working in KAJ, while around 15% of them have their family members working in Pani Samiti.

It is interesting to note here that over 70% of the respondents are aware about the various activities conducted by these institutions for the prevention of salinity ingression in their villages. Most of them found village level institutions to be helpful. On an average 16% of respondents specifically mentioned their awareness and knowledge about techniques used for efficient use of groundwater.

5 WORKING OF INSTITUTIONS IN THE SELECTED VILLAGES

In order to have an idea about capacity building through various activities carried out by village level institutions, it is important to understand the working of village level institutions and the type of participation taking placed by villagers.

When asked about the frequency of meeting, most of the respondents were of an opinion that meetings of these associations are held once (more than 60%) or twice in a month. Attendance at these meetings is fairly regular. Virol has the highest number of respondents (92%) attending meeting regularly, which is followed by Kukaswada with 82% and Farangta with 79%. The important observation here is that the regular turnout in these meeting is not merely for their presence, but members actively participate in discussion and give their opinion during the meetings (over 80%). On an average, one-fourth of the respondents have contributed their opinion in decision-making and another one-fourth of them has participated very actively in discussion on any issue. The success of these institutions stated by the villagers is because of the direct public participation, through the regular public meetings.

Table 5 Awareness about the work carried out through Pani Samiti, KAJ and KASS.

Village	*Helpful %*	*Different programmes %*	*Organize public meeting %*	*Village development works %*	*Efficient use of ground water %*	*No response %*
Kukaswada	80.65	9.68	3.23	4.5	16.67	0
Virol	8.7	7.69	23.08	15.38	20	25.15
Farangta	7.14	35.71	7.14	8.56	14.29	27.16

Note: Total percentage is not equal to hundred because of the multiple answers.

Table 6 Frequency of meetings.

	Every 15 days %	*Every month %*	*After 1.5 month %*	*After 2 months %*	*As per requirement %*	*No response %*
Kukaswada	12.9	45.2	0	3.2	0	38.7
Virol	7.7	92.3	0	0	0	0
Farangta	7.1	71.4	0	0	7.1	14.3

Source: Fieldwork, 2004

Table 7 Regular attendance at the meetings.

Village	*Yes %*	*No %*	*No response %*	*Total*
Kukaswada	81.9	12.9	5.2	100
Virol	92.3	7.7	0.0	100
Farangta	78.6	21.4	0.0	100

Source: Fieldwork, 2004

Table 8 Role-played by members in various meetings (%).

Village	*Attendance %*	*Listening %*	*Active participation in decision%*	*Opinion %*	*Participation in discussion %*	*No response %*
Kukaswada	5.13	23.08	20.51	10.26	41.03	0
Virol	6.67	13.33	26.67	36.67	16.67	0
Farangta	2.94	8.82	29.41	29.41	23.53	5.88

Source: Fieldwork, 2004

The proper meeting turns into good learning. Training given to the stakeholders in various meetings plays an important role in capacity building in the process of institutional development. The training programmes, rallies and meetings have benefited the residents to a great extent. On an average 45% households have learnt to prevent salinity ingression. People have realized the importance of water. Inspired by the local institutions, they have started adopting horticultural crops and using micro irrigation devices for efficient use of water. Drip irrigation technique for agriculture has gained importance in Kukaswada villages. Sprinklers have been adopted by many farmers in Virol and Farangta. It was found from the survey that many people want more knowledge about the various techniques, which they learnt during their training.

Table 9 Learning from the meetings and trainings.

Village	*Well Recharge* %	*Saving Water* %	*Use of Rain Water* %	*Efforts to prevent salinity* %	*Obtain useful Information* %	*Drip Irrigation and Sprinklers* %
Kukaswada	0	21.4	21.4	42.9	14.3	42.9
Virol	15.38	9.1	0	72.7	18.2	3.4
Farangta	7.14	50	12.5	19.64	25	2.3

*Note: Total percentage is not equal to hundred because of the multiple answers.

In short, the analysis shows that the activities carried out by institutions have helped villagers to become aware and sensitive towards the increase of salinity ingression and to use groundwater in efficient manner. Villagers have realized the value of water. They have also learned the techniques for water conservation and management. However, they need more and more information regarding the various techniques.

6 WATER MANAGEMENT IN AGRICULTURE

As mentioned earlier, salinity ingression due to overwithdrawal of groundwater has been the major problem of this region. In order to control withdrawal of groundwater it is very important that the water is used in a wise manner. Water management in agricultural usage becomes important in this regard. Local level institutions KAJ and KASH promote horticultural crops and encourage farmers to adopt micro irrigation practices such as drip irrigation and sprinkler as technical devices for saving water used in agriculture.

It is important to understand the cropping pattern in order to see the impact of horticultural crop promoting strategy. The following table shows that the selected villages have different cropping patterns. Kukaswada has more area under horticultural crops, while Virol and Sangawada have started cultivation of Chikoo. Different cropping patterns in all the selected villages make variation in overall figures of gross draft of groundwater. Adopting horticultural plants is rather difficult and there is need to change the mindset of farmer. Return from cash crops like groundnut is more in short term when compared to that of Chikoo plantation as Chikoo plantation starts giving fruits only after five years. However, over a period of time the return increases in Chikoo plantation. Different research studies and surveys have proved that the returns from Chikoo plantation over a period of twenty years are much more than that of groundnut. (Mishra, 2002; Lodhia, 2004).

Table 10 Village-wise percentage of area under different crops.

Village	*Groundnut* %	*Wheat* %	*Bajra/Juwar* %	*Coconut* %	*Chikoo* %	*Banana* %	*Grand Total* %
Kukaswada	31.2	9.4	16.7	67.4	17.4	13.0	100
Virol	89.7	55.2	34.5	0.0	6.0	3.4	100
Farangta	97.7	58.1	34.9	2.3	1.0	0.0	100

Source: Fieldwork, 2004

Horticultural crops require less water. This is reflected in case of gross draft calculated for each crop. Groundnut is mostly rain fed crop. However, the frequent droughts and vagaries of rainfall forces farmers to irrigate groundnut from well water. Wheat is a winter crop, which requires more water for irrigation. An analysis of gross draft for different crops in these villages shows that water requirement for horticultural crops are less than half of water requirement in groundnut or wheat. Thus, promoting horticultural crops can drastically reduce withdrawal of groundwater.

Using micro irrigation techniques for horticultural crops further declines the withdrawal of groundwater. The following table shows that the difference in gross draft of farmers with and without using micro irrigation devices for every crop is very high. It shows that the quantity of water draft for irrigation can be declined by 33% in groundnut, 67% in coconut and 49% in banana by using micro irrigation techniques.

Table 11 Average gross draft for different crops with and without using micro irrigation devices.

Indicators	*Groundnut m^3*	*Wheat m^3*	*Bajra m^3*	*Coconut m^3*	*Chikoo m^3*	*Banana m^3*
Average Gross Draft for sample (Without drip or sprinkler)	10563	15554	2372	6364	5491	646
Average Gross Draft (With drip or sprinkler)	7093	NA	NA	3098	NA	330
Average Gross Draft per *Vigha* (Without drip or sprinkler)	3143	5025	846	2077	631	430
Average Gross Draft per *Vigha* (With drip or sprinkler)	315	NA	NA	253	NA	132

Source: Fieldwork, 2004

It is important to see how the micro irrigation techniques help in conservation of water in our study area. This is explained by using indicators like pumping hours and drafting of water. The following table shows the average pumping hours of wells owned by farmers.

Table 12 Average hours of pumping for farmers.

Village	*With micro irrigation*	*Without micro irrigation*	*Total Average*
Kukaswada	14	93	90
Virol	136	240	142
Farangta	133	239	222
Total	**139**	**199**	**140**

Source: Fieldwork, 2004

Table 9 shows a big gap in average hours of pumping between farmers with and without using micro irrigation devices. Pumping hours for the farmers who are using micro irrigation devices are less. The ratio of pumping hours between beneficiaries and non-beneficiaries of micro irrigation devices is 1:7 for Kukaswada, 1:1.7 for Virol and 1: 1.8 for Farangta. Majority of the area in Kukaswada taluka is under horticultural crops. As horticultural crops require less water, the average pumping hours in Kukaswada is less. Further, the difference in pumping hours between farmers with and without drip irrigation techniques in Kukaswada is quite visible showing the effectiveness of micro irrigation techniques in horticultural plantation. On the other hand, sprinklers are widely accepted for groundnut crop in Virol and Farangta. This shows the effectiveness of the technical devices of micro irrigation in reducing pumping hours.

It is important to note here that pumping hours depend on other four variables, namely, capacity of pump, depth of groundwater, cropping pattern and area under irrigation. Pumping hours can be less when capacity of pump is high or the depth of groundwater is less. All the three villages are located near coastline. Therefore, the groundwater level is not much deep and the variation is less among coastal villages of Mangrol taluka starting from 15 m to 25 m. Secondly, the range of capacity of pump used by farmers in this region varies from 5 HP to 10 HP. In order to eliminate the impact of both these variables, (depth of groundwater and capacity of pump) per *Vigha* (Unit of area) gross draft is calculated.

The effect of micro irrigation devices is quite effective when average gross draft per unit of land (*Vigha*) is calculated. The following tables 10 and 11 show that gross draft per *Vigha* of land is much less when micro irrigation technique is used. The difference is quite high in Farangta as beneficiaries of micro irrigation devices withdraws only 284 cubic metres to irrigate one *Vigha* of land as against of 2150 cubic metres groundwater withdrawal by non beneficiaries. In case of Virol, non-beneficiaries withdraw five times more water than that of beneficiaries to irrigate one *Vigha* of land. However, the reason for such a big difference in these two villages is due to different cropping patterns of beneficiaries and non-beneficiaries. Most of the beneficiaries grow horticultural crops, which require less water. Irrigating such horticultural crops with drip irrigation device drastically reduces the use of water resources. In case of Kukaswada, the difference is less as many non-beneficiaries also cultivate horticultural crops, which require less water.

Table 13 Average gross draft per unit of land.

Village	*With micro irrigation* m^3	*Without micro irrigation* m^3
Kukaswada	150	195
Virol	345	1592
Farangta	284	2150

Source: Fieldwork, 2004

Table 14 Average gross draft per farmer.

Village	*With micro irrigation* m^3	*Without micro irrigation* m^3	*Total Average* m^3	*Reduction in gross draft by using micro irrigation devices %*
Kukaswada	1956	6303	6103	69.0
Virol	6905	10694	10441	35.4
Farangta	7281	19460	17430	62.6

Source: Fieldwork, 2004

It is important to calculate how much reduction occurs in gross draft per farmer when both the strategies are adopted together. Gross draft per farmer has reduced by 69% in Kukaswada, 35% in Virol and 63% in Farangta for the farmers who have adopted micro irrigation devices.

Furthermore, Kukaswada is showing less withdrawal of water per farmer compared to other two villages; as most of the farmers in Kukaswada have adopted horticultural crops. Once again the table confirms that farmers using micro irrigation devices for irrigating land are withdrawing less water.

In short, overall groundwater withdrawal can be controlled by using modern irrigation techniques in a vast geographical area. Such type of water management in agriculture can help in maintaining balance of groundwater resource of the region.

7 CONCLUSIONS

- Activities carried out by village level institutions helped the villagers to become aware and sensitive towards the increase of salinity ingression. Villagers have realized the value of water. They have also learnt the techniques for water conservation and management. However, they need more and more information regarding the various techniques.
- In order to reduce the withdrawal of groundwater, the role of village level institutions is quite significant. The work of KAJ, KASS and Pani Samiti has started showing their positive effects through various indicators. Adopting horticultural crops and using drip irrigation or sprinklers have drastically reduced withdrawal of groundwater in these villages. With more and more farmers taking the benefits of such devices, it will make a positive impact on larger scale by improving water resources of the region and protecting the livelihood of local people.
- Finally, there is no doubt about it that the control on withdrawal of groundwater along with recharge at a large scale is the only solution to regenerate the declined and deteriorated groundwater resources. This will ultimately result in preventing salinity ingression and protecting and maintaining the current level of livelihood in this region.

References

AKRSP(I). 2002. *Annual progress report 2001*, Aga Khan Rural Support Programme (India), Ahmedabad.

Barber W. 1989. *Mission report on groundwater management stud.* Aga Khan Rural Support Programme (India). Junagadh. Gujarat.

GWRDC. 1998. *Report of the committee on estimation of groundwater resource and irrigation potential in Gujarat state.* Narmada and Water Resources Department Government of Gujarat, December 1998.

Lodhia S. & Singh H. 2004. *Baseline survey for the SCALE programme of AKRSP(I).* A report prepared for AKRSP(I). Centre for development alternatives, Ahmedabad, Gujarat.

Mishra T. 2002. *Feasibility study of Chikoo cooperative in Gadu programme area.* Institute of Rural Management, Anand, August 2002.

Srinivas M., *et al.* 2004. *Community Management of Groundwater resources in rural India: A study of Bhanavas, Samrapur and Nana Kothasana villages of Mahesana district.* VIKSAT. Ahmedabad.

Arsenic Hot Spots Detected in Patna District, Bihar (India) – A Serious Health Hazard for Estimated Human Population of 5.5 Lakhs

7

Ashok K. Ghosh, Shatrunjay K. Singh, Nupur Bose, N.P. Roy, Sushant K. Singh, Ajay K. Upadhyaya, Shailendra Kumar and Amardip Singh

Department of Environment and Water Management, A.N.College, (Magadh University), Boring Road, Patna-800013, India

Tel. No. :91-612-2282387, E-mail: ashok.ghosh51@gmail.com

ABSTRACT

Globally, arsenic contamination of groundwater is a serious emerging health hazard. The permissible limit for arsenic content in drinking water is 10 µg/L, as per WHO Guidelines. However, this study team has detected arsenic content in groundwater in the State of Bihar, that is much higher [almost three times more] than that found in the Bengal Basin. In modern parlance, arsenic is viewed as being synonymous with 'toxicity', with fatal implications for the affected populace.

The purpose of this study was to determine the existence and intensity of Arsenic Contamination of Groundwater resources in Patna, the most heavily populated district of Bihar, located along the River Ganges. The findings will aid in mitigation program for the arsenic affected population.

The methodology involves use of Field test kits for initial detection, followed by testing of the confirmed samples for arsenic content by Atomic Absorption Spectrophotometer.

This study has detected arsenic contamination in many of the public hand pumps that supply water for drinking and domestic purposes to the rural population of Patna. Clinical symptoms of srsenic poisoning among the affected population has also been detected. The concentration of this contaminant varies within a 5 km. belt, on the southern bank of the Ganga. The highest value of Arsenic content detected in the study area is 724 µg/L. Three epicenters of arsenic hot spots have been detected in Patna District. Distinct groundwater arsenic iso-lines with tapering arsenic concentrations have been detected for the first time in the study of arsenic contamination in groundwater in the region. There also exist intervening stretches of Arsenic-free aquifers. The arsenic contamination in the groundwater of the study area is alarmingly high. An estimated population of 5.5 lakhs is potentially exposed to serious health risks, with possible rise in morbidity in the next 5 -10 years.

Keywords: Arsenic contamination, toxicity, Atomic Absorption Spectrophotometer, groundwater, arsenic aquifers.

1 INTRODUCTION

W.H.O. (World Development Report,1999-2000) has published Guidelines for Drinking Water Quality in which a contemporary value for an acceptable maximum level of arsenic was set at 10 µg/l/ for safe water. National standards range from 7 µg/l in Australia to 50 µg/l in Vietnam, Cambodia and Bangladesh (Fazal *et.al.*, 2001) . Naturally occurring Arsenic, as a water quality issue, began to attract international attention in the early decade of the nineties, when widespread chronic arsenic poisoning cases became apparent in Bangladesh and later in West Bengal, India (Chadha *et.al.*, 1999: Chakrovorty et.al, 1994). Later, the discovery of arseniferrous aquifers in Simaria-Ojhapatti village under Shahpur Block in Bhojpur, Bihar, located further upstream of the Bengal basin in 2003, raised apprehensions about the greater spatial spread of this dreaded contamination along the Ganga Valley.

Arsenic in drinking water poses the greatest threat to public health (Chakrovorty, 1989). Arsenic is highly toxic carcinogen (Category I) and also a mutagen / teratogen (harming foetus). Long-term oral exposure via drinking water can cause cancer of the skin (Astolfi *et.al.*, 1981)' lungs, urinary bladder, and kidneys. With long term exposure the first changes are usually seen in the skin pigments (indicator of arsenic poisoning), then hyperkeratosis. Symptoms of chronic arsenic poisoning can take five to 15 years to appear and are apparently influenced by nutrition and general health standards (Chaudhary *et.al.*, 2000). People tend to overlook or even forget the exposure. Increased risks of lung and bladder cancer and of skin lesions have been observed at arsenic concentrations of less than 0.05 mg / L of drinking water. High risk of prostrate cancer has been reported in Victoria, Australia where high levels (400 µg/L) of arsenic occur in groundwater. Some studies have reported hypertensive and cardiovascular diseases, diabetes and reproductive affects of arsenic poisoning.

Arsenic occurs widely in aquifers of deltaic sediments (Acharya *et.al.*, 1999), near zones of orogeny, and in deep sandy aquifer layers (AusAid Guidelines, 2004) as fluvial deposits. In 2002, Prof. D. Chakraborty of the School of Environmental Studies, Jadavpur University, detected high arsenic levels similar to the Bengal contamination readings. This called for further detection of Arsenic infestation in the surrounding aquifers near the Ganga (Rajiv Gandhi National Drinking Water Mission, 1998-99), within Bihar. The cause of this contamination is still subject to hypotheses.

The study area of this research incorporates Patna District, being justified by its close proximity to the Bhojpur arsenic contaminations. The two districts are separated by the River Sone which meets the Ganga on the northern boundaries of both Bhojpur and Patna. The study area of Patna district was confined to an approximate belt of 10 kms. along the southern bank of River Ganga, based on the findings of previous studies stating the greater possibilities of contamination in newer fluvial deposits (Fazal *et.al.*, 2001; Morgan *et.al.*, 1980).

2 METHODOLOGY

This research team formulated the Protocol, duly approved by Govt. of Bihar, Govt. of U.P., and UNICEF, for detection of arsenic over large areas. This Protocol is specifically intended for use by all those involved with arsenic detection in hand pumps using field test kits. The efficacy of the FTK results were tested in Sriram Institute of Industrial Research, New Delhi, where a positive correlation of over 6.0 was noted between them and AAS results of the same water samples. The methodology incorporated in the Protocol has the following five significant components:

- All the public hand pumps of the study area were tested for Arsenic content by trained field workers through Field Test Kits manufactured by CHEM-IN Corporation, Pune.
- Each public hand pump tested was marked, with a unique Identification Code at the time of testing. The 13-digit code was derived from the Census 2001. This information identified the State, District, block and village, and finally the serial number of the Hand pump.
- Bihar Plains consist of sandy soils, where recent research has shown that presence of silica often masks the signal of arsenic from many chemicals (AusAid Guidelines, 2004), thereby "creating a false optimum" of arsenic content lower than actually existing. Hence, 10% random verification of field test kit results was done through Atomic Absorption Spectroscopy.
- All field test results above 40 mg/L were retested using AAS (APHA - 1996).
- Marking of hand pumps were done to inform the public of their status (safe/unsafe). The color codes applied were - Blue for FTK readings below 40 µg/L; yellow for FTK readings 40- 79 µg/L; and, Red for FTK readings of 80 µg/L and above.
- There is an increasing need to map the level of arsenic concentration, trends of arsenic flow and temporal changes occurring in its concentration levels. Recording the locations of Arsenic-affected hand pumps, using Global Positioning System (GPS) units (Basu and Sil, 2004), was done, followed by mapping of the arsenic occurrences. The other references used were Block Maps, Topographical Maps.

3 ANALYSIS OF F.T.K. DATA

Of the total of 4989 public hand pumps surveyed and sampled by Field Test Kits till date, 1060 hand pumps were contaminated with arsenic. Taking into account the permissible limit of 10 µg/L set by WHO(WHO,1993), the table below reveals massive contamination in Maner, Danapur, Bakhtiarpur, and Barh blocks. These arsenic hotspots, forming the first group, were found to be highly concentrated in the Diara region. It would not be out of place to mention that there is a continuity of occurrence of ground water arsenic, from Bhojpur district in the west to Patna in the east. The arsenic readings tapered down towards Mokama to comparatively lower concentrations.

The second group of blocks, registering a fair presence of arsenic above 10 µg/L included Mokama, Sampatchak, Patna, Fatuha and Phulwari. The general areal extent of this contamination is in close proximity to the right bank of river Ganga, except in Sampatchak, where arsenic concentrations of 50 µg/L low. and below were spread over the entire block area . While most of the blocks had a considerable number of "safe" or BDL [below detectable limit] marked hand pumps, in Maner maximum contaminated hand pumps were identified. Except Bihta, Naubatpur, Ghoswari and Daniawan all the blocks of Patna district had contamination levels of above 10 μg/L to 50 μg/L, that urgently requires regular monitoring.

The following table is the summary of FTK data of the Study Area in Patna District.

Table 1 Status of Total No. Of Samples Patna District FTK data [BDL=Below Detection Limit]

S.No.	*Name of Block*	*BDL*	*1 to 10 mg*	*11 to 50 μg/L*	*51 to 100 μg/L*	*100+ μg/L*	*Total*
1.	**Maner**	555	38	188	100	76	957
2.	**Bihta**	240	—	—	—	—	240
3.	**Naubatpur**	369	—	—	—	—	369
4.	**Sampatchak**	123	38	32	—	—	193
5.	**Phulwari**	324	21	06	—	—	351
6.	**Danapur**	168	12	38	24	11	253
7.	**Patna**	406	16	04	—	—	426
8.	**Fatwah**	173	31	05	—	—	209
9.	**Mokamah**	290	61	37	—	—	388
10.	**Ghoswari**	18	—	—	—	—	18
11.	**Khusrupur**	188	26	02	—	—	216
12.	**Daniyawan**	155	—	—	—	—	155
13.	**Bakhtiarpur**	515	82	50	14	37	698
14.	**Pandarak**	178	38	17	—	—	233
15.	**Barh**	227	26	17	06	07	283
	Total	**3929**	**389**	**396**	**144**	**131**	4989

4 MAPPING OF ARSENIC HOTSPOTS IN PATNA DISTRICT, BIHAR

The base map was prepared on the basis of FTK data of 4989 water samples to identify the Arsenic hot spots identified in Patna District. A total of 171 villages were surveyed in Patna. This map was prepared manually with the help of block maps. Map scales differed among all the block maps. Hence repetitive enlargement and reduction of the respective block maps were done. Again, distortion of the river channel could not be eliminated manually in this map.

To obtain the base map of the study area on same scale, Topographical Sheets, and G.P.S. Readings were integrated with inputs from the block level maps. This map depicts the spatial spread of arsenic contaminated ground water being tapped for human consumption. Every block had some degree of arsenic contamination, with Maner having more than 40 %, and Danapur and Sampatchak having 30-40% contaminated samples. Sampatchak was the single block with diffuse spread of arsenic contaminated hand pumps – low intensity arsenic contamination existed in almost every village, and the highest reading went upto 50 μg/L. In Patna block, due to access of deeper safe aquifers in urban areas, less than 10 % of the samples yielded more than 10 μg/L arsenic content. The remaining blocks had 10 to 40 % contaminated water samples.

The highest arsenic contaminations of each affected village, marked with coloured dots for studying the intensity of the contamination, were located on G.P.S. In this map, 4 clusters of hotspots were noted – Maner cluster, Danapur Cluster, Bakhtiarpur Cluster and Barh Cluster. The red spots, indicate arsenic levels of > 110 μg/L in all the clusters. Some hand pumps with below 10 μg/L and even "nil" contamination were also found. All these hotspots were in the Diara or floodplains that are subject to annual inundation. Mokama and Pandarak had frequent occurrences of 50 μg/L too. Gradual decreases in the number of highest readings were noted from west to east. Also, arsenic concentrations tapered down towards south, supporting the hypothesis of arsenopyrite oxidation of Holocene sedimentary deposits (Chadha & Sinha Roy, 1999; Rajiv Gandhi Drinking Water Mission). This oxidation has been attributed to excessive withdrawal of ground water by the inhabitants, both for domestic purpose and agricultural use.

As one moved away from the riverbank, the frequency of arsenic contaminated sources decreased. Tests were also extended southward along the right bank of river Sone in order to seek spread of contamination. However, all the samples collected from the Sone bank further south of its confluence with the Ganga, tested negative. It can be safely concluded that the menace of Arsenic in ground water exists in Patna district, parallel to the Ganga riverbank, in a broken belt from Maner in the west to Mokama in the east. These findings need further regular monitoring to detect future changes in its extent and intensity.

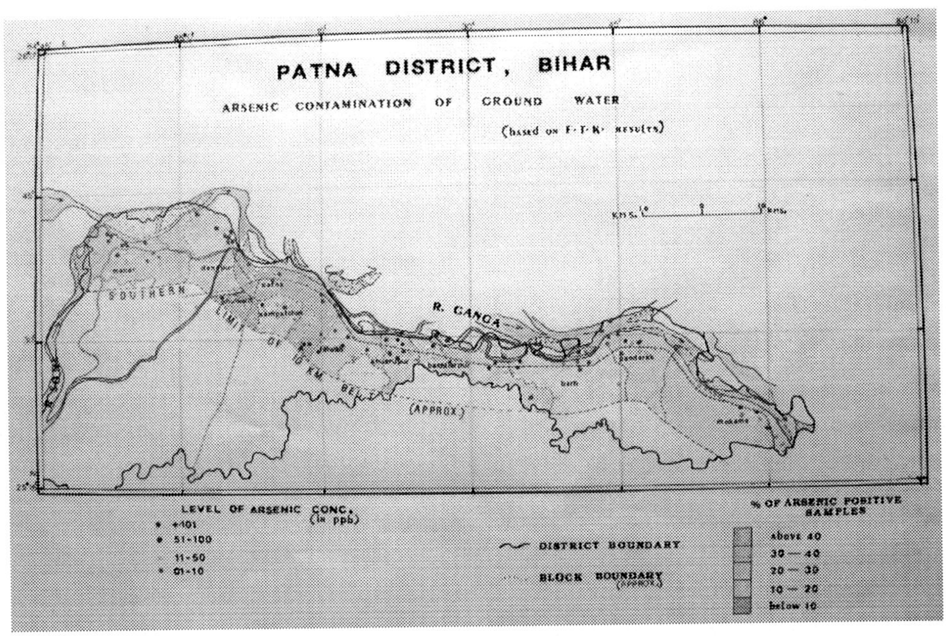

Map 1 Arsenic Contaminated Drinking Water hotspots, Patna District [based on F.T.K. Data and G.P.S. Fixes]

The following G.P.S. coordinates were obtained for the AAS tested high arsenic contaminated villages in Maner, Danapur, Bakhtiyarpur and Barh for mapping procedures.

5 ANALYSIS OF AAS RESULTS

Out of 6101 samples, 457 samples were retested by AAS-HG method at Sriram Institute of Industrial Research, New Delhi. Of the 457 samples tested there, 390 samples were confirmed

Table 2 GPS Data-Latitude, Longitude, and Altitude of As. Hot Spots, Patna District

Name	*As. Cont. (μg/L)*	*Location*	*Latitude*	*Longitude*	*Altitude (m)*
Hotspot 1	378 μg/L	Zirakhantola, Maner	N25°29.092'	E84°25.778'	44
Hotspot 2	148 μg/L	Ratantola, Maner	N25°38.940'	E84°54.388'	36
Hotspot 3	370 μg/L	Panapur, Danapur	N25°40.310'	E85°04.401'	37
Hotspot 4	288 μg/L	Ramnagar Village, Maner	N25°40.909'	E84°52.691'	39
Hotspot 5	250 μg/L	Ramnagar PO, River Bank, Maner	N25°40.928'	E84°52.766'	37
Hotspot 6	203μg/L	Badantola Temple, Maner	N25°40.960'	E84°52.570'	43
Hotspot 7	724 μg/L	Pundev Singh, Naya Tola, Maner	N25°40.997'	E84°52.288'	41
Hotspot 8	328 μg/L	Purana Tola, Maner	N25°40.980'	E84°52.238'	39
Hotspot 9	179 μg/L	Dhajwa Tola, Maner	N25°40.634'	E84°52.141'	41
Hotspot 10	340 μg/L	Primary School, Satana, Maner	N25°40.613'	E84°52.303'	35
Hotspot 11	278 μg/L	Krishna Mandir, Saat Aana, Maner	N25°40.654'	E84°52.431'	26
Hotspot 12	214 μg/L	Dudhaila, Maner	N25°40.472'	E84°52.496'	36
Hotspot 13	378 μg/L	Hathitola, Maner	N25°39.972'	E84°52.556'	40
Hotspot 14	484 μg/L	Malahibanda, Barh	N25°28.623'	E85°40.701	29
Hotspot 15	553 μg/L	Gyaspur Mahaji, Bakhtiyar pur	N25°29.656'	E85°26.9966	35
Hotspot 16	452 μg/L	Kasimchak, Danapur	N25°39.150'	E85°02.500'	35
Hotspot 17	409 μg/L	Harshamchak, Danapur	N25°41.012'	E85°03.050'	34

with arsenic content of over 10 μg/L, out of which 272 samples tested had more than 50 μg/L arsenic. The highest arsenic content detected by AAS Method was 724 μg/L, this sample being collected from Naikatola Village, Maner block, in the hand pump located near the house of Mr. Pundev Singh.

In Danapur block the highest As. Level detected by AAS method was 450 μg/L at Kasimchak, from the hand pump located near the house of Mr. Sitaram Rai. The next highest reading of 409 μg/L in this block, was from the hand pump near the house of Mr. Laxman Rai of Harshamchak village, Danapur block.

In Bakhtiarpur, the highest arsenic level detected by AAS method was 553 μg/L in Ghiaspur Mahazi, outside the house of Ashok Mahto, followed by 538 μg/L in Kala Diara, outside the house of Faudar Paswan. In Barh, the highest arsenic level read was 484 μg/L in Malahi Banda, near the house of Balaj Mahto; and then 441 μg/L near the house of Mohan Mahto.

Table 3 Arsenic contaminated villages confirmed by AAS-HG method in Patna District.

Adminstrative Block	*Arsenic affected Villages*
Maner	Naikatola, Purankatola, Hathitola, Ramnagar, Badantola, Ratantola, Satanna, Zirakhantola, Dhajwatola, Dudhaila.
Danapur	Kasimchak, Bishunpur, Harshamchak, Rampur, Panapur
Bakhtiyarpur	Gyaspur Mahazi, Kala Diara.
Barh	Malahi Banda

AAS-HG Tests revealed four concentrations of high arsenic levels. The resultant pattern is highly indicative of the presence of arsenic rich rock layer within or very close to the major aquifer [depth- 60 to 110 feet] in this area, from where maximum water is withdrawn. This finding needs to be further assessed by possible digging of boreholes, for geochemical analysis, in the peak value spots identified in the map. The space between the Maner and Danapur Clusters was uninhabited, but it fell within the zone of potential arseniferrous aquifers. The peculiarity of this metalloid in Patna is that it is non-existent in open wells, and is absent on, and, beyond the older alluvium of the southern levee.

Table -3 reveals a tapering down of the number of samples in the successive higher arsenic value groups. 27.35% of the AAS-tested 457 samples fall within the 50-99 µg/L group, and 25.38% have arsenic content of between 11-49 µg/L. Uncontaminated samples were only 10.5%, while another 4.6% samples were within the permissible limit of below 10 µg/L. All the four critical clusters had large numbers of contaminations between 11-99 µg/L.

Table 4 Number of A.A.S.-HG tested drinking water samples in Patna District, Bihar

As. Content	*Maner*	*Danapur*	*Barh*	*Bakhtiarpur*	*Mokama*	*Other Blocks*	*Total*
BDL	04	01	NIL	NIL	21	22	48
1 – 10 µg/L	04	NIL	NIL	06	08	03	21
11-49 µg/L	72	12	11	16	05	NIL	116
50-99 µg/L	92	11	05	17	NIL	NIL	125
100-199 µg/L	70	04	04	12	NIL	NIL	90
200-299 µg/L	24	01	02	03	NIL	NIL	30
300-399 µg/L	05	01	NIL	06	NIL	NIL	12
400-499 µg/L	01	01	02	07	NIL	NIL	11
500-599 µg/L	NIL	NIL	NIL	03	NIL	NIL	03
600 µg/L & above	01	NIL	NIL	NIL	NIL	NIL	01
TOTAL	273	31	24	70	34	25	457

The findings of this study also revealed the quantum of population being exposed to arsenic contaminated hand pumps [Table-4]. Taking into account the efficacy of field test kits, 32.27% of the total population of the study area were at risk. The critical figure of affected population in Sampatchak [82.63%] was attributed to widespread contamination of 50 µg/L and less of the drinking water sources. Classic symptoms of arsenic poisoning were visible in all the four worst affected blocks, although clinical investigations are yet to be done. The highest Arsenic contaminated aquifers of the fertile, densely populated villages of the Diara lands of Maner, coincided with a large number of children with skin discolorations. There were common complaints of eczemas not responding to medicines. In the remaining three important blocks, visible symptoms were commonly confined to skin problems and gastro-intestinal symptoms. Epidemiological studies state that the external symptoms of arsenic poisoning appears after at least 5 years of continuous ingestion of arsenic, by which time difficulties arise in treatment of symptoms. The general population of Bihar has a wide youthful base, hence children in the contamination zones are the most vulnerable to arsenic related health problems.

6 RECOMMENDATIONS

- A comprehensive study for arsenic contamination in ground water is urgently required all along the banks of Ganga.

Table 5 Population at Risk

Sl.No.	Name of Block Contaminated Samples[%]	As. [Census 2001]	Total Population	Population Affected	% of Affected Population
1.	Maner	42.00	201345	109957	54.61
2.	Sampatchak	36.26	76154	62930	82.63
3.	Phulwari	07.69	191005	54148	28.34
4.	Danapur	33.59	325457	37189	11.42
5.	Patna[rural]	4.69	66409	1101	01.65
6.	Fatwah	17.22	153470	32863	21.41
7.	Mokama	25.25	168036	74581	44.38
8.	Khusrupur	12.96	80428	20000	24.86
9.	Bakhtiarpir	26.21	172531	110474	64.03
10.	Pandarak	23.60	123538	14449	11.69
11.	Barh	19.78	162381	37647	23.18
	TOTAL	21.24	1720754	555339	32.27

- All hand pumps confirmed as having more than 50 µg/L. Arsenic content be capped immediately.
- Identification of arsenic-free aquifers be carried out in the affected villages identified by this study.
- Supply of low-cost, community-based arsenic filters be provided to the affected villages.
- Usage of dug wells, with proper water treatment.
- Through bioremediation by "arsenic munchers"[microbes and ferns].
- Clinical investigation be initiated in affected villages by a team of doctors.
- Provisions for alternate safe sources of drinking water.
- Intensive awareness campaign should be initiated in arsenic affected villages.

Acknowledgements

This Research Project was funded by UNICEF.

References

The World Bank .1999 . Entering the 21st. Century, World Development Report, 1999-2000, Oxford University Press, N.Y.

WHO Guidelines for drinking water quality. V. I. Recommendations Geneva:WHO, 1993:41-2.

Fazal, A., Kawachi, T., & Ichion. 2001. "Extent and Severity of Groundwater Arsenic contamination in Bangladesh", International Water Resources Association, *Water International*, Vol. 26, No. 3, 370-379, Sept.

Chadha D.K. and Sinha Roy S.P.1999. "High Incidence of Arsenic in Groundwater in West Bengal", CGWB, Ministry of Water Resources, Faridabad, India, July 1999 Report.

Chakraborty, D., Samanta, G. et al. 1994 ."Arsenic in Groundwater In Six Districts of West Bengal, India: Biggest arsenic Calamity in the World" . Environ. Geochemistry and Health, 18, 5-15.

Chakraborty, D. 1989. "Arsenic Orphans", Banabithi, Environmental Special Issue: 7-16., Dept. of Environment and Forest, Govt. of West Bengal, India.

Astolfi, E., Maccagno, A., et al, 1981, "Relation between Arsenic in Drinking Water and Skin Cancer", Biological Trace Element Research, 3, 133-143.

Chaudhary, U.K., Biswas, B. K., Roy Choudhury, T., et al. 2000. "Ground water Arsenic Contamination in Bangladesh and West Bengal, India", Environmental Health Perspective, 108, 388-397.

Acharya, S. K., Chakrabarty, P., et al, 1999. "Arsenic Poisoning in Ganges Delta [Brief Communication], *Nature,* 401: 55.

Rajiv Gandhi National Drinking Water Mission.1998-99. http://rural.nic.in/book98-99/chapter7.pdf

Fazal, A., Kawachi, T., & Ichion, E., 2001. "Validity of the Latest Research Findings on Causes of Groundwater Arsenic Contamination in Bangladesh", International Water Resources Association, *Water International,* Vol 26, No. 2, 380-389.

Morgan, J.P., McIntire, W.G. 1980. Quaternary geology of the Bengal Basin, East Pakistan, and India, Bulletin of Geological Society of America, 70, 319-342.

Standard Methods for the Examination of Water and Wastewater. 1996. published jointly by American Public Health Association [APHA], American Water Works Association and Water Pollution Control Federation.

Basu, B. & Sil, S.2004.,"Arsenic Mapping for North 24 Parganas District of West Bengal- using GIS and Remote Sensing Technology". www.gisdevelpment.net/application/environment/water/pdf/ 2004pdf.

Multiple Statistical Analysis for Groundwater Quality Assessment in Marudaiyar Basin, Tamil Nadu

8

S. Balaselvakumar[1], K. Kumaraswamy[1] and R. Arthur James[2]
[1]Department of Geography, [2]Department of Marine Science,
Bharathidasan University, Tiruchirapalli – 620 024
*E-mail: james_fse@yahoo.com

ABSTRACT

Understanding the quality of groundwater is necessary to find the suitability of groundwater for various purposes and also to know the extent of contamination. Multiple statistical analysis technique has been applied to find out the quality and factors controlling groundwater geochemistry of Marudaiyar basin. Two seasons of data (pre- and post-monsoon) with 32 locations of the entire basin were used for this assessment. The geochemical study denotes that the blocks located along south-eastern, central and central eastern part of the basin bear good water quality. Based on the study, major ion chemistry is controlled by natural weathering factor followed by carbonate and anthropogenic activities.

Keywords: Groundwater geochemistry, PCA and factor analysis, Marudaiyar basin.

1 INTRODUCTION

The chemical composition of the natural waters is governed primarily by the natural environmental factors to which the water is exposed in the hydrological cycle. The composition of the natural waters is influenced by the type and amount of soluble products of rock weathering and decomposition and by the terrain traversed by that water (Ramesh and Anbu, 1996). Hence, major ion chemistry assessment has been a prerequisite for hydrological investigations. Groundwater chemistry is a very important tool and plays a significant role in determining the water quality for various uses of man. The quality of required groundwater supply depends on its purpose; thus, needs for drinking water, industrial and irrigation purposes that vary widely. One should be very keen in understanding the water chemistry before applying it for various uses especially in agriculture sector to achieve maximum benefit. The quality of groundwater is described by its physical, chemical and microbiological properties and are always interlinked. Multiple statistical analysis gives interlinking behaviors of various mechanisms and facilitate rapid monitoring methods. Kumar (1993) studied correlation among water quality parameters of groundwater samples from different parts of Indian and developed linear regression equations. In the present study, an attempt has been made to study the groundwater quality using various statistical methods to find out the nature of mechanism and other controlling factors for the Marudaiyar basin.

1.1 Study Area

Marudaiyar sub-basin, the study area is located in the central part of Tamil Nadu State concerning an areal extent of 623 km^2. It is geographically located between the latitudes 11°02′-11°15′ N and the longitudes 78°48′-79°15′ E (Fig. 1). The area is composed of series of a plains, valley bottoms, undulating uplands and broken chains of eastern-ghats viz., Pachamalai. The average height of Pachamalai hill is 100 m. But few of its peak above 1020 m above msl. The elevation of the basin ranges from 250 – 400 m. The Marudaiyar sub-basin has its origin from the Pachamalai hills, and its flow in the south-eastern direction, passing through the Perambalur, Kunnam, Ariyalur Udaiyarpalayam and Lalgudi taluks of Perambalur and Tiruchirappalli District before joining the Coleroon Rivers. On the northern side the study area

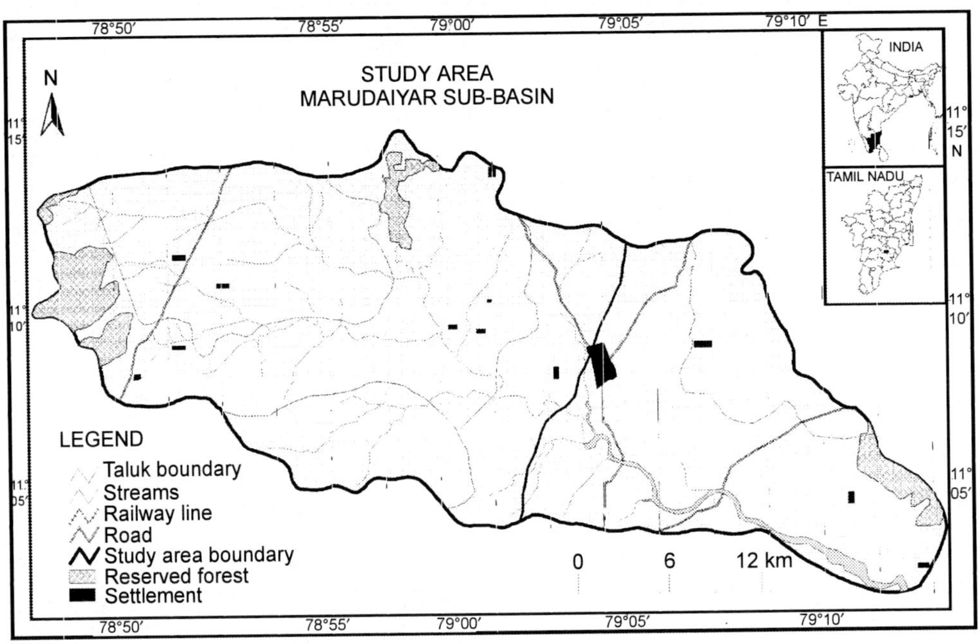

Fig. 1 Study area of Marudaiyar basin, Tamil Nadu.

is bounded by Chinnar basin and on the eastern side, it is bounded by Udiayarpalayam minor basin. On the southern and western side it is located by Nambiyar and Swedhanadhi minor basin respectively. In the study area, the average annual rainfall ranges from 750 to 1000 mm. The area comes under the influence of both the south-west monsoon (June – September) and north-east monsoon (October – December) due to geographic difference. Geologically, the river basin is mainly occupied by Archaean group of rocks composed of gneisses and charnockites. The Marudaiyar river basin, especially in its eastern part is composed almost entirely of sedimentary rocks especially calcareous rocks (limestone, gypsecous sandstone, calcareous sandstone, etc). Alluvium, deposited by the Marudaiyar river is found to occur in the downstream area especially in areas where the river joins in the Coleroon River.

1.2 Database and Methodology

Thirty-two observation samples were collected from the representative bore wells from the entire study area for two seasons viz. pre-monsoon and post-monsoon. The collected water samples were analysed for major anions and cations viz. Ca^{2+}, Mg^{2+}, Na^{+}, K^{+}, HCO_3^-, CO_3^{2-}, SO_4^{2-}, Cl^- and NO_3^- using standard procedure (APHA 2001). The statistical methods have been used to support the assessment programme in the study area. The principal component analysis on different chemical constituents present in the groundwater chemistry was assessed followed by Lawrence and Upchurch, (1982); Jayakumar, (1993); Gupta and Subramanian (1998); Babu and Sankara Pitchaiah, 2004).

1.3 Results and Discussion

Table 1 gives the overall water quality type of the groundwater samples collected in the study area. The calculated TDS ranges from a minimum of 239 mg/L to the maximum of 2889 mg/L showing the heterogenic nature of distribution in the study area. According to the data, more than 69% of the study area is well within the permissible limit of the WHO drinking water standard. Statistical package (SPSS) was used to perform symmetrical correlation matrix between major ions species. Tables 2 and 3 show the symmetrical correlation matrix of variables from which the interrelationships among the individual ions in the groundwater of the study area are determined. It is observed that the highest correlation exists between Ca^{2+} vs. Mg^{2+} and Cl^- vs. Na^+, during pre-monsoon and Ca^{2+} vs. Mg^{2+} and Na^+ vs. SO_4^{2-} during post-monsoon. A preliminary quantitative assessment of the principal components has been carried out using this matrix and pictograms have been developed to show the contribution of different variables. All the absolute values of the correlation coefficients greater than 0.5 have been taken into consideration; otherwise all the ten elements are extractable within two components. From the pictograms it could be seen that a group of variables like chloride are closely correlating with SO_4^{2-} during pre-monsoon season, while Ca^{2+} is closely correlated with magnesium and sulphate during post-monsoon season.

1.4 Principal Component Analysis

In the principal component analysis (PCA), the correlations between the variables are treated as similarity measures. The correlation matrix was computed with eigenvalues and percent of trace. Then the eigenvectors were calculated to bring out the principal axis matrix where the

Table 1 Water quality of the study area.

S.No.	pH	Total Hardness (as $CaCO_3$) mg/L	TDS (mg/L)	SAR (epm)	Geochemical type
1.	7.4	145	250	1.12	$CaHCO_3$
2.	7.5	220	1309	4.58	NaCl
3.	7.5	160	287	0.06	$CaHCO_3$
4.	7.4	180	249	0.09	$CaHCO_3$
5.	7.8	565	825	0.06	CaCl
6.	7.6	265	402	0.06	CaCl
7.	7.3	300	1232	3.00	NaCl
8.	7.7	790	1374	0.15	CaCl
9.	7.7	295	405	0.93	CaCl
10.	7.9	390	663	0.03	CaCl
11.	7.6	840	2889	0.93	NaCl
12.	7.3	175	415	2.12	CaCl
13.	7.8	490	1503	1.08	NaCl
14.	7.6	305	909	0.45	NaCl
15.	7.1	335	793	0.06	NaCl
16.	7.6	265	450	1.23	CaCl
17.	7.9	230	1544	3.17	NaCl
18.	7.3	105	239	0.15	$NaHCO_3$
19.	7.4	610	1224	3.68	$CaHCO_3$
20.	7.5	500	1285	0.63	NaCl
21.	7.4	235	651	0.09	$NaHCO_3$
22.	7.8	150	304	0.04	$CaHCO_3$
23.	7.2	370	1118	0.16	NaCl
24.	7.6	740	1323	0.23	CaCl
25.	7.6	60	585	0.34	$NaHCO_3$
26.	7.4	425	716	0.14	CaCl
27.	7.5	410	929	1.09	NaCl
28.	7.6	370	949	0.16	NaCl
29.	7.4	175	390	0.24	NaCl
30.	7.4	385	871	0.29	NaCl
31.	7.5	185	294	0.06	NaCl
32.	7.5	220	489	0.18	NaCl

variables loaded in the components were brought out. The principal component loading matrix are presented in Tables 4 and 5.

The component I of pre-monsoon is weighed by Mg^{2+}, Na^{+}, K^{+}, SO_4^{2-} and Cl^{-}, which contributes 46% of the sample variance. The concentration of ions indirectly reflects the water interaction with rocks and soil types. Ca^{2+}, Mg^{2+}, Na^{+} and SO_4^{2-} area represented in post-monsoon which contributes to 56% of the distributions of chloride ion concentration indirectly reflect the water interaction with rocks, permeability variations of aquifer and inter-transmissivity of rocks (Hem, 1970; Jeyakumar 1993). The addition of Ca^{2+} ions confirms the lithological control over the groundwater in the study area. The sources of Ca^{2+} are attributed to dissolution of carbonate minerals in the aquifer.

Table 2 Symmetrical correlation matrixes for pre-monsoon season.

	Ca^{2+}	Mg^{2+}	Na^{+}	K^{+}	HCO_3^-	CO_3^{2-}	SO_4^{2-}	Cl^-	NO_3^-	EC
Ca	1.00									
Mg	0.39	1.00								
Na	0.23	0.60	1.00							
K	0.31	0.11	0.25	1.00						
HCO_3	0.06	0.14	0.23	0.45	1.00					
CO_3	−0.16	−0.18	0.27	0.34	0.54	1.00				
SO_4	0.18	0.59	0.89	0.22	0.05	0.16	1.00			
Cl	0.53	0.78	0.84	0.33	0.05	−0.09	0.78	1.00		
NO_3	0.26	0.43	0.08	0.22	−0.21	−0.28	0.09	0.28	1.00	
EC	0.42	0.76	0.89	0.46	0.23	0.15	0.87	0.92	0.37	1.00

Table 3 Symmetrical correlation matrixes for post-monsoon season.

	Ca	Mg	Na	K	HCO_3	CO_3	SO_4	Cl	NO_3	EC
Ca	1.00									
Mg	0.91	1.00								
Na	0.81	0.80	1.00							
K	0.05	0.05	0.09	1.00						
HCO_3	0.02	0.13	0.32	0.17	1.00					
CO_3	−0.23	−0.23	−0.03	0.16	0.42	1.00				
SO_4	0.86	0.83	0.95	0.07	0.21	−0.12	1.00			
Cl	0.93	0.92	0.93	0.13	0.14	−0.22	0.91	1.00		
NO_3	0.05	0.26	0.09	0.48	0.03	0.01	−0.01	0.15	1.00	
EC	0.89	0.90	0.96	0.22	0.26	−0.09	0.94	0.97	0.23	1.00

Table 4 Principal component matrix for pre-monsoon.

Sl. No.	Variables	P.C. I	P.C. II	P.C. III
1.	EC	0.98		
2.	Ca			0.49
3.	Mg	0.79		
4.	Na	0.89		
5.	K	0.45		0.63
6.	HCO_3		0.75	
7.	CO_3		0.87	
8.	SO_4	0.85		
9.	Cl	0.95		
10.	NO_3			0.49
11.	Eigen Values	4.64	1.99	1.28
12.	% Trace	46.42	19.97	12.79

The Component II pre and post-monsoon is represented by carbonate and bicarbonate ions with the variance of 19 and 16%. The sources of carbonate and bicarbonate are mostly from decayed organic matter and sulphate reducing bacteria. The source from weathering of silicate

Table 5 Principal component matrix for post-monsoon.

Sl. No.	*Variables*	*P.C. I*	*P.C. II*	*P.C. III*
1.	EC	0.99		
2.	Ca	0.93		
3.	Mg	0.94		
4.	Na	0.95		
5.	K		0.68	
6.	HCO_3		0.65	
7.	CO_3		0.68	
8.	SO_4	0.95		
9.	Cl	0.98		
10.	NO_3			0.71
11.	Eigen Values	5.61	1.67	1.32
12.	% Trace	56.14	16.74	13.22

minerals is relatively possible and also it is used as a measure of eviction of groundwater flow and transmissivity gradient (Kumaraswamy *et al.*, 1996). Third component of pre and post-monsoon represented by calcium, potassium and nitrate ion with the variance of 12 and 13%. The sources of nitrate in groundwater are from decaying organic matter, intense agricultural activity and usage of nitrate fertilizers, sewage waste, etc. Potassium is found in relatively small concentrations in most of the samples because of its being reconstituted into insoluble secondary minerals formed in the process of weathering and absorbed in the clay minerals.

The score of the three principal components with reference to post and pre-monsoon seasons water samples of the study area have been considered, since the cumulative contribution of these three factors account to nearly 94%. The positive score made by individual samples with reference to these three principal components can be utilized to project the chemical characteristics of groundwater, by way of developing a model. Any type of developmental planning related to agriculture, industry or public water supply could utilize such a model for deterministic purposes.

1.5 Factor Analysis

Factor analysis deals the raw data standardized to have mean of zero and standard deviation of 1 giving a matrix "*R*" containing correlation coefficient among "*m*" variables. The matrix "*R*" is produced when a matrix of raw data (x), having m columns and n rows, is multiplied with its transpose (x'). Principal factor matrix is nothing more than the eigen vectors of a correlation or variance-covariance matrix. Communality provides an index to the efficiency of reduced set of factors. Varimax rotation moves each factor axis to positions so that projections from each variable onto the factor axes are either near the extremities (± 1) or near the origin (0). Thus, interpretation in terms of original variables made easier (Gupta and Subramanian 1998). Now the varimax rotation is performed to maximize the variance. By the process of manipulation or adjusting the factor axis, it is possible to achieve a simple and pragmatically more meaningful factor solution. Then, the commonalties are inserted in the diagonal of the correlation matrix and the extracted factors are based only on the common variance as original variables sharing with all other variables. Finally, interpretable rotated factor matrix and varimax factor scores are computed. The rotated factor matrix of both the periods are presented in Tables 6 and 7.

The factor I of pre- and post-monsoon seasons was represented by calcium, magnesium, sodium, sulphate and chloride with the variance of sample 40 and 55%. The water mass of the post monsoon represents Na^+, Cl^-, and SO_4^{2-} facies type. Further Tables 6 and 7 clearly indicate that the TDS factor influences the overall hydrochemistry of the groundwater. It represents the total ion, which presents in the dissolved forms and hence show the affinity by natural way of forming together consequently controls the major ion chemistry. Additionally, if we compare the components behavior between pre-monsoon and post-monsoon, the Rotated Factor I (RFI) value clearly explains very good record towards post-monsoon (0.93 to 0.98) season owing dissolution of the existing major ions, which tend to move together after the monsoon season.

Table 6 Rotated factor loading matrix for pre-monsoon.

Sl. No.	*Variables*	*R.F. I*	*R.F. II*	*R.F. III*	*Communality*
1.	EC	0.89			0.98
2.	Ca			0.69	0.55
3.	Mg	0.73			0.72
4.	Na	0.96			0.96
5.	K		0.72	0.47	0.77
6.	HCO_3		0.83		0.70
7.	CO_3		0.78		0.77
8.	SO_4	0.94			0.89
9.	Cl	0.89			0.93
10.	NO_3			0.77	0.64
11.	Eigen-values	4.02	1.99	1.91	
12.	% Trace	40.20	19.93	19.08	

Table 7 Rotated factor loading matrix for post-monsoon.

Sl. No.	*Variables*	*R.F. I*	*R.F. II*	*R.F. III*	*Communality*
1.	EC	0.97			0.99
2.	Ca	0.94			0.90
3.	Mg	0.93			0.89
4.	Na	0.95			0.94
5.	K			0.82	0.72
6.	HCO_3		0.84		0.75
7.	CO_3		0.82		0.72
8.	SO_4	0.96			0.93
9.	Cl	0.98			0.97
10.	NO_3			0.88	0.79
11.	Eigen-values	5.56	1.53	1.52	
12.	% Trace	55.62	15.25	15.25	

Though the pre-monsoon season value also has the good affinity, the individual components are slightly (0.72 to 0.95) lower than post-monsoon season. This kind of prediction proves further the groundwater percolation after the monsoon as well as the major ions coexistence in the groundwater hydro-geochemistry. Similarly, for factor II and III also correctly explain the above-mentioned factor and support the carbonate weathering and fertilizer input in the groundwater. The Rotated Factor II values are slightly higher in the post-monsoon than pre-monsoon (0.82-0.84) and also the higher Rotated Factor III (0.82-0.88) values when compared to the pre-monsoon season. From this study it can be concluded that major ion chemistry is controlled by natural weathering (TDS factor) factor followed by carbonate and anthropogenic activities (fertilizer factor). Thus, one can understand the significance of groundwater hydro-geochemistry and other factor, which influences by means of natural and anthropogenic activities.

The Component II pre and post-monsoon is represented by carbonate, bicarbonate and potassium ions with the variance of 19 and 15%. The sources of carbonate and bicarbonate are mostly from decayed organic matter and oxidation of organic carbon, produce high levels of carbonic acid, resulting in the formation of HCO_3^-. Also the kaolinisation of K-feldspar and consequent production of bicarbonate ions explains this factor. The chemical reaction may be expressed as: 2K-Feldspar + $2CO_2$ + $11H_2O$ = Kaolinite + $2K^+$ + $2HCO_3^-$ + $2H_4SiO_4$. The HCO_3^- ion, which makes up considerable amount in total anion composition, is responsible for the high TDS concentration in the Marudaiyar basin. Further potassium is easily absorbed in the clay minerals and the secondary products of water-rock interaction.

1.6 Multiple Regression Analysis

In Multiple Regression, any observed variable could be considered a function of any other variables measured on the same samples. However, in regression model, independent variables upon dependent variables can be expressed as.

$$Y = b_0 + b_1\,X1 + b_2\,X2 + b_3\,X3 + \ldots + b_n\,Xn$$

where,

Y = Dependent Variable;

b_0 = Constant;

b_1 to b_n = are the coefficients and

$X1$ to Xn = are the independent variables

In our study the TDS values mainly depend on the concentration of major ions. The substitution of the regression coefficient, obtained each variable, exhibits the functional relationship between the dependent (TDS) and independent (major ions) variables which are listed in Tables 8 and 9.

Table 8 Multiple regression analysis for pre-monsoon.

Source of Variation	Sum of Squares	Degrees of Freedom	Mean Squares	F Values
Regression	23428140	9	2603126.7	1603.5
Deviation	35715	22	1623.4	
Total variation	23463855	31		
Goodness of Fit		= 0.998		

Table 9 Multiple regression analysis for post-monsoon.

Source of Variation	*Sum of Squares*	*Degrees of Freedom*	*Mean Squares*	*F Values*
Regression	53840155	9	5982239.5	2774.1
Deviation	47442	22	2156.5	
Total variation	53887597	31		
Goodness of Fit		= 0.999		

The correlation coefficient for both periods shows the value around 0.99, is clear that there exists a strong positive relationship between the dependent variable TDS and the independent variables i.e., major ions. The goodness of fit for pre and post-monsoon are further indicating that the assumption made is correct and validated for the present analysis.

2 CONCLUSION

Based on the multiple statistical analysis in the groundwater samples of Marudaiyar basin, it shows three components of loading, namely, natural weathering, followed by carbonate factor and anthropogenic activities. The dominating ions are sodium, chloride, magnesium followed by carbonate, bicarbonate, nitrate and sulphate for both the periods. All over the study area, the geochemical modeling denotes that the blocks located along south-eastern, central and central eastern part of the basin are bearing good quality groundwater.

References

APHA (2001) "Standard methods for the examination of water and wastewater 25th edition, New York, pp. 1200.

Babu, M.N and Sankara Pitchaiah, P., (2004): "Chemical Nature of Brahmanapalle-Vaemula Mine Water, Cuddapah District, Andhra Pradesh - Multivariate Statistical Approach", *Journal of Applied Hydrology*, Vol. 17, pp. 79 - 85.

Chebotarev, I., (1935): "Metromorphism of Natural waters in the Crust of Weathering", *Geochemica Cosmochimica Acta*, Vol. 8, pp. 22 - 48.

Dawchy, D.R. and Feth, J.H., (1967): "Applications of Factor Analysis in Study of Chemistry & Groundwater Quality in Mogave Valley, California", *Water Resource Research* No. 3, pp. 505 - 510.

Gupta, L and Subramanian, V (1998) "Geochemical factors controlling the chemical nature of water and sediments in the Gomti river, India" Environmental Geology, Vol.24, pp.1 - 26.

Hem, J.D., (1970): "Study and Interpretation of the Chemical Characteristics of Natural Water", (2nd Edition), *Geological Survey Water-Supply* Paper 1473. United States Department of the Interior, United States Government Printing Office, Washington, D.C.

Hitchen, B., Billings, G.K and Elovan, J.E., (1971): "Geochemistry and Origin of Formation Waters in the Western Canada Sedimentary Basin–III, Controlling Chemical Composition", *Geochemica Casmochemica Acta*, No. 35, pp. 567 - 598.

Jayakumar, R., (1993): "Deterministic and Predictive Models for Characteristics in Saline and Alkaline Soils", In *Geographical Perspectives on Irrigation* (eds.) by Ramkumar, Rawat Publication, Jaipur, pp. 69 - 80.

Kumaraswamy, K., Jayakumar, R. Kamaraj, M. and Unnikrishnan, K., (1996): "Multivariate Hydrogeochemical Investigation in Vaippar Basin, South India, *Transactions of the Institute of Indian Geographers*, Vol.18, No.2, pp.79 - 85.

Lawrence, F.W. and Upchurch, S.B., (1982): "Identification of Recharge Areas Using Geochemical Factor Analysis", *Groundwater*, Vol. 20, No.6, pp.680 - 687.

Ramesh, R and Anbu, M. (1996) "Chemical methods for environmental analysis" Macmilan India Ltd., Chennai, pp. 210.

Razack, M and Dazy, J., (1990): "Hydrogeochemical Characterisation of Groundwater Mixing Sedimentary and Metamorphic Reservoirs with Combined use of Piper's Principal and Factor Analysis", *Journal of Hydrology*, Vol. 114, pp. 371 - 393.

Seyhan, E., Nienhmis, P.R and I. Simmers., (1987):_"Spatial Analysis of Hydrogeological Sub-systmes", *Journal of Geological Society of India*, Vol. 29, pp. 38 - 56.

Vajrappa, H.C and Rajadhan Singh, N., (2004): "Factor Analysis of Hydrochemical Data of Suvarnamukhi Sub-basin of Arkavathi River, Bangalore District, Karnataka", *Journal of Applied Hydrology*, Vol. 17, pp. 32 - 37.

Physico-chemical Characteristics and Determination of Cu(II), Mn(II), Fe(II), Zn(II), Ni(II) and Pb(II) in the Effluent from the Sites of Electroplating Industries of Agra City, India

9

Monika Singh, Anjali Mishra and Susan Verghese
School of Chemical Sciences, Department of Chemistry, St. John's College, Agra, India

ABSTRACT

Waste water from four electroplating sites of Agra city was assessed for physico-chemical characteristics and heavy metal concentration. A comparative study revealed that each sample was highest in particular metals concentration and some samples were higher in organic pollution than the others. Site AB_1 sample was highest in organic pollution (BOD 285.14 mg/L and COD 580 mg/L). AB_2 sample was highest in iron (46.70 mg/L) and zinc (10.79 mg/L) levels. AB_4 site sample was highest in copper (3.68 mg/L) and manganese (56.80 mg/L) concentration. All these heavy metals have severe adverse effect on human, plant and animal when present beyond tolerance limit. As the waste water finally reaches the Yamuna, the paper draws attention towards the contribution of electro-plating industries and units in contamination of this river and need of more adequate and efficient treatment of waste water let off from such industries.

1 INTRODUCTION

Metal pollution is the product of rapid increase in population, industrialization, urbanization and mechanization that serve to provide the necessities of population. The metallic wastes invariably discharged into environment results in water pollution (Dixit *et al.*, 2003). With the rapid development, the evolution of metal-based industries has led to the contamination of environment with heavy metals as these metal ions are non-biodegradable and toxicity produced by them is slow and long lasting (Gupta *et al.*, 2001). Increased concentration of heavy metals in aqueous environment is capable of causing phytotoxicity, bioconcentration and biomagnification by organisms (Joshi 2000).

In India there are over 50,000 large, medium and small electroplating units mostly scattered in the urban areas. Most large units are captive to some other large industrial units. Medium units cater to selected business groups as feeder units and small tiny units mostly carry out job plating. Unlike captive and feeder units, small-scale job plating units suffer from a lot of uncertainties and most of the operations are done manually. Due to variations in day-to-day operations, standardization of plating operations is a bit difficult. As most of the small units are in tiny residential areas as family owned business, resource and space constraints do not permit effluent treatment (Upadhyay *et al.*, 1992).

Electroplating normally involves three basic steps pre-treatment, electroplating and post-treatment in which acids, alkalis, and salts of Cd, Cr, Cu, Ag, Ni and Zn etc. are used at each step, the process involves thorough cleaning and rinsing with water and finally the articles are dried before packing. After a few days or weeks of use the plating bath contents and the acids used in pretreatment get contaminated and need to be disposed of, for maintaining the quality of plating. Typical waste generation sources are: Drag out losses, concentration liquid wastes, solid wastes, fumes and particulates. Drag out is the loss of bath liquid during transfer of work-piece from one bath to another. Costly and hazardous chemicals used in plating baths are lost in the rinse water or on the floor.

Bath solutions need to be replaced when their quality deteriorates due to contamination by dissolved metals and insoluble salts. All of these concentrated spent acid baths, spent alkali baths, and spent passivation dip baths are generally disposed of into drains. Rinse water solution usually has a lower concentration of chemicals and metals but its large volume contributes to a significant load (Rai *et al.* 1998).

This paper deals with the physico-chemical characteristics and heavy metal concentrations in waste water from four electroplating sites of Agra city and as this waste water finally reaches river Yamuna, the paper draws attention towards the contribution in contamination of Yamuna water caused by such industries and units.

2 MATERIALS AND METHOD

Waste water from four electroplating sites AB_1, AB_2, AB_3 and AB_4 of Agra city (Fig. 1) were assessed for heavy metals Cu (II), Mn (II), Fe (II), Zn (II), Ni (II) and Pb (II) concentrations and other physico-chemical characteristics (BOD, COD, alkalinity, pH, conductivity). One major electroplating industry was also among the sampling site, which is believed to contribute largely in electroplating waste of Agra. Two sites were small electroplating shops in main city and one electroplating sampling site was in the exterior of the city.

Grab samples from the four electroplating sites were collected in plastic bottles of one liter capacity for specific parameters and were properly labelled with the site name, date of collection, time, sample number, etc. Temperature, colour, and odour were noted at the time of collection and pH, conductivity within two hours from collection were measured at the laboratory directly after arrival.

Chemical and physical analysis of sample water was done following the procedure recommended by APHA 1992, pH, and conductivity are measured with the help of a NIG 333 digital pH meter and NDC 736 digital conductivity meter after calibration of the respective instruments. Alkalinity is determined by titration with phenolphthalein or methyl orange.

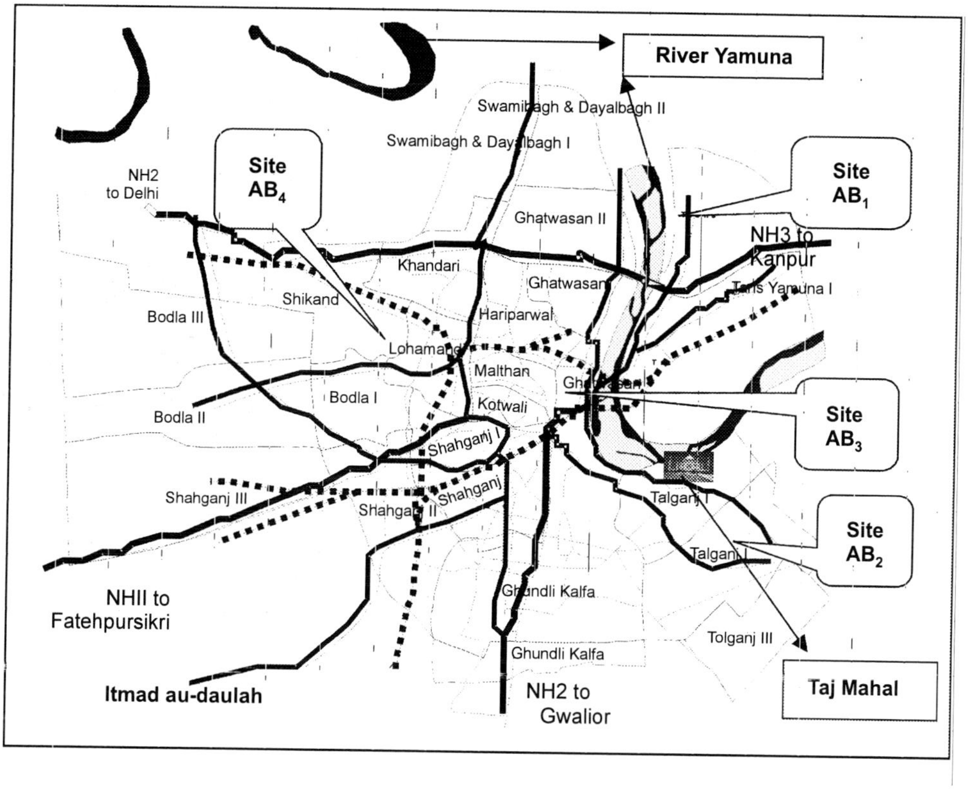

Fig. 1 City map of Agra showing the study area and the location of different sites.

BOD (biological oxygen demand), COD (chemical oxygen demand) is determined by titration method. The metal analysis was performed on an Atomic Absorption Spectrometer (Perkin-Elmer A. Analyst 100), following the condition of operation for the instrument. For the metals analysis Atomic Absorption Spectrometer uses acetylene and air as fuel and oxidant respectively. The standard solution of each metal (Cu, Mn, Pb, Fe, Zn and Ni) was made using analytical grade reagents for calibration purpose, samples were filtered and digested with nitric acid before determination of the metal concentration. Wavelengths used for particular metals were as follows: Copper (324.8 nm), Manganese (279.5 nm), Lead (283.3 nm), Iron (248.3 nm), Zinc (213.9 nm), and Nickel (232.0 nm).

3 RESULTS AND DISCUSSION

Physico-chemical characteristics of waste water collected from different electroplating sites are given in Table 1. pH varies considerably in all the four sites. Waste water sample of site AB_1 has the pH value of 13.27, and was the most alkaline sample water, pH above 7 is due to presence of sufficient quantity of carbonates. Site AB_2 waste water was highly acidic among the four samples, value of pH being 4.16, pH value of AB_3 and AB_4 site were almost close to each other values being 5.66 and 5.09 respectively. Thus, pH values in decreasing order were as follows: $AB_1 > AB_3 > AB_4 > AB_2$. Tolerance limits of pH for industrial effluents discharge into inland surface waters (IS: 2490-1974) is 5.5 – 9.0. AB_1 pH value is above 9.0 and two site AB_2 and AB_4 pH value is below 5.5.

Site AB_4 has the highest conductivity, value being 3.97 μS/cm. Site AB_2 has least conductance, value being 0.08 μS/cm. Site AB_3 conductance was 2.72 μS/cm and site AB_1 conductance was 0.46 μS/cm. Conductance in the decreasing order was as follows: $AB_4 > AB_3 > AB_1 > AB_2$.

Table 1 Physico-chemical characteristics of waste water from electroplating sites of Agra city.

S.No.	Parameters	AB_1	AB_2	AB_3	AB_4
1.	Temperature	30°C	30°C	25°C	25°C
2.	Colour	Colourless	Colourless	Green	Green
3.	Odour	Odourless	Odourless	Detergent like	Detergent like
4.	pH	13.27	4.16	5.66	5.09
5.	Conductivity (μS/cm)	0.46	0.08	2.72	3.97
6.	Alkalinity (mg/L)	650	3425	1587	1327
7.	COD (mg/L)	580	88	86	92
8.	BOD (mg/L)	285.14	170	54.85	56.06

Biological oxygen demand (BOD) is an index of organic pollution in water, it is the rate of removal of oxygen by microorganisms in aerobic degradation of dissolved or even particulate organic matter in water. All the four sites have different BOD values. Site AB_1 has the highest BOD, value being 285.14 mg/L. Site AB_2 BOD value being 170 mg/L, AB_3 and AB_4 site sample BOD values were close to each other, values being 54.85 mg/L and 56.06 mg/L respectively. BOD values in decreasing order were as follows: $AB_1 > AB_2 > AB_4 > AB_3$. Tolerance limit of BOD for industrial effluents discharged into inland surface waters (IS:2490-1974) is 30 mg/L and in all the site samples BOD value crosses this value of 30 mg/L prescribed by Indian standard for discharge into inland surface waters.

Chemical oxygen demand (COD), the measure of oxygen required in oxidizing the organic compounds present in water by means of chemical reactions involving oxidizing substances such as potassium dichromate and potassium permanganate showed different values in all the waste water samples, site AB_1 COD value was the highest among the four site its value being 580 mg/L which crossed the tolerance limit for industrial effluents discharged into inland surface waters (IS: 2490-1974) which is 250 mg/L, site AB_2 and AB_3 COD values were close to each other 88 mg/L and 86 mg/L respectively. Site AB_4 COD value was 92 mg/L, site AB_2, AB_3 and AB_4 COD values were below the ISI standards for discharge of industrial effluents into inland surface waters. COD values in decreasing order were as follows: $AB_1 > AB_4 > AB_2 > AB_3$. The results of the analyzed metals, copper, manganese, lead, iron, nickel and zinc in waste water samples from the four sites are presented in Table 2.

Copper is industrial health hazard, excess of copper in human body is toxic and causes hypertension, sporadic fever, uremia, coma and even death. Copper also produces pathological changes in brain tissue. Copper concentration was highest in sites AB_4 value being 3.68 mg/L, least in site AB_1 0.24 mg/L, 1.40 mg/L and 2.93 mg/L in site AB_2 and AB_3 respectively, copper in decreasing order of concentration was as follows: $AB_4 > AB_3 > AB_2 > AB_1$. Site AB_4 exceeded the tolerance limit for industrial effluents discharged into inland surface

Table 2 Metal concentration in waste water from electroplating sites of Agra city.

S.No.	Metal	Concentration (mg/L)			
		AB_1	AB_2	AB_3	AB_4
1.	Cu	0.24	1.4	2.9	3.7
2.	Mn	ND	3.2	55.1	56.80
3.	Fe	0.08	46.7	0.36	0.3
4.	Zn	0.01	10.8	6.5	6.7
5.	Ni	16.4	7.3	31.9	31.9
6.	Pb	ND*	ND*	ND*	ND*

*ND: Non-detectable

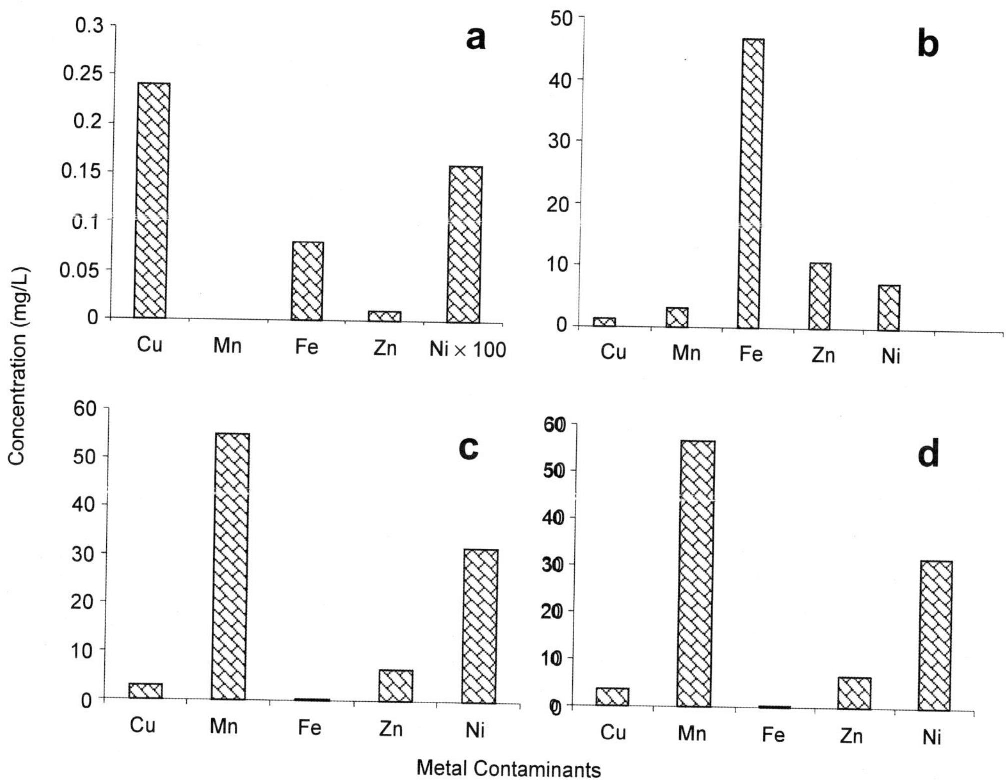

Fig. 2 Representing metal concentrations in electroplating sites: a) AB_1; b) AB_2; c) AB_3 and d) AB_4.

waters (IS: 2490-1974) which is 3.0 mg/L and copper concentration in rest three sites' samples was within this tolerance limit.

Manganese concentration above 100 mg/L is toxic to man and mammals both. It causes growth retardation, fever, muscular fatigue and even blindness. Permanganate manganese is most toxic, MnO_4^- ion is less toxic, Mn^{2+} ion is three times more toxic than Mn^{3+} ion, manganese concentration was highest in site AB_4, value being 56.8 mg/L and was non-detectable in site AB_1, site AB_2 and AB_3 concentration was 3.18 mg/L and 55.1 mg/L respectively. Manganese in decreasing concentration was as follows: $AB_4 > AB_3 > AB_2 > AB_1$.

More than 10 mg/kg level of iron causes rapid increase in pulse rates, congestion of blood vessels, hypertension and drowsiness. 3.0 mg/L is the waste water discharge standard for iron prescribed by Central Pollution Control Board (CPCB), Ministry of Environment and Forest, July (2000). Highest concentration of iron was found in site AB_2 (46.7 mg/L), least in site AB_1 (0.08 mg/L). Concentration in site AB_3 and AB_4 was 10.36 mg/L and 0.28 mg/L respectively. Concentration in decreasing order of concentration was as follows: $AB_2 > AB_3 > AB_4 > AB_1$.

Zinc salt's heavy doses (165 mg) in continuation causes vomiting, renal damage, and cramps. Discharge standard as per (CPCB; 2000) for zinc is 5.0 mg/L. Zinc concentration was highest in site AB_2 (10.79 mg/L); least in site AB_1 (0.01 mg/L). Sites AB_3 and AB_4 had almost same concentration level, value being 6.46 mg/L and 6.68 mg/L respectively. Zinc in decreasing order of concentration was as follows: $AB_2 > AB_4 > AB_3 > AB_1$.

Nickel concentration more than 30 mg causes change in muscle, lung, liver, brain and kidney and also can result in cancer. It also causes tremor, paralysis and even death. Nickel concentration was highest in site AB_4 (31.9 mg/L) and concentration level was almost same in site AB_3, value being 31.90 mg/L. Concentration in site AB_1 and AB_2 was 16.5 mg/L and 7.3 mg/L respectively. Nickel in decreasing order of concentration was as follows: $AB_4 > AB_3 > AB_1 > AB_2$. The nickel concentration in all the sites exceeded the tolerance limits for industrial effluent discharge into inland surface waters which is 3.0 mg/L (IS: 2490-1974) and (CPCB 2000).

Lead concentration above 800 mg in humans causes toxicity, mild anaemia, brain damage, loss of appetite, vomiting, convulsions uncoordinated body movement and stupor eventually producing coma and death. Lead toxicity is due to concentration of lead diffusion in to soft tissues. Discharge standard as per (CPCB 2000) for lead is 0.1 mg/L. Table 2 shows that the lead concentration was non-detectable in all the four sites.

Highest BOD and COD values were found in site AB_1 but it had least metal concentration among the four sites. Highest concentration of iron and zinc was found in site AB_2. Highest copper and manganese concentration was found in site AB_4. Highest concentration of nickel was found in site AB_1.

4 CONCLUSION

It was thus, concluded that AB_1 waste water sample was highly alkaline and highly polluted with organic matter. AB_2 was acidic, second highest in BOD value with high iron and zinc concentration. AB_3 sample was acidic, organically polluted, with high concentration of zinc, nickel and iron. AB_4 sample was acidic, with high conductivity, copper, manganese highest in concentration, and zinc, nickel second highest in concentration. The data was analyzed statistically for correlation values between different metals among all the sites. It is revealed from the correlation value that the '*r*' value between copper and manganese is 0.955, between iron and zinc is 0.722 and between zinc and nickel is –0.157.

In all these electroplating units, certainly there is a need to treat the waste water for certain heavy metals and for chemical and organic pollution before it is discharged into open drains. As these effluents from electroplating units are highly corrosive due to the presence of acids and toxic metals, there discharge directly into rivers (Yamuna river in case of Agra) without neutralization decreases the pH of river water and results in mass mortality of aquatic flora and fauna. When discharged into the sewers they attack concrete structure, hydrolyze soaps present in the sewage, liberates fatty acids which form floating scum that sticks to objects and clog the sewers and inhibit or kill the microorganisms that take part in the purification of sewage.

References

APHA, AWWA & WEF. 1992. Standard Methods for Examination of Water and Wastewater. 18th ed. New York: American Public Health Association.

Dixit, R.C., Verma, S.R., Nitnaware, V. & Thacker, N.P. 2003. Heavy metals contamination in surface and groundwater supply of an urban city. *Ind. J. Environ. Health* 45 (2):107 – 112.

Gupta, K.V., Gupta , M., & Sharma, S. 2001. Process development for the removal of lead and chromiu m from aqueous solutions using red mud an aluminium industry waste. *Water Res* 35, 1125 – 1134.

Joshi, S. 2000. Ecotechnological treatment for the industrial waste water containing heavy metals. *J. IAEM* 27: 98 – 102.

Rai, A.K. & Kumar, S. 1999. Treatment of chromium bearing waste water by adsorption on brick kiln ash and fly ash. *Ind. J. Environ. Hlth* 41: 65 – 73.

Upadhyay, Y.D., Upadhyay, S.N. & Haribabu, E. 1992. Removal of chromium (VI) by fly ash. *Chem. Environ.Res* 1 (3): 289.

10

Nutrient Chemistry of Groundwater of NCT-Delhi, India

Manish Kumar[1], AL. Ramanathan[1], Bhawna Sharma[1] and M.S. Rao[2]
[1]School of Environmental Sciences, Jawaharlal Nehru University, New Delhi-110067, India
[2]National Institute of Hydrology, Roorkee, Uttarakhand, India

ABSTRACT

Present study has been carried out to assess and understand nutrient concentration (nitrate, phosphate and potassium) and its spatial and seasonal variations in the groundwater of National Capital Territory of Delhi, India, in 2003. The concentration of nutrients in groundwater acts as an indicator to identify the nature and influence of agricultural and urban runoff on the shallow subsurface environment. Results of the study suggested that leaching from the landfill sites and solid waste disposal sites along with agricultural activities, including application of fertilizers, soil mineralization processes and irrigation return flow, are major processes regulating the nutrients chemistry in the groundwater of this region. Salinity and nitrate contamination are the two major problems in the area which is alarming for drinking purpose. The seasonal water level fluctuation shows that rising water level increases nutrients concentration in groundwater due to the agriculture related activities. The result also indicates a different source of origin for the nitrate and potassium and not a common source for their origin as it was thought earlier. Few groundwater samples in the northern part and eastern part of the study area exhibited very high phosphate concentration. This may be due to local factors such as discharge of agricultural runoff or due to mixing of sewage with the shallow groundwater table and leachates from the landfill sites.

1 INTRODUCTION

With the exponential growth in human population each and every natural resource is going under stress with different intensity simultaneously. During the last century agricultural dependence on inorganic fertilizers increased to meet the food and fiber demand of the world population that grew by some 5 billion during the period. Recent projections indicate that fertilizer requirements will reach 180 million metric nutrient tons per year by 2030—a 30% increase from the current 138 MT fertilizer was an integral part of technological trinity—seed, water, and fertilizer—responsible for bringing about the "Green Revolution" that helped many densely populated countries, including India, China, and Indonesia, achieve self-sufficiency in a short span of 20 to 25 years. Since the 1960s, global cereal production has more than doubled, and more than 40% of the world's dietary protein is derived from nitrogen fertilizers (Smil, 1999). But all this eventually leads to imbalance in nutrient load in groundwater.

Studies on nutrient concentration in groundwater have been widely reported during recent decades from all parts of the world. The major nutrients derived from agricultural activities are nitrate, phosphate and potassium. The nitrate molecule is highly mobile and plays a role in pollution resulting from agricultural activities (Hem 1985; Guimera 1998; Hudak 2000; Nolan 2001; Oren *et al.*, 2004). Other nutrients applied to the cropland, such as phosphate and potassium, can also dissolve and move from the surface towards the water table through the vadose zone. The contribution of anthropogenic activities to major ion composition in groundwater is limited in comparison to nutrients. But in the case of nutrients, anthropogenic contributions, such as agricultural and domestic activities as well as solid waste disposal act do play a major role. As other than fertilizer landfill leachates, human, plant and animal wastes are also some other major contributors in total input of nutrients. Hantzsche and Finnemore (1992) described the various processes by which soil microorganisms and nitrogen interact to form nitrates. Organic nitrogen (manure, plant waste, etc.) can be easily converted to the ammonium ion through the process of ammonification. The ammonium is subsequently oxidized to nitrate by nitrification (Freeze and Cherry 1979). Walter *et al.* (1975) reported that the nitrogenous fertilizers that are applied are converted into mobile nitrates by natural processes irrespective of their form or composition. The rate of leaching is governed by the soil properties and the amount of water present in the soil system.

Helen et al. (2005) examined the pattern in nitrate and phosphorus sources, loads and concentrations in a groundwater-dominated lowland catchment, the Hampshire Avon and concluded that the groundwater chemistry is modified by the impact of point source sewage effluents. Thomas et al. (1982) described the importance of annual nutrient loading estimates on the stream water in order to know the groundwater contribution to the total water budget of lake. Ritter (1988) reported that there is a considerable amount of research data showing that excessive fertilization will not increase crop yields but will only increase nitrogen and phosphorus concentration in surface water and groundwater. Exner and Spalding (1979) reported contamination of groundwater due to fertilization and irrigation of the sandy soils. Further, they concluded that 50% of the applied nitrogen fertilizer infiltrates into the groundwater reservoir in this area. Several researchers studied the effect of agrochemicals on groundwater quality (Boyer and Pasquarell 1995; Hamilton and Helsel 1995; Stuart *et al.*, 1995; Kraft *et al.*, 1999). Driescher and Gelbrecht (1993) reported that agricultural land use is one of the sources of phosphorus contamination of an aquifer in the catchment area of the lower river spree, Germany. Phosphate ions in groundwater of the alluvial aquifer of Eskisehir Plain, Turkey, are mainly caused by agricultural activities (Kacaroglu and Gunay 1997).

India supports more than 16% of the world's population with only 4% of the world's fresh water resources (Singh, 2003). Although agriculture sector in this country has been major user of water, share of water allocated to irrigation is likely to be decreased by 10 – 15% in next two decades (CWC, 2000). In India, several studies have reported groundwater contamination by nitrate due to agricultural activities (Mukherjee and Pandey 1994; Chandu *et al.*, 1995; Prasad 1998).

Groundwater is used for irrigating nearly two-thirds of the area under irrigation in the Delhi area. Agriculture return flow also makes groundwater richer in nutrients concentration. Present study is carried out to understand the nutrient chemistry of groundwater and its spatial and seasonal variation. Previous study done by Datta and others (1997) on the impact of fertilizer use on the groundwater quality of Delhi by using ^{18}O isotopes not included phosphate. Therefore, an attempt has been made to evaluate the real nutrient status of the groundwater and with monsoon effect on its concentration in the capital city of India.

2 STUDY AREA

The National Capital Territory of Delhi has an area of 1483 sq km. The area lies between 28°24′17″ - 28°53′00″N and 76°50′24″ - 77°20′37″E (Fig. 1). Delhi region is a part of Indo-Gangetic Alluvial Plains, at an elevation ranging from 198 m to 220 m above mean sea level. Lithologically, the area is transacted by a quartzite ridge, a prolongation of Aravalli Hills extending along the southern border of Delhi and ending to the north on the west bank of the Yamuna river. The Ridge acts as a water divide between eastern and western parts.

The climate is of semiarid nature due to marked diurnal differences of the temperature, high saturation deficit and low moderate/rainfall. The climate is markedly periodic and is characterized by dry and gradually increasing hot seasons between March and June, a dry and cold winter October to February and the warm, monsoon period from July to September. The average rainfall (1954 – 2004) is 721 mm and the annual evaporation is about 2565 mm. The mean minimum and maximum temperatures are 18.7° and 30.5° C, respectively, with daily maximum temperatures during hottest months commonly exceeding 42.2° C.

Physiographically, the region can be divided into four major divisions:

(a) The Delhi Ridge: A prolongation of the Aravalli hills consisting of quartzite rocks and extends from southern part of the territory to western bank of the Yamuna for about 35 km. The alluvial formations overlying the quartzitic bed rock have different nature on either side of the ridge.

(b) The nearly closed Chattarpur alluvial basin covering an area about 48 km^2 that is occupied by the alluvium derived from the adjacent quartzite ridge.

(c) Alluvial plains on eastern and western sides of the ridge.

(d) Yamuna flood plain deposits. These are of recent origin also termed as Newer alluvium.

The Newer alluvium is characterized by the absence of permanent vegetation due to periodic flooding and lack of kanker. The thickness of alluvium overlying the bedrock increases in the direction away from ridge and reaches to 300 m or more in western parts of Nazafgarh and Kanjhawla block and in the northern part of Alipur block. The thickness of alluvium in the areas of the Yamuna in the east of the ridge is about 165 m.

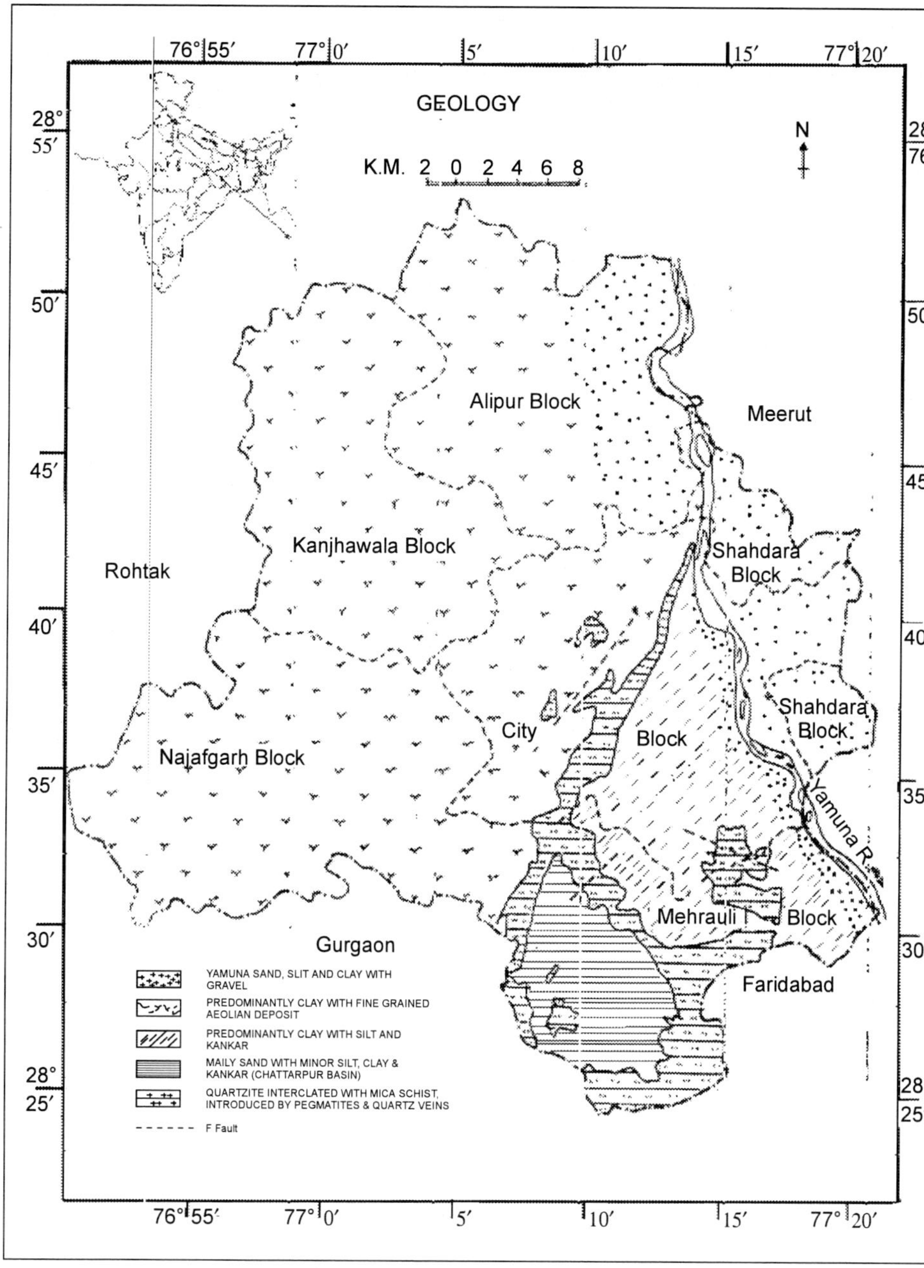

Fig. 1 Location and general geological map of NCT-Delhi, India.

3 METHODOLOGY

To understand the general variation in groundwater chemistry over the study area, a well inventory survey was carried out during March 2003 and electrical conductivity (EC) and pH were measured. A global positioning system (GPS) was used for location and elevation reading. This was supported by topographic sheets made available from Geological Survey of India. These data were used to select the representative wells and hand pumps for groundwater sampling. Sampling wells were selected in such a way that they represent different geological formations as well as land-use pattern at varying topography of this area. Thirty groundwater samples (Fig. 2) were collected during May and November 2003 in order to cover the monsoon impact on nutrient status.

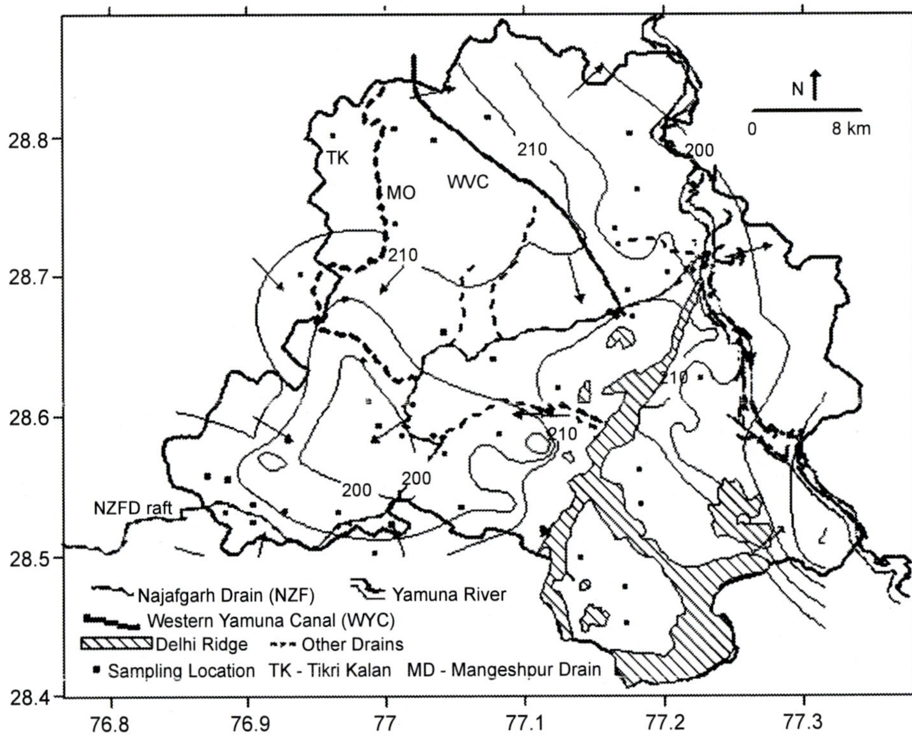

Fig. 2 Water table elevation, flow movement with sampling location.

Samples were collected in clean polyethylene bottles. At the time of sampling, sampling bottles were thoroughly rinsed 2 – 3 times with groundwater to be sampled. In the case of bore wells and hand pumps, the water samples were collected after pumping the water for 10 minutes. In the case of open wells, water samples were collected 30 cm below the water level using a depth sampler. This was done to remove groundwater stored in the well itself and to obtain presentative samples. *In situ* measurements included EC, pH, DO, ORP, temperature and bicarbonate were measured using a portable field kit and titration methods.

Samples collected were brought to the laboratory and were filtered using 0.45 μm millipore filter paper and acidified with nitric acid (Ultrapure Merck) for cation analyses and HBO_3 acid was used as preservative for nitrate analysis (Kumar, 2004). For anion analyses, these samples were stored below 4°C. K^+ was analyzed on flame photometer and heavy metals were analyzed using Schimadzu AAS. The chemical analysis was carried out as per standard procedure given in APHA (1995). Nitrate analysis was performed by using Brucine Method. Apart from the collection and analysis of water samples, other data were collected from the Central Ground Water Board (CGWB) and Central Pollution Control Board (CPCB), New Delhi.

4 RESULT AND DISCUSSION

The average, minimum and maximum values for each water quality parameter analyzed with their standard deviation for the both the pre-monsoon and the post-monsoon seasons is given in Table 1 with the percentage of samples exceeding permissible limit for the drinking water given by WHO (1993). The pH values for different lithology groundwater indicate alkaline in nature and their low values in the post-monsoon indicate dilution due to influx of rainwater of lower alkalinity. In general, EC is found to be high all along the western border of the study area. Along the west, on moving from north to south, EC increases from 3000 μS/cm and reaches to a maximum value of 37,500 μS/cm at the Tikri Kalan near Tikri border and thereafter it gets diluted on moving to the south towards Nazafgarh depression zone where it reaches 4000 μS/cm, indicating, a salinity mound close to the middle of the western region near the Tikri Kalan. The Cl^- concentration that measured at this site was 2628.9 mg/l which is also the highest among all the measured sites. In addition to this general trend, local points of high salinity are also seen in the study area. The highest salinity in the location close to the river Yamuna is observed at the site near Bhalswa lake. It is a low lying area located ~5 km west of the river Yamuna in the north Delhi and the EC measured at this site is 11,400 μS/cm. Such a

Table 1 Chemical parameters with their monthly range and average value.

Parameter		*May*	*June*	*July*	*August*	*September*	*October*	*November*
pH	R.	6.5-9.9	6.7-9.3	6.8-8.5	6.81-8.68	6.85-8.71	6.87-8.68	6.87-8.6
	Av	7.32	7.26	7.46	7.56	7.6	7.26	7.26
EC	R.	453-37500	512-38464	413-27525	453-17679	453-12650	1392-10000	341-9684
(μS/cm)	Av	3074.9	3123	2857	3074.9	3074.9	2982	2654
Na^+ (mg/L)	R.	0.6-490.5	0.9-512	2-604.6	4-948.3	6-1025.5	11-1374	9-1257
	Av	108.1	118	126	164	178	189	186
K^+ (mg/L)	R.	BD-107	BD-105	0.21-169	0.38-225	0.41-237	0.39-258	0.31-215
	Av	10.71	11.31	12.37	12.88	11.77	9.44	9.28
HCO_3^-	R.	28-220	26-220	68-542	81-831	95-915	120-1050	78-846
(mg/L)	Av	88.57	78.35	110.7	320.5	340.6	362.44	295
Cl^- (mg/L)	R.	5.67-2629	9-2832	11-3294	12-3312	14-3645	14-3935	14-3754
	Av	177	212	367	557	621	699	643
NO_3^- (mg/L)	R.	9.29-109	12-104	8-219	6-264	5-287	4.06-300	4.29-289
	Av	34.32	37.56	43.21	43.56	43.56	37.73	37.56
PO_4^{3-}(mg/L)	R.	0.03-2.34	0.03-3.13	0.03-4.45	0.12-2.45	0.03-1.64	0.10-0.93	0.09-1.34
	Av	0.31	0.36	0.42	0.45	0.48	0.38	0.36
Fe (mg/L)	R.	0.11-23.08	0.06-25.12	BD-11.8	BD-9.21	BD-1.02	BD-0.86	BD-2.12
	Av	1.73	1.54	0.87	0.43	0.32	0.21	0.25
Zn (mg/L)	R.	0.01-1.96	0.02-2.96	BD-2.12	BD-1.85	BD-1.62	BD-2.06	BD-2.09
	Av	0.18	0.58	0.54	0.54	0.34	0.23	0.24

high salinity at the proximity of perennial-river suspects the origin of its salinity to be from local origin. One of the dominant freshwater recharging source in the north Delhi is observed to be Western Yamuna Canal (WYC) (Sharma *et al.*, 2004). Water samples collected all along the canal up to its meeting point with Nazafgarh drain were found to be fresh (EC is ~500 in 100 ft depth ~800 at 50 ft depth) indicating canal to be the predominant recharge source in the north Delhi along its stretch. In general, a correlation appears between the potentiometric surface and groundwater salinity. Groundwater in high land areas of Ridge and the canal appears fresh and with the approach of low laying area towards the west and south-west the salinity increases.

Dissolution of carbonates and reaction of silicates with carbonic acid accounts for the addition of HCO_3^- to the groundwater. In post monsoon HCO_3^- is dominating anion at many places over Cl^-. The concentration of phosphate was relatively higher in drains and also river water near Okhla Barrage (Kumar 2004). The high PO_4^{3-} concentration is attributed to input from the domestic and industrial wastewater. A few groundwater samples of northern and eastern part of the study area exhibited very high phosphate concentration. This may be due to local factors such as discharge of agriculture runoff or due to mixing of sewage with the shallow groundwater table and leachates from the landfill sites.

The common sources of nitrate in water are the leachates from the landfill sites whether in use or abandoned, local agricultural fields, domestic sewage, etc. (Fig. 3). However, because of laminar flow condition and due to difference in specific gravity and viscosity between nitrate

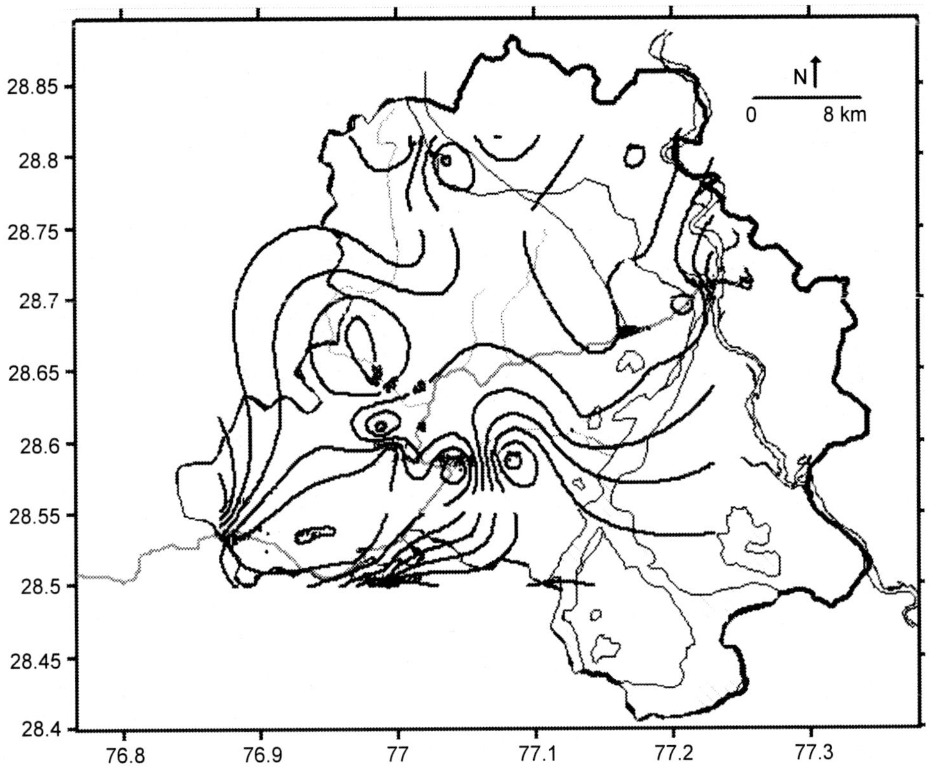

Fig. 3 Distribution map of nitrate (mg/L).

slugs and the natural freshwater, some diffusion bound to occur. But no such natural sources are reported to exist in the area for NO_3^- (Handa, 1988). Delhi soil and the alluvium have significant amounts of illite, which may fix K from water. Hence, certain groundwater in Delhi has low K^+ values (Subramanian and Saxena, 1983). In order to investigate fertilizer input as a probable common source of potassium and nitrate to the groundwater as was thought in the earlier studies (Datta et al., 1997), a cross plot between NO_3^- and K^+ was examined (Fig. 4). The plot indicates that an excess nitrate above a median value is normally associated with a low potassium content (<20 mg/L). This indicates multiple source of origin for the nitrate as against the common source for their origin as was thought earlier. Around 41% samples of pre-monsoon and 23% of post-monsoon are having more nitrate than that of permissible limit. Other than Fe rest of the heavy metals are either in between desirable and permissible limit or below desirable limits.

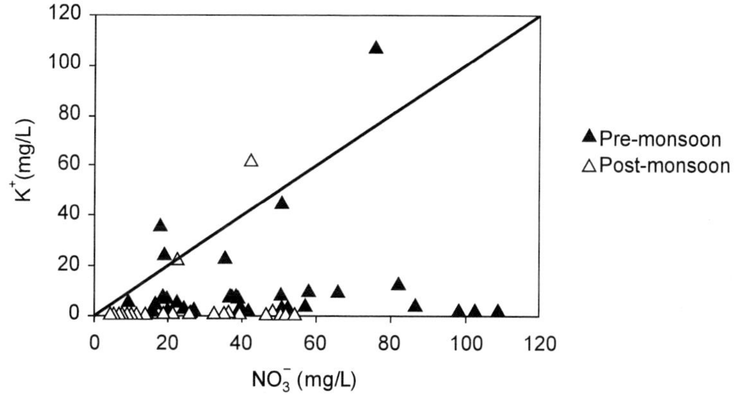

Fig. 4 Scatter Diagram of NO_3^- vs. K^+ in groundwater of Delhi.

The average value of chloride in the post-monsoon is much more than that of in pre-monsoon that is perhaps due to the rising water table in the post-monsoon period which dissolves more saline salts from the soils and increases the salinity of water after monsoon (Ramesam 1982; Ballukraya and Ravi 1999). The chloride concentration was found higher in the area covered with sand dunes, especially western and northern part of the study area which are mainly covered with the sand dunes. High chloride weathering of ridge-material also contributed to the salinity problem in the area. Previous studies also reported high Cl^- level as high as 1680 ppm (Paliwal and Yadav 1976). In this sand dunes region rainwater may react with the evaporite deposits to enhance the topsoil with Cl^-, HCO_3^- and Na (Subramanian and Saxena 1983). Beside these, anthropogenic sources such as domestic wastewater, industrial discharge etc. the large lateral variation of chloride concentration indicate recharge and discharge zones of lateral flow regime. Local recharge to the unconfined aquifer is more dominant than recharge from lateral flow (Datta and Tyagi 1996). The chloride ion is also considered to be conservative in groundwater of the Delhi Area (Datta et al., 1996). Groundwater is under-saturated with respect to NaCl. Therefore, physical processes such as mixing with another aquifer of different Cl^- concentration, change in evaporation rate during recharge either spatially or temporally, may result in changes in Cl^- content of groundwater.

Na^+ ion is showing higher values because it behaves like conservative element, i.e. not used up in biological process or clay mineral formation. (Subramanian and Saxena, 1983). The concentration of Mn^{2+} for all sampling seasons are showing higher value in the eastern part of the study area which may be attributed to anthropogenic activity such as waste discharge from industrial units. The groundwater samples are showing relatively higher concentration of Zn^{2+} especially in the south and western part of study area, which may be attributed to runoff from the agriculture field. Other than Fe and Mn rest of the heavy metals are either in between desirable and permissible limit or below desirable limits.

5 SEASONAL VARIATION IN NUTRIENT CHEMISTRY

The seasonal variations in groundwater chemistry in any area are principally due to variations in groundwater recharge, pumping, well lithology and geochemical reactions. The composition of the infiltrated water depends on rainfall, soil environment, agricultural use and thickness of the unsaturated zone (Scheytt, 1997). In the study area, rainfall recharge generally occurs during the monsoon period from July to September and evaporation is very high from February to July when the temperature rises to 42°C (Fig. 5). Groundwater concentration of nitrate increases in monsoon probably due to two main reasons first due to water level rise during the monsoon and second due to more leaching from landfill sites and recharge with industrial contaminant (Fig. 6). Similar results were reported elsewhere (Scheytt, 1997; Rao,

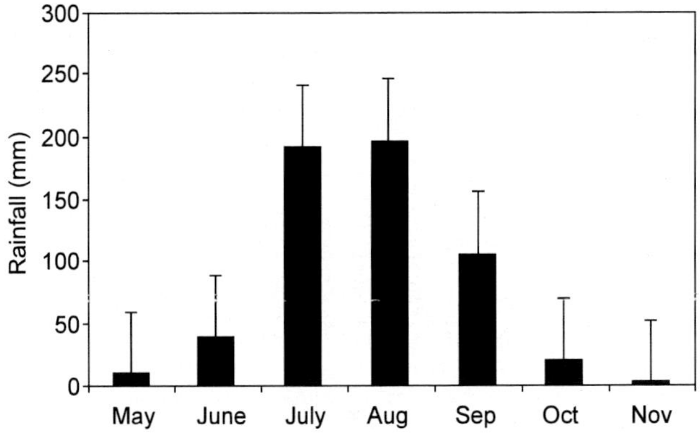

Fig. 5 Bar diagram of monthly precipitation during sampling period.

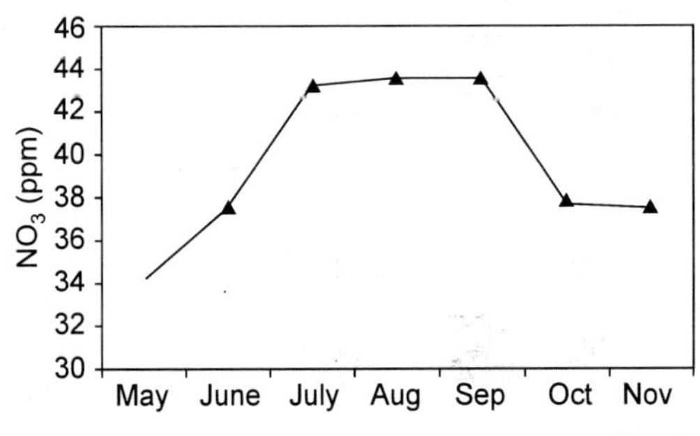

Fig. 6 Temporal variation in nitrate concentration.

1998). During monsoon rainfall, the groundwater is rapidly recharged through the agricultural lands and hence, there is an increased propensity for nitrate leaching. When the water level rises dissolution of nitrates concentrated in the capillary and vadose zones during the preceding dry season takes place. Such an observation was also made by Williams *et al.* (1998). Rao (1998) reported that high nitrate content in groundwater was contributed by clayey soil in the lower Vamasadhara River basin, India. Nitrate is concentrated in the vadose zone because of its large ionic size and it is likely to be retained in the clayey soil (Degens and Chillingar 1967; Klimas and Paukstys 1993). Hence, the accumulation of nitrate due to agricultural activities in the vadose zone and its flushing during rainfall are responsible for its concentration in the groundwater of this area. In addition, mineralization of soil nitrogen usually continues after harvest, causing accumulation of nitrate in the soil zone, which is flushed by the recharging water (Rajmohan 2001). Although there is not much difference in the average value of nitrate but the maximum value observed do increase threefold (Table 1). Phosphate concentrations in groundwater show seasonal variations more or less similar to nitrate levels (Fig. 7). The concentration of phosphate increases during rainfall because of the flushing of phosphate adsorbed during the preceding dry season. In the case of potassium, most of the samples have concentrations below detection limits (Fig. 8) perhaps due to flushing effect of the monsoon. However, locations with substantial potassium concentration behave like nitrate and phosphate. Higher concentrations are observed at shallow water level conditions and suggest

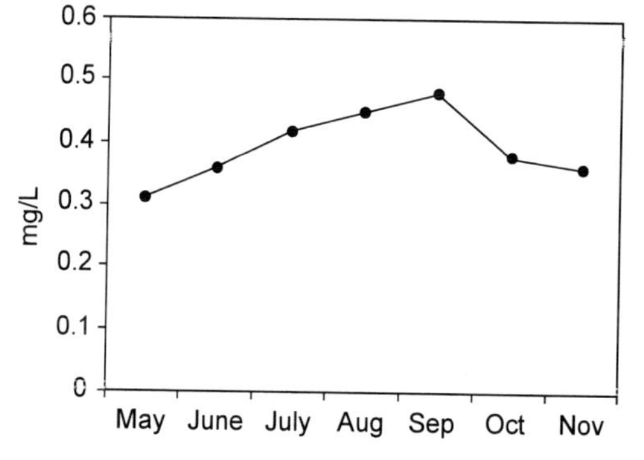

Fig. 7 Temporal variation of phosphate concentration.

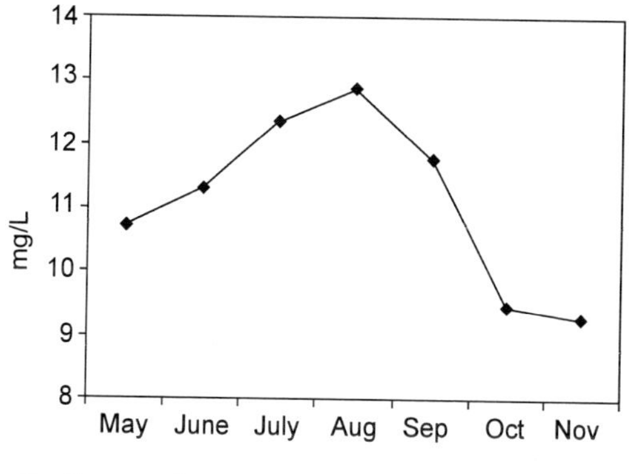

Fig. 8 Temporal variation of potassium concentration.

that rising water table might dissolve the nutrients adsorbed in the aquifer materials and increase their concentrations in groundwater.

Hence, seasonal water level fluctuation is an important factor controlling the nutrients concentration in groundwater of the study area.

6 CONCLUSION

Results of the study suggested that leaching from the landfill sites and solid waste disposal sites along with agricultural activities, including application of fertilizers, soil mineralization processes and irrigation return flow, are major processes regulating the nutrients chemistry in the groundwater of this region. Salinity and nitrate contamination are the two major problems in the area which is alarming for drinking purpose. The seasonal water level fluctuation shows that rising water level increases nutrients concentration in groundwater due to the agriculture related activities. The result also indicates a different source of origin for the nitrate and potassium and not a common source for their origin as it was thought earlier. Few groundwater samples in the northern part and eastern part of the study area exhibited very high phosphate concentration. This may be due to local factors such as discharge of agriculture runoff or due to mixing of sewage with the shallow groundwater table and leachates from the landfill sites.

Acknowledgements

Author (MK) thanks Council of Scientific and Industrial Research (India) for financial grant through a junior research fellowship. The authors also thank to the Head, Nuclear Division, National Institute of Hydrology, Roorkee for allowing availing the analysis facility in his department. The authors also acknowledge the Department of Science and Technology (DST), under Government of India.

References

APHA (1995) Standard methods for the examinations of water and waste water, 17th edn. APHA, Washington DC, USA.

Boyer DG, Pasquarell GC (1995) Nitrate concentrations in karst springs in an extensively grazes area. Water Resour Bull 31:729–736.

Ballukraya PN, Ravi R (1999) Characterisation of groundwater in the unconfined aquifer of Chennai City, India. J Geol Soc India 54:1–11.

Chandu SN, Subbarao NV, Raviprakash S (1995) Suitability of groundwater for domestic and irrigational purposes in some parts of Jhansi District, UP. Bhujal News 10:12–18.

Datta PS, Bhattacharya SK, Tyagi SK (1996) ^{18}O studies on recharge of phreatic aquifers and groundwater flow-paths of mixing in the Delhi area. J. Hydrol 176:25–36.

Datta PS, Tyagi SK (1996) Major ion chemistry of groundwater in Delhi area: Chemical weathering processes and groundwater flow regime. J Geol. Soc. India. 47:179–188.

Datta PS, Deb DL, Tyagi SK (1997) Assessment of groundwater contamination from fertilizers in the Delhi area based on ^{18}O, NO_3^- and K^+ composition. J. Conta Hydrol, 27:249–262.

Degens ET, Chillingar GV (1967) Diagenesis of subsurface waters. In: Larson G, Chillingar GV (eds) Diagenesis in sediments. Developments in sedimentalogy. Elsevier, Amsterdam.

Driescher E, Gelbrecht J (1993) Assessing the diffuse phosphorus input from subsurface to surface water in the catchment area of the lower river spree (Germany). Water Sci Technol 28:337- 347.

Exner ME, Spalding RF (1979) Evolution of contaminated groundwater in Halt country, Nebraska. Water Resour Res 15:139–147.

Freeze RA, Cherry JA (1979) Groundwater.Prentice-Hall, New Jersey.

Guimera J (1998) Anomalously high nitrate concentrations in groundwater. Groundwater 36:275–282.

Hamilton PA, Helsel DR (1995) Effects of agriculture on groundwater quality in five regions of the United States. Groundwater 33:217–226.

Handa BK (1988) Content of potassium in groundwater in India. Fertilizer News 33(11): 15–27.

Hantzsche NN, Finnemore EJ (1992) Predicting groundwater nitrate-nitrogen impacts. Groundwater 30:490–499.

Helen P. Jarviea, Colin Neala, Paul J.A.Withersb, Chris Wescott, Richard M. Acornley (2005) Nutrient hydrochemistry for a groundwater-dominated catchment: The Hampshire Avon, UK, Sci. of the Total Env.344 (2005) 143–158.

Hem JD (1985) Study and interpretation of the chemical characteristics of natural waters, US Geological Survey Water-Supply paper 2254:263.

Hudak PF (2000) Regional trends in nitrate content of Texas groundwater. J Hydrol 228:37–47.

Kacaroglu F, Gunay G (1997) Impacts of human activities on groundwater quality of an alluvial aquifer: a case study of the Eskesehir plain, Turkey. Hydrogeol J 5:60–69.

Klimas A, Paukstys B (1993) Nitrate contamination of groundwater in the Republic of Lithuania. NGU Bull 424:75–85.

Kraft GS, Stites W, Mechenich DJ (1999) Impacts of irrigated vegetable agriculture on a Humid north–Central US Sand plain aquifer. Groundwater 37:572–580.

Kumar M (2004) An integrated hydrogeochemical and isotopic study of NCR-Delhi, India. M.phil Dissertation, Jawaharlal Nehru University, N.Delhi, India (Unpublished).

Mukherjee S, Pandey DS (1994) Nitrate pollution in groundwater at Jaunpur and its environs, Uttar Pradesh. Bhu-jal News 9:22–25.

Nolan BT (2001) Relating nitrogen sources and aquifer susceptibility to nitrate in shallow groundwaters of the United States. Groundwater 39:290–299.

Oren O, Yechieli Y, Bo¨ hlke JK, Dody A (2004) Contamination of groundwater under cultivated fields in an arid environment, central Arava Valley, Israel. J Hydol 290:312–328.

Paliwal KV, Yadav BL (1976)Irrigation water quality and crop management in U.T. of Delhi. Water Tech. Centre, IARI, New Delhi, Res. Bull., 9:1-45.

Prasad R (1998) Fertilizer urea, food security, health and the environments. Curr Sci 75:667–683.

Public Work Department (PWD) (2000) Groundwater perspectives: a profile of Kancheepuram district, Tamil Nadu. 220 pp

Rajmohan N (2001) Hydrogeochemical studies in a part of Palar and Cheyyar river basin, Tamil Nadu, India. PhD Thesis, Anna University, Chennai.

Ramesam V (1982) Geochemistry of groundwater from a typical hard rock terrain. J Geol. Soc. India 23:201-204.

Rao NS (1998) Impact of clayey soils on nitrate pollution in the groundwater of the lower Vamsadhara river basin, India. Hydrol Sci J 43:701–714.

Ritter WF (1988) Reducing impacts of non point source pollution from agriculture: a review. J Environ Sci Health 7:645–667.

Scheytt T (1997) Seasonal variations in groundwater chemistry near Lake Belau, Schleswig-holstein, Northern Germany. Hydrogeol J 5:86–95.

Sharma B, Kumar M, Ramanathan AL (2004) Impact of Surface Water on the Groundwater Quality in and around Nazafgarh Drain, Delhi. In:Ramesh R, Ramachandran S (ed) Freshwater Management. Cap. Pub. Co. Delhi. pp 70–86.

Smil, V. (1999) "Long-Range Perspectives on Inorganic Fertilizers in Global Agriculture," Second Travis P. Hignett Memorial Lecture, IFDC Lecture Series LS-2, IFDC, Muscle Shoals, Alabama, U.S.A.

Stuart MA, Rich FJ, Bishop GA (1995) Survey of nitrate contamination in shallow domestic drinking water wells of the inner coastal plain of Georgia. Groundwater 33:284–290 829.

Subramanian V, Saxena K (1983) Hydrogeology of groundwater in Delhi region of India, Relation of water quality and quantity (Proceedings of the Hamberg symposium IAHS publication no. 146.

Thomas D. Brock T, David Robert Lee Z, David Janes L And David Winek L (1982) Groundwater seepage as a nutrient source to a drainage lake; Lake Mendota, Wisconsin *Water Res.* 16. 1255-1263.

Walter MF, Bubenzer GD, Converse JC (1975) Predicting vertical movement of manurial nitrogen in soil. Trans Am Soc Agric Eng 18:100–105.

Williams AE, Johnson JA, Lund LJ, Kabala ZJ (1998) Spatial and temporal variations in nitrate contamination of a rural aquifer, California. J Environ Qual 27:1147–1157.

World Health Organization (WHO) (1993) Guidelines for drinking water quality, vol. 1, 2nd ed. Recommendations, Geneva.

Study on Age-old Non-conventional Adit/Tunnel Wells in the Lateritic Terrain of Kasargod District, Kerala

11

N.C. Nayak[1], D.S. Thambi and K. Balakrishnan[2]
[1]Central Ground Water Board (SER), Bhubaneswar, Orissa, India
[2]Central Ground Water Board (SWCR), Tiruvananthapuram, Kerala, India

ABSTRACT

Tunnel wells are special kind of groundwater abstraction structures prevalent in Kasargod district of Kerala state. They are horizontal wells driven into the laterite plateaus and hills for tens of meters. No lifting device like bucket and rope or pump is necessary for harnessing groundwater as it drains under gravity. Each tunnel well is unique in its structure as the availability of groundwater leads to its modification and specialization. Most of them yield 0.1 to 0.3 l/s during summer, which is sufficient for the domestic requirement. However, few good yielding wells of 1 to 3 l/s discharge are being used for irrigation. In general, the rainy season discharge is 10 to 15 times that of the summer yield. Though these structures are the best feasible in the area, they lead to the excessive draining of the aquifer which render it unproductive during summer seasons. This necessitates the need for groundwater conservation by suitably modifying these structures.

1 INTRODUCTION

Kasargod, the northernmost district of the Kerala state has an aerial extent of around 1961 km^2. Most part of the district is marked by rugged topography with steeply sloping hills separated by deep cut valleys. Though the district receives very high annual rainfall of around 3294 mm, it experiences severe water scarcity in summer months. Laterite serves as the main aquifer system in the area, which is very thick at the summit and moderate along the slope of the hills. Dug wells at the summit and slope of the hills and plateau are deep (15 to 25 m bgl). Most of them either go dry or need frequent deepening at regular intervals to fetch sufficient water during summer months. Moreover, adjacent dug wells showing contrasting water levels and their fluctuations, add another dimension to the problem. Thus to avoid such gamble and to take advantage of the inhomogenity and anisotropism of the lateritic aquifer system, tunnel wells are common.

Known popularly as "Surangam" in Malayalam and Sanskrit, they are rectangular wells driven horizontally into the lateritic plateau and hills for tens of meters. Structurally they resemble with Qantas (Beaumont, 1971) found in the arid regions of South-west Asia and North Africa extending from Afghanistan to Morocco, running underground through miles and are in use since ancient times. Probably the expertise in such structures had come to India with the Arabic Muslims who got settled in the Malabar Coast at the end of seventh century (Francis Day). Studies have been conducted in and around Bedadka area of the district, which falls in Survey of India Toposheet Number 49 L/15 and 49 P/3. Three areas have been chosen around Kutiyanam, Munnad, Jayapuram and Periya village where almost all the water requirements for domestic and irrigation purposes are met by these tunnel wells. As many as 35 such tunnel wells have been studied in detail.

Generally, a tunnel well starts as a trench across the slope of the hill for a few meters (Fig. 1). After getting a height it transforms into a tunnel. It is an unlined structure in laterite, which is strong enough to withstand against caving. The average width and height of the tunnels are 0.4 to 0.8 m and 1.6 to 2.0 m respectively. The length varies from 10 to 100 m depending on availability of groundwater and the water requirement. Most of them are situated on higher elevation so that water flows under gravity due to the elevation difference between the tunnel and the collection point. The hydrostatic pressure head is used for irrigation and domestic water supply purposes (Fig. 2).

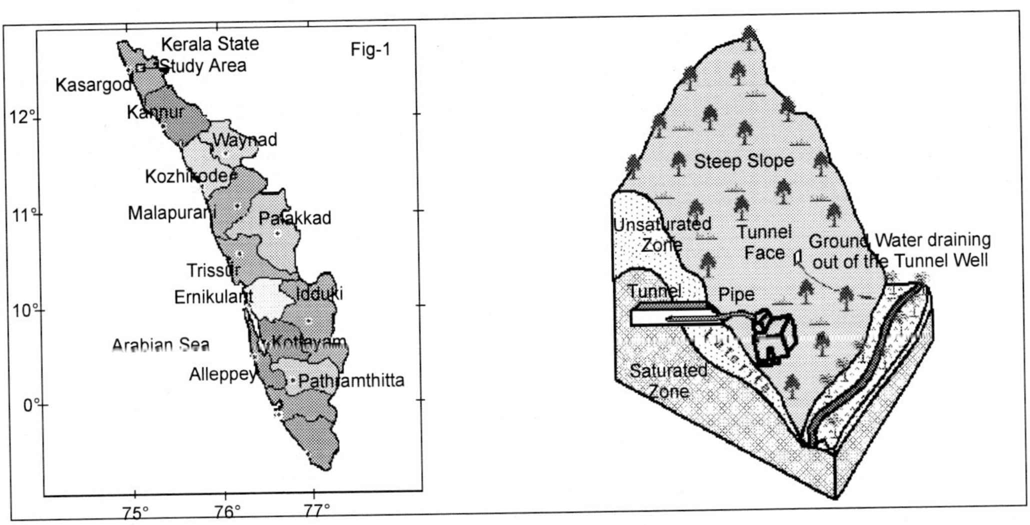

Figs. 1 and 2 Study area and a schematic sketch of a common tunnel well.

2 CLASSIFICATION OF TUNNEL WELLS

Functionally tunnel wells can be divided into two categories, tributaries and distributaries. Tributary tunnels are driven in the saturated zone of laterite where they serve as groundwater harvesting and storage structures. The distributaries are generally driven in the unsaturated lateritic portion to act as conduits for the groundwater collected in the tributaries for further use. Both the structures are complementary to each other. A tributary can serve as conduit for stored groundwater while a distributary may yield groundwater. Due to the rise of water table in rainy season, all the tunnels yield whereas in summer season, only a part of the tributary yields water (Fig. 3). Though tunnel system is a complex one, however it can be expressed as one of the following types or a combination of two or more of them. A brief summary of these types are as follows:

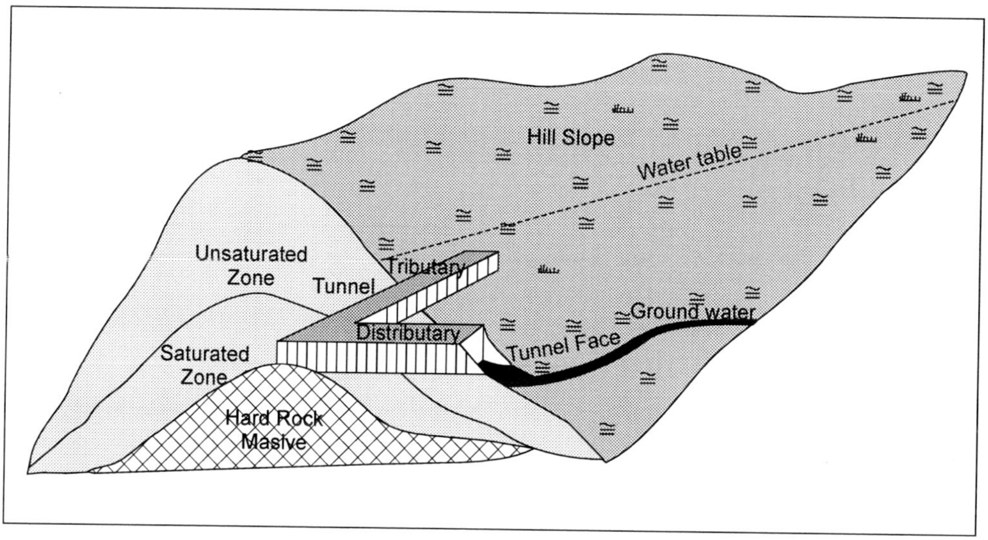

Fig. 3 Types of tunnel wells based on their function.

2.1 Single Tunnel

They are simplest structures, which serve both as tributary and distributary (Fig. 2). In rainy season most part of the tunnel section yields groundwater whereas in summer only it's working face yield. This is commonly seen in homogeneous lateritic areas with thick lateritic cover along the slope.

2.2 Criss-cross/Diversion Tunnels

It is a tunnel system with numerous tunnels arranged in different orientations, out of which some are dry and others yielding. It is commonly seen in heterogeneous laterite where remnant big boulders of massive rocks are encountered or when the core of a lateritic hill is a massive rock. Since the chances of getting groundwater are limited, tunnels in different directions are driven in order to tap a potential groundwater bearing zone (Fig. 4).

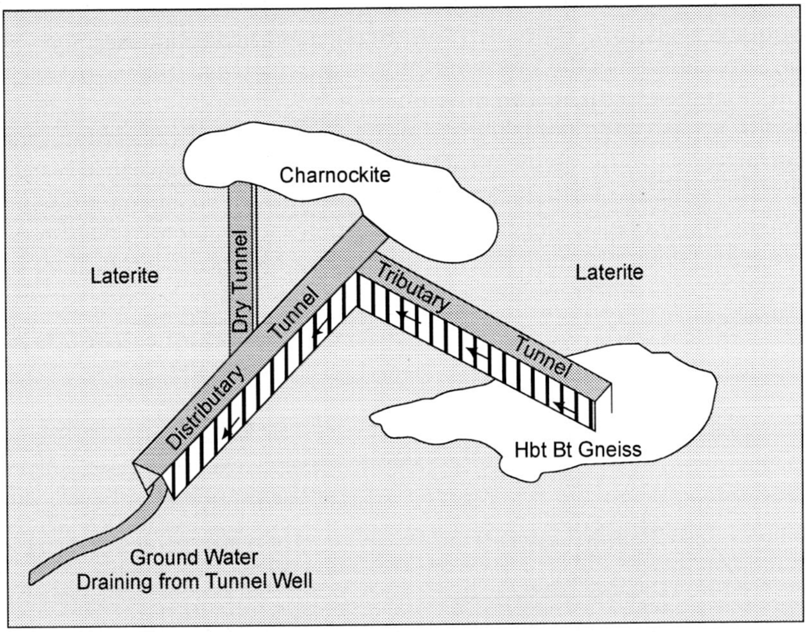

Fig. 4 Modification and ramification of tunnel Systems due to inhomogeneity.

2.3 Tunnel System Starting from a Dug Well

It works as a conduit to the groundwater stored in a dug well, which is tapping a potential groundwater rich zone but having the elevational or locational disadvantage such as higher elevation and away from the habitation. The system works as tributary and serves the purpose of one or a group of houses with water flowing under gravity. This structure is preferred in areas where laterites are yielding at the summit of lateritic hills and plateau and remains dry or having very deep water level along the slope (Fig. 5).

2.4 Tunnel System Ending at a Dug Well/Storage Tank

Most of the tunnels in the system act as tributaries and the dug well and storage tank serve as groundwater storage structures. They are commonly used for irrigation purposes where laterites are yielding more (Fig. 5 & 6).

2.5 Tunnel System with a Dug Well in Between

In this system the tunnels on the up slope side of the dug well are tributaries contributing groundwater to the dug well. Several distributaries, which are starting from the dug well, serve as supply lines for domestic and irrigation purposes (Fig. 7).

2.6 Yield from Tunnel Wells

The discharge from a tunnel well depends on its length inside the saturated lateritic zone. In rainy season when most part of the tunnel is inside the saturated laterite, the yield is maximum.

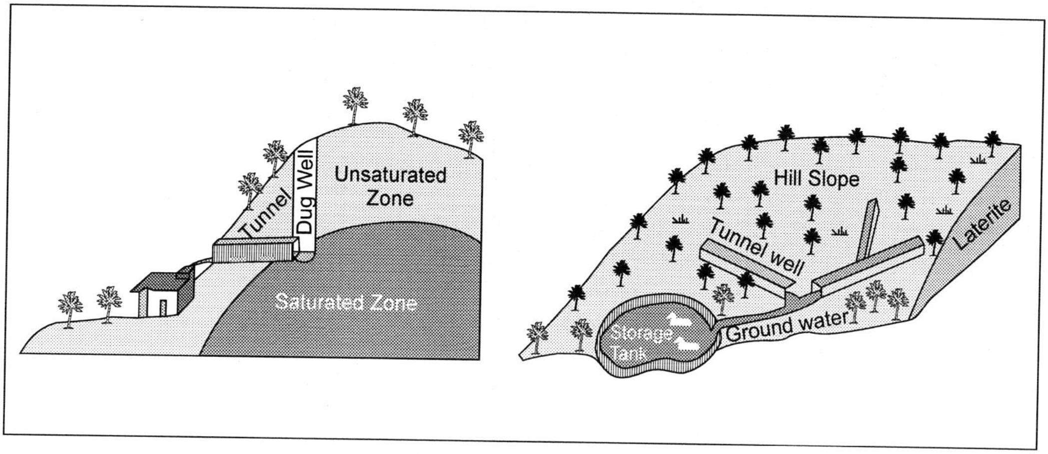

Figs. 5 and 6 Tunnel systems starting from a dug well and ending at a storage pond respectively.

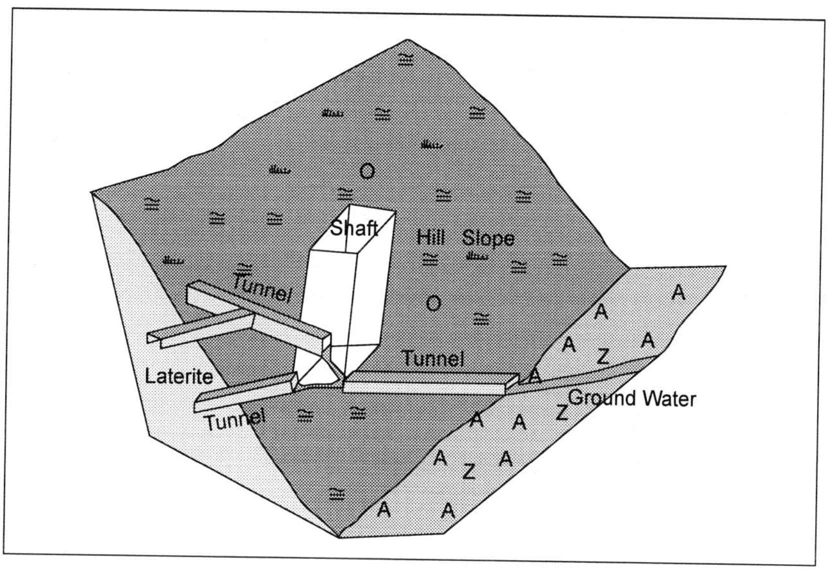

Fig. 7 Tunnel system with a dug well in between.

As months progress, it drastically reduces and at peak summer it is just enough for domestic use. On an average, the rainy season discharge is 10 to 15 times more than that of the summer discharge. Presuming the tunnel face as a rectangular notch, the height of the flowing groundwater above the base of the tunnel can give a fair estimate of the discharge from the tunnel at any point of time.

3 JUSTIFICATION FOR SUCH STRUCTURES IN THE AREA

3.1 Physiography and Climate

Physiographically the study area forms a part of the midland region of Kerala with ground elevation ranges from 10 to 300 m above msl. The area is characterized by an undulating terrain with small steeply sloping hillocks separated by deep cut valleys. Lateritic plateau dotted with low hills are common sight.

The normal annual rainfall is very high (3294 mm) with the monsoon rainfall between June to September contributes about 85% of the total annual rainfall. The year to year variability of rainfall is around 18% and the frequency of incidence of deficient rainfall (I.M.D classification) is only 11% which indicates the consistency of rainfall in the area.

3.2 Hydrogeology

Laterite (Buchanan, 1807), the product of intense subaerial rock weathering formed as capping on the hard rock, forms the main aquifer system in the study area. Groundwater occurs under water table condition in shallow dug wells that are existing in lateritic trough and valley fills. It is under semiconfined condition in deep dug wells commonly seen in hill summits and lateritic plateau. The laterite is very thick (15 to 30 m) and porous and is developed on Hypersthene Granite (Charnockites), (Krishnan 1982), Hornblende-Biotite Granite and Granite Gneiss. A generalized cross section of a laterite hill depicting the different geological horizons in laterite and the hydrogeological condition is shown in Fig. 8.

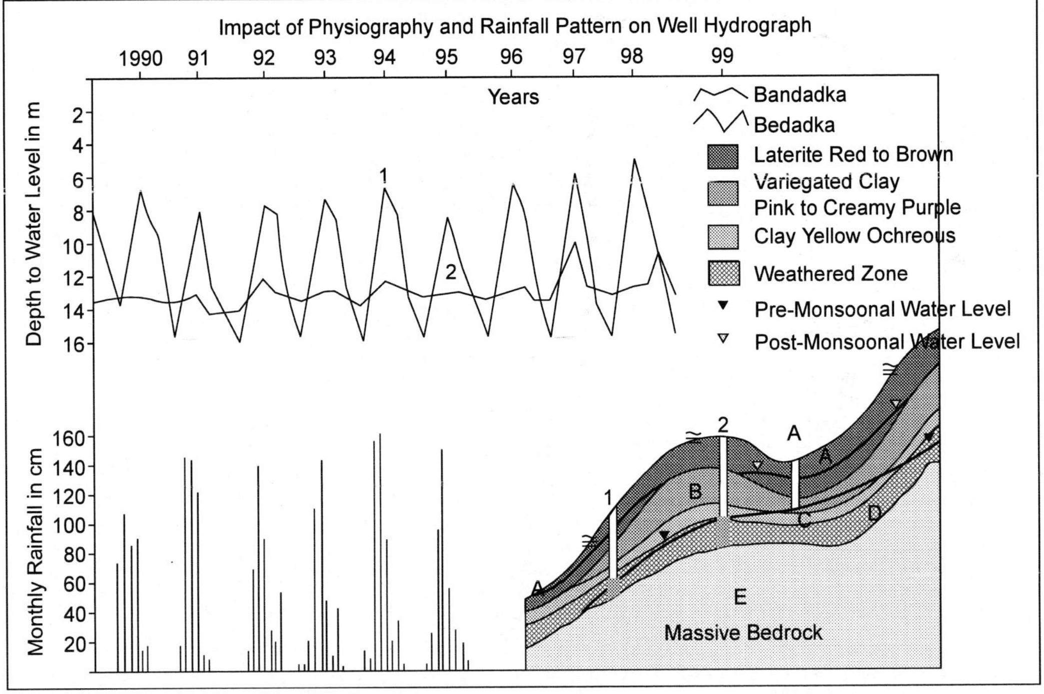

Fig. 8 Impact of physiography, rainfall and lithology on well hydrograph.

- A = Red–brown laterite, porous with very hard black colored lateritic oxidized zone developed on surface, generally 4 to 8 m thick.
- B = Thick variegated clay of pink to creamy purple colour. Spotted appearance is due to the presence of comparatively less oxidized pockets in an otherwise highly oxidized mass, generally 4 to 7 meters thick.
- C = Yellow clay sometimes ocherous, generally 2 to 4 m thick, grading to white clay down depth. Irregular 5 to 20 cm thick impervious and discontinuous very hard Goethitic bands are observed which are responsible for perched water table condition in some dug wells. These Goethitic bands are commonly confined to water level fluctuation zone in laterite, which is indicative of repeated oxidation and reduction leading to its formation. These structures are more prevalent in laterites developed on Hypersthene Granite (Charnockite) and Hornblende - Biotite Gneiss than on and Garnetiferous Sillimanite Gneiss (Khondalite), (Krishnan 1982).
- D = Weathered zone developed on Charnockite, white in color, generally 2 to 5 m thick, characterized by free Quartz embedded on Kaolin. Kaolin in the form of Pseudo morph of feldspathic Rhombs is commonly observed.
- E = Fractured Charnockite

Physiography and the structure of laterite are the main controlling factors for the success of a dug well in this area. This is the reason for nearby dug wells showing totally contrasting groundwater level and fluctuation. The Monitoring Well at Bedadka and Bandadka are best examples (Hydrograph – 1 and 2 in Fig. 8).

The Bedadka Monitoring Station, located at the emergence point of a valley with a moderate slope, shows high fluctuation in water level and remains dry in summer. The well is getting groundwater from the top lateritic aquifer (A zone) in rainy season, which gets dry in summer. Since the bottom weathered zone (D zone) is thin and non-productive during summer, the well is dry during most part of summer. But the Bandadka well is situated on a lateritic plateau is tapping the deep weathered zone (D zone) which is thick and productive. Thus, the well never gets dry during summer and not responds much to the monsoonal recharge, as the groundwater exists under semi-confined condition.

The success of a dug well depends on the following favorable condition:

- The top lateritic zone (A zone in Fig. 8) should be thick enough to retain sufficient groundwater for summer months.
- The clay zone should not be very thick for the stability of the well and the economy of its construction.
- The weathered zone (D zone in Fig. 8) must be thick enough and must get sufficient recharge from the upslope side to yield groundwater in summer.
- The well should not end up against massive rock.

Therefore, in order to avoid failure of deep dug wells and to take the advantage of the slope and the aquifer, tunnel wells are preferred, which are undoubtedly the best solution for the water problem in the area.

4 LIMITATIONS

The major serious problem with this system is the excessive draining of the aquifer leading to the wastage of most of the groundwater reserve in the first two to three months after peak recharge, as there is no control over the system. The example enumerated below can show its dimension. A single good tunnel well yields around 10 liters per second just after peak recharge which is equivalent to 864 m^3/day, since it is a continuous flow for 24 hours. Presuming a hydrogeologically closed basin having one square kilometer area and showing water level fluctuation between pre and post-monsoon as 4 meters and specific yield of the lateritic aquifer as 0.03 (G.E.C Norm, GWREM, 1997), the groundwater reserve can be calculated as 120,000 m^3. Even though the discharge from a tunnel well used to reduce exponentially with time, eight to ten such wells will prove sufficient to drain the whole aquifer.

5 REMEDIAL MEASURES

As there is no substitute for tunnel wells in such adverse condition, modification of the structure is the feasible answer. A close observation of the structure can indicate the following zones in the tunnel:

- The outer zone which is dry for the whole year (OA in Fig. 9).
- The innermost zone of the tunnel, which is yielding the required discharge (PC in Fig. 9 if groundwater requirement is 1.5 L/sec).
- The remaining part of the tunnel (AC in Fig. 9) yields copious water in rainy season and subsequently goes dry. This part is responsible for the excessive draining of the aquifer and is to be sealed off by plastering of the walls.

The method advocated above can reduce the excessive discharge from the aquifer. However, in case of lengthy and complex tunnel wells this proves uneconomical. In such cases the outlet of the tunnel well (tunnel face) is to be sealed off and the required groundwater for domestic and irrigational purposes can be obtained by an outlet pipe.

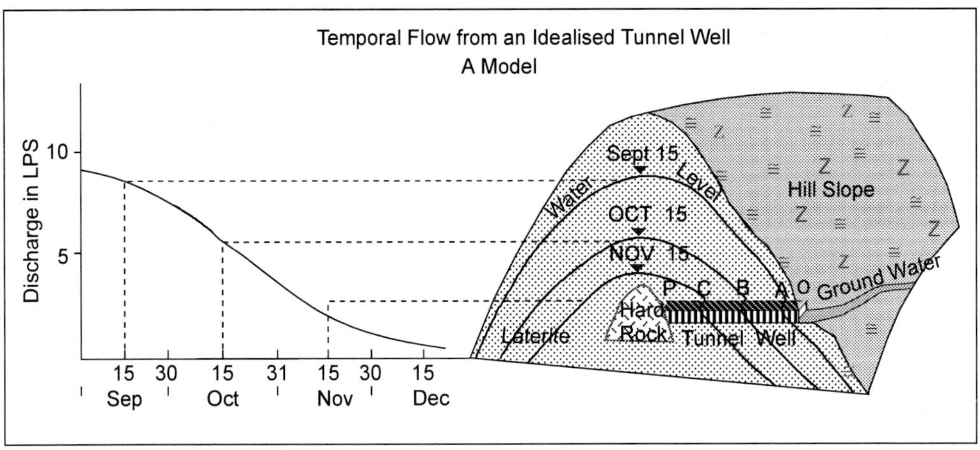

Fig. 9 Temporal variation in flow from an Idealized tunnel well

6 CONCLUSION

To meet the growing need of groundwater in the years to come, more tunnel wells will be added, which may further deteriorate the groundwater situation in the area. As villagers are ignorant of the ill effects of these structures, they must be educated through mass awareness campaign. User agencies and various research organizations can take up such tunnel well modification in small watershed on experimental basis to study its effectiveness. Since not much expenditure and skills are involved, villagers once educated will come forward to adopt it. Government institutions like National Bank of Rural Development (NABARD) and Cooperative Banks and Gram Panchayats should motivate people of the district towards the same by giving some incentives.

Acknowledgements

The authors are thankful to Dr. Saleem Romani, Chairman, Central Ground Water Board for his kind permission for the preparation and presentation of this paper. Thanks are due to Shri S. Chakladar, Regional Director, C.G.W.B (South Eastern Region), Bhubaneswar for his kind support during the paper preparation work. Thanks to the C.G.W.B (Kerala Region) for providing the concerned report and the necessary data. Lastly sincere thank goes to Late Sri P. Laxmi Narayanan, Superintendent Hydrogeologist for his constant encouragement and support during the study.

References

Beaumont, P. Quanat systems in Iran, Bull. Intl. Assoc. Sci. Hydrogeology. V. 16. pp. 39–50. 1971.

Buchanan, F. 1807. A journey from Madras through the countries of Mysore, Canara and Malabar. East India Co., London, 2: 440–441.

Ground Water Resource Estimation Methodology. 1997. Central Ground Water Board. Ministry of Water Resources. New Delhi. pp. 53.

Krishnan, M.S. 1982. Geology of India & Burma. CBS Pub. India. pp. 95–100.

Methods for Estimation of Replenishable Groundwater Resources Adopted in India

12

Rana Chatterjee[1] and B.M. Jha[2]
[1]Central Ground Water Board, New Delhi, India
[2]Central Ground Water Board, Faridabad, India

ABSTRACT

The availability of groundwater is non-uniform in space and time. Hence precise estimation of groundwater resources is essential for planning its development. National Water Policy, 2002 of India also stresses the need for periodical assessment of groundwater resources on scientific basis. In India, extensive and intensive studies on evaluation of groundwater resources were undertaken through multi-disciplinary groundwater balance projects in varied hydrogeological conditions across the country. In these projects, various techniques like soil moisture and surface water balance, chemical and isotope methods, water level fluctuation, numerical modeling etc. were applied for recharge estimations. In 1979, a methodology was evolved for state-wise estimation of groundwater resources. The methodology was subsequently revised in 1984 and 1997. This paper discusses various techniques for assessment of groundwater resources applied in India and suggests strategies for future refinements in the estimation.

1 INTRODUCTION

The definition of groundwater resource, which can be exploited for utilization, has evolved with time and also varied in space depending on the demographic and socio-economic situations and agro-climatic conditions of different countries. Todd (1959) defines the safe yield of a groundwater basin as the amount of water that can be withdrawn from it annually without producing an undesired result. Any withdrawal in excess of safe yield is an overdraft. More recently FAO in its document on Water Reports 23 (FAO, 2003) has defined 'Exploitable water resource' as the 'water resources considered to be available for development under specific economic and environmental conditions. The computation of exploitable water resources considers factors such as dependability of flow, extractable groundwater and minimum flow required for non-consumptive use'. In India, National Water Policy, 2002 (Ministry of Water Resources, 2002) stresses that 'exploitation of groundwater resources should be so regulated as not to exceed the recharging possibilities, as also to ensure social equity.' The exploitable quantity of groundwater, which is recharged annually, is termed as annually replenishable groundwater resources.

2 METHODS USED FOR ASSESSMENT OF GROUNDWATER RESOURCES IN VARIOUS PROJECTS/STUDIES IN INDIA

Estimation of groundwater resources were initiated in India since nineteen seventies with the advent of projects on detailed water balance studies. These studies were undertaken to assess the complete water resource in an area on basin basis adopting multi-disciplinary approach. The water balance studies involve application of a combination of techniques, which can be broadly classified into three groups on the basis of the hydrologic sources or zones from which the data are obtained, namely surface water, unsaturated zone and saturated zone (Scalon et. al. 2002). These techniques alongwith examples from Indian case studies are enumerated in the following paragraphs. These studies have facilitated in the latter years in evolving the groundwater resource estimation methodologies and various norms for recharge estimations, which were used for resource estimation of the entire country.

2.1 Techniques based on Surface Water Studies

2.1.1 Estimation from baseflow

In watersheds with gaining streams, groundwater recharge can be estimated from stream hydrograph separation. Use of baseflow discharge to estimate recharge is based on a water budget approach in which recharge is equated to discharge (Scalon et. al., 2002). Recession-curve displacement method (Rorabough, 1964) is a commonly used approach. A study on the relation of rainfall infiltration factor to the landform using stream hydrograph separation method in Srikakulam district of Andhra Pradesh (Karanth, 1987) indicated that overall infiltration decreases as the terrain becomes more rugged, being about 36% in plain areas and less than 2% in hilly areas. Separation of baseflow in a hydrograph has also been done by solving two simultaneous equations using the chemical quality (Ca/Na ratio) of water in Vedavati river basin project, Karnataka (Central Ground Water Board, 1977). It was found that 39% of the total flow of Vedavati river constituted groundwater runoff during a period of one week.

2.1.2 Water balance method

The general water balance equation using surface water gains or losses has been applied to estimate the recharge in Vedavati river basin project (Baweja & Karanth, 1980).

Gross Recharge = Rainfall – ET losses – Surface runoff – Change in surface water storage + Ground water draft

Gross Recharge in this case has been found to be 14% of rainfall. Water balance method was used to study groundwater recharge in Luni basin, Rajasthan (Srivastava et. al., 1978). Taking daily rainfall in the catchment area as input and daily average evapotranspiration and surface runoff as the total water output, the hydrologic balance of the basin has been studied and groundwater recharge potential estimated on daily basis. It was calculated that during three wet years studied, an average of about 49 mm i.e. 3062 million cubic metre of rainwater is available as groundwater recharge (including runoff). Whereas during the two dry years when rainfall was not only scanty but also spread over, no groundwater recharge took place.

2.2 Techniques based on Unsaturated Zone Studies

2.2.1 *Lysimeter*

Lysimeter consists of containers filled with disturbed or undisturbed soil, with or without vegetation, that are hydrogeologically isolated from the surrounding soil, for purposes of measuring the components of water balance (Scalon et al., 2002). In different parts of the country, Indian Metereological Department conducted lysimetric studies for quantification of evapo-transpiration for major crops in different climatic zones. Central Ground Water Board has used these ET data for various water balance projects like Ghaggar basin project, in Haryana and Punjab (UNDP, 1985). ET losses in the project area varied from 1052 mm for sugarcane in rainy season of June-October to 287 mm for wheat crop during November-May. Techniques similar to lysimeter, commonly known as 'drum culture', have been employed in Andhra Pradesh and Tamil Nadu (Baweja and Karanth, 1980). The method involves raising crops in two drums of standard size, one with bottom closed and another with bottom open in representative agricultural plots. The rate of infiltration due to applied irrigation in the paddy fields in three representative basins in Andhra Pradesh as determined by drum culture varied from 13 mm/day to 28 mm/day.

2.2.2 *Soil moisture storage change*

Infiltration occurring at the land surface can be estimated by the soil moisture balance approach. Soil moisture budgeting, taking into account evapotranspirational abstraction from precipitation, provides a measure of moisture available for runoff and infiltration (Karanth, 1987). Soil moisture budget technique was applied in Ghaggar project, Haryana and Punjab (UNDP, 1985). Studies showed that in arid zone, there is no contribution to groundwater from mean seasonal rainfall. Only if the rainfall in a given year is more than double of mean rainfall, it will contribute to groundwater. During the study period, in the wet year, rainfall's contribution to groundwater recharge was about 8%. Similarly, in semi-arid zone, rainfall contribute to groundwater recharge only with above normal rainfall. About 3 to 15% of rainfall contributes to groundwater. In sub-humid and humid zones, rainfall contributes to groundwater in each year of mean or above mean rainfall. In sub-humid zone, contribution is 24% and in humid zone contribution in the study period was noted on an average to be 15%. The method of soil moisture budget has been extensively used in several other water balance studies in India as well (Baweja & Karanth, 1980) e.g. in climatic water balance study in UNDP, Phase II, Rajasthan, water balance studies in uncropped and cropped deep vertisols and on deep and medium alfisols, Andhra Pradesh, ICRISAT, Hyderabad, studies in typical watersheds in Basaltic terrain, Maharashtra by GSDA, water balance studies in basalts and sedimentaries in Karnataka by Department of Mines and Geology, Karnataka.

2.2.3 *Unsaturated flow modeling*

Unsaturated flow modeling is used to estimate deep drainage below the root zone or recharge in response to meteorological factors like rainfall (Scalon et. al., 2002). National Institute of Hydrology, Roorkee used soil moisture modeling approach to develop a numerical model based on one-dimensional Richard's equation for estimation of irrigation return flow in unsaturated zone (National Institute of Hydrology, 2004). Soil moisture measurements and finite element unsaturated flow modeling carried out in red loamy soil near Bangalore, Karnataka by Indian Institute of Science (Ground Water Resource Estimation Committee, 1997) indicated that the combined recharge factor for rainfall and groundwater irrigation is 12 to 16%.

2.3 Techniques based on Saturated Zone

The following techniques are used in both unsaturated – saturated and saturated zones to estimate groundwater recharge.

2.3.1 *Water table fluctuation method*

The water table fluctuation (WTF) method is based on the premises that rise in groundwater levels in unconfined aquifers are due to recharge water arriving at the water table (Scanlon, 2002). The fluctuations in water table indicate changes in groundwater storage. These in conjunction with specific yield are utilized to obtain the quantum of groundwater recharge. This method has been by far the most commonly used method in all the water balance projects undertaken in India. Water budget of the Noyil basin in Tamil Nadu was done in CGWB-SIDA assisted groundwater project (Raju et. al. 1983). The project area is covered with high-grade metamorphic rocks. The change in groundwater storage obtained through water level fluctuation method and the corresponding annual groundwater recharge for three consecutive years (1976-78) are as follows — –69 mcm: 422 mcm, 345 mcm: 837 mcm, –81 mcm: 611 mcm. In the water balance studies carried out in the upper Yamuna basin (Central Ground Water Board, 2000), which is covered with alluvium, groundwater recharge estimated through empirical method are compared with those estimated through water level fluctuation method. It was seen that in the years of above normal rainfall, the recharge worked out by empirical method were under estimations. Therefore recharge figures obtained by water level fluctuation method were preferred. The water level fluctuation method also forms the base of groundwater resource estimation methodology adopted for the country.

2.3.2 *Chloride mass balance (CMB) method*

The Chloride mass balance (CMB) approach was applied in the arid zones in India to estimate recharge e.g. in the Lathi basin, Western Rajasthan (Karanth et. al., 1980). The average chloride content in rainfall in the Lathi basin is 4 mg/L. The chloride content in the groundwater in the recharge area of the Lathi basin ranges from 100 to 150 μg/L. From these, the rainfall infiltration was worked out as 3 to 4 %. The values obtained compared well with estimations by other methods.

2.3.3 *Radioactive tracers*

These include both historical as well as applied tracers. Historical tracers result from human activities or events in the past, such as contaminant spills or atmospheric nuclear testing. These historical tracers or event markers are used to estimate recharge rates during past 50 years. In applied tracers technique, chemical or isotopic tracers are applied as a pulse at the soil surface or at some depth within the soil profile to estimate recharge (Scalon et. al. 2002). Studies have been made by National Geophysical Research Institute (Ground Water Resource Estimation Committee, 1997) to estimate recharge from rainfall by both historical and applied tracer techniques using Tritium isotopes. The studies are carried out at a number of sites all over the country in different agroclimatic and hydrogeological regions. Uttar Pradesh Irrigation Research Institute, Roorkee used Tritium tagging method (applied tracers) to find out recharge due to rainfall as well as due to applied irrigation in the command areas of Uttar Pradesh (Ground Water Resource Estimation Committee, 1997). Rainfall recharge varies from 18 to 24% and recharge from applied irrigation varied from 22 to 64%. National Institute of Hydrology, Roorkee (Ground Water Resource Estimation Committee, 1997) found out the rainfall recharge factor by tritium tracer studies in various hydrogeological setup like alluvial areas of Indo-Gangetic plains of Northern India (17 – 22%), alluvial areas of Gujarat (3.3 – 13%), granitic terrain (8.3 – 6.1%), Gondwana sandstone (5.4 -15.5%), hard rock areas of Maner basin, Andhra Pradesh (7.6%) and Rajasthan – sand (8.0%), sedimentaries (1.7 to 3.0%), crystalline (1.5%).

2.3.4 *Groundwater dating using tracers*

The estimation of the elapsed time since the recharge to groundwater has taken place is of importance in groundwater resource evaluation. Tritium, Silicon-32 and Radiocarbon are widely used for dating of groundwater. ^{32}Si and ^{14}C measurements on groundwater samples from confined aquifers in the Mehsana district of Gujarat indicated a residence time of 500 years. Similar studies in Jaisalmer district of Rajasthan yielded a residence time of 500 to 3000 years. It was also noted that confined aquifers receive significant recharges of fresh water as leakage through confining beds above (Karanth, 1987).

2.3.5 *Numerical modeling*

Early estimates of groundwater recharge were made by graphical analysis of flow nets for both unconfined and confined aquifers. However this approach has largely been replaced by groundwater flow models. Groundwater calibration or inversion is used to predict recharge rates from information on hydraulic heads, hydraulic conductivity and other parameters (Scalon et. al., 2002). In Vedavati river basin project in Karnataka and Andhra Pradesh (Ground Water Resource Estimation Committee, 1997), finite difference method was adopted with simultaneous estimation of specific yield and recharge factor by model calibration. Results show specific yield values vary from 0.01 to 0.04 and rainfall recharge factor vary from 3 to 15%. Similar studies were also taken up (Ground Water Resource Estimation Committee, 1997) by Indian Institute of Science and Narmada Planning Agency, Madhya Pradesh in Deccan trap basalts, Vindhyan and alluvial formations in Omkareshwar Composite Command, Bargi Diversion project in Sone and Narmada valleys, Tank commands in various areas of Madhya Pradesh. Finite difference numerical modeling technique was also applied in Chitradurga district, Karnataka to get specific yield (0.005 – 0.035) and rainfall recharge factor (1.5 – 20%) values (Ground Water Resource Estimation Committee, 1997). Finite difference method was used in the UNDP assisted project for the Mehsana area, Gujarat (Ground Water Resource Estimation Committee, 1997) for estimating the recharge from rainfall in the deeper aquifers so that remedial measures like restricting withdrawal and artificial recharge could be adopted to restrict water level decline. A numerical modeling study was undertaken in Upper Ganga Canal Irrigation system, Western Uttar Pradesh by National Institute of Hydrology, Roorkee

(Ground Water Estimation Committee, 1984) for studying and inferring suitable strategies for the development of surface water system and in conjunctive utilization of surface water and ground water for irrigation purpose. Finite difference method was adopted for simulation of the groundwater system.

2.4 Scope of Application of Techniques

Various techniques used for recharge estimations in India have indicated that the efficacy of the method depends to a large extent on availability of data and applicability of a methodology in the particular study area. Water table fluctuation technique is the most widely applied method because of the simplicity in application and also since it reflects the actual field manifestation of groundwater recharge. The other advantage of this method is that it can be easily applied on a regional scale because of availability of data from groundwater monitoring program of Central and State agencies. Techniques like CMB and isotope methods have shown good results in arid climatic conditions. Soil moisture balance method has also been successfully applied in several studies. However, the latter methods require adequate infrastructure facilities and considerable data on meteorological, hydrologic and soil characteristics on watershed basis which may not be readily available with state level groundwater organizations (Ground Water Resource Estimation Committee, 1997). Groundwater modeling studies though a useful tool for recharge estimations but it can be better applied with adequate and reliable database. Considering the present availability of database with the state organizations, not all the methods of recharge estimations described in the previous paragraphs are feasible for application on regional basis. However, these can be applied in selected areas to generate norms on various recharge parameters.

3 GROUND WATER ESTIMATION METHODOLOGY FOR STATE-WISE ESTIMATION

For the purpose of state-wise quantitative assessment of groundwater resources, methodologies have been evolved, which can be uniformly applied throughout the country, using database easily generated by the central and state agencies and the results of which matches fairly well with the actual field conditions. The present methodology, which is commonly known as GEC (Groundwater Estimation Committee), is used for assessment of groundwater resource of states for planning groundwater development and management strategies. Assessments are done jointly by CGWB, the central agency and respective state groundwater agencies.

3.1 Concept of GEC

Seasonal rise and fall in water table are observed due recharge and discharge of the aquifers (Das, 2006). The groundwater which is available in this zone of water level fluctuation is called as dynamic groundwater resources or annual replenishable groundwater resources. Below the zone of water table fluctuation, the groundwater which is available in the perennially saturated portion of the aquifer is called as static or in-storage groundwater resources. Groundwater resources in the unconfined aquifers are computed based on groundwater storage concept using specific yield of the rock formations. In the areas where groundwater occurs in confined aquifers, groundwater resources are computed by determining the rate of flow of groundwater in aquifers, using transmissivity/ hydraulic conductivity of the rock formations.

3.2 Background

In 1979, Ground Water Over Exploitation Committee evolved a methodology for groundwater estimation, which was to be uniformly followed by the states. This committee recommended definite norms for groundwater resources computations. It also suggested that in areas where adequate data is available and in all cases where groundwater development is high, water level fluctuation method (same as WTF described in 2.3.1) be applied for groundwater resource estimation. The Gross groundwater recharge included annual recharge from rainfall and other sources excluding recycled water from groundwater irrigation (Agriculture Refinance Development Corporation, 1979). Based on the methodology and norms recommended by the committee, the Gross Groundwater Recharge was computed as 460 billion cubic meter (bcm) and the Net Recharge as 320 bcm (Central Ground Water Board, 2006).

The Over-exploitation Committee, 1979 methodology was subsequently modified to 'Ground Water Estimation Methodology, 1984 (GEC-1984), which recommended estimation of rainfall recharge by two approaches – rainfall infiltration factor method based on norms on rainfall-recharge relations arrived at in various projects mentioned in previous paragraphs and water level fluctuation method using specific yield of the aquifers. Norms on recharge through other sources like canal seepage, return flow from irrigation, seepage from tanks etc. were also recommended by the committee (Ground Water Estimation Committee, 1984). The total groundwater resources included annual recharge from rainfall and other sources. The methodology suggested that about 15% of this annual resources be allocated for domestic, industrial and other uses as well as for committed baseflow of river. The rest of the resource is available for irrigation. Prior to Nineteen Nineties, the focus on groundwater development in the country was predominantly for irrigation purpose, hence the status of groundwater development was determined solely in terms of its utilization in irrigation sector. The groundwater resources of the States and the country were re-estimated based on GEC-1984 for the base year 1988 to 1993. The total replenishable ground water in India was estimated as about 432 bcm. The ground water resource available for irrigation purpose was about 361 bcm (Central Ground Water Board, 1995).

Changing scenario in groundwater utilization pattern, uptodate findings of groundwater assessment studies and enhancement in the database especially on water level led to the revision of the methodology. The revised methodology is called as Ground Water Resource Estimation Methodology - 1997 or commonly as GEC–1997. The revised methodology made a major shift in the concept of groundwater resource by including the recycled component of groundwater irrigation as a component of recharge. The status of groundwater development is now computed for the gross draft from irrigation, domestic and industrial sectors taken together. Also, long-term trend of water level became one of the criteria for categorization of assessment unit apart from the stage of groundwater development. Few other changes were made in GEC-1997 based on the experiences of the previous estimate and findings of new studies (Ground Water Resource Estimation Committee, 1997).

4 GROUND WATER RESOURCE ESTIMATION METHODOLOGY-1997 (GEC-1997)

4.1 Principle

GEC-1997 involves estimation of season-wise groundwater recharge from rainfall and other sources including canal seepage, irrigation, water bodies etc., determination of present status of groundwater utilization and categorization of assessment units based on groundwater utilization status and water level trend.

The assessment year starts from the onset of monsoon of one calendar year to the end of the summer season of the next calendar year. In major part of the country where South West monsoon is predominant, the assessment year starts from late May to late June of one calendar year to May / June of the next calendar year. In areas with North East monsoon, it is October of one calendar year to September of next calendar year.

Groundwater recharge during monsoon season is estimated using water level fluctuation technique and empirical norms and during non-monsoon season using empirical norms only. Monsoon season is June/July to September/October. However, water level fluctuation is taken between the deepest water level in April/May, prior to monsoon and November, one month after the cession of monsoon to account for the baseflow, which occurs immediately following the monsoon period but may not be utilized for the groundwater development as per the present practice. Figure 1 depicts seasonal fluctuation of groundwater level used for recharge estimation during monsoon season.

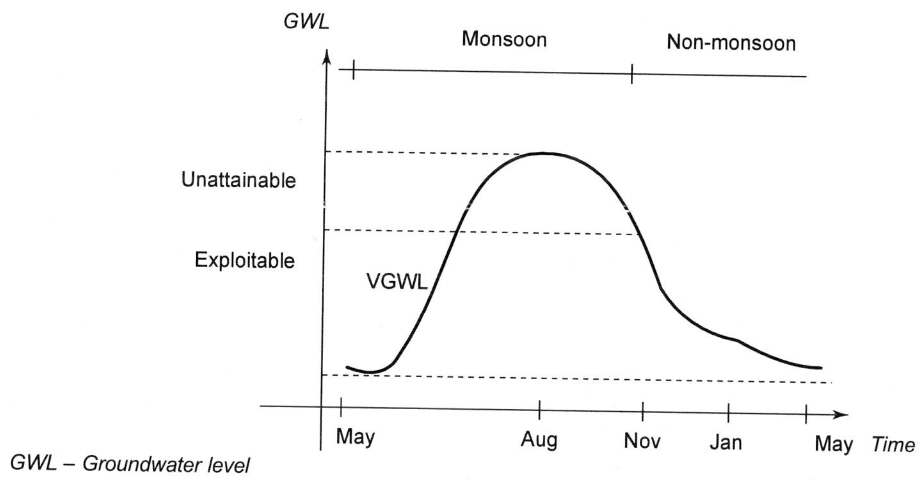

Fig. 1 Typical hydrograph showing seasonal fluctuation of groundwater level used for recharge estimation during monsoon season

4.2 Steps for Computations

(i) Assessment unit – block in alluvial areas and watershed in hard rock areas. Sub-units – command (area irrigated through major/medium irrigation schemes) and non-command areas.

(ii) Recharge during monsoon:

(a) Determination of monsoon recharge using water level fluctuation method ~ Change in Storage + Groundwater draft during monsoon.

(b) Recharge from other sources during monsoon (Canal seepage + return flow from irrigation + recharge from water bodies + recharge from water conservation structures).

(c) Monsoon rainfall recharge — Recharge from other sources during monsoon (at 'b') to be subtracted from monsoon recharge (at 'a').

(d) Normalization monsoon rainfall recharge (at 'c').

(e) Determination of monsoon rainfall recharge using Rainfall Infiltration Factor method.

(f) Determination of normal monsoon rainfall recharge - comparison between Rrf (normal, WLF method) – Rrf (normal, RIF method) – PD (Percent difference)

(g) Total recharge during monsoon season for normal monsoon season - rainfall recharge (normal) (arrived at 'f') + recharge from other sources (arrived at 'b').

(iii) Recharge during non-monsoon ~ Rainfall Recharge during non-monsoon season + recharge from other sources during non-monsoon season.

(iv) *Annual replenishable groundwater resource* – monsoon recharge (iii-g) + non-monsoon recharge (iv).

(v) Allocation for natural discharge during non-monsoon period :

5% of annual replenishable groundwater resources – for natural discharge (baseflow and subsurface inflow/outflow) during non-monsoon season. In case, monsoon rainfall recharge is arrived at using RIF method, 10% of annual resource to be deducted.

(vi) Net annual ground water availability – Annual replenishable groundwater resources – allocation for natural discharge during non-monsoon period.

(vii) Determination of the status of groundwater development

(a) Stage of GW development (%) ~ (annual groundwater draft /net annual groundwater available) * 100.

(b) Determination of the trend of groundwater level for last 10 years.

(c) Categorization of the assessment unit based on stage of GW development and trend of water level into – 'Safe', 'Semi-critical', 'Critical' and 'Over-exploited'.

4.3 Governing Equations for Recharge Estimations

- *Estimation of recharge during monsoon season (Water table fluctuation technique)*

$$R = h \times Sy \times A + D_G \quad \text{...(i)}$$

where,

R = Recharge during monsoon

S = Change in storage ($S = h \times Sy \times A$)

D_G = Gross draft during monsoon season

h = water level fluctuation between pre-monsoon and post monsoon

Sy = Specific yield

A = Area of assessment

- *Normalization of rainfall recharge*

$$R_{rfi} = R - R_c - R_{sw} - R_{gw} - R_{wc} - R_t \quad \text{...(ii)}$$

$$= h \times Sy \times A + D_G - R_c - R_{sw} - R_{gw} - R_{wc} - R_t \quad \text{...(iii)}$$

Where,

R_{rfi} = Rainfall Recharge during monsoon season for i^{th} particular year

R_c = Recharge from canal seepage during monsoon season (in command areas) for i^{th} particular year

R_{sw} = Recharge from surface water irrigation during monsoon season (in command areas) for i^{th} particular year

R_{gw} = Recharge from groundwater irrigation during monsoon season for i^{th} particular year

R_{wc} = Recharge from water conservation structure during monsoon season for i^{th} particular year

R_t = Recharge from tanks during monsoon season for i^{th} particular year

Rainfall recharge is normalized for estimating recharge corresponding to the normal monsoon rainfall using following linear relationship between recharge and rainfall- (a) $R = a\,r$, Where, R is rainfall recharge during monsoon season, r is monsoon season rainfall and a is constant

Or (*b*) $R = a\,r + b$, Where, R is rainfall recharge, r is rainfall and a and b are constants

- *Rainfall Infiltration factor approach (RIF)*

$$R_{rf} = f \times A \times \text{Normal rainfall during monsoon season} \quad \text{...(iv)}$$

Where,

R_{rf} = Rainfall recharge during monsoon

f = rainfall infiltration factor

A = area of assessment unit

- *Percentage difference*

$$PD = [(R_{rf}(\text{normal, WLF method}) - R_{rf}(\text{normal, } RIF \text{ method}))/ R_{rf}(\text{normal, } RIF \text{ method})]*100 \quad \text{...(v)}$$

Where,

PD = Difference in WLF & RIF estimate expressed as a percentage of RIF estimate

If,

(a) if PD is within ± 20%, Rrf(normal) = R_{rf}(wlf)

(b) if PD is <-20%, Rrf(normal) = 0.8 * R_{rf}(RIF)

(c) if PD is>20%, Rrf(normal) = 1.2 * R_{rf}(RIF)

- *Total recharge during the monsoon season for* ***normal*** *monsoon season*

$$R\text{ (normal)} = R_{rf}\text{ (normal)} + R_c + R_{sw} + R_{gw} + R_{wc} + R_t \quad \text{...(vi)}$$

Where,

R (normal) = Total recharge during normal monsoon season

R_{rf} (normal) = rainfall recharge during monsoon season for normal monsoon

- *Recharge estimation of non-monsoon recharge*

  Rainfall recharge – Using RIF approach

  Recharge from other sources – Using norms

  Total recharge is the sum of rainfall recharge and recharge from other sources
- *Annual recharge – recharge during monsoon + recharge during non-monsoon.*

5 DYNAMIC GROUNDWATER RESOURCES OF INDIA

Groundwater resource of India has been estimated based on GEC'97 for the base year, 2004 (Central Ground Water Board, 2006).The Annual Replenishable Ground Water Resource for the entire country is 433 billion cubic metre (bcm). Figure 2 presents the over-all scenario of groundwater resource and development potential of the country. The overall contribution of rainfall to country's Annual Replenishable Ground Water Resource is 67% and the share of other sources taken together is 33%. South-west monsoon being the most prevalent contributor of rainfall in the country, about 73% of country's Annual Replenishable Ground Water Recharge takes place during the Kharif period of cultivation. Keeping an allocation for 34 bcm for natural discharge, the Net Ground Water Available for utilization for the entire country is 399 bcm. The total Annual Ground Water Draft is 231 bcm out of which 213 bcm is for Irrigation use and 18 bcm is for Domestic and Industrial use. In general, the irrigation sector remains the main consumer of groundwater (92% of total annual groundwater draft for all uses).

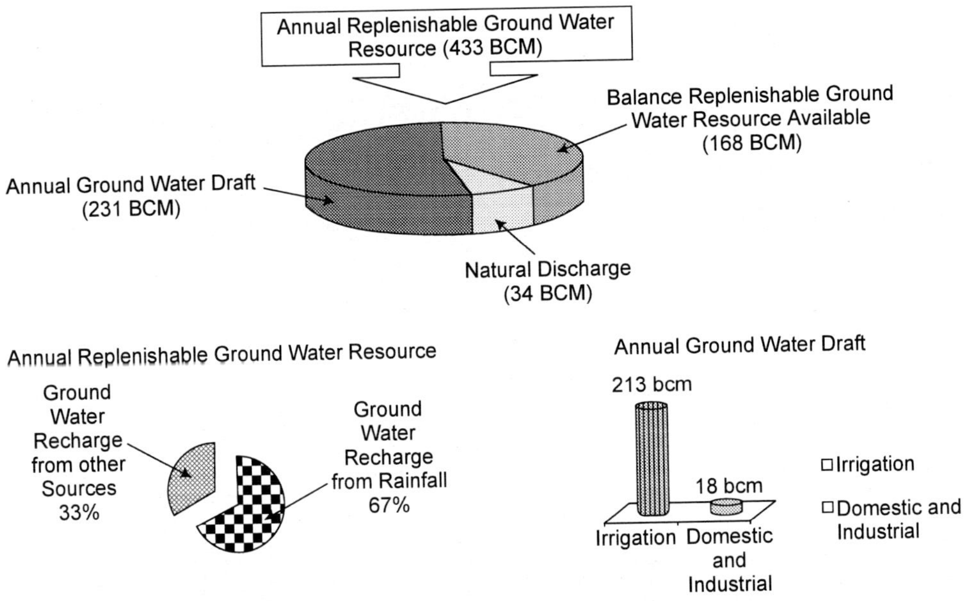

Fig. 2 Groundwater resources availability and utilization in India (from Dynamic Ground Water Resources of India, CGWB, 2006).

The overall stage of groundwater development in the country is 58%. The status of groundwater development is quite high in the states of Delhi, Haryana, Punjab and Rajasthan and UT of Daman & Diu and Pondicherry, where the overall Stage of Groundwater Development is more than 100% which implies that in the these states the average annual groundwater consumption is more than average annual groundwater availability. In the peninsular states of Andhra Pradesh, Gujarat, Karnataka, Kerala, Madhya Pradesh, Maharashtra and Tamil Nadu, the general scarcity of sustainable water supply has led to over-stress on groundwater regime. Since most of the area in this region is occupied by hard rock terrains, limited availability of groundwater has resulted in growing number of over-exploited Blocks/ Mandals/ Taluks. The average stage of groundwater development in these states are also high. Uttar Pradesh and Uttaranchal also have high stage of groundwater development. In the states of Bihar, Orissa, West Bengal and the valley areas of Assam and Tripura in the north east, there is ample scope for groundwater development since the stage of groundwater development in these states are lower and there are few Blocks/Taluks in Critical/Over-exploited category. In the states of Chhattisgarh, Goa, Himachal Pradesh, Jammu & Kashmir, Jharkhand, North Eastern States of Arunachal Pradesh, Manipur, Meghalaya, Mizoram, Nagaland and UTs of Andaman & Nicobar Islands, Dadra & Nagar Haveli, though the stage of groundwater development is lower, however considering the prevailing hydrogeological situations, the groundwater development should be taken up with caution on site-specific basis.

Out of 5723 numbers of assessed administrative units (Blocks/ Taluks/ Mandals/ Districts), 4078 units are 'Safe', 550 units are 'Semi-critical', 226 units are 'Critical', 839 units are 'Over-exploited' and 30 units are 'Saline'. Number of Over-Exploited and Critical administrative units are significantly higher (more than 15% of the total assessed units) in Andhra Pradesh (where categorization was done upto sub-unit level i.e. within Mandal - command and non-command-wise), Delhi, Gujarat, Haryana, Karnataka, Punjab, Rajasthan and Tamil Nadu and also the UTs of Daman & Diu and Pondicherry (figure 3).

The summary of dynamic ground water resource estimates of India is given in table– 1.

Table 1 Dynamic ground water resources of India (As on 2004)

Annual replenishable ground water resource	433 bcm/yr
Natural discharge during non-monsoon season	34 bcm/yr
Net annual ground water availability	399 bcm/yr
Annual ground water draft	231 bcm/yr
Projected demand for domestic & industrial uses	29 bcm/yr
Ground water availability for future irrigation	161 bcm/yr
Stage of ground water development	58%

Source: Dynamic Ground Water Resources of India, CGWB, 2006

6 SUGGESTIONS ON REFINEMENTS IN ASSESSMENTS

Groundwater resources estimation is a continuing process. Following suggestions are made regarding future refinements in the resources assessments.

- *Assessment units with Hydrologic/ hydrogeologic boundaries :* The groundwater balance equations suggested in GEC-1997 can be better applied in assessment units with

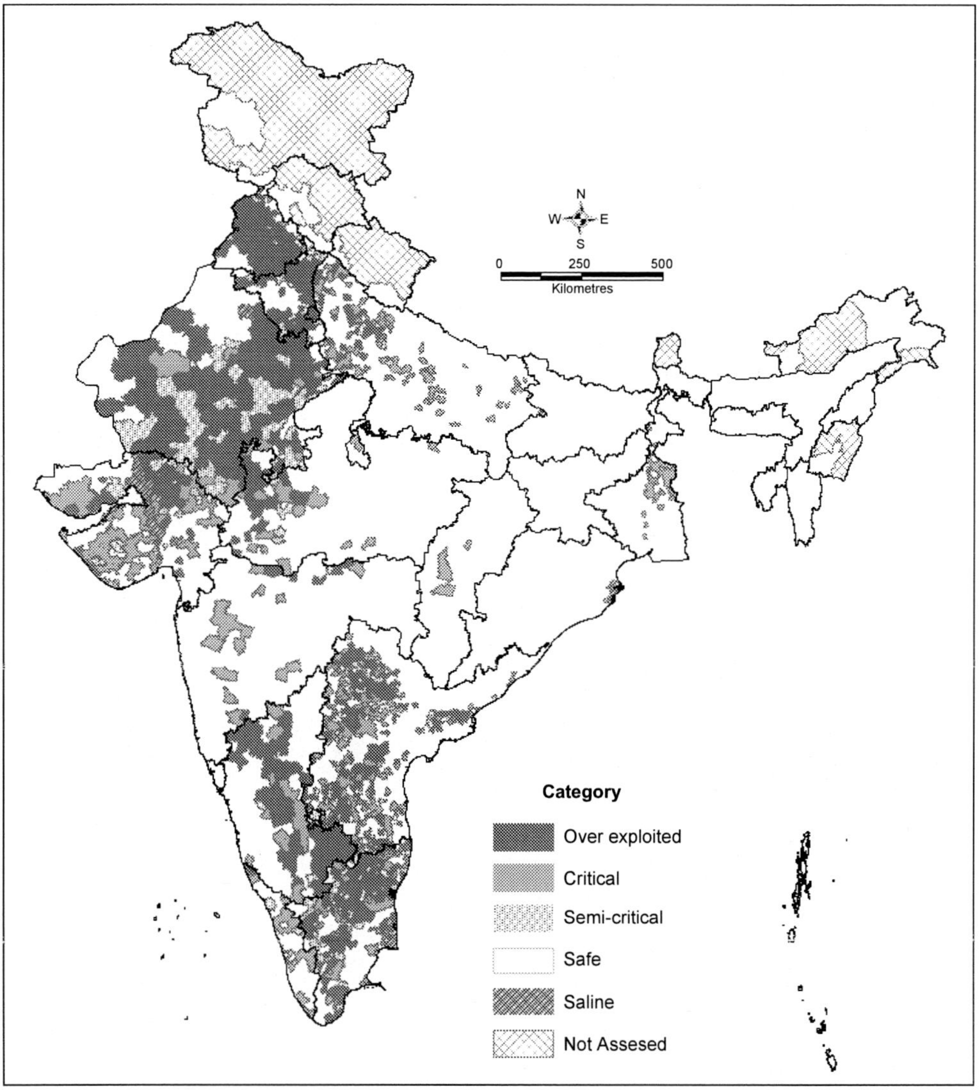

Fig. 3 Categorization of Blocks/ Mandals/Taluka as on 2004 (from Dynamic Ground Water Re sources of India, CGWB, 2006)

hydrologic/hydrogeologic boundaries. In case of predominantly hard rock states, the assessment unit should be Watershed. This would require re-organization of database in watershed-wise and sub-unit wise (canal command/non-command) formats.

- *Strengthening of database on norms for resources estimation :* Refinement in groundwater resources estimation would require strengthening of database on norms of various parameters used for recharge estimation like specific yield, rainfall infiltration factor, infiltration and seepage factors of irrigation, canal, water bodies, water conservation structures, baseflow and natural loss etc. These are to be achieved through special R&D studies on groundwater balance, irrigation projects, artificial recharge etc.

- *Strengthening the database on groundwater draft* : As per the Third Census on Minor Irrigation (2000-01), there are about 18.5 million groundwater structures in the country of which about 16.4 million are in use. However, most of these structures are privately owned and there is no metering system installed in these structures by which the exact amount of water being withdrawn by these structures can be measured. In absence of direct method of estimation of quantity of water being withdrawn, the groundwater draft has to be estimated by indirect means. This brings in uncertainties in groundwater draft estimations. Therefore, more than one method of draft estimation needs to be applied to arrive at a realistic figure matching the field situations. The recommended methods are : unit draft method, cropping pattern method, pattern of electricity consumption, remote sensing method in non-command areas. Existing mechanism for data collection at state level needs to be streamlined for this purpose.
- *Assessment of additional groundwater resources :* In the hilly terrains, where due to steep hydraulic gradient, annually replenished groundwater could not be retained for utilization for the larger period of the year, assessment of water available through alternate sources like spring discharges needs to be done. In the shallow water table areas, where inspite of regular groundwater extraction, the water level is not declining, the *sustainable yield* (Kalf & Woolley, 2005) of the aquifers is to be determined.
- *Assessment of groundwater availability and potentiality of the aquifer systems* : The present paper has mainly dealt with the concept of groundwater recharge. Since the stress on groundwater regime is ever increasing due to accelerating demand of water, there is need for assessment of groundwater availability and development potential of aquifers at all depths. The first step in this regard would be systematic aquifer mapping and determination of hydraulic characteristics of the individual aquifers within the regional aquifer systems of the country.

7 CONCLUSION

Groundwater resource computations done by GEC'97 presents the generalized assessment of the exploitable groundwater resource and status of its utilization. This first level of qunatification should be followed by micro-level studies, wherein besides precise estimation of groundwater resource, various options for future groundwater development and management have to be evaluated. Since detailed investigations are time-consuming process, areas for micro-level studies need to be prioritized. Over-exploited and critical areas would be the natural preference in the priority list. Areas with quality problem can also be taken up on priority for detailed groundwater assessment.

Applicability of a technique for quantifying groundwater recharge depends on several factors like agro-climatic conditions, spatial scale of study, hydrogeological setup etc. Each technique has its own uncertainties, which could influence the result of recharge estimate. Therefore it is recommended that multiple techniques should be used in order to reduce uncertainties and increase confidence in recharge estimates.

Present estimation based on GEC'97 is done as on March, 2004. There is a need for periodical reassessment of groundwater resource of the country due to rapidly changing groundwater scenario. Estimation methodology also requires further refinements which would include enhancement and improvement in database and a holistic approach on groundwater resource by including assessment of additional sources like spring discharge, sustainable yield in shallow water table areas and assessment of groundwater potential of regional aquifer systems.

References

Agriculture Refinance Development Corporation, 1979. Report of the Ground Water Over-Exploitation Committee.

Baweja, B.K. & Karanth, K.R., 1980. *Groundwater recharge estimations in India.* New Delhi: Central Ground Water Board.

Central Ground Water Board., 1987. Hydrogeology and groundwater balance of the Vedavati river basin in parts of Karnataka and Andhra Pradesh - an interim report on concept, methodology and tentative findings. Unpublished report.

Central Ground Water Board., 1995. *Ground water resources of India.* New Delhi: Central Ground Water Board, Government of India.

Central Ground Water Board., 2000. Water balance studies in upper Yamuna basin - terminal report - project findings and recommendations. Chandigarh: Central Ground Water Board.

Central Ground Water Board., 2006. *Dynamic ground water resources of India (As on March, 2004).* New Delhi: Central Ground Water Board, Government of India.

Das, S. 2006. Groundwater overexploitation and importance of water management in India - vision 2025. Tenth IGC Foundation Lectures. The Indian Geological Congress, Roorkee.

FAO., 2003., *Water Reports 23: Review of world water resources by country.* Rome: FAO.

Kalf, Frans R.P. & Woolley, D.R., 2005. Applicability and methodology of determining sustainable yield in groundwater systems. *Hydrogeology Journal* 13: 295-312.

Karanth, K.R., Mehta, M., Jati, K.K. 1980. A reappraisal of ground water recharge in the Lathi basin, Rajasthan. New Delhi: Central Ground Water Board.

Karanth, K.R. 1987. *Ground water assessment, development and management.* New Delhi: Tata McGraw-Hill Publishing Company Limited.

Ground Water Estimation Committee., 1984. *Ground water estimation methodology.* New Delhi: Ministry of Irrigation, Government of India.

Ground Water Resource Estimation Committee. 1997. *Ground water resource estimation methodology – 1997.* New Delhi: Ministry of Water Resources, Government of India.

Ministry of Water Resources, Government of India. 2002. *National water policy.* New Delhi: Ministry of Water Resources, Government of India.

National Institute of Hydrology, 2004. *Implementable technologies for water resources development.* Roorkee: National Institute of Hydrology.

Raju, K.C.B. *et. al.* 1983. Ground water resources of Noyil, Ponnani and Vattamalai Karai basins. New Delhi: Central Ground Water Board.

Rorabough, M.I. 1964. Estimating changes in bank storage and groundwater contribution to streamflow. *Int Assoc Sci Hydro Publ* 63: 432-441.

Scanlon, B.R., Healy, R.W. & Cook, P.G. 2002. Choosing appropriate techniques for quantifying groundwater recharge. *Hydrogeology Journal* 10(1): 18-39.

Srivastava, K.K., Kapoor, V.K., Abbi, S.D.S., 1978. Potential evapotranspiration loss and ground water recharge in Luni catchment in West Rajasthan.

Todd, D.K. 1959. Ground water hydrology. New York: John Wiley & Sons.

UNDP. 1985. Ground water studies in the Ghaggar river basin in Punjab, Haryana and Rajasthan. Final technical report, volume I - Surface and Ground water resources, New York: UNDP.

13

Assessment of Groundwater Reserves: A Case Study

Ramesh Chand, R.S.K. Srinivasulu, V.S. Singh and N.C. Mondal
National Geophysical Research Institute, Hyderabad, Andhra Pradesh, India

ABSTRACT

Assessment of groundwater reserves to a basin in granite, having an area of 6100 km^2 in Andhra Pradesh, India, has been measured by artificial tritium tagging method. Tracer was injected at 54 selected locations before the onset of monsoon covering various soil types. Annual replenishment of groundwater reserves of the drought-prone area, due to percolation of fraction of the annual precipitation, has been quantitatively estimated. Vertical soil profiles, down to a maximum depth of 290 cm were collected. Soil moisture content and tritium concentration (T.U.) were measured for every 10 cm section of the soil moisture profile. Moisture content and tritium concentration of the soil cores collected after monsoon were measured in the laboratory. Tracer displacement was estimated by using the variation in tritium concentration vs. moisture with depth and the same is measured for estimation of recharge. The recharge values calculated from moisture content and displacement of the tracer peak in the study basin were found to vary from -2.90 cm to 21.0 cm. The mean recharge value for the entire basin was computed as 2.74 cm, i.e. 4.45% of the average annual precipitation. This gives an annual groundwater reserves in the study area of about 140 MCM.

1 INTRODUCTION

The study area is located in Anantapur, Cuddapah and Kurnool districts of Andhra Pradesh (A.P.), India. The climate is semi-arid and the basin is chronically drought affected due to frequent failure of monsoon in the last few decades. The surface water resources are limited to the quantum supplied through Tungabhadra canal in the basin. There were curtailments in this supply too, owing to depletion in reservoir's storage. As such large areas of the basin depend on the groundwater supply.

The chief source of recharge to groundwater in this basin is by direct precipitation due to annual rainfall. Due to non-uniform recharge condition and precipitation, large fluctuation in groundwater levels are often observed at many places. It has, therefore, become essential that the annual input or recharge to various aquifers of the basin is quantified and the data is used for planning and optimal utilization of the groundwater resources.

Recharge due to precipitation in the basin which together form major part of Rayalseema region of A.P. was measured by tritium injection method for the hydrological cycle. The results of recharge at various tritium-injected sites are represented in this research paper.

1.1 Location

Major parts of the basin are located in Anantapur district of A.P. Small portion at the tail ends falls in Cuddapah district of A.P. in the lower reaches and Kolar district of Karnataka in the upper reaches (Fig. 1). The basin is bounded within Latitudes 13°35′ to 14°50′ N and Longitudes 77°30′ to 78°15′ E. The basin occupies as area of 6100 km^2 comprising two sub-basins, namely, Maduleru and Kushavati, which are the major streams of Chitravai river.

1.2 Physiography and Drainage Pattern

Physiographically, the basin is characterized by four main units namely, Denudational hill ranges with steep slopes, undulating topography with moderate slopes, pediplains and alluvial areas. The basin slopes from south to north with the maximum elevation in the extreme southern end in Karnataka of about 900 m and minimum of 200 m at the northern extremity near the mouth of Chitravati river. A major portion of the basin forms a plain ranging in elevation from 600 to 300 m from south to north with hill ranges and hillocks of relatively small relief scattered over the area. The alluvium is seen along the course of Chitravati, Maduleru and Kushavati river sections and also in the vicinity of river lets. Alluvium ranges in thickness from few meters to a maximum of 30 m in the river section of Chitravati. This river is the tributary of river Pennar and originates near Yaligere in Karnataka and flows in a northernly direction. Kushavati and Maduleru are the two major streams of Chitravati river. The drainage pattern of the Chitravati basin is dendrite and parallel to sub-parallel. The basin has two major tanks, namely, Dharmavaram and Kottacheruvu, which are interconnected. The overflows of these tanks drain the Chitravati river. There are a number of major irrigation tanks in the basin. Figure 1 shows the drainage map of Chitravati river basin.

1.3 Climate

The basin receives both SW and NE monsoons but does not enjoy the full benefit of as it is situated at the coast in semi-arid area of Rayalseema region of A.P. The average annual rainfall varies between 550 mm and 800 mm. The major portion (60%) of the rainfall comes from SW monsoon during the period June to September and a minor amount of rainfall (30%) is due to NE monsoon during the period October to November. The rest of the rainfall is scattered in the other periods of the year. Few observations show that isolated showers tend to produce flashy runoff in this semi-arid area where the catchments have few storage possibilities. Evapotranspiration in this basin is also high. The average rainfall is 615 mm/year.

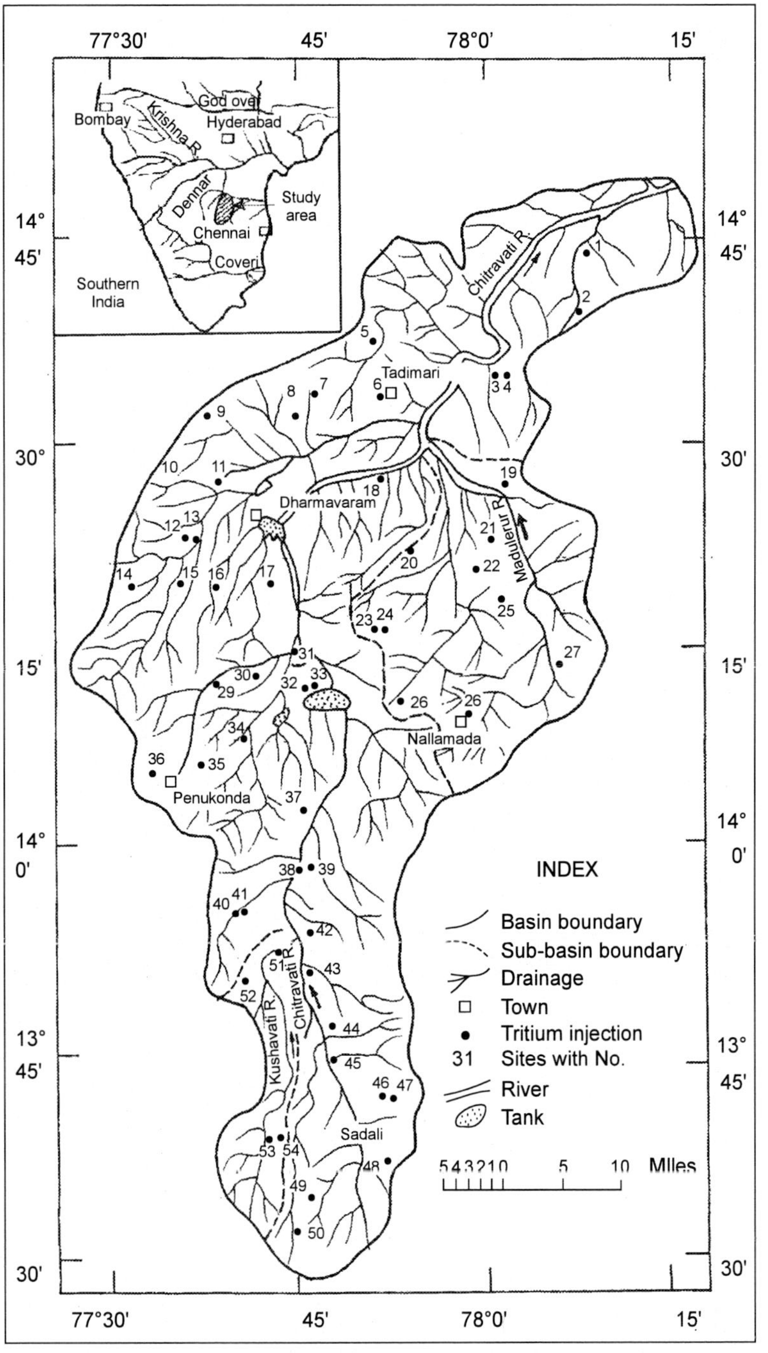

Fig. 1 Location of injection sites and drainage pattern.

1.4 Soil Types

The chief soil types of this basin are red loamy and black cotton soils. The maximum thickness of the black cotton is as much as 9.0 m in Pennar plain. Red soils are thin as compared with black soil. They range in thickness from few cm to 2.5 m. Alluvium is restricted to the banks of rivers and streams. It generally comprises sands, gravels, silt and clay having 5–10 m thickness.

1.5 Geology and Hydrogeological Setting

The geological formation of the Chitravati river basin can broadly be divided into two groups. An older group of metamorphic rocks belongs to Archaean and a younger group of sedimentary rocks belongs to Cuddapah super group of Proterozic age. The general stratigraphic succession of this basin is given by Raju *et al.*, (1979). The geological map of the basin is shown in Fig. 2. Groundwater in this region occurs in the intergranular space of the weathered rocks and in the joints and fractures criss-crossing the rocks. Dolerites also hold water under weathered and fractured conditions. Long and relatively unweathered dolerite dykes act as barriers for groundwater and hence groundwater accumulates on one side of the dyke as against the poor occurrence on the other side depending upon the direction of groundwater flow in the region which, in turn, is largely guided by physiographic consideration.

2 RECHARGE MEASUREMENTS

Artificial tritium injection experiments have been up in Chitravati river basin with the aim of calculating average recharge and total input to the shallow groundwater reserves of the basin due to monsoon precipitation.

Tritium (H^{3+}), a radioactive isotope of hydrogen, is a soft β-emitter having mass 3 and a half-life of 12.26 years. It exists in the form of water molecules HTO and as such it is an ideal tracer for studying movement of water.

The use of tritium as a tracer in recharge measurement involves injection of tritiated water at a certain depth in the soil column and study of the vertical movement that this tracer undergoes during the next hydrological cycles. This application of tritium in recharge measurements is based on an assumption of the Piston-Flow model, proposed by Zimmermann *et al.*, (1967 a, b) and Munnich (1968), assumes that the percolating soil moisture moves downward in discrete layers and any addition of a fresh layer of moisture at the surface, would push down an equal amount of water immediately below and so on, till the last such layer in the unsaturated zone will be added to the saturated regime or the water table. The principal steps involved and other field and laboratory procedures of the tritium injection method for calculating recharge have been described by Athavale *et al.* (1978), (1980), (1983), Ramesh (2005) and Chand *et al.* (2004).

2.1 Tritium Injections

Tritiated water was injected at 54 selected sites in the study area in the month of July before the commencement of monsoon. The depth of injection was 60 cm. Duplicate sets of injections were made at 8 selected sites of the area. The adjacent injections were separated by a distance of 3 m. Tritium injection technique used the results from these sites in studying the reproducibility of

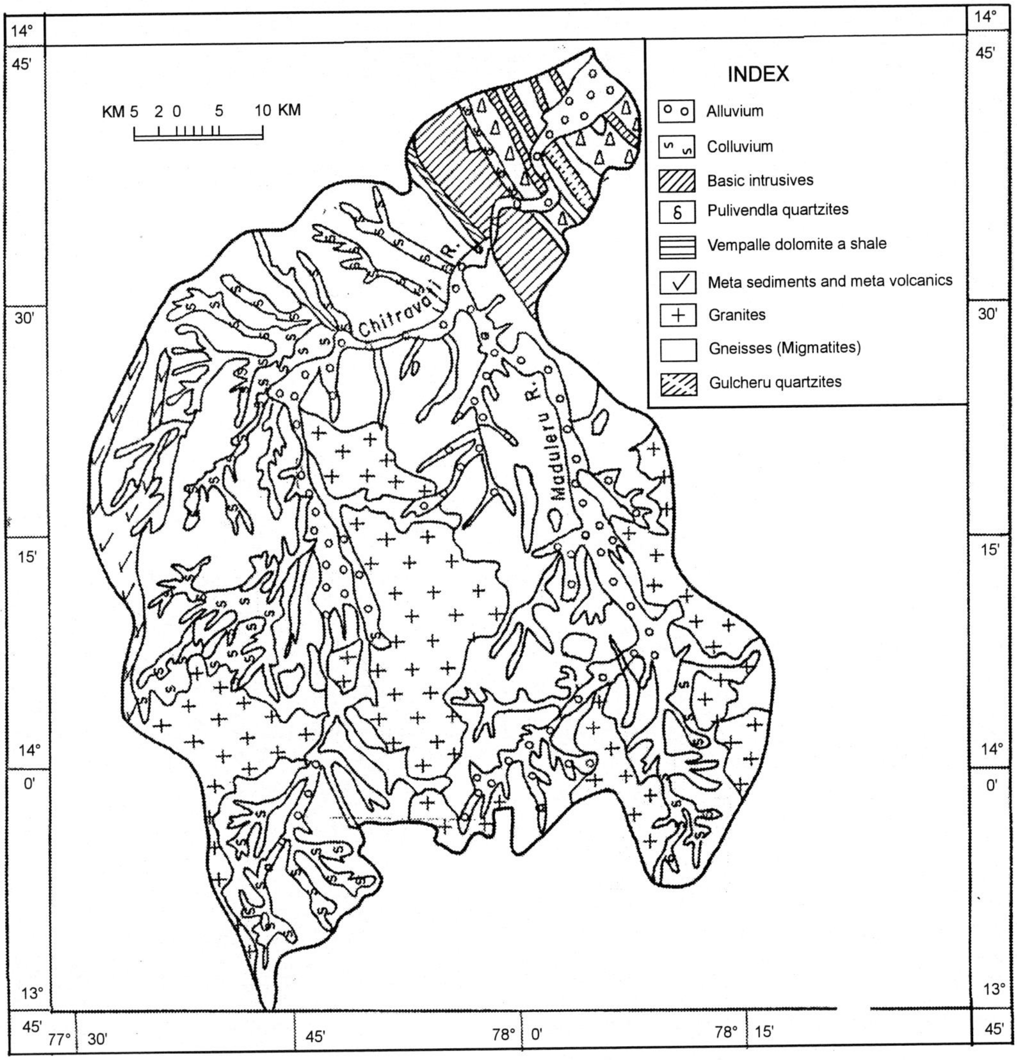

Fig. 2 Geology map.

recharge measurements. The location of tritium-injected sites in the area is shown in Figure 1. Tritiated water (2.5 ml) having an activity of 10 µci/ml was injected at selected depth in a hole made by using drive rod. All the holes were filled back with local soil. Thus, 125 µci of tritium was injected at each site.

2.2 Soil Core Collection

Vertical soil core profiles were collected from the injected sites with a 10 cm sampling intervals in the month of December. The maximum depth of collection was 2.9 m. The samples were weighed in the field and sealed in air-tight polythene bags.

2.3 Laboratory Measurements

In the laboratory, the soil samples were reweighed and the moisture content of the samples was determined by using torsion moisture balance, heated by an infrared lamp. The soil samples were subjected to partial vacuum distillation for extracting soil moisture. 4 ml of the distillate was mixed with 10 ml of Instagel (scintillator cocktail) in low potash glass vials (capacity 20 ml) and the tritium activity of the sample was counted on manually operated scintillation counter (ECIL make).

The recharge in the case of each site was calculated by determining the peak/centre of gravity of the tritium versus depth profiles. The mean displacement of the tracer from the depth of injection depth and the peak/center of gravity. The net recharge was taken as the height of an imaginary water column over a square centimeter in the soil profile interval between the depth of injection and the depth of peak/centre of gravity. This figure was computed from the moisture concentration, tracer displacement and wet bulk density, as described by Athavale *et al.*, (1980). Figs. 3-4 show some typical tritium profiles of Chitravati river basin.

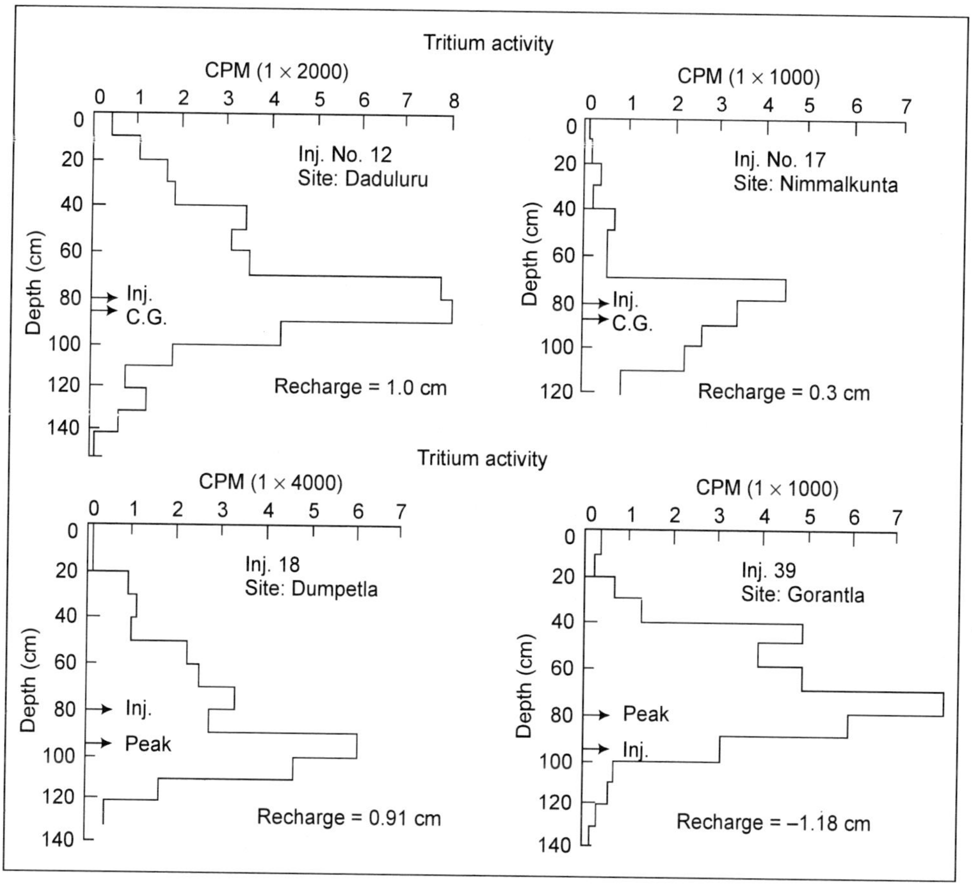

Fig. 3 Typical tritium profiles at Daduluru, Nimmalkunta, Dumpetla and Gorantla villages.

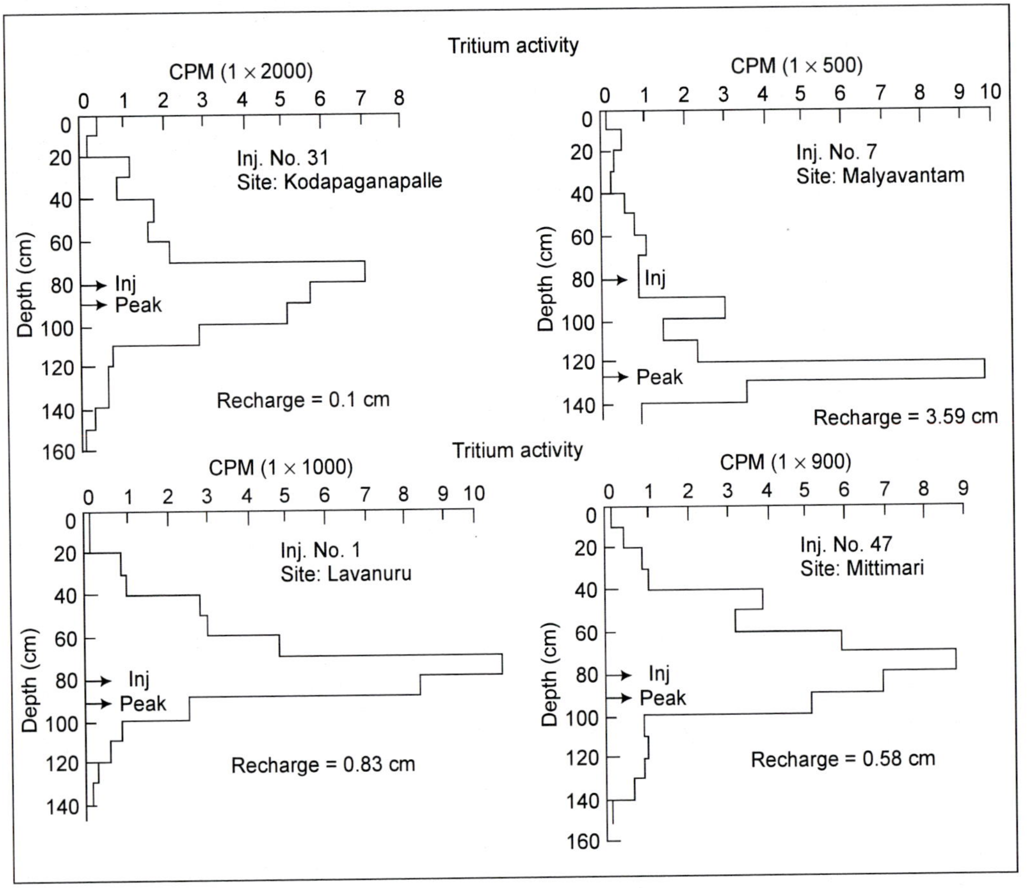

Fig. 4 Typical tritium profiles at Kodapaganapalle, Malyavantam, Lavanuru and Mittimari villages.

3 RESULTS AND DISCUSSIONS

The recharge values for 54 sites in Chitravati basin range from –29 mm to 210 mm. The arithmetic mean of all the recharge values is 27.47 mm with a standard deviation of ± 42.93 mm.

The negative recharge value indicates that the side has come under discharge zones in the basin. Maduleru and Kushavati are sub-basins of the main Chitravati basin (Figure 1). The average recharge values obtained from 9 and 4 sites located in these sub-basins are 23 mm and 44 mm respectively. Table 1 presents the recharge values estimated for each site in Chitravati river basin.

Raju *et al.*, (1979) have, on the basis of field studies, indicated that greater than 5% slope cannot hold surface water for sufficient time to allow the percolation and so the area above 5% slope can be neglected for calculating the effective recharge area in a basin. Using this criterion, the effective infiltration areas of Chitravati basin is calculated as 5378 km^2.

The total input to the groundwater reserves of this basin is estimated by multiplying the effective infiltration area with the respective average recharge value. In total 140 (MCM) in

Table 1 Recharge due to monsoon precipitation at Chitravati basin (Total 46 sites, Mean = 34.1 mm).

Inj. no.	*Site name*	*Depth of soil samples collected (cm)*	*Recharge (mm)*	*Soil type*
1.	Lavanuru	240	8.30	Loamy
2.	Kuttapalle	250	38.20	Black clay
3.	Dontellapalle	260	9.90	Clay loam
4.	Dontellapalle	180	8.00	Clay loam
5.	Kunukuntla	190	1.60	Loam
6.	Tadimarri	170	52.60	Sandy
7.	Malyavantam	210	35.90	Sandy
8.	Yangadagarpally	200	65.80	Sandy
9.	Sapatla	190	6.50	Sandy loam
10.	Maruru	180	−3.00	Loam
11.	Kunuturu	130	−4.00	Sandy loam
12.	Daduluru	200	10.00	Clay loam
13.	Daduluru	200	9.00	Clay loam
14.	Polepalle	180	0.30	Black clay
15.	Kuntimaddi	180	53.00	Sandy loam
16.	Nagasamudram	200	8.40	Black clay
17.	Nimmalkunta	120	3.00	Sandy loam
18.	Dumpetla	170	9.10	Sandy loam
19.	Dorigallu	200	7.30	Sandy loam
20.	Krishnapuram	260	23.60	Loam
21.	Nagamagubbalu	230	−29.00	Clay loam
22.	Mudigubba	170	4.60	Loam
23.	Siddarampuram	290	16.00	Loam
24.	Siddarampuram	250	14.00	Loam
25.	Arunagiri	210	28.00	Murrum
26.	Narasimhapalle	220	149.10	Black clay
27.	Patnam	160	−8.40	Loam
28.	Nallamada	200	22.00	Loam
29.	Gutturu	260	41.00	Loam
30.	Markamtapalle	170	81.20	Sandy loam
31.	Kodapaganipalle	170	1.00	Sandy loam
32.	Kottacheruvu	200	19.00	Sandy loam
33.	Kottacheruvu	200	18.00	Sandy loam
34.	Hovaldarpally	200	12.30	Sandy loam
35.	Ramapuram	150	7.60	Sandy loam
36.	Penukonda	260	21.90	Sandy loam
37.	Tariavapalle	210	60.50	Sandy
38.	Gorantla	240	−12.10	Clay loam
39.	Gorantla	270	−11.80	Clay loam
40.	Pulagurapally	190	148.00	Sandy loam
41.	Pulagurapally	220	210.00	Sandy loam
42.	Bhumagarpally	210	−8.70	Loam

Table: contd.

43.	Morasapally	270	−6.70	Loam
44.	Bagepally	250	39.00	Loam
45.	Huldolpally	190	79.00	Sandy loam
46.	Mittemarri	240	4.80	Red sandy
47.	Mittemarri	190	5.80	Red sandy
48.	Sadali	250	17.40	Black clay
49.	Aruru	210	1.50	Loam
50.	Yaligere	220	16.40	Clayey
51.	Kodulur	210	50.70	Clay loam
52.	Chillmatturu	180	15.90	Loam
53.	Bangampally	200	68.00	Sandy loam
54.	Bangampally	170	64.00	Sandy loam

Chitravati basin has been recharged to groundwater of the near surface aquifers due to monsoon. This is equivalent to average net recharge of 0.026 MCM over one km^2 area in Chitravati basin.

Table 2 shows the comparison of recharge values of 8 duplicate sites in Chitravati basin. Assuming that the change in soil condition at a distance of 3.0 m is negligible, the difference in recharge values obtained at the two adjacent points can be regarded as a measure of the cumulative error in the injection technique of measurement of recharge. It can be seen that out of 8 duplicate sites, 7 sites show a deviation of +–15% from the mean and 1 site shows a deviation of less than 25% from the mean.

Table 2 Study of reproducibility of recharge measurements in the area.

S. No.	*Inj. No.*	*Site name*	*Recharge (mm)*		*Mean recharge (mm)*	*Deviation from Mean (%)*
			Site1	*Site 2*		
1.	3,4	Dontelapalle	9.9	8.0	8.95	14.97
2.	12,13	Daduluru	10.0	9.0	9.5	7.40
3.	23,24	Siddarampuram	16.0	14.0	15.0	9.4
4.	32,33	Kotacheruvu	19.0	18.0	18.5	13.07
5.	38,39	Gorantla	−12.1	−11.8	−11.95	1.77
6.	40,41	Pulagurupalle	148	210	179	24.49
7.	46,47	Mittemarri	4.8	5.8	5.3	13.34
8.	53,54	Banganpally	68.0	64.0	66.0	4.28

4 CONCLUSIONS

The recharge estimates for 54 sites in Chitravati river basin vary from –29 mm to 210 mm with a mean value of 27.4 mm. This is 4.45% of the monsoon rainfall of 615 mm during monsoon period. These measured recharge values show a wide scatter ranging which is due to variation in pattern and magnitude of rainfall, change in soil type and its heterogeneity. The total annual input to groundwater reserves of Chitravati river basin due to monsoons over an effective infiltration area of 5378 is 140 MCM. The reproducibility of recharge measurements obtained by tritium injection method has been found to be +15% from the mean. 8 sites out of 54 selected gave negative values of recharge indicating discharge zones in this basin. The sites found, through tritium injection studies, to be giving high recharge values, can be considered favorable for implementation of schemes of artificial recharges.

References

Athavale, R.N. Murti, C.S. & Chand, R. 1978. Estimation of recharge to the phreatic aquifers of lower Maner basin by using the tritium injection method. Tech. Report No. GH-9-NHI, Hyderabad.

Athavale, R.N. Murti, C.S. & Chand, R. 1980. Estimation of recharge to phreatic Aquifers of Lower Maner basin by using the tritium injection method. *Journal of Hydrology*. 45, 185–202.

Athavale, R.N. Ramesh Chand & Rangarajan, R.1983. Groundwater recharge estimates for two basins in the Deccan Trap basalt formation. *Hydrological Sciences-Journal*. 28, 4, 12/1983.

Chand, R. Chandra, S. Rao, V.A. Singh, V.S. & Jain, S.C. 2004. Estimation of natural recharge and its dependency on sub-surface geoelectric parameters. *Journal of Hydrology*. 299 (2004) pp. 67–83.

Munnich, K.O. 1968. Moisture movement measurement by Isotope tagging. In: Guidebook on Nuclear techniques in hydrology, 112-117, IAEA, Vienna.

Raju, K.C.C. Karemuddin, M. & Prabhakar Rao, P. 1979. Operation Anatapur, p. 39. Miscellaneous publication No. 47, GSI, Govt. of India.

Ramesh, C. Hodlur, G.K. Ravi Prakash, M. Mondal, N.C. & Singh, V.S. 2005. Reliable natural recharge estimates in granite terrain. Current Science. V. 88, No.5, pp.821–824.

Zimmermann, U. Ehhalt, D. & Mummich, D.O. 1967a. Soil water movement and evapotranspiration. change in the isotope composition of water. In: *Isotopes in hydrology*. 567–586, IAEA, Vienna.

Zimmermann, U. Munnich, K.O. & Roether, W. 1967b. Downward movement of soil moisture traced by means of hydrogen isotopes. AGU Geophysical Monograph 11, 28–36.

Groundwater Quality and Drastic Vulnerability Assessment of the Aquifers in Aizawl City, Mizoram

14

L. Sailo[1], C. Mahanta[2]

[1]Indian Institute of Technology Guwahati, Assam, India. sailo@iitg.ernet.in

[2]Indian Institute of Technology Guwahati, Assam, India. chandan@iitg.ernet.in

ABSTRACT

Due to scarcity of potable water, groundwater quality is becoming an acute concern in the Aizawl city, the capital of hill city Mizoram in the Northeast India. In this study, the basic groundwater chemistry of the region was studied using statistical analysis, spatial distribution of major ions and vulnerability of the groundwater to pollution was assessed using DRASTIC. For statistical analysis, Principal Component Analysis (PCA) was used, which is one of the favorite tools in environmetrics for data compression and information extraction. Thus, the present work aimed at estimating aquifer vulnerability by applying the DRASTIC model and utilizing sensitivity analysis to evaluate the relative importance of the model parameters for aquifer vulnerability in the Aizawl city. An important objective was to demonstrate the combined use of the DRASTIC and geographical information system (GIS) as an effective method for groundwater pollution risk assessment. The DRASTIC shows highest vulnerability in the Aizawl west region, where the depth to water table parameters inflicts largest impact on the vulnerability.

1 INTRODUCTION

1.1 Study Area

Aizawl city, the capital of the State of Mizoram, is located at the intersection of 23° 44″ N latitude and 92° 44″ E longitude and is contained in the Survey of India toposheet No. 84 A/10, 84 A/14 & 84 A/9 as shown in Figure 1. The population of *Aizawl* city is 2, 29,714 (2001 census). It is located at the top of a ridge where the main sources of water are generally streams, spring and

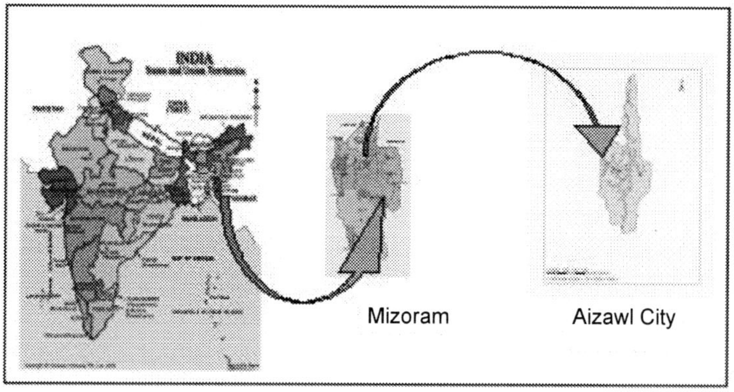

Fig. 1 location of Aizawl City

rainfall, thus, faces acute shortage of water during dry season (Central Ground Water Board, 2004). Due to shortage of piped or tap water, about 20% of the population depends entirely on water from tubewells, springs, streams and rainwater for drinking and cooking purposes, which causes incidence of water-borne gastrointestinal diseases like cholera (Town & Country Planning 2004). As many of the springs are reportedly drying up lately, next resort for the people is pumping of groundwater. As water demand is rising, besides an increasing depletion of reserve, the quality aspect of groundwater is assuming greater importance with time.

The climate of Aizawl city is characterized by its coolness, relatively high humidity nearly all throughout the year and abundant rainfall. It receives ample rains during the summer with an annual average 209 cm. This heavy rainfall causes groundwater recharge and rejuvenates the streams and springs. It is observed that the areas under the influence of faulting/jointing get saturated with water easily. As the city area is devoid of vegetation, runoff is predominant feature and much of the precipitation water flows as runoff. Groundwater reserve of the area has been estimated to be about 1400 MCM, which is based on precipitation and only negligible part is being used through handpumps and tube-wells (Central Ground Water Board India, 2004). The average per capita supply of water in the areas served is 30 liter per capita per day (lpcd) against 135 lpcd Indian Standard (Town & Country Planning 2004).

Groundwater quality and pollution problems have become a major global concern. So, many studies and researches have been carried out on the chemical evolution of the groundwater and geochemical modeling of the groundwater (Cronin, A.A., *et al.*, (2005), Guler, C. and Thyne, D.G., (2003), Allen and Matsuo (2002), Ali, R., *et al.*, (2002), Robins, N.S., (2001), Edmunds, W.M., *et al.*, (2001), Zilberbrand, M., *et al.*, (2001), Sikdar, P.K., *et al.*, (2000) Taylor, R.G. and Howard, K.W.F (1994), Davies, J., (1994)). Water quality related studies were done with the PCA (Goonetilleke, *et al.*, (2005), Kunwar P S., Malik, *et al.*, (2004), Sarbu, C., Pop, H.F., (2004), Vazquez *et al.*, (2003) Farnham, I.M., *et al.*, (2002)). And the DRASTIC pollution potential of the groundwater (Babiker, I.S., *et al.*, (2005), Al-Adamat., *et al.*, (2003), Renduo Zhang *et al.*, (1996)).

1.2 Hydrogeology

The detailed hydrogeology of the Aizawl city is much controlled by the lithology and the secondary structures present in the rocks. The general lithology consists of a sequence of siltstone, shale and sandstone, wherein the porosity of the rocks is very limited. This has significantly reduced the water yielding capacity of the aquifers (Lalfakzuala & Chamliana, J., 2003). The hill slopes contain a range of springs and seepages, while a good number of them are perennial and tapped for local water supply. The spring's discharges in most cases reduce severely starting from December-January (Central Ground Water Board India, 2004).

In Aizawl North the formation encountered was well bedded fine grained sandstone with siltstone having a dip of 35°-75° easterly. Which resulted in a number of successful tube wells along the lower levels having good discharges and a shallow static water levels ranging from 0-12 meters below the ground level.

Aizawl East is bounded by two parallel faults and longitudinal faults. The highly surface built up areas largely disrupt the natural recharge of groundwater, and hence, further development of groundwater could be undertaken only to limited scale.

In Aizawl West the longitudinal strike were much disturbed. Frequent changes of altitudes have been noticed in the area. However, high angle of dips of formation, with numerous joints system and fracture planes supplemented by longitudinal lineaments enhances the potential of groundwater development.

Aizawl South is another structural block bounded by faults. In the lower levels, development of groundwater potential increases, as there is a decrease in dips, in some places reversal of dips against the slope. Numerous spring discharges have been observed at different altitudes.

2 METHODOLOGY

2.1 Collection of Data

Fifty samples were collected during August 2004, followed by a second set during April 2005. The sample collection was done covering as many as localities possible, to get a comprehensive idea about the overall water quality of the study area. The samples were classified into four different groups namely, Aizawl North (N), Aizawl East (E), Aizawl West (W), Aizawl South (S). These classifications were made keeping in view the geological formations, geographical locations.

The water quality reports of the groundwater were collected from the Geology and Mining Wing, Directorate of Industries and "Water Analysis Report" Water Testing Laboratory, Public Health Engineering Department, Government of Mizoram. The rainfall data from the year of records (1986) were collected from the Agriculture Department, Government of Mizoram.

2.2 Data Analysis

The testing of the four parameters viz. Temperature, pH, Electrical Conductivity and Alkalinity were performed in the laboratory of Public Health Engineering Dept., Tuikhuahtlang, Aizawl, while in the field. All the samples were then preserved and tested for the rest of the parameters viz. sodium, potassium, calcium, iron, nitrate, fluoride, phosphate, and sulphate in laboratory of IIT Guwahati. All the analyses were carried out according to the standard methods of American Public Health Association (APHA, 1998).

2.3 Principal Component Analysis (PCA)

PCA, a multivariate statistical data analysis, is also known as eigenvector analysis, eigenvector decomposition or Ksarhunen-Loeve expansion. It is a technique which reduces a set of raw data into a number of principal components that retain the most variance within the original data in order to identify possible patterns or clusters between objects and variables (Goonetilleke, *et al.*, 2005, Brereton 2003). Since patterns in data can be hard to find in data of high dimension, where the luxury of graphical representation is not available, PCA is a powerful tool for analyzing data. With PCA, a set of correlated measured variables is transformed into a set of uncorrelated principal components. PCA takes the form (Brereton 2003):

$$X = T.P + E \qquad ...(2.1)$$

where, X = original data matrix, T = Score matrix, P = Loading matrix, E = Error matrix.

The principal components are described using two lower dimensional data matrices (scores and loadings) that capture the underlying patterns within the original data (Farnham I.M *et al.*, 2002). The principal components are obtained through eigenanalysis of the correlation or covariance matrix. Each eigenvalue/eigenvector pair describes a principal component (Brereton, 2003, Farnham M.I *et al.*, 2002). The first principal component is the combination of variables that explains the greatest amount of variation. The second principal component defines the next largest amount of variation and is independent to the first principal component. There can be as many possible principal components as there are variables (Brereton, 2003). In this case only first two principal components were used.

A software "Multivariate Analysis Add-in for Excel, Version 1.3" (Brereton, 2003), was used for undertaking the multivariate data analysis. This software was selected due to its versatility, ease of use and superior data handling capabilities. The water quality concentration data as mg/L were arranged into matrix using the software, for each study area. The columns in the matrix that was developed, defined the parameters and the rows, and the site location of the tube-wells.

2.4 DRASTIC Method: Evaluating Groundwater Pollution Potential

Groundwater vulnerability to contamination can be defined as the tendency or likelihood for contamination to reach a specified point in groundwater system after introduction at some location above the uppermost aquifer (Aller, *et al.*, 1987). A numerical rating scheme, called DRASTIC, has been developed for evaluating the potential for groundwater pollution in a given area based on their hydrogeological settings (Aller, *et al.*, 1987). The DRASTIC sensitivity model functions on the basis of the following linear combination equation:

$$\text{Pollution Potential} = D_rD_w + R_rR_w + A_rA_w + S_rS_w + T_rT_w + I_rI_w + C_rC_w \qquad ...(2.2)$$

The acronym of DRASTIC is derived from the seven factors in the rating scheme where, D is the Depth to groundwater, R the net recharge rate, A the aquifer media (geology), S the soil media (texture), T the topography (slope), I the impact of the vadose zone, C the hydraulic conductivity of the aquifer, and the subscripts r and w denote the rating and the weight, respectively. Ratings reflect the relative significance of classes (1-10) within each of the seven parameters, while weights (1-5) provide an indication of relative parameter influence within the equation (Table 1). The weights from the modified version called "Pesticide DRASTIC" would be more appropriate for studies related to nitrate groundwater pollution (Aller, *et al.*, 1987). All variables are dimensionless.

Table 1 Assigned Importance Weight for Factors in Pesticide DRASTIC

Factor	*importance weight*
Depth to Groundwater (*D*)	5
Net Recharge (*R*)	4
Aquifer Media (*A*)	3
Soil Media (*S*)	5
Topography (*T*)	3
Impact of Vadose Zone (*I*)	4
Hydraulic Conductivity (*C*)	2

Based upon the importance weights and typical ratings as described above, the range of possible pollution potential pesticide DRASTIC, it is 29 to 256. Conceptually, the greater the score for an area, the greater is the potential for groundwater pollution. There are no standard subdivisions of scores; however, the colors of the spectrum were chosen to show relative vulnerability to pollution. The contour has been plotted and the darker shades (red, orange, and yellow) indicate areas with potentially greatest vulnerability; the cool colors (blue, indigo, and violet) indicate areas of lower vulnerability to pollution.

The aquifer sensitivity and groundwater vulnerability mapping procedures carried out in this study incorporated the use of a Geographic Information System (GIS). A GIS is a computerized mapping and spatial data analysis system, which enables the manipulation and analysis of spatially referenced information to describe the relationship between landscape features (Zhang R, *et al.*, 1996). Though not originally designed as a GIS-based tool, the DRASTIC model lends itself to such an implementation (Babikar I.S., *et al*, 2005). GIS applications of the DRASTIC model (Babiker, I.S., *et al*, (2005), Al-Adamat., *et al*, (2003), Zhang. R., *et al*. (1996)) have been widely documented in the literature. Several types of data were used to construct thematic layers of the seven model parameters. A summary of the data types, sources and usages is available in Table 2.

Table 2 Data used for constructing the seven parameters used for DRASTIC model

Data Type	*Source*	*Used to Produce*
Borehole Data	Hydrological Mapping of Aizawl City, Geology and Mining Wing, Govt. of Mizoram	D
Geology	-do-	A & I
Soil	Geology and Mining Wing, and Master plan 2002, Town and Country Planning, Govt. of Mizoram.	S
Hydraulic Conductivity	- do -	C
Rainfall Data	Department of Agriculture & Minor Irrigation, Govt. of Mizoram.	R
Topography	Contour Map 1:2000, Dept. of Land Revenue, Master plan 2002, Town & Country Planning, Govt. of Mizoram	T & R

The GIS was used in a number of procedures, including: (i) converting hardcopy map information into a digital format, (ii) creating a contour map of distribution of major ions of groundwater, (iii) assigning sensitivity rating values to mapped attribute values._Using the Kriging method, water quality information could be interpolated to areas where data are either scarce or are not available. Key issues included:

- The unavailability of reliable data;
- The unavailability of temporal groundwater information;
- A limited amount of data therefore existed for the groundwater quality operational area and the extension of this database using kriging was proposed.

As there is no simple technique to determine which method of ordinary kriging is suited to each chemical variable, each method (i.e. Gaussian, linear, spherical, circular and exponential) was evaluated to determine the model of "best fit" (Davis 1986). A choice was made when an almost exact fit of the appropriate model appears with a cellsize of 50m using Gaussian model.

3 RESULTS AND DISCUSSIONS

3.1 Univariate Data Analysis

The mean of the datasets for the study area is shown in Table 3. The number of groundwater data points analyzed in the region is given in the parenthesis.

Table 3 Mean of the datasets

Parameters	*pH*	*EC*	*Alkalinity*	*Na*	*K*	*Ca*	*Mg*	*Cl*	SO_4	NO_3	*F*
North (11)	6.4	393	109	40	4	81	10	51	39	11	0.3
East (5)	6.8	498	110	46	6	86	8	54	98	16	0.3
West (20)	6.7	444	113	31	5	68	10	52	76	5	0.4
South (14)	6.4	443	143	34	3	78	13	48	22	6	0.2

Based on the data given in Table 3, the following can be discerned:

- From the four regions, the higher concentrations of major ions are found in the region of Aizawl East and Aizawl North for most of the parameters. This indicates a high variability of groundwater quality in an urban region thus underlying the difficulties in developing urban water quality predictive models.
- The nitrate concentration, which is highest in the region of Aizawl East, may be due to high residential development, and high usage of fertilizer for gardening.
- The sulphate concentration is highest in the Aizawl West region with the highest standard deviation (160); there may be some other point sources and geological setting leading to a higher variability.

3.2 Multivariate Data Analysis

Subsequent to the univariate study, multivariate techniques were applied to identify linkages between various chemical constituents, parameters and correlations with the land use. In the PCA undertaken, the transformed data for pattern recognition and for the identification of correlations between selected variables were interpreted as:

- The closer to the origin indicate lower intensity.
- Closely clustered objects are very similar.
- Object diametrically opposites are having an opposite nature.

Using Principal Component Analysis (PCA) the two graphs plotted are with respect to the two principal components (PC1 and PC2) are shown, the Score and a Loading plots are shown in Figure 2 to 9.

Various conclusions derived from Loading plot of PCA for *Aizawl North* (Fig. 2) are given below:

- Na^+, K^+ and Ca^{2+} are significantly related with EC. (closely related with EC)
- Ca^{2+} is significantly correlated with Alkalinity and with SO_4^{2-}.
- Cl^- is weakly correlated with Na^+ and K^+.
- It is found that NO_3^- and Mg^{2+} do not have a correlation with any chemical species.

From Score plot of *Aizawl North* (Fig. 3) it was seen that:

- A close correlation between N8, N10 and N5, N6 was observed. In these sites the parameters do not vary much according to chemical constituents of ions.
- N2 and N4 are diametrically opposite with one another. Here, N4 is having the highest chemical composition in *Aizawl North,* whereas N2 has a very low concentration of ions in that region.

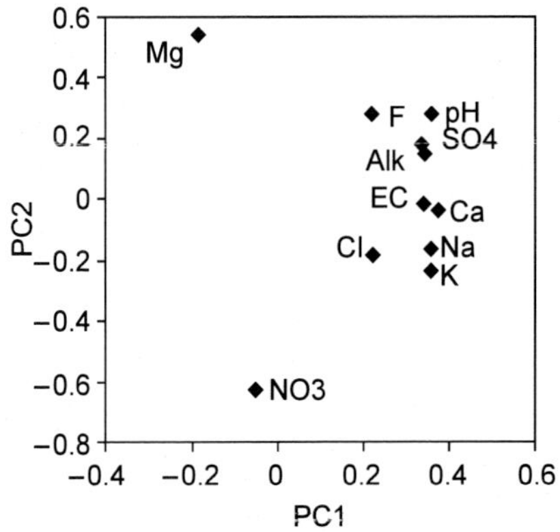

Fig. 2 Loading plot of Aizawl North.

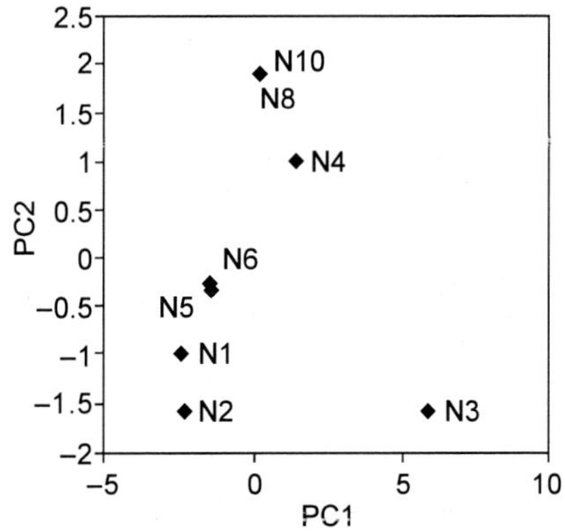

Fig. 3 Score plot of Aizawl North.

From the Aizawl East (Fig. 4), various features were observed:

- A strong correlation was observed between SO_4^{2-} with Na^+ and K^+.
- A strong inverse correlation was observed between Alkalinity and K^+ and, NO_3^- with F^- as the chemical species are found to be diagonally opposite to each other.
- EC is weakly correlated with Mg^{2+} and K^+.

From Score plot of Aizawl East (Fig. 5)

- E1 and E3 are significantly correlated.
- E1 is diametrically opposite to E5, which shows the opposite nature of both the sites.

In the Aizawl West (Fig. 6) it was found that:

- A close correlation was found between EC and Ca^{2+} and, also among Na^+, K^+ and NO_3^-.
- A significant correlation was observed between SO_4^{2-} and Ca^{2+}. It was noticed that Alkalinity and Mg^{2+} has less variation and almost independent of the other species.

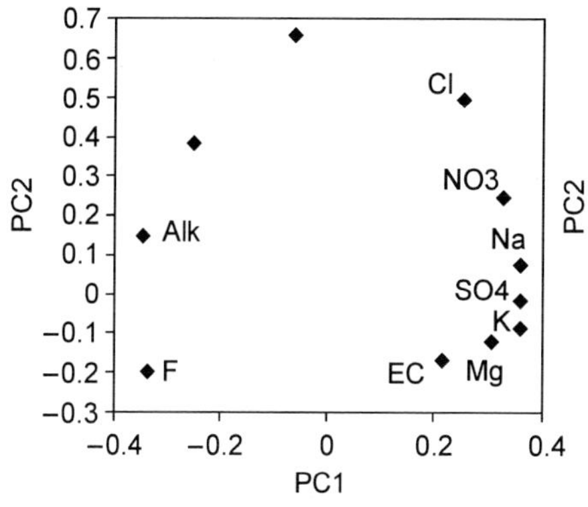

Fig. 4 Loading plot of Aizawl East.

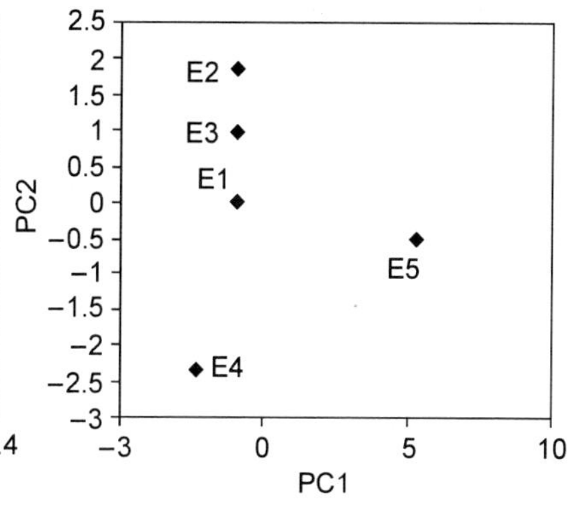

Fig. 5 Score plot of Aizawl East.

From Score plot (Fig. 7) it was seen that:

- Most of the samples were found to cluster near the center; it shows that parameters variation of the sites is of not much difference.
- Diametrically opposite natures were found between W15 and W16, W17 and W20, and among W2 with W8 and W9.

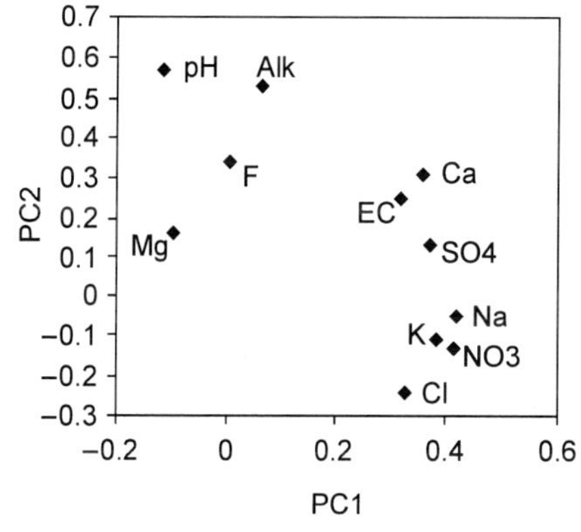

Fig. 6 Loading plot of Aizawl West.

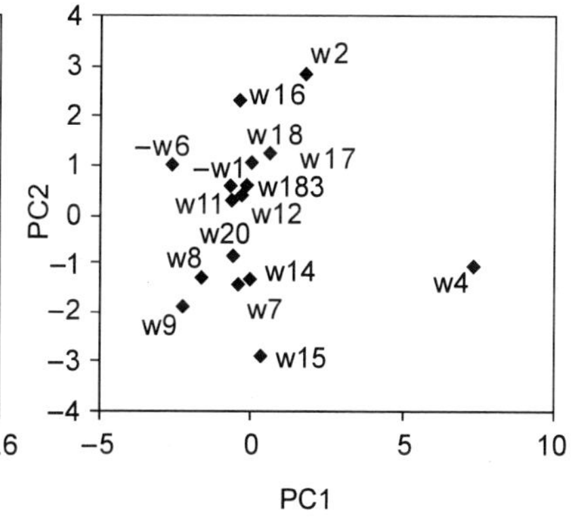

Fig. 7 Score plot of Aizawl West.

It was observed that in Aizawl South (Fig. 8) from the loading plot:

- The Na^+ and Cl^- is closely correlated with one another and with EC.
- Alkalinity and Ca^{2+} are highly correlated with each other.
- EC is in weak correlation with Ca^{2+}, Na^+ and Cl^-.
- A diagonally opposite K^+ and Cl^- are inversely correlated with each other.

In *Aizawl South* from the score plot (Fig. 9):

- A close correlation was seen between S3 and S4.
- A significant relation was noticed among S1, S2, S5 and S6.

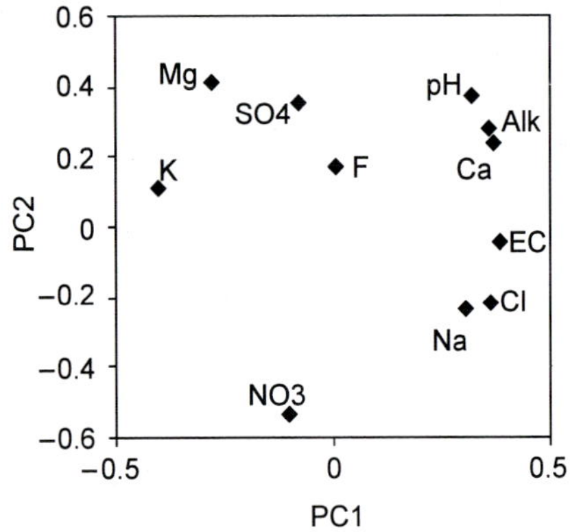

Fig. 8 Loading plot of Aizawl South.

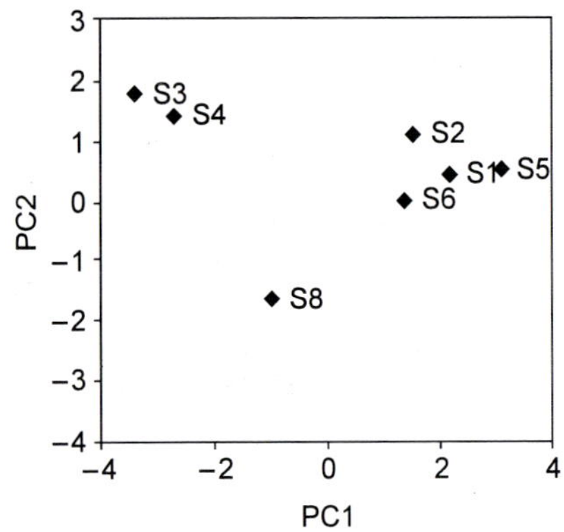

Fig. 9 Score plot of Aizawl South.

The following are the various parameters of groundwater quality in the Aizawl city in relation to the land use, aquifer type and anthropogenic sources encountered in the study area.

3.2.1 pH

From the measured pH of the samples it was found that the average value is 6.6, which is slightly acidic to neutral due to greater extent of oxidation in shallow aquifer and results in increase the rate of dissolution of silicate minerals (Robins N.S., 2002). The higher standard deviation of pH (1.2) due to possible point sources input from the organic/inorganic waste in the Aizawl North.

3.2.2 Electrical conductivity

From the PCA plots it can be seen that EC closely related with Ca^{2+}, Na^{+}, K^{+} and Cl^{-} in the entire region. This high concentration of dissolved salts in the groundwater from sedimentary formations as a result of dissolution of major cations like Ca^{2++}, Na^{+}, Mg^{2+} from the geological formations may be the source of increased EC. Higher mean value of EC was found in the region of Aizawl East where the concentrations of chemical constituents are higher.

3.2.3 Calcium and magnesium

It was found that 38% of the water samples were having a hardness value greater than 300 mg/L of $CaCO_3$, which is characterized as very hard water. Scaling of cooking pots through $CaCO_3$ deposition and a salty taste are two commonly observed byproducts of utilizing chemical hard water. Calcium and magnesium are abundant in rocks and soil, particularly limestone and dolomites anthropogenic inputs of these chemicals are small and essentially insignificant (A Kas *et al.*, 2005, Minnesota Pollution Control Agency, 1999).

From the PCA plots, it was found that Ca^{2+} is closely correlated with HCO_3^- and SO_4^{2-} which may be due to deposition of Calcite ($CaCO_3$) and gypsum ($CaSO_4$) (Guler, C. & Thyne D.G 2004) in the region. The Mg^{2+} has a very less variation and has a weak correlation with K^{+} and EC in Aizawl East.

3.2.4 Sodium

From the PCA plots, Na^+ is closely correlated with EC, K^+, Cl^- and Ca^{2+} for the entire region. Sodium is comparatively low and varies within a small range. This is ascribed to adsorption on mineral surface of high cation exchange capacities such as clay (A Kass *et al.*, 2005, Eby G.N., 2004, Evangelic V.P 1998). Significant increase suggests the silicate weathering and/or dissolution of soil salts (Na_2SO_4) in Aizawl East. The analogous distribution of chloride, sodium and EC indicates common source of these ions such as leaching of saline residues in the soil due to climatic conditions and to a limited scale may be due to anthropogenic activities.

3.2.5 Potassium

Potassium (K) is strongly adsorbable and relatively immobile in soil and subsoil (A Kas et al., 2005, Evangelic V.P, 1998). The background potassium: sodium ratio (K:Na) of soiled water and other wastes derived from plant organic matter is considerably greater than 0.4, whereas the ratio in septic tank effluent is less than 0.2. Consequently a K:Na ratio greater than 0.4 can be used to indicate contamination by plant organic matter, usually in farmyards, occasionally landfill sites (from the breakdown of paper). However, a K:Na ratio lower than 0.4 does not indicate that farmyard wastes are not the source of contamination (or that a septic tank is the cause), as K is less mobile than Na (Daly, 1996). The five sites where the K:Na concentration is greater than 0.4 are W1, W6, W7, W23, S3. Except W1 all the other sampling sites were located on the periphery or outskirts, where wastes derived from plant organic matter is suspected.

3.2.6 Chloride

It was observed that Cl^- is significantly correlated with the Na^+, which shows it may be from same source with same mobility and residence time. It may be due to the dissolution of Halite (NaCl) (Guler, C. & Thyne D.G 2004, Evangelic V.P 1998, Neil P.O., 1993). Chlorides are found to be generally high in the region highly inhabitant areas. The principle source of chloride in uncontaminated groundwater is rainfall and so in any region, depending on the distance from the sea and evapotranspiration, chloride levels in groundwater will be fairly constant (Neil P.O., 1993). Human activities are another source of inorganic substances in groundwater. Chloride, like nitrate, is a mobile anion. Because sewage is such a rich source of chloride, a high level or significant increase in ions may give rise to suspicions of pollution from sewage (EPA, 2001).

3.2.7 Phosphorous

The phosphorous was found only in those samples where iron is present in the samples. The abundance of phosphorous in natural water is controlled both by solubility of the phosphorous containing minerals and adsorption. Phosphorous phases that contain iron [e.g. vivianite, $Fe_3(PO_4)_2.H_2O$, and strengite, $FePO_4.2H_2O$] has a greatest solubility at intermediate pH values (Eby, 2004). The phosphate levels in the well water < 2 mg/L coincided with regions of thick sandy formations or red sandy soils. They ascribed the low phosphorous abundance for waters associated with these units to the very low adsorption characteristics of sandy material (Eby. G.N., 2004).

3.2.8 Sulphate

The SO_4^{2-} is significantly correlated with Ca^{2+} and Na^+. Significant increase suggests the silicate weathering and/or dissolution of soil salts (Na_2SO_4) (Guler, C. & Thyne D.G 2004) and deposition of gypsum ($CaSO_4$) in the entire region.

3.2.9 *Nitrate*

The major potential sources which have been identified in the Aizawl city are from the septic tank (pit latrine) apart from minor sources such as rainwater, domestic animal wastes and possible rare occurrence of nitrate mineral in host rock. The two samples viz. N2 and E5 found to surpass the WHO nitrate limit (45 mg/L) in the region, N2 is located adjacent to the weekly marketplace, and E5 is also located near the main business and market area. Similar link to probable contaminant sources have been identified in Nigeria where the presence of cattle market were implicated in elevating the nitrate concentration in adjacent wells (Edmund W.M, 1996).

The natural soil profile has long been recognized as an effective system for disposal and purification of human sewage by the removal of feacal micro-organism and the breakdown of many chemical compounds due to bacteriological action and filtration (Connelly & Taussig, 1994). In Aizawl city on-site sanitation is usually practiced as there is no sewage treatment plant so sewages are directly disposed of in to the side-drains. When effluent enters the unsaturated zone, pore clogging eventually develops at the infiltration surface and forms a crust. This clogging allows effective filtration and purification mechanism to proceed. The thicker the unsaturated the more effective contaminant removal will be. Therefore, newly established septic tanks system and pit latrines are at a greater risk of polluting groundwater until the crust builds up sufficiently in the pits to reduced infiltration.

Where the underlying rocks is mostly of fractured and with faults as in the Aizawl North and Aizawl West regions, it provides direct access through this open fractures from the pit latrines into the groundwater without any natural treatment through the unsaturated zone. Where the latrines are subjected to extreme loading (for example in schools and other public places) contamination is greater risk compared to other latrines where loading is not severe (Connelly & Taussig, 1994).

3.2.10 *Iron*

Usually a high level of iron was reported by some local residents near the tube-wells S3, S4, N5, W15 and W13 but out of these samples, only S3, N5 and W15 were having a concentration beyond the permissible limit (0.3 mg/L). Iron is present in significant amounts in soils and rocks, principally in insoluble form (Eby. G.N, 2004, Evangelic V.P 1998). Taste is not usually noticeable at iron concentrations below 0.3 mg/L. Laundry and sanitaryware will stain at Fe concentration above 0.3 mg/L (Woods & Wright, 2003).

Although naturally present in groundwater in some areas, they can also be good indicators of contamination by organic materials. Since organic pollution can lead to serious deoxygenation of groundwater and provide reducing conditions to bring metal into solution, marked increase in levels of iron above background, can be considered indicative of such pollution (Daly, 1996). However, the prevalence of corroded borehole rising mains throughout indicates that galvanized steel pipes in boreholes acts a secondary source of trace metals. Similar link to probable contaminant sources have been identified in Uganda, where corroded rising mains throughout acts as a secondary source of trace metal (Taylor and Howard, 1994).

3.2.11 *Fluoride*

Fluoride is relatively rare element in sedimentary rocks, except where concentrated in specific areas of mineralization. The fluoride concentration in the Aizawl city was quite low i.e. mean concentration is 0.33 mg/L. This may be because of the sedimentary rocks formation of the region. As the major ion is calcium, so it is believed to have saturated and thus formed the mineral fluorite (CaF_2), which leads to lower concentration of fluoride in the groundwater.

3.3 Nitrate Vulnerability Assessment Using DRASTIC

Drastic Vulnerability Index was calculated according to the given equation. The score for the study area is with an average value of 124 with maximum value of 154 for sample N1. Conceptually, the greater the score for an area, the greater is the potential for groundwater pollution.

It is found from the vulnerability mapping (Fig. 10), that Aizawl West (western boundary), Aizawl North and Aizawl East (near the boundary) show higher vulnerability to the groundwater pollution by nitrate due the reason that the geology of the formation is Upper Bhuban i.e. arenaceous compared to argillaceous formation. And, due to the prominent faults and fractures encountered on the north zone and western scarp, chances of subsurface infiltration is more.

The above vulnerability mapping (Fig. 10) can be correlated with the nitrate concentration (Fig. 11) of the groundwater. It was found that the Aizawl West is having a highest vulnerability of nitrate pollution but the nitrate concentration was found to be lowest. This may be because of the less population density and largely vegetated land cover, where the pollution because of the anthropogenic source is the lowest. But, in Aizawl East, Aizawl North the nitrate concentration was found to be quite high, and have a high vulnerability for pollution due to nitrate. The origin of high concentration may be due to the geology of the area have high impact where anthropogenic sources like domestic waste, septic tanks and fertilizers also exist. Aizawl East in particular already has the highest concentration of nitrate, inferring the likely sources from the main market, septic tanks and domestic waste discharged from this highly populated zone.

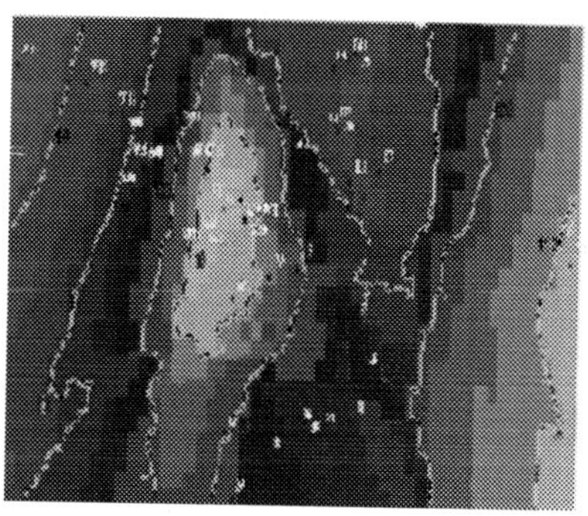

Fig. 10 DRASTIC Vulnerability mapping of the Aizawl City.

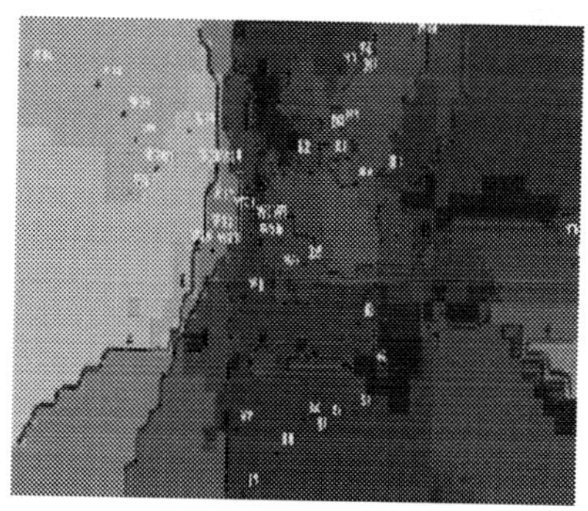

Fig. 11 Spatial distribution of Nitrate in the Aizawl City.

3.3.1 Sensitivity of DRASTIC model

The statistical summary of the seven rated parameter maps used to compute DRASTIC index is provided in Table 4. The highest risk of contamination (mean of 7.8) of groundwater in Aizawl city originates from depth of groundwater parameter. The aquifer type and soil media imply moderate risks of contamination. As the formation is siltstone and sandstone where conductivity is very low, so hydraulic conductivity imposes lowest risks of contamination to the groundwater. The recharge net, soil media and topography are moderately variable, while the hydraulic conductivity of the area is very low. The low variability of the parameters implies smaller contribution to the vulnerability index across the study area.

Table 4 Statistical Summary of DRASTIC parameters

Parameters	*D*	*R*	*A*	*S*	*T*	*I*	*C*
Maximum	10	6	7	7	5	5	1
Minimum	5	3	6	3	1	2	2
Mean	7.8	4	6.6	4	3	4	1.3
SD	1.5	1.4	0.5	1.4	1.4	1	0.7
CV (%)	19	36	8	33	47	28	36

Table 5 Summary of correlation analysis results between seven DRASTIC parameters

Factor	*Importance weight*
Aquifer Type (A) and Impact of Vadose Zone (I)	0.7
Aquifer Type (A) and Soil Media (S)	0.65
Aquifer Media (A) and Hydraulic Conductivity (C)	0.6
Soil Media (S) and Impact of Vadose Zone (I)	0.5
Net Recharge (R) and Topography (T)	0.4

The rank-correlation analysis is shown in Table 4 between the seven parameters of DRASTIC. It indicates that strong correlation exists between Aquifer Type and Impact of Vadose Zone (r = 0.7), Aquifer Type with the Soil Media (r = 0.65). This relation can be attributed to the procedure followed in order to construct layers from subsurface geology map, and the uniform and unconfined nature of the Aizawl city aquifer. Aquifer type and hydraulic conductivity (r = 0.6) may be because of the locations of the faults and dips in the region where arenaceous soils are found viz. in the region of Aizawl West and Aizawl North. Moderate correlation existed between Recharge Net and Topography (r = 0.43), Soil Media and Impact of Vadose Zone (r = 0.56). The former can be explained from the facts that, the aquifers are in hilly areas and the slope of the topography plays role in recharge. The latter is due to the different soil media in the urban land cover.

4 CONCLUSIONS

The study of groundwater quality in the Aizawl city was carried out and the various results of groundwater composition, statistical analysis and its chemical evolution and DRASTIC vulnerability mapping were used in drawing the following conclusions:

- The composition of the groundwater was found to be more mineralized in the Aizawl East core city area, where business and residential buildings and offices cover the major portion of the area.
- The groundwater composition was found to be relatively of immature chemical composition (Ca-HCO_3).

- The DRASTIC vulnerability map indicated that the Aizawl West (western boundary), Aizawl North and Aizawl East (near the boundary) show higher vulnerability to the groundwater pollution because of nitrate due to arenaceous formation and prominent fractured and faults encountered, where anthropogenic sources exist.
- The depth to water table parameter inflicts largest impact on the intrinsic vulnerability of the Aizawl aquifer. While Aquifer media (6.6) parameter has a moderate impact on the vulnerability.
- DRASTIC provides the users with a measure of relative groundwater vulnerability to pollution and, therefore, may be one of many criteria in siting or other decisions, but should not be the sole criteria.

References

Al-Adamat., R.A.N., Foster, I.D.L. and Baban, S.M.J., (2003) "Groundwater vulnerability and risk mapping for the Basaltic aquifer of the Azraq basin of Jordan using GIS, Remote sensing and DRASTIC" *Applied Geography* 23:303-324.

Ali R., Salama, R., Pollock, D. and Bates, L., (2002) "*Geochemical Interaction between Groundwater and Soil, Groundwater Recycling and Evaporation in ORIA*" Technical Report, CSIRO Land and Water, Perth.

Allen, D.M. and Matsuo, G.P. (2002) "*Results of the Groundwater Geochemistry Study on Hornby Island, British Columbia*" Final Technical Report. Earth Sciences, Simon Fraser University.

Aller L,Bennett T, Lehr J.H, Petty R.J, Hackett G,(1987 a) "*A Standardized System for Evaluating Groundwater Pollution Potential Using Hydrogeologic Settings*" USEPA , Ada Oklahoma.

Arizona Department of Environmental Quality (1997) "*Ambient Groundwater Quality of the Prescott Active Management Area: An ADEQ 1997-1998 Baseline Study*"

Ashley, R.P. and Burley,M.J, (1996) "Controls on the occurrence of fluoride in groundwater in the Rift Valley of Ethiopia". In: H. Nash and G.J.H McCall (editors), *Groundwater Quality,* Chapman & Hall publication, London.

A. Kass, I. Gavrieli, Y. Yechieli, A. Vengosh, A. Starinky (2005) "The impact of freshwater and wastewater irrigation on chemistry of groundwater: a case study from the Isreali Coasta Aquifer", *Journal of Hydrology* 300(2005): 314-331

Babiker, I.S., Mohamad, M.A.A., Hiyama, T. and Kato, K., (2005) "A GIS-based DRASTIC model for assessing aquifer vulnerability in Kakamigahara Heights, Gifu Prefecture, central Japan" *Science of the Total Environment* 345:127-140.

Brereton R.G., (2002) "*Chemometrics: Data Analysis for the Laboratory and Chemical Plant*" Wiley publishers.

British Geological Survey (2001) "*Groundwater quality: Southern India*", "*Groundwater quality: Ethiopia*" National Environmental Research Council (NERC), United Kingdom.

Canter L.W., (1997) "*Nitrates in Groundwater*", Lewis Publishers USA.

Central Ground Water Board, (2004) "*Groundwater Resources of Mizoram*" CGWB, North East Region, Ministry of Water Resource, Guwahati.

Connelly R J and Taussig D. (1996) "Nitrate contamination of groundwater in the Kutama and Sinthumule districts of Venda, South Africa". In: H. Nash and G.J.H McCall (Ed), *Groundwater Quality,* Chapman & Hall publication. London.

Cronin, A.A., Barth, J.A.C., Elliot, T., Kalin R.m., (2005) "Recharge velocity and geochemical evolution for the permo-Triassic Sherwood Sandstone, Northern Ireland" *Journal of Hydrology.* In press xx:1-17.

Davies, J. (1996) "The hydrochemistry of alluvial aquifer in central Bangladesh" In: H. Nash and G.J.H McCall (Ed), *Groundwater Quality,* Chapman & Hall publication, London.

Department of Agriculture & Minor Irrigation (2003) "*Daily & Monthly Rainfall Record*" Government of Mizoram.

Eby G.N., (2004) "*Principles of Environmental Chemistry*" Brooks/Cole-Thomson publishers.

Edmunds, W.M., (1996) "Characterization of groundwater in semi-arid and arid zones using minor elements" In: H. Nash and G.J.H McCall (Ed) *Groundwater Quality*, Chapman & Hall publication, London.

Edmunds, W.M., Rivera, C.J.J., Cardona, A., (2002) "Geochemical evolution of groundwater beneath Mexico City" *Journal of Hydrology* 258:1-24.

EPA, 2001, "*National Primary Drinking Water Regulations; Arsenic and Clarifications to Compliance and New Source Contaminants Monitoring*".

Evangelic V.P (1998) "*Environmental Soil and water Chemistry – Principle and Application*" A Wiley-Interscience Publication.

Foster, S.S.D., 1996 "Groundwater for development an overview quality constraint" In: H. Nash and G.J.H McCall (Ed) *Groundwater Quality*, Chapman & Hall publication, London.

Goonetilleke, A., Thomas, E., Ginn, S. and Gilbert, D., (2005), "Understanding the role of land use in urban stormwater uality management", *Journal of Environmental management* 74:31-42.

Guler, C. and Thyne D.G (2004) " Hydrologic and geologic factors controlling surface and groundwater chemistry in Indian Wells-Owens Valley area, southeastern California, USA" *Journal of Hydrology* 285:177-198.

Hoven, S.J.V., Solomon, D.K. and Moline, G.R., (2005) "Natural spatial and temporal variations in groundwater chemistry in fractured, sedimentary rocks: scale and implications for solute transport" *Applied Geochemistry* 20:861-873.

Kevin H., (2005). "*HYDROGEOLOGY: principles and practice*" Blackwell Publishing.

Kunwar P S., Malik, A., Mohan, D., Sinha, S. and Singh, V.K., (2004) "Chemometric data analysis of pollutants in wastewater- a case tudy" *Analytica Chimica Acta*. In press.

Lalfakzuala and Chamliana, J (2003)"*Report on Detail Hydrological Mapping of Aizawl City (core area)*" Geology and Mining Wing, Directorate of Industries, Government of Mizoram.

Lerner D.L., (2004) "*Urban Groundwater Pollution*" A.A. Balkema Publishers.

Litchfield, N., Craw, D. and DOONS, P.O., (2002) "Geology and geochemistry of groundwater within the Taieri Basin, east Otago, New Zealand" *New Zealand Journal of Geology & Geophysics* 45:481-497.

Mazlum, N., et al., (1999) "Interpretation of Water Quality Data by Principal Components Analysis" *Tr. J. of Engineering and Environmental Science* 23., 19-26.

Minnesota Pollution Control Agency, (1999) "*Barium, Beryllium, Calcium and Magnesium in Minnesota's Groundwater*", Environmental Outcome Division, Groundwater Monitoring and Assessment Program.

Neil, P.O., (1993) "Environmental Chemistry" Chapman & Hall Publication. London.

Public Health Engineering Department (2004) "*Water Analysis Report*" Water Testing Laboratory, Government of Mizoram.

Robins, N.S., (2002) "Groundwater quality in Scotland: major ion chemistry of the key groundwater bodies" *The Science of Total Environment* 294:41-56.

Sikdar, P.K., Sarkar, S.S and Palchoudhury, S., (2001) "Geochemical evolution of groundwater in the quaternary aquifer of Calcutta and Howrah, India" *Journal of Asian Earth Science* 19: 579-594.

Taylor, R G. and Howard, K.W.F (1996) "Groundwater quality in rural Uganda: hydrochemical considerations for the development of aquifer within the basement complex of Africa". In: H. Nash and G.J.H McCall (Ed) *Groundwater Quality*, Chapman & Hall publication, London.

Town and Country Planning (2004) "*Master Plan For Aizawl 2002*" Local Administrative Department. And Approved by the State Govt. Under Mizoram Urban & Regional Development Act, 1990. Amended in 1996.

Woods L. and Wright G., (2003) "An assessment of the quality of public groundwater supply sources in County Wicklow" *Groundwater quality report,* Geological Survey of Ireland.

Zhang, R.,Hmerlinck, J.D., Gloss P.S. and Munn L., (1996) "Determination of nonpoint-source pollution using GIS and numerical models" *Journal of Environmental Quality* 25(3): 411-418.

Zilberbrand M., Rosenthal, E. and Schachnai, E., (2001) "Impact of urbanization on hydrochemical evolution of groundwater and on unsaturated-zone gas composition in the coastal city of Tel Aviv, Israel" *Journal of Contaminant Hydrology* 50:175-208.

Problem of Selenium Toxicity in Nawanshahr and Hoshiarpur Districts of Punjab State, India

15

K.P. Singh[1], K.S. Dhillon[2], S.K. Dhillon[2] and N.S. Tiwana[3]
[1]Geology Department, Panjab University, Chandigarh, India.
[2]Department of Soils, Punjab Agricultural University, Ludhiana, India.
[3]Punjab State Council for Science & Technology, Chandigarh, India.

ABSTRACT

Selenium toxicity in groundwater, soils and plants has been observed in parts of Nawanshahr and Hoshiarpur districts of Punjab state in India. The problem is prevalent only in about 1000 hectares of area covering about 5-6 villages, namely: Mehadpur, Jainpur, Barwa, Rakkar, Simbli and Nizarpur. Selenium content of soils from toxic regions ranges from 0.23 to 4.55 mg/kg and is present upto 180 cm depth of soil profile. Concentration of selenium in groundwater ranges from 2.5 µg/L to 65.5 µg/L. Groundwater pumped from shallow tube wells (upto 36 m) contains 2-3 times more selenium than that from deep tube wells (more than 76 m depth). Higher concentration of selenium has affected the human and animal health. In the present paper, effects of selenium toxicity on human and animal health and the remedial measures to tackle the problem are discussed along with the possible causes of selenium toxicity.

1 INTRODUCTION

Selenium (Se), depending upon its concentration, can be beneficial or toxic to plants, animals and humans. Selenium toxicity in soils, groundwater and crops has been observed in parts of Nawanshahar and Hoshiarpur districts of Punjab state which lie in the northern parts of India (Fig. 1). The area is underlain by Indo-Gangetic alluvial plains and rocks of Shiwalik formation are exposed in the NE parts. Geomorphologically, it lies in Sutlej-Beas interfluves and forms parts of plains adjoining piedmont area. The area is primarily an undulating plain traversed by number of streams or choes (seasonal streams) originating from the Shiwalik hills. Study of remotely sensed data (IRS and Landsat TM data) reveals presence of palaeo channels of river Sutlej close to the area.

Higher concentration of selenium in soils and groundwater and crops has affected the human and animal health. The livestock fed on fodders containing large amounts of selenium suffer from selenium poisoning. Human population also has adverse effects on the health. In the present paper, the entire problem is discussed along with the remedial measures.

2 SELENIUM IN SOILS

Toxic soils containing selenium more than 0.5 mg/kg cover more than 1000 hectares and are located in villages Panam, Nazarpur, Simbly, Barwa, Jainpur, Menhdpur, Rakkara Dhahan and Bhanmajra (Fig. 1). These sites are sporadically distributed in the piedmont plains and adjoining alluvial areas.

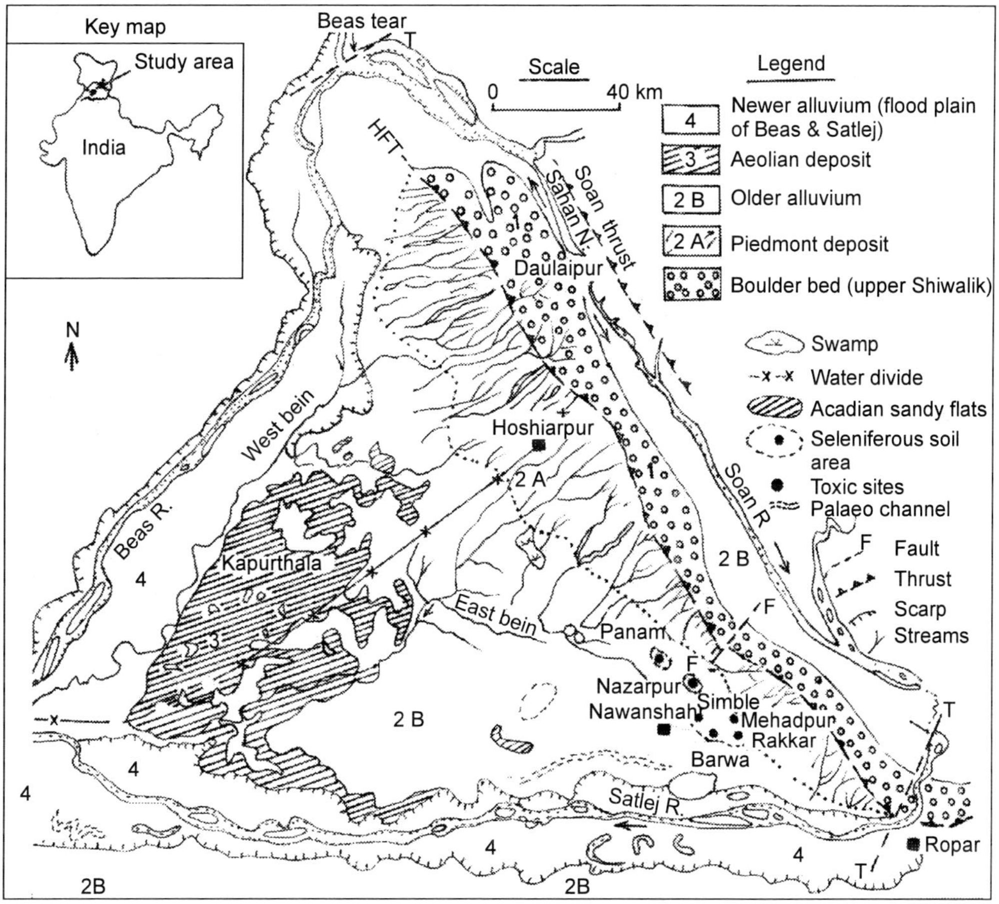

Fig. 1 Geological and geomorphological map of area (Modified after Parsad and Kar, 2004).

Total Se content of soils from toxic region ranges from 0.23 to 4.55 mg/kg. The soils are alkaline in reaction, calcareous, silty clay loam in texture and contain appreciable quantities of selenate-Se followed by selenite-Se. Application of phosphatic fertilizer may affect Se status of soils because phosphate ions can easily desorb selenium over a large pH range. Desorbed Se could be easily oxidized to selenate and in turn could easily leach out due to its higher solubility. In the entire piedmont plain ouf of 184 surface soil samples collected, 18% were found to contain Se more than 0.5 mg/kg (Dhillon *et al.*, 2004).

3 SELENIUM IN GROUNDWATER

In the seleniferous region of Punjab, groundwater constitutes the only source of irrigation and drinking water. Groundwater in top aquifers occurs under phreatic conditions whereas deeper aquifers of more than 100 m depth are under semi confined-confined conditions. Depth to groundwater is variable and ranges from 10 to 15 m bgl. Regional groundwater flow is from NE to SW direction. Groundwater is alkaline with pH ranging from 7.8 to 8.8 and electrical conductivity ranging from 0.32 to 0.78 dS m^{-1}. Bicarbonate is the dominant anion followed by chloride or carbonate and sulphate ions. Sodium is dominant cation followed by calcium, magnesium and potassium (Dhillon and Dhillon, 2003). Selenium content of groundwater ranges from 2.5 to 69.5 µg/L. The maximum permissible concentration of Se for drinking and irrigation water is 10 µg/L and 20 µg/L respectively. Groundwater from shallow tube wells (upto 36 m depth) contained 2–3 times more Se from deep tube wells (Dhillon and Dhillon, 2003). Selenium content of groundwater from shallow tubewells (upto 36m) is highly variable even within a small village as revealed by the results from Barwa (0.66–8.95 µg/L), Mehandpur (0.61–13.80 µg/L) and Jainpur (1.05–35.13 µg/L) villages.

Results of deep exploration carried out by Central Ground Water Board, Govt. of India recently in June, 2004 at Mehadpur village indicate that groundwater pumped from deeper zones (97-103 m depth) contains 3.5 µg/L Se and is free from Se toxicity (Director's personal communication).

4 SELENIUM IN CROPS

Soils containing more than 0.5 mg Se/kg are able to supply sufficient Se to produce vegetation containing more than 5 mg Se^{-1} which is toxic (Dhillon *et al.*, 2004). Typical Se toxicity symptoms appear on the wheat plants grown over Se toxic soils having snow white or papery-white chlorosis with pink coloration at the lower surface of the leaf. Wheat plants exhibiting varying degrees of toxicity symptoms contain Se ranging from 107.5 to 262.4 mg kg^{-1} in shoots and from 29.2 to 66.5 mg/kg in roots (Dhillon *et al.*, 2003). It was observed that leaves containing <100 mg Se/kg remained lush green in colour. The results of 134 fodder samples reported by Dhillon *et al.* (2004) indicate 10% of them as toxic with Se ranging from 5–10 mg/kg; 58% as safe with Se less than 2 mg/kg and 27% as potential toxic (Se between 5 and 10 mg/kg) and 5% as highly toxic with Se more than 10 mg kg^{1}.

5 EFFECTS ON HUMAN AND ANIMAL HEALTH

Selenium toxicity has been observed on humans and animals of the area consuming toxic crops. Typical symptoms in humans are (i) Loss of hair from body, particularly head

(ii) Malformation of fingers as well as toe nails (iii) Swollen joints (iv) Severe headache (v) Nausea and loss of appetite, and (vi) Progressive deterioration of general health. Selenium content of hair and nails of affected persons was higher than of healthy persons (Dhillon *et al.*, 2004). It has been observed that leaving the region temporarily for 3-4 months resulted in remarkable recovery from selenium toxicity when source of food is changed from seleniferous to non-seleniferous, thereby reducing the selenium intake. Results of house to house survey carried out recently by health department indicate that in Nawanshahar district out of total population of 1459 (of effected 6 villages), 59 persons were critically affected. Similarly, in Hoshiarpur district 472 persons were affected critically by Se toxicity, out of total population of 3958 (of 3 affected villages, namely, Simli, Panam and Nazarpur.

In animals selenium toxicity, leads to (i) cracks in the hoofs and their gradual detachment (ii) shedding of horn cornium (iii) loss of hair, (iv) necrosis of tails, etc. House to house survey carried out by animal husbandry department indicates that in Hoshiarpur district, out of total population of 894 animals in Nazarpur and Simbli, 119 were affected with selenium toxicity.

6 SOURCES OF SELENIUM

Exact nature of parent material of seleniferous soils in Punjab is not known but convincing evidence is available that Se has been transported by rainwater through seasonal rivulets from nearby hills of Shiwalik range and deposited in low lying areas (Dhillon and Dhillon, 1991). The toxic sites are located at the dead ends of the seasonal rivulets (Fig. 1). The parent material in seleniferous region is derived from upper Shiwaliks that are composed of conglomerates of variable compositions (granite, basalt, limestone) which are derived from metamorphic terrain of Himalayas (Karunakaran and Rao, 1979). Further, tectonic control by Himalayan Frontal Fault (HFF) and Nawanshahar fault (Parsad and Kar, 2004) might have helped the selenium mobility, from the source rocks to the site of deposition which was perhaps a paleo bog (swampy) area. However, detailed investigations of the Shiwalik hills and the adjoining Kandi area for soils and groundwater is required to be done to know the regional hydrogeological and hydrochemical environment in relation to selenium toxicity.

7 REMEDIAL MEASURES

The remedial measures suggested here mainly aim to reduce the entry of Se into the food chain in the affected region:

(i) Groundwater from deeper wells (>100 m depth) should be used for irrigation and drinking water as deeper aquifers are free from Se toxicity.

(ii) Cropping pattern requiring less irrigation needs to be adopted (Rice-wheat to be replaced by maize-wheat) in the area.

(iii) In case of multicut fodders, avoid feeding first cut to animals as it contains 2–3 times more Se than the following cuts.

(iv) Apply gypsum 0.8 to 1 ton/ha every alternate year before sowing fodder crops since it helps in reducing Se absorption upto 60–70%.

(v) Oat, maize and sorghum should be grown as fodder because of their lower Se absorption capacity, compared to other crops.

(vi) Toxic top soil should be replaced from the soils of another area and removed soil should be used for brick kiln purpose.

(vii) Education and awareness programs to be launched in the area.

The above measures should help in reducing the effects of Se toxicity.

8 CONCLUSIONS

Problem of selenium toxicity in soils, groundwater and plants is prevalent in Nawanshahr and Hoshiarpur districts of Punjab in India. The affected area covers about 1000 hectares and 6-8 villages. Seleniferous soils got developed at the dead end of rivulets which brought selenium rich material from Shiwalik hills. Himalayan Frontal Fault and Nawanshahar fault (Parsad and Kar, 2004) might have helped in the Se mobility from Shiwaliks to the localized deposition. Subsequently, it leached down into the top aquifer system and when crops grew under such waters and soils rich in Se, Se moved into the crops and food chain thereby affecting human and animal health in the area. In order to reduce entry of Se in food chain, it is recommended to use groundwater drawn from deeper aquifers (>100 m depth) which is free from Se hazard; use crops requiring less water i.e. maize-wheat instead of rice-wheat and to replace top toxic soils by the soils from other non-toxic areas.

References

Dhillon, K.S and Dhillon, S.K. 1991. Selenium toxicity in soils, plants and animals in some parts of Punjab, India. Int. J. Environ. Stud., 37, 15–24.

Dhillon, K.S and Dhillon, S.K. 2003. Quality, of undergroundwater and its contribution towards selenium enrichment of the soil-plant system for a seleniferous region of north west India. J. Hydrol., 272, 120–130.

Dhillon, K.S; Dhillon, S.K. and Singh, K.P. 2003. Status Paper on Selenium Toxicity in Punjab. Punjab State Council for Science & Technology, Chandigarh. 1-14. (PSCST/SP-I-WRM 2003).

Dhillon, K.S; Dhillon, S.K. and Singh, K.P. 2004. Selenium toxicity in Nawanshahr and Hoshiarpur districts, Punjab, India. IGCP-454 Workshop on Medical Geology, Nagpur, India. Feb. 2004; 93–94 (Abst.).

Dhillon, K.S; Dhillon, S.K; Minhas, N and Randhawa, S.S. 2004. Distribution of seleniferous soils in piedmont plain adjoining Siwalik range in eastern Punjab. Bull. Ind. Geol. Assoc., 37, 59–66.

Karunakaran, C and Rao, A.R 1979. Status of exploration for hydrocarbons in the Himalayan region–contribution to stratigraphy and structure. Miscellaneous publications, Geological Survey of India, No. 41, part IV, 1-66.

Parsad, S and Kar, S.K. 2004. Geological Control of selenium concentration in the soils of Punjab plains, India. Proceedings of IGCP 454 Workshop on Medical Geology organized by G.S.I. Nagpur, 306–313.

Impact of Watershed Development in Groundwater Augmentation in India

16

R.K. Sivanappan
International Consultant, 14, Bharathi Park, 4th Cross Road, Coimbatore–641 043.

ABSTRACT

The importance of groundwater to meet the growing demand of our nation for its ever increasing agricultural, industrial and domestic growth needs little emphasis. Though the overall extraction of groundwater is only about 35% in India, it is more than 80% in southern states and in Tamil Nadu the groundwater level is more than 300 m in some parts and the water mining is continuing as there is no law to control the extraction.

Watershed development programme is taken up by the states and central government in a big way in order to augment the groundwater for the last three decades on watershed basis. Soil and water conservation and water harvesting measures are implemented by various departments and NGOs under this programme with the objective to use the land and water resources to get maximum benefit with the involvement of people living in the watershed mainly to recharge the groundwater in order to maintain the water table for sustainable agricultural production. In India thousands of crores of rupees are spent every year by various agencies for watershed development works to achieve the objectives.

This paper gives the details of the impact studies carried out by about ten researchers/ scientists on watershed development works mainly in recharging the groundwater in different parts of the country and how this has augmented the groundwater.

1 INTRODUCTION

Watershed is considered a unit of planning and development for all the activities. The viable area for implementing any development may be about 500 ha (micro watershed). The government of India has decided as a policy to take up any development work, since 1975, on watershed basis. Unfortunately in practice, even now this concept is not yet clearly understood by the field staff and the development work is carried out in a haphazard manner in various departments. In fact, all the activities (by various departments) should be planned and executed in the selected watershed to have any impact. Even the selection of watershed should also be done after prioritising so that which one should be taken first can be identified. This is applicable whether it is purely an agricultural watershed or purely a forest watershed or mixed lands.

2 WATERSHED DEVELOPMENT AND MANAGEMENT

The watershed management is an integration of technologies within the boundaries of drainage areas for optimum development of land, water, plant resources (forest) to meet the basic and minimum needs of the people living in that watershed on sustained basis. The associated watershed factors are geology, soils, size, shape of watershed, topography, drainage, land cover, rainfall (surface and groundwater), socio-economic conditions of the people, etc. The problems and need of the treatment of watersheds are bound to vary from watershed to watershed, as the parameters mentioned above are not the same in all the watersheds. Therefore, the planning and activities of watershed development and management varies depending on the location specific aspects. Similarly, in the degraded forest areas, a viable micro watershed can be identified/selected and with the help/cooperation or involvement of people living in and around the area, forest development/regeneration and other watershed development works are taken up. The project has conceptualized a unique blend of forest development, watershed development and rural development. This three-pronged strategy gives equal focus on regeneration of forests, hydrological improvement particularly the groundwater and poverty alleviation and community development to uplift the socio-economic condition of the interface villages.

The government of India through various departments and institutions, namely, department of land resources, agriculture, environment and forest, rural development, NABARD/CAPART, etc., have launched the following watershed development programmes under different names/programmes in the last 2-3 decades.

1. Integrated wasteland development programmes (IWDP).
2. Comprehensive watershed development programme (CWDP).
3. Drought prone area programmes (DPAP).
4. Desert development programme (DDP).
5. Western Ghats development programme (WGDP).
6. Hill area development programme (HADP).
7. National watershed development programme in rainfed area (NAWDPRA).

Apart from this, the external funding agencies are also financing the development programme on watershed basis in agricultural and forest watersheds. The agencies who are helping the programmes are,

1. Japan Bank for international cooperation for forest development in Tamil Nadu and Karnataka (JBIC).
2. Department for international development, UK government (DFID).

3. Ford foundation.
4. USAID.
5. SIDA, DANIDA, etc.

3 OBJECTIVES OF WATERSHED DEVELOPMENT

Whoever may fund the watershed development programmes, the main objectives are to use the land and water resources to get the maximum benefit with involvement of the people living in the watershed. The various main activities involved in the watershed development programme are:

1. Land development.
2. Water resources development.
3. Agricultural development/forest development.
4. Increase the livelihood of the rural people including landless and labourers.

In India, thousands of crores of rupees are spent every year by various agencies, but not implemented in an integrated manner by all the departments on a watershed basis in a saturated way. If all the money allotted in the various programmes are pooled and the development works are undertaken under one nodal agency involving all the line departments on watershed basis, the nation can become a developed nation before 2020.

Unfortunately, still many people in the top administration who have a say in approving the watershed development programmes are not convinced about the impact and benefit of these programmes. In this paper, the impact of watershed development works carried out in many states were gathered and their fundings are detailed. In the watershed development/waste land development/forest development programmes, major amount is allotted to land development and water resources development, especially for recharging groundwater.

4 THE RESULTS OF THE IMPACT STUDIES

The impact studies carried out by the researches/scientists in various parts of the country are analysed and the findings are detailed below, especially how it has improved the groundwater potential and increased the water table for a sustainable use.

The impact studies of watershed development programmes have indicated the following:

1. The results of the study conducted by WTC/TNAU, Coimbatore have indicated that by construction of water conservation measures such as percolation ponds, check dams, farm ponds and renovation of tanks have enhanced the storage capacity in the watershed to store the excess run-off, which in turn has increased the groundwater recharge. In Coimbatore district, the data revealed that the water levels in the wells have increased from 0.5 m to 4.4 m. The farmers in the watershed reported that the water conservation measures are very useful in conserving rainwater and recharge the groundwater which reduced the water scarcity during most part of the year. Due to this, the cropping intensity has increased from 100% to 134% and in many cases the productivity of major crops has increased from 31.7% for sorghum to 127.3% for maize.

   On the basis of the data collected from the sample farmers it is found that the water level in the open dug wells has risen in the range of 2.5 to 3.5 m and 2 to 3 m in Kattampatti in Kodamgipalayam in Coimbatore district of Tamil Nadu.

2. Due to construction of percolation pond and check dam under Western Ghats Development Programmes (WGDP) by the Department of Agricultural Engineering, Government of Tamil Nadu, the groundwater level has raised to 65′ from 160′ within 2 to 3 years in Theni District of Tamil Nadu as reported by the evaluation team of the project (Sivanappan, 2000).
3. The study conducted in various watersheds in the country by Centre for Soil and Water Conservation Research and Training Institute, Ooty, has demonstrated the impact on recharge in watershed management projects such as water table rise, increased perenniality of wells, increased recuperation, increased irrigated area and crop diversification in their different case studies. Based on the water table data from sample survey of observation wells and piezometers and interview of farmers, it was found that the water table level has risen in the range of 0.8 m to 2.0 m in various DPAP watersheds and IWDP watersheds in Coimbatore district. It was also observed that recharge of wells away from the percolation pond and the influence of these percolation ponds could be generally observed upto a distance of about 500 m. (Sikka, 2001).
4. Studies conducted in Kallambella watershed development project in Karnataka state has revealed that there is a direct benefit of improvement in the groundwater level which reduced much of the burden of the government regarding the drinking water supply.

   Further the study of irrigation bore wells in 8 villages of watershed area and 7 villages outside watershed area has revealed that within the watershed area there is no bore well failed, in the outside watershed area about 7 bore wells failed. Finally, it was concluded that the treated watershed area has impact on groundwater recharge measure. The water harvesting structure across the Nalla (stream) harvests the water and increases the infiltration rate resulting in the recharge of groundwater (Narayanaswamy, 2000).
5. The soil and water conservation works executed in Kandi watershed area in Punjab has been instrumental in reversing the falling water table to raising water table in the district of Hoshiarpur and the study has recommended the need for taking of watershed management works in the upper reaches of declining water table areas on priority basis (Khepar, 2001).
6. The studies made by the National Institute of Hydrology, Roorkee have revealed that the construction of percolation pond, a significant recharge influence is observed in the downstream side of these ponds up to a distance of about 350 m in Uralipatti in Dindigul district, Tamil Nadu (Tyagi, 2001). From the trace technique used to find out the effect of recharge from the percolation pond in Pappankulam village in Coimbatore district, Tamil Nadu, it is found that the zone of influence extend to about 750 m in the downstream side. It was also reported that there is about 3 m raise in water level and according to the farmers the duration of pumping water has been increased considerably (Kanagasabai *et al.*, 2002).
7. Present emphasis on groundwater depletion needs the strategies for artificial recharge of aquifers. Soil and water conservation measures adopted on watershed basis have definite impact on replenishment of groundwater. This is evident from the results obtained from several watershed management projects studied by CSWCRTI, Dehradun and its regional research centre on development of groundwater resources (Sastry, 2001).

8. The study conducted by Agricultural Engineering Department, Government of Tamil Nadu in the priority watershed of Lower Bhavani catchment in Mettupalayam Taluk of Coimbatore district showed that there is increase in the groundwater recharge due to the construction of percolation pond. There has been an increase of groundwater from 16% to as high as 188% with a general trend of increase around 50%. The percentage of increase in the number of wells ranged from 8 to 50% after the construction of percolation pond. The increase in the irrigated area ranged from 21 to 170%, which is clear indication of the increased groundwater availability. In a study conducted to observe the effect of contour bunding in groundwater recharge in Theni district of Tamil Nadu, an increase of 14.5% was observed in the water yield of wells after bunding (Agricultural Engineering Department, 1996- 97).
9. The study conducted by the minor irrigation department of Andhra Pradesh, on the impact of percolation tanks and check dams, the structures are very useful in drought affected areas. The farmers are able to reap the benefits from the immediate subsequent years. The percolation ponds are quite effective in augmenting the groundwater. A number of wells have been benefited by recharge under the percolation pond. The water level has been considerably raised due to construction of percolation pond. The study conducted by the College of Agriculture, Indore, Madya Pradesh has revealed that various soil and water conservation measures in the watershed enhance groundwater recharge and ultimately increased crop intensity and crop productivity (Ranada *et al.*, 1999).

5 CONCLUSION

From the above observations, it is clear that the impact of watershed development works, namely, construction of percolation pond, check dam, soil conservation measures like contour bunding, compartmental bunding, vegetative/stone barriers helped in augmenting groundwater in the watershed and the farmers in the watershed raised crops in more areas under irrigation than before the implementation of the watershed projects. This is necessary, since our's is a monsoonal country where the rainfall is erratic, uneven, and failure occurs once in 3 or 4 years. The watershed development work will help in storing and recharging the groundwater when the rainfall is more and the stored/conserved water can be used in the subsequent drought years. Therefore, much attention should be paid to take up integrated/ comprehensive watershed development programmes in the country to mitigate water scarcity especially in drought years.

References

Agriculture Engineering, Department, Goverment of Tamil Nadu. 1996-97, Project appraisal cum evaluation report, Kundah and lower Bhavani River Valley Project, Chennai.

Kanagasabai S. *et al.*, 2002, Studies on the influence of percolation pond on groundwater recharge- Tracer technique.

Khepar S. D 2001, Impact on soil & water conservation works on groundwater regime in Kandi area of Punjab, Journal of soil and water conservation, Vol. 45.

Narayanaswamy KM, 2000, Kallam bella watershed development project, watershed development programme, Tumkur, Karnataka.

Ranada D. H. *et al.*, 1999, Increasing groundwater availability through soil conservation measures, Indian Journal of soil conservation 27 (3), 258–260.

Sastry P. R. et al., 2001, Augumenting ground water resources through integrated watershed management, Journal of soil and water conservation, Vol. 45, 67–71.

Sikka A. K. (2001), Augumentation and management of ground water resources: Participatary watershed management as a policy and strategy.

Sivanappan R. K. *et al.*, 2000, Perspective plan for Western Ghats development, Planning & Development Deptt, Govt of Tamil Nadu, Chennai.

Tyagi J. V., 2001, Effectiveness of percolation ponds for artificial recharge – a case study, Journal of soil and water conservation, Vol. 45, pp 53–66.

Predictive Model for Aquiferous Units and Quality of Underground Water in Onitsha Metropolis, Anambra Basin, Nigeria

17

O.A. Ehinola and W.O. Nneli
Department of Geology, University of Ibadan, Ibadan, Nigeria

ABSTRACT

Onitsha metropolitan Area (300 sq km) is located at the confluence of river Niger and river Anambra, in the Anambra basin. This is a highly populated centre which includes the present Onitsha township that lies within the southeastern part of Nigeria. The field mapping exercise involved water sample collection, lithologic description and GPS readings, while the laboratory analysis involved major and trace element determination. The component of the Aquiferous units and depth to water table was characterized using boreholes, sp and resistivity logs and GIS software. Also the underground water distribution and flow pattern, topographic driving force and water table were produced in 3D Models. This is to identify the aquiferous units and to ascertain the quality of underground water from boreholes in the study area. The borehole result shows that the study area consists of 3 aquiferous units system. The flow pattern of the underground water is in east-west direction with topography playing a major role as main contributing factor. The trace element concentration (i.e. Ba, Ca, Na, Cu, Fe, K, Mg, Mn, Si, Zn, & Sr) shows higher values than the guideline level. This indicates that most of the boreholes in the study area are not potable. Natural sources, lack of planning in this area and awkward sitting of wells close to roads, drainage system, waste disposal sites, sewage tanks and latrines as well as industrial effluent may have aggravated the level of trace element concentration. The information presented in this study could be employed in the borehole management and evaluation in the area and other similar city in Nigeria.

1 INTRODUCTION

Groundwater resources are dynamic in nature and are noted to be the largest available source of fresh water lying beneath the ground. It has become crucial not only for targeting of groundwater potential zones, but also monitoring and conserving it is very importance (Das, D., 2001). The most visible evidence of this is the human health problems associated with discharge of pathogens, domestic waste, industrial effluent etc. into drinking water resources (Burke and Beltran, 2001).

Nigeria is one of the largest countries in Africa (1,924,000 km^2. in area) and the most populous. The low service level of water accounts for why water-borne diseases are prevalent sometimes to epidemic scale and food shortages become critical. Within the last fifteen years or thereabout, the government has become partially aware of the significance of water resources development and some attention has been focused on water development and management (Hadidu, 1990).

Hydrological models have been specifically used to explain the interrelationship of the various hydrological phenomena and to solve various hydrological and water resource problems (Ayoade, 1988).

2 HYDROGEOLOGICAL SETTING

The Onitsha metropolitan area lies within the Nkisi-Idemili watershed, and covers an area of about 300 km^2 and falls within latitude 6° 04′N and 6° 15′ N and longitude 6° 45′ E and 6° 53′ E on the geographical map (Fig. 1). The elevation ranges from about 32 m above sea level from east to coastal part of river Niger to about 100 m above sea level from the central half to the east extremity. Generally, the area is drained by the Anambra, Nkisi and Idemili rivers all of which discharge into the Niger .Nkisi and Idemili Rivers flow in an E-W direction while the Anambra River flows in a S-W direction.

The Onitsha area is located within the tropics and has two distinct seasons which are wet and dry. The wet season is from about April to October while dry season is from about November to March. Available rainfall record has shown an average rainfall of about 2000mm (2 m) most of which falls with the month of April to October (Nwankwo and Eche, 1990). The 1959-62 rainfall record of the area shows that precipitation is greatest from June to October. During this period, the mean rainfall varies form 18 mm to 330 mm except in August when it is as low as 75 mm. Between June and part of October much water is discharged through run off and infiltration. Local violent storms are greatest in June (180 mm), July (260 mm), and September (330 mm).Also this period is characterized by moderate temperatures (20°C) and high relative humidity. The months of November to February have scanty rains, higher temperatures and low humidity (Ogbukagu, 1976). The high temperature, humidity and precipitation of the area favor quick plant growth and hence the vegetation is mainly tropical rainforest type in the virgin areas but this has generally been degraded by human activities (mainly urbanization) in the study area.

The Onitsha metropolis is entirely underlain by Bende-Ameki Formation (Fig. 3), which consists of sequence of interbedded units of Sand, friable Sandstones, Sandy Clay, Shale and occasional lignite seams. They were deposited during the regression of the sea in early Eocene. The beds have low to moderate dip angles of about 10° to 20° to the west (Reyment, 1965),and lie conformably on a thick shale sequence (Imo Shale). Patches of alluvial deposits consisting of sand and gravel overlie the Bende–Ameki formation along River Niger banks (Nwankwo and Eche, 1990).

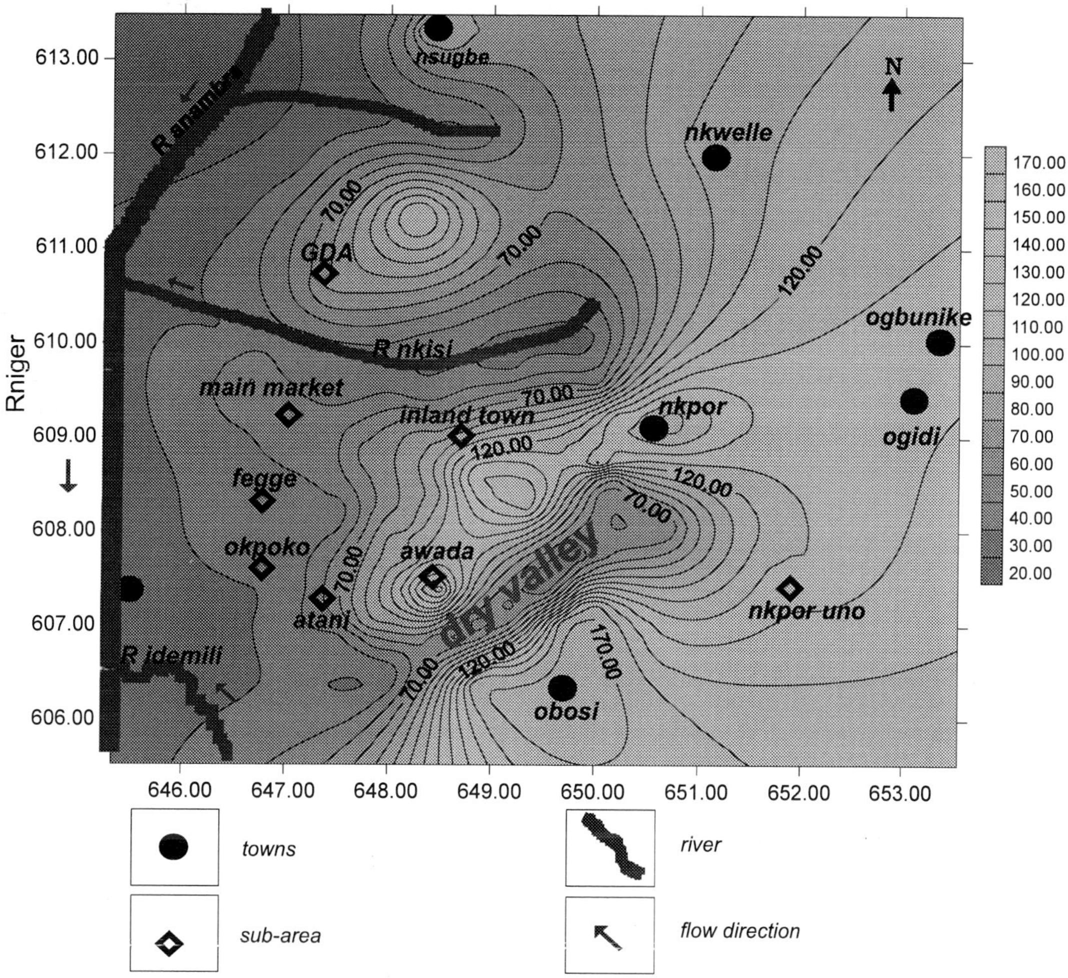

Fig. 1 Location map and rivers in the study area.

3 METHODOLOGY

The study area was mapped using the Garmin E-trek Venture Global Positioning System (GPS). This was used in the determination of the longitude, latitude, elevation of the location of the boreholes for the map production.

Water samples were collected from 10 bored wells in the study area and were analyzed using Inductively Couple Plasma-Optical Emission Spectroscopy (ICP-OES) technique. Up to 36 trace elements were analyzed and compared with the reporting limit. The trace element analysis was carried out at the Activation laboratories Ltd, Ontario, Canada.

Hydrogeologic survey of the Area was carried out by the determination of the well depths of different boreholes located in Onitsha metropolis. These depths were obtained from the drillers of these boreholes and in some cases owners of the boreholes. 74 borehole depth of water wells were collected for a comprehensive map production (models) of the study area (Fig. 1).

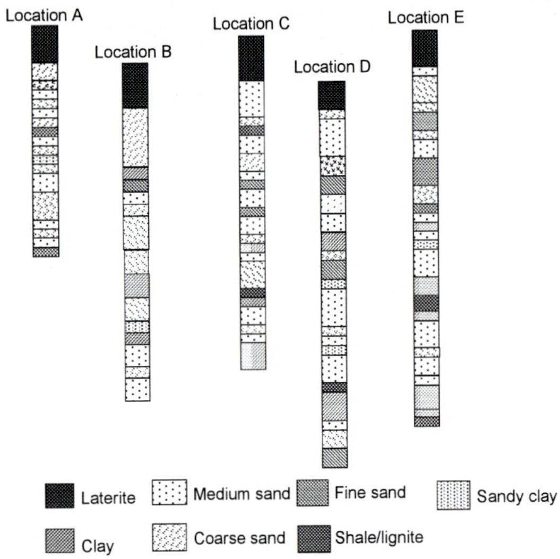

Fig. 2 Lithological cross section in the study area

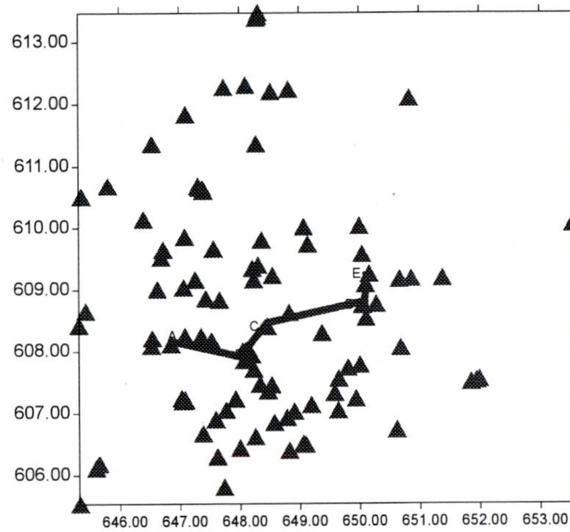

Fig. 3 Location map of wells and lithlogic cross section.

The component of the Aquiferous units and depth to water table was characterized using lithologic logs, well logs and GIS software. Also results obtained from trace element analysis of the water samples were plotted to generate models for the trace element concentration in the study area. Also plotted using the GIS package was the hydrological cycle of the study area, topographic driving force of the study area, regional underground water flow pattern and water table of the area.

4 RESULT AND INTERPRETATION

4.1 Aquifer System

Information from the lithologic and geophysical logs reveals a hydrostratigraphic sequence of the study area and well to well log correlation (Figs 2 and 3). The sequence show three major and lateral extensive contour aquifer units that alternates with clay shale and lignite seam aquitards to form a multi aquifer system. The geophysical logs differentiated the permeable formation from non-permeable zones. Aquifer 1 and 2 belong to the Benin formation (Miocene to Recent) while the third aquifer belonged to the ogwashi-asaba formation (Oligocene to Miocene) [Adegoke, 1969]. The first aquifer occurred at a depth of 20 m while the second and third aquifers varied due to topographic difference in different areas and was predicted using the models. Field studies have shown that determination of the third aquifer which is the regional aquifer units based on well cuttings has resulted in dry holes in some areas due to the uniqueness of the well cuttings down the holes. Geophysical logs have clearly showed the aquifer units and thus it is very important that geophysical surveying and well logging are applied for successful ground water exploration in the study area. The model was also able to show that the 3rd aquifer would be the best for ground water exploration due to farness from

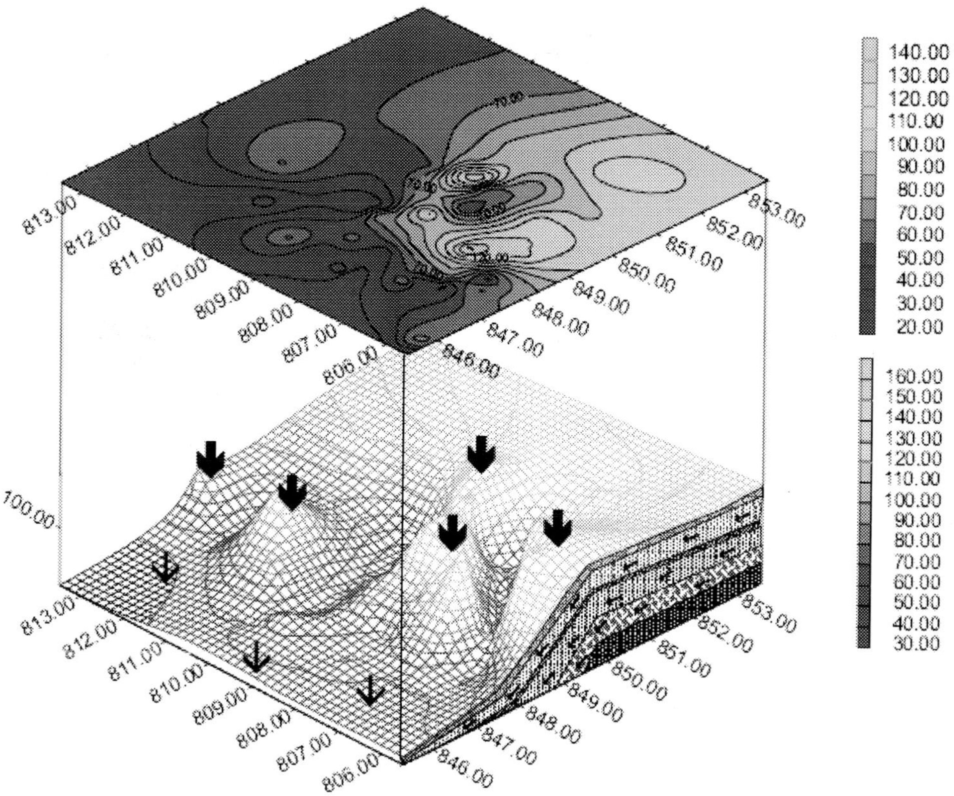

Fig. 4 Control of topography in the groundwater flow in the study area

contaminants on the ground surface. The quality of water sampled from neighboring Imo area has been found to be affected based on geology and human activities in the study area resulting in high concentration of trace elements and other problems (Ezeigbo, 1989). Lithologic succession in boreholes located in the study area have shown 3 major and lateral extensive units that alternates with clay-shale aquitards to form a multiaquifer system (Nwankwo and Eche, 1990).

The borehole depth ranged from 30 m to 152 m within the study area. Borehole depths were lower at the coastal areas to the river Niger while the highest water table depth was located in high land areas like Obosi, Awada and Nkpor areas. Deep aquifers were observed in topographically high areas while shallow aquifers were observed in the topographically low areas. This is highly significant in the underground water quality evaluation and supply information for reconnaissance survey before borehole exploration. The flow pattern of the underground water using the well depths is shown in (Fig. 4). The model represented the natural flow field of the study area, which is an east-west subsurface base flow. Though not yet studied it could be predicted that combined with natural processes from water cycle, the net recharge into the aquifer system and the subsurface flow of ground water has resulted in the recharge of the rivers in the study area. This could be supported by the large ground water discharge of $11.23 m^2/s$ (9.70×10^5 m^3/day) across the drainage basin (Nwankwo and Eche 1990).

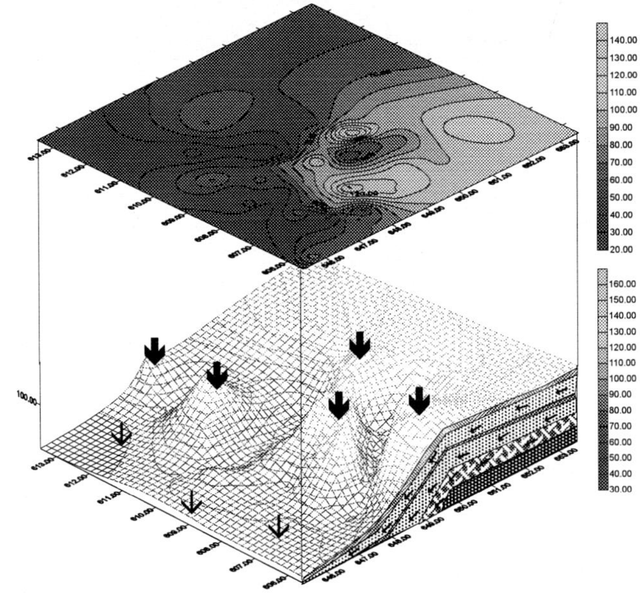

Fig. 5 Topographic driving force in the study area

The elevation difference of outcrops or the outcrop and the sub crop, provides the topographic drive for ground water movement (Chamberlin 1885). Gravity difference, line of least resistance in ground water movement (King 1899) and strata inclination (Larone and Gvitzman 1997) as observed in the lithologic succession within the study area. A 3D model of the topographic drive of the study area (Fig. 5) shows how the slope of the land is an important factor in determining whether the contaminant released will become runoff or infiltrate the aquifer. With a slow slope and subsurface geology, the contaminant is less likely to become runoff and therefore more likely to infiltrate the aquifer.

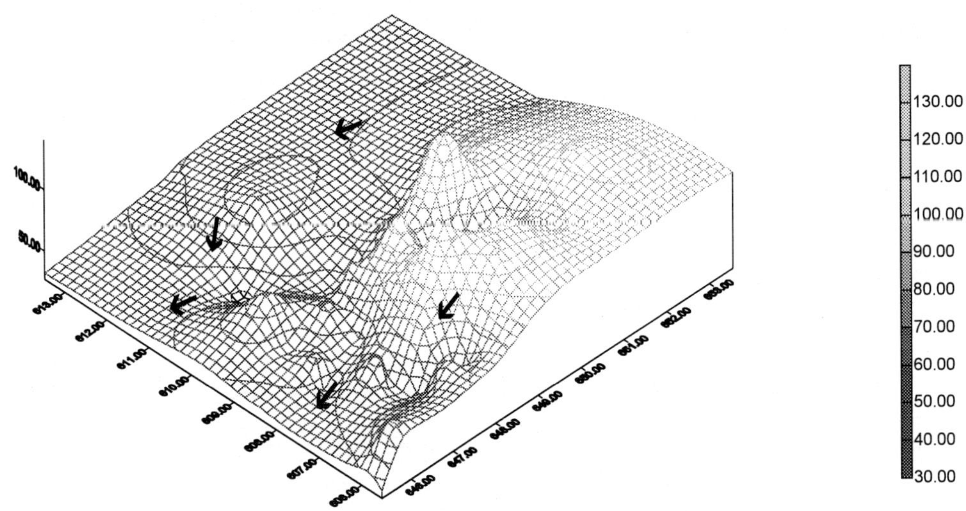

Fig. 6 Underground water flow pattern in the study area

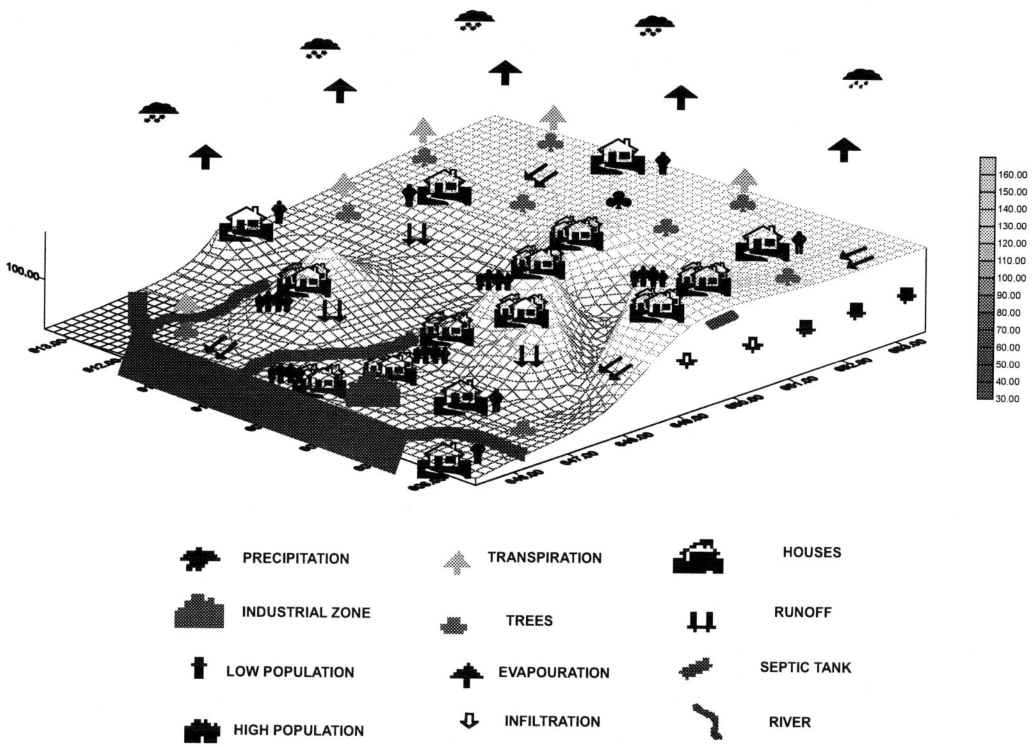

Fig. 7 Hypothetical model showing the hydrological cycle in the study area.

The distribution of the underground water in the study area using the concept of water cycle is presented in (Fig. 6). This could also be studied based on the annual rainfall of the area and the lithostratigraphic sequence of the Area. The influence of precipitation, infiltration interflow, baseflow, evaporation, transpiration was noted to influence the distribution of groundwater and this is important in assessing the groundwater contamination source.

4.2 Trace Element Concentration

In this study, groundwater values of Ag, Al, As, B, Ba, Be, B1, Ca, Cd, Ce, Co, Cr, Cu, Fe, K, L1, Mg, Mn, Mo, Wa, Ni, P, Pb, Sb, Te, T1, TL, U, V, W, Zn, S, Sn, Sr, Y, were determined . However, trace element analysis showed that metal concentrations of Li, B , Tl, Sn, As, B,P, Ag, Cd, Cr, Mo, W, V Se,Te and S were safe as they were below the reporting limit in the study area. The model maps generated and plots of trace element level indicates excessive level of Ba, Ca, Co, Cu, Fe, K, Mg, Mn, Si, and Sr (Table 1 and Figs 7 a to v) above the reporting limit (Activation laboratories Ltd, Ontario, Canada) in the area.

These trace elements are highly significant and have been noted to constitute public health challenges that require urgent attention and additional study in the study area.

The concentration of Barium range from -20 μg/l to 26 μg/L with 9 out of 10 samples exceeding the reporting limit at 20 μg/L (Fig 7 a and b). The source of Barium is said to be geological in origin due to its erosion from natural deposits (EPA, 2002). Barium can cause toxic

Table 1 Results of Trace Elements Concentration

S/N	Latitude	Longitude	Elevation (m)	Water depth (m)	Ba µg/L	Cu µg/L	Mn mg/L	Fe mg/L	Zn µg/L	Ca µg/L	Mg mg/L	Na mg/L	Si mg/L	K mg/L	Sr µg/L
1.	648.276	611.359	114.605	72.248	40	10	0.05	0.03	115	2.8	0.5	6.1	6.3	0.8	20
2.	646.724	609.643	22.25	72.248	–20	8	0.03	–0.01	355	4.1	0.4	1.7	4.7	0.5	–10
3.	646.851	608.115	42.367	39.624	120	30	0.06	0.05	40	1.4	0.7	14.8	6.2	2	20
4.	646.627	609.006	38.71	38.71	80	12	0.22	0.04	705	8.8	0.6	9.9	11.7	1.6	40
5.	648.21	609.344	3.962	35.052	140	10	0.48	–0.01	40	10.1	4.1	29.8	5.6	18.1	60
6.	645.318	608.424	29.566	36.576	120	42	0.46	0.02	100	14.1	2.1	28.6	5.8	19	50
7.	653.549	610.062	136.55	64.008	260	126	0.13	0.02	210	22.3	3.4	40.2	5.5	25	70
8.	649.367	608.229	160.934	47.72	120	66	0.52	0.04	120	7.8	2.4	43	5.9	8.8	50
9.	650.107	608.543	108.814	64.008	40	6	0.03	0.03	145	2.6	0.4	2.1	4.8	0.9	–10
10.	648.07	607.852	95.402	88.392	40	–2	0.26	–0.01	1160	13.5	2.0	7.9	8.5	2.8	40

effect to the heart, blood vessels and nerves (Dorota Z. Haman and Del B. Bottcher, 1986).It also leads to increase in blood pressure (USEPA, 2002).The concentration of copper ranges from -2 µg/L to 126µg/L with all the samples exceeding the reporting limit at 2 µg/L (Fig 7 c and d) .Source of copper concentration could be said to be of natural source as it is frequently found in surface water and ground water due to erosion of natural deposits (USEPA, 2002). As a metal used extensively by modern industrial civilization, cases of dissolution from plumbing fixtures and water pipes is known. Organic and inorganic compounds of copper have been used in agricultural pesticide sprays (Hem, J.D., 1985). These may have also resulted in its dispersal in the environment. Because of its low toxicity the interim copper limit is based on taste and is set at 1 mg/L. However in areas of high concentration, Short term exposure leads to gastrointestinal distress, and Long term exposure leads to Liver or kidney damage (USEPA, 2002).The concentration of manganese ranges from 0.03mg/L to 1.13mg/L with all the samples exceeding the reporting limit at 0.01 mg/L (Fig 7e and f). Source of the element could be natural as found in surface water and ground water survey and anthropogenic based on concentration of Tea, nuts, cereals and dried fruits in waste. Manganese is a known mutagen (Beckam, Milvran and loeb, 1985).The accumulation of Mn causes hepatic encephalopathy (Layrargues et al., 1998). Moreover, the chronic ingestion of Mn in drinking water is associated with neurological damage (kondakis et al., 1989).The concentration of zinc range from 40 µg/L to 1160 µg/L with all the samples exceeding the reporting limit at 5 µg/L(Fig 7g and h). Sources of Zn in the environment are usually from natural and anthropogenic source including from industrial effluents (Klein *et al.*, 1974), domestic effluents (Preuss and Kollmann,1974), urban storm water runoff (Field and Lager, 1975) and spoil heaps(Heitfeld and Schottler, 1973). Rarely, if the zinc concentrations in water approach 40 rag/L, human toxicity may occur, characterized by symptoms of irritability, muscular stiffness and pain, loss of appetite, and nausea (National Academy of Sciences-National Research Council. 1977). The concentrations of calcium ranged from 1.4 mg/L to 22.3 mg/L with all the samples exceeding the reporting limit at 0.1 mg/L (Fig 7i and j). Its source could be said to be from natural. Excess calcium may produce deficiencies in magnesium and potassium. Its abundance in groundwater leads to most common problem associated with groundwater which is hardness. Hard water has not been shown to cause health problems but can be a nuisance as it may cause soap curds and deposits to form on pipes and plumbing fixtures. The concentration of magnesium ranges from 0.4mg/L to 4.1mg/L with all the samples exceeding the reporting limit of 0.1mg/l(Fig 7k and l). Sources of Mg in the environment are usually from natural source to the groundwater. Any excess amount is quickly expelled by the body. No upper limit has been set for this metal in drinking water. It can, however, create a problem for people with kidney disease. They may develop toxic reactions to high levels of magnesium, including muscle weakness, coma, hypertension, and confusion (Dorota Z. Haman and Del B. Bottcher, 1986).The concentration of iron range from -0.01 µg/L to 0.05 µg/L with 7 out of 10 samples exceeding the reporting limit of 0.01 µg/L(Fig 7m and n). Its source is natural in nature based on its minimal concentration. Iron causes a variety of different nuisance problems (Wilson et al., 2000). They usually do not present a health hazard when they are present in the household water supply. The presence of iron in drinking water may increase the hazard of pathogenic organisms, since most of these organisms need iron to grow (Haman & Bottcher, 1986). The concentration of sodium ranged form 1.7 mg/L to 43.0 mg/L with all the samples exceeding the reporting limit of 0.1 mg/L(Fig 7o and p). It is present in most natural waters. Sewage and industrial effluents are also sources of sodium (Hem, J.D., 1985), Water with high Na varies blood pressure slightly and this is of particular concern in people with heart, liver, kidney and other disease where salt intake has to be restricted (WHO, 1996).The concentration of silica ranged from 4.7mg/L to 11.7mg/L with

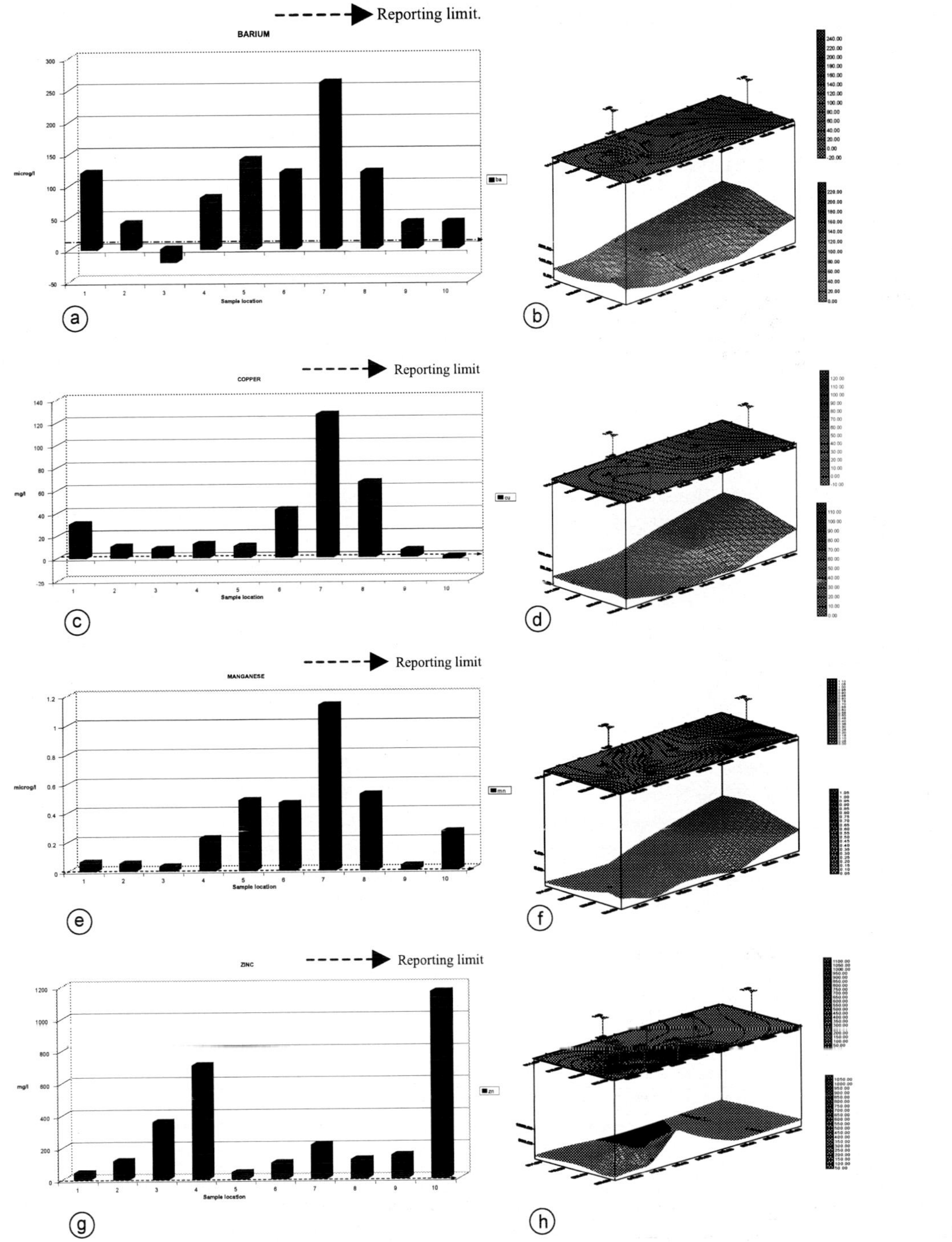

Fig. 8 (a) Levels of barium in groundwater. (b) Barium map. (c) Levels of copper in groundwater. (d) Copper map. (e) Levels of manganese in groundwater. (f) Manganese map. (g) Levels of zinc in groundwater. (h) Zinc map.

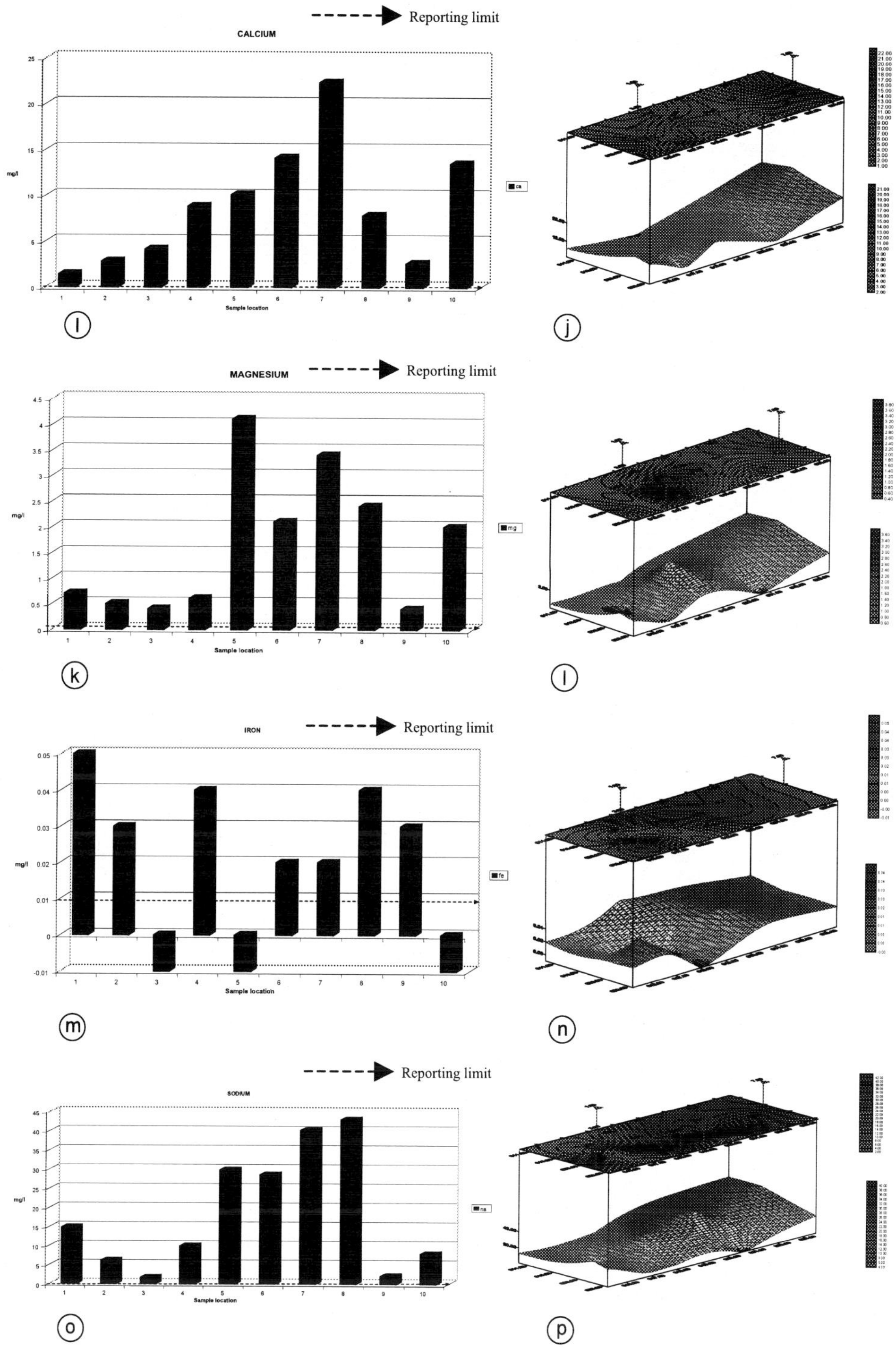

Fig. 9 (i) Levels of calcium in groundwater. (j) Calcium map. (k) Levels of magnesium in groundwater. (l) Magnesium map. (m) Levels of iron in groundwater. (n) Iron map. (o) Levels of sodium in groundwater. (p) Sodium map.

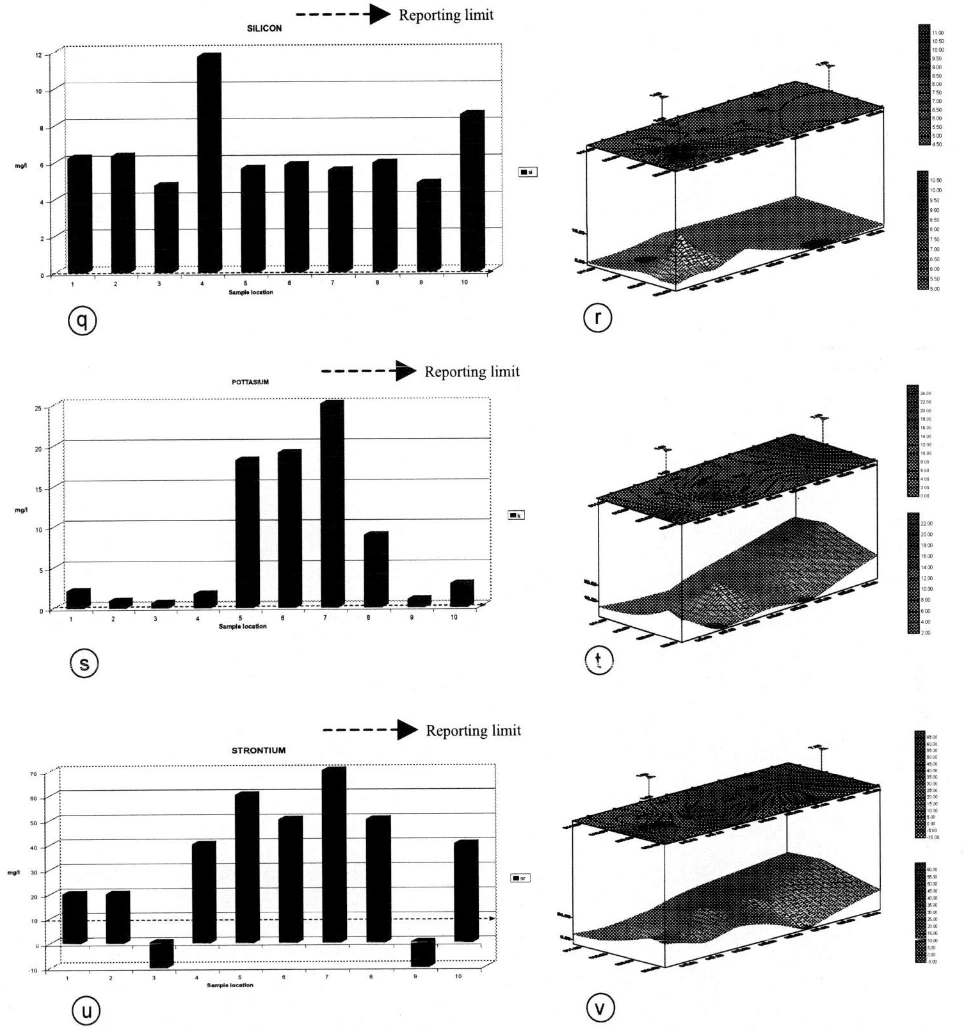

Fig. 10 (q) Levels of Silicon in groundwater. (r) Silicon map. (s) Levels of Potassium in groundwater. (t) Potassium map. (u) Levels of Strontium in groundwater. (v) Strontium map.

all the samples exceeding the reporting limit of 0.17mg/L (Fig. 7q and r). The elevated levels of silicon in the groundwater samples could be said to be due to the bedrock of the study area as it constitutes the bulk of the grains in most sandstones (Hem, 1985). It causes no harmful affect, but large amount can cause scaling in pipes that impact water flow and it can interfere with iron and manganese removal. The concentration of potassium ranging from 0.5mg/L to 25mg/L with all the samples exceeding the reporting limit being 0.1mg/L (Fig. 7s and t). it is fairly abundant in sediment, but concentration present in solution in natural waters are normally no more than 10mg/L. The source of potassium in the area is both natural and anthropogenic, it is a component of sewage, as it is essential in metabolism, and it is always in animal metabolic waste(Hem, 1985). Concentration exceeding 2000 mg/L may be harmful to the nervous and digestive system. The concentration of Strontium ranges from 10 ug/L to 70ug/L with 8 out of 10 samples exceeding the reporting limit at 10ug/L (Fig. 7u and v). The source of strontium in the study area could not be ascertained. It is particularly dangerous for growing children as it is easily deposited in the bones and is believed to induce bone cancer and leukemia.

5 CONCLUSION

Three major aquifer units occur in the stratigraphic sequence that underlies the study area. Ample amount of rainfall, the general undulating topography over most areas and the relatively large outcrops of permeable rock units and soil causes adequate infiltration. The underground water flow was in the East-West direction. The well depths ranged from 30m to 152m within the study area. The distribution of underground water in the study area was illustrated using the concept of water cycle. It showed the natural and anthropogenic source contribution to ground water quality in the study area. The groundwater is contaminated with Ba, Cu, Mn, Zn, Ca, Mg, Fe, Na, Si, K and Sr with natural source of imputes existing. The inhabitants who depend solely on groundwater retrieved from hand dug wells for portable supply are endangered. From field observation, sitting of wells close to roads, drainage system, waste disposal sites, sewage tanks and latrines has been found to affect the chemical quality of groundwater in terms of their excessive concentration. This has a significant impact on the public health within the area as it has affected crop yield, animal and human health.

In addition to yield information, this study shows several order type of data which can be used to aid in the development of a water supply. GIS has helped to make the result of a complicated model more clearly through visual representation, thus providing an applicable tool for decision makers. Current improvements in analytical methods has allowed for detection of trace elements at very low concentration in groundwater samples. Consequently groundwater supply which is considered to be pure has been found to contain various contaminants both from natural and anthropogenic sources. Protection of groundwater quantity and quality can best be accomplished by controlling potential contaminant sources and by managing land uses in prime recharge areas, using the knowledge of local geology and groundwater direction as shown in this study.

Another means of reducing the chances of groundwater contamination from residential areas is to limit the allowable housing density. This study should be considered as an introduction to some specific water problems one might encounter in the study area (especially water exploration) and how one should go about identifying and solving them. In view of these groundwater quality problems, a new water supply should be tested before use and old sources should be periodically checked for changes. Community supplies are very important in the study area as it is safer than individual system.

References

Amber Wilson, Kathleen Parrott and Blake Ross., 2000, Iron and Manganese in Household, Virginia Cooperative Extension, *Virginia* state university, Water Publication Number 356-478.

Ayoade,J.,O., 1988, tropical hydrology and water resources. Published by Macmillan publisher's ltd.london and Basingstoke 6:113-130, 8:169-170, 10:242.

Brian, J., Skinner, and Stephen, C., porter, 2000, The Dynamic Earth. An introduction to physical geology (forth Edition) printed by von Hoffmann press Inc copyright by John Wiley & sons Inc pp281-305.

Das, D., 2001, GIS application in hydrogeologic studies, GIS development. Department of Environmental Science, University of Kalyani, W.B. - 741235, India, ddas@klyuniv.ernet.in or GISdevelopment.net

Domenico, A. P and Schwartz, W. F., 1998, physical and chemical hydrogeology, Second Edition, John Wiley and Sons, Inc.

Dorota Z. Haman and Del B. Bottcher,1986 Circular 703, Florida Cooperative Extension Service, Institute of Food and Agricultural Sciences, University of Florida.

Edward, J., Calabresi, Andrew, T., Canada and carol Sacco (1985). Trace elements and public health. Ann. Rev. Public health. 1985 6: 131-146.

Edward, A., Keller, 2000, Environmental Geology 8th Edition Prentice Hall Inc. pp 291-368.

Ehinola, O. A., 2002, Hydrochemical characteristics of ground water in parts of the Basement complex of south western Nigeria. Journal of Mining and Geology 38(2): 25-133.

Ehinola, O. A, Oladunjoye M.A, Adabanija M.A and Coker T.J (2005). Quality of underground water and the health Hazards implication in Agbowo Area of Ibadan, south Western Nigeria science focus, Lautech, Ogbomoso (in press).

Ezeigbo, H. I., 1989, Ground water quality in parts of Imo state Nigeria Journal of mining and geology vol. 25 No 1&2 1989.

Emenike, E.A., 2001, Geophysical exploration for ground water in a sedimentary Environment. A case study from Nanka over Nanka formation in Anambra basin south Eastern Nigeria published in Global journal of pure and applied sciences Vol. 7 January 2001.

Handidu, J. A., 1990, National growth, water demand and supply strategies in Nigeria in the 1960s, journ.of the Nigerian association of hydro geologists (NAH), Vol. 2.1.

Hem, J.D., 1985 Study and interpretation of chemical characteristics of natural water: U.S. Geological Survey Water-Supply Paper 2254,

Kenneth, E., Foote and Margaret, lynch, 1995, Geographic information system: context, concepts and definitions. The Geographers project, Department of Geography. The University of Colorado at Boulder.

Larrone, L., and Gvirtzman, H., 1997, Subsurface river like Ground master flow patterns at the Judean Desert, the Hebrew University of Jerusalem.

Morrison, G., fatoki, O.S., Linders, and Lumdeth, C., 2003,Determination of heavy metal concentrations and metal finger print 0f sewage sludge from Eastern cape province South Africa by inductively coupled plasma - mass spectrometry (ICP - Ms) and laser Ablation - Inductively coupled plasma - mass spectrometry (LIA - ICP - MS): water, air and soil pollution 152. 111- 127, 2004 Kluwer academic publishers printed in Netherlands.

Nwankwo, G. I., and ECHE, B. C, 1990. Hydrological evaluation of the Greater - Onitsha water scheme, Anambra state, journal of mining and geology vol 26 NO2 1990

Ogbukagu, I. N., 1976.soil erosion in the northern parts of the Awka-Orlu uplands, Nigeria, journal of mining and geology vol 13(2):6-9.

Olade, M. A., 1975, Evolution of Nigeria, Benue through Aulagogen) a tectonic model. Geol mag Wl 122 pp 575 - 583.

Olade, M .A. 1979. The Abakaliki pyroclastic of southern Benue through Nigeria, their petrology and Nigeria mining and geosciences society vol 16. No 1 P 17-25.

Onuoha, K. M., 1986, Basin Subsidence, sediment decompaction and burial history modeling technique: Applicability to the Anambra basin. Nigerian association of petroleum Exploration (Nape) Bull vol 02/01 (pp 6-17).

Reyment, R. A. 1965. Aspect of the Geology of Nigeria. University of Ibadan Press, Nigeria. 145p.

Schlumberger, 1974, log interpretation volume II applications.No2 P 13.

Seth H. Frisbie, Richard Ortega, Donald M. Maynard and Bibudhendra Sarkar ,The health effects of arsenic and other toxic metals in Bangladesh's drinking water. SOS-arsenic.net.

US Environmental Protection Agency 2002, List of Contaminants & their MCLs, EPA 816-F-02-013

WHO (1996), Guidelines for Drinking-water Quality, Volume 2: Health Criteria and Other Supporting Information, 2nd ed. Geneva, Switzerland: WHO; 152, 162, 213, 266, 278, 279.

18

Methods of Artificial Recharge of Groundwater

P. G. Gaikwad[1] and Deepak Khare[2]
[1]N.D.M.V.P. Samaj's College of Engineering, Nasik, Maharashtra, India
[2]Indian Institute of Technology, Roorkee, India

ABSTRACT

Groundwater is the largest source of freshwater available on earth, which is exploited to satisfy our needs for domestic, agricultural and industrial purposes. Surface water sources are limited in nature and open land for natural recharge is also decreasing day by day. More abstraction of groundwater in comparison to natural recharge results in depletion of groundwater storage. More storage of water behind dams and especially in aquifers via artificial recharge is necessary to save water in times of water surplus for use in times of water shortage. Limited natural rainfall recharge and increased water use throughout the country calls for conservation as well as augmentation by artificial recharge. Variety of methods have been developed to recharge the groundwater such as direct surface recharge, combination of surface and subsurface recharge and indirect recharge.

1 INTRODUCTION

1.1 Need of Artificial Recharge

Water is going to be the most serious problem that India will be facing in 21st century. Though India is blessed with good rainfall, large amounts of water is lost through run-off and it cannot be impounded on the surface for want of suitable sites for storage. A hungry country like India cannot afford to lose this amount of water, which could be made available for domestic, agricultural and industrial purposes. Population growth and high living standards will cause ever-increasing demands for good quality water for drinking and industrial use. At the same time more and more irrigation water will be required to meet increasing demands of food for growing populations. In certain high demand areas groundwater development has already reached a critical stage—demand outstripping the supply.

Overdevelopment of the groundwater resources results in declining groundwater levels, shortage in water supply, intrusion of saline water in coastal areas and increased pumping lifts necessitating deepening of groundwater structures and seriously affecting socio-economic conditions. In such areas there is urgent need for augmenting of groundwater resources. Object is to ensure water for drinking and protective irrigation in rainfed cultivated area. Moreover, nature's precious gift—rainwater needs to be shared on an equitable basis to ensure social stability. The non-availability of sites for surface storage has led to considerable quantities of water being lost through surface runoff. Vast storage space is available in the groundwater reservoirs.

Non-conventional methods for utilisation of water such as through inter-basin transfers, artificial recharge of groundwater and desalination of brackish or sea water as well as traditional water conservation practices like rainwater harvesting, including rooftop rainwater harvesting, need to be practiced to further increase the utilisable water resources.

1.2 Scope of Artificial Recharge

It is possible to harness the large percentage of excess run-off going as waste to sea and store it in the available groundwater reservoirs by creating recharge structures. Rainwater harvesting to augment water resources is not new in India. Percolation ponds (village ponds) and rooftop rainwater harvesting has been practiced from a very long time.

Artificial recharge of groundwater can be achieved with surface infiltration systems (basin spreading facilities) wells, trenches, pits, shafts, stream channel modification or augmentation, soil aquifer treatment (SAT) systems, aquifer storage and recovery (ASR) wells, vadose zone wells, combination of surface and subsurface recharge, induced recharge, incidental recharge, etc. (Pliakas 2005).

1.3 Definitions of Artificial Recharge

Artificial recharge is a planned activity of man whereby rainwater as well as surface water from streams and lakes is made to infiltrate the ground commonly at rates and in quantities many times in excess of natural recharge, giving corresponding increase in the magnitude of the safe yield (Huisman and Olsthoorn 1983).

Artificial recharge systems are engineered systems where surface water is put on or in the ground for infiltration or subsequent movement to aquifer to augment groundwater resources.

Artificial recharge is the process of replenishing groundwater by augmenting the natural infiltration of rainwater or surface water into underground formations through various methods designed depending upon topographic, geologic and soil conditions.

1.4 Purposes of Artificial Recharge

The basic purpose of artificial recharge, as already defined above is to augment the amount of available groundwater, however some other purposes include:

1. To reduce, stop or even reverse declining level of groundwater.
2. To protect underground freshwater in coastal aquifer against saltwater intrusion from the ocean.
3. To store surface water including flood or other surface water, imported water and reclaimed wastewater for future use.
4. To improve water quality by removing suspended solids by filtration through ground.
5. To store water to reduce the cost of pumping and piping.
6. Reduce or stop significant land subsidence.
7. Provide a localized subsurface distribution system for established wells.

2 FACTORS TO BE CONSIDERED FOR ARTIFICIAL RECHARGE

Artificial recharge of groundwater is expected to increase worldwide as population rises, while artificial recharge resulting in conjunctive use of surface water and groundwater and long-term underground storage or water banking is preferred where possible (Bouwer, 2002). Before undertaking the artificial recharge project, following factors should be taken into consideration:

1. Quantity of the available surface water that can be recharged into aquifer.
2. Quality of the water to be recharged, as it determines the resulting quality of groundwater.
3. Resulting water quality (reactions with native water and aquifer materials).
4. Clogging potential of the recharge system.
5. Underground storage space available.
6. Depth of underground storage space.
7. Transmission characteristics of soil strata.
8. Application method (infiltration or injection).
9. Cost of the project.

3 METHODS OF ARTIFICIAL RECHARGE

A variety of methods have been developed to recharge groundwater and although no two projects are identical, most use variations or combinations of direct-surface, direct subsurface or indirect techniques. The specific methods in these categories are listed below:

Direct surface recharge

- Flooding
- Ditch and furrow system
- Basin recharge
- Stream channel modification
- Stream augmentation, and
- Overirrigation

Direct subsurface recharge

- Natural openings
- Pits and shafts
- Reverse drainage
- Wells
- Aquifer storage recovery (ASR) wells
- Soil aquifer treatment (SAT)

Combination of surface and subsurface recharge

- Subsurface drainage i.e. collector with wells
- Basins with pits, shafts or wells.

Indirect recharge

- Induced recharge from surface water sources
- Aquifer modification

3.1 Direct Surface Recharge

Direct surface recharge techniques are among the simplest, oldest and most widely applied methods of artificial recharge. In these methods water moves from land surface to the aquifer by percolating through soil. Direct surface recharge methods can be grouped into several categories such as: flooding, ditch and furrow method, basin recharge, stream channel modification, stream augmentation and overirrigation.

Flooding

Recharge by flooding can be done only on land having a 1% to 3% slope. The objective is to spread water over a large area in a thin film that travels slowly down hill without disturbing the soil. Water is spread over the land surface from several distribution points as shown in Fig. 1. to obtain an even application. Embankments or ditches may bound the system to localize infiltration or to protect adjacent land. Excess water may be collected at the system's topographic low point for disposal.

The biggest problem with the flooding technique is containments. Other problems are related to large area required and evaporation. The method's greatest advantage is relatively low cost of construction and maintenance.

Ditch and furrow system

In this method, a source stream provides recharge water that is passed through closely spaced, shallow, flat-bottomed ditches or furrow. Most ditch and furrow systems have one of three patterns: lateral, dendritic or contour.

Lateral systems

These characteristically have one or more main supply canals from which smaller ditches protrude at right angles. Gates at the head of these systems control flow rates (Fig. 2). Furrow

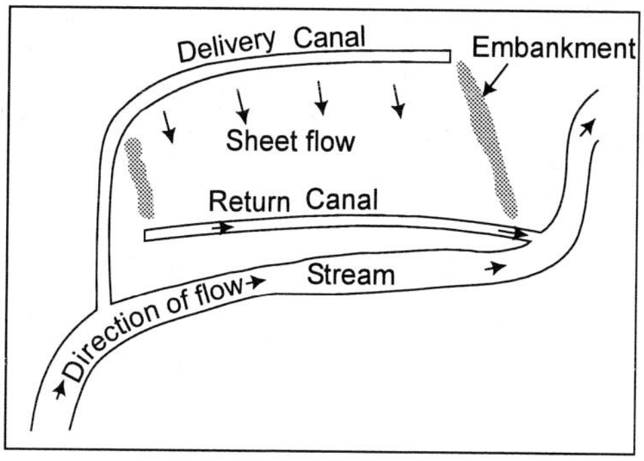

Fig. 1 Direct-surface recharge by flooding.

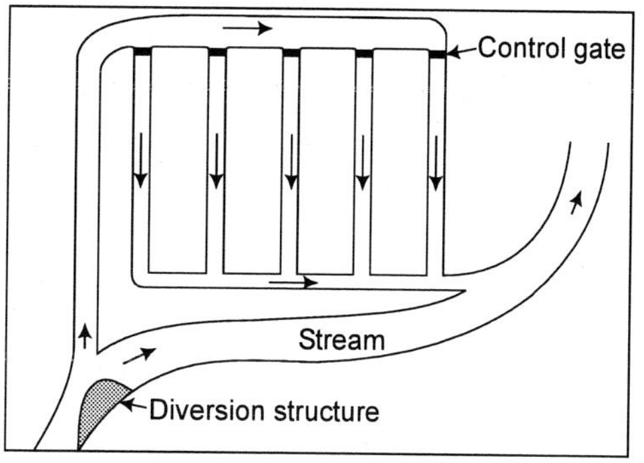

Fig. 2 Lateral ditch pattern.

depth depends on topography but rarely exceeds that necessary to maintain a uniform velocity with maximum wetted surface. Most systems divert stream water into main canals, pass the water through a series of lateral furrows, and collect run-off in a canal further down slope that routes the water back to the source stream.

Dendritic systems

These divert flow from a main canal to series of successively smaller ditches and gates control the flow to each series of ditches. The bifurcation of ditches continues until virtually all water has infiltrated; terminal collection ditches are optional (Fig. 3).

Contour systems

These spread water through a ditch or ditches that follow the contour of the land. A switchback is made wherever the ditch approaches the limit of the spreading area. In effect, the ditch traverses the spreading area repeatedly and at the lowest point, returns the water to the source stream. (Fig. 4).

Ditch and furrow system is particularly advantageous where recharge water contains high loads of suspended sediment. Generally, system flow rates are sufficient to carry a large percentage of foreign materials through the systems and back into the source stream.

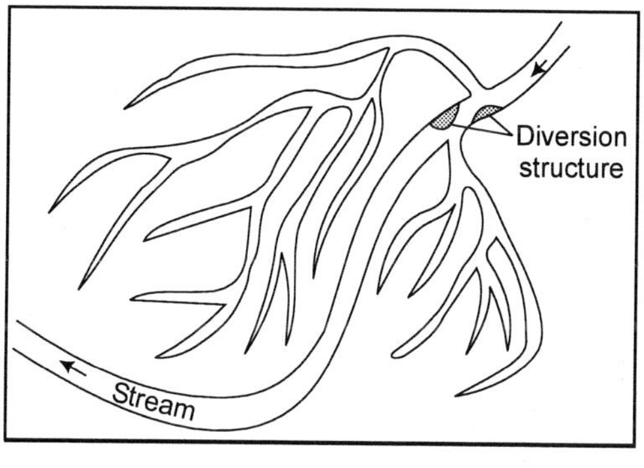

Fig. 3 Dendritic ditch pattern.

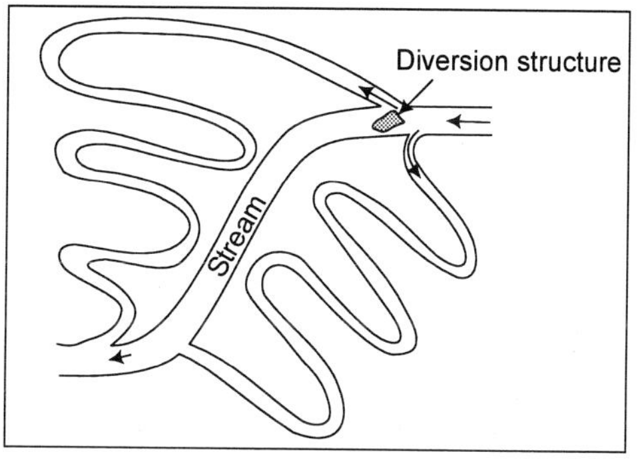

Fig. 4 Contour ditch pattern.

Basin recharge

Basins are probably the most favoured method of recharge because they allow efficient use of space and requires only simple maintenance. Basins are either excavated or are enclosed by dikes or levees. Basin geometry is flexible, allowing construction to be adapted to the terrain. Basins may be constructed individually, such as in small drainage areas to collect urban run-off, or in the series for infiltration of stream water or storm water as shown in the Fig. 5. Use of multiple basins for infiltration of stream water provides several advantages: the storage capability allows a longer time for recharge, the upstream basins act as clarifiers for those below, and the ability to bypass the basins permits periodic maintenance (such as scrapping) to restore infiltration rates. In flat areas, basins are more costly to construct because natural landform containments cannot be used; basins in such areas are commonly long, straight, and narrow and are constructed side by side.

The advantages of basins include:

1. Expected flows can generally be accommodated by constructing basins of appropriate size.
2. Intermittent floodwater can be stored for later infiltration.
3. Clogging can be easily mitigated through basin construction techniques or operational procedures.
4. Land is used efficiently.

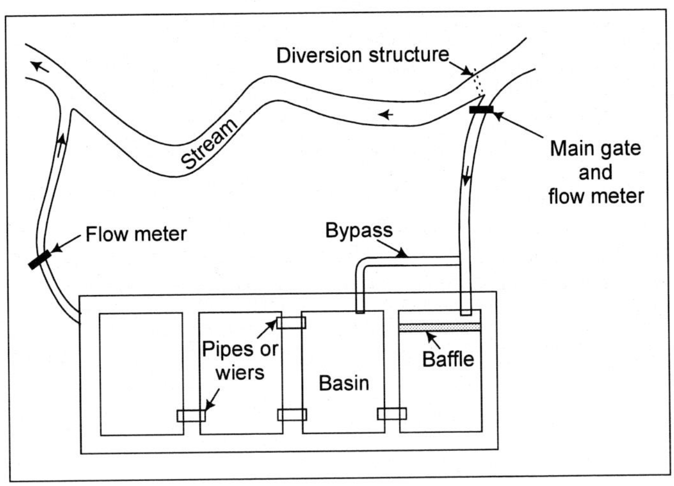

Fig. 5 Basin recharge.

Stream Channel Modification

Stream channel modification entails construction of check dams across stream floodplains. Above the dams, basin-like impoundments enhance recharge by increasing the wetted area and driving head while detaining water for recharge, below the dams, recharge is enhanced through exposure of more than just the original stream channel to water (Fig. 6). Because stream water flowing at velocities greater than the stream bed infiltration rate is essentially lost,

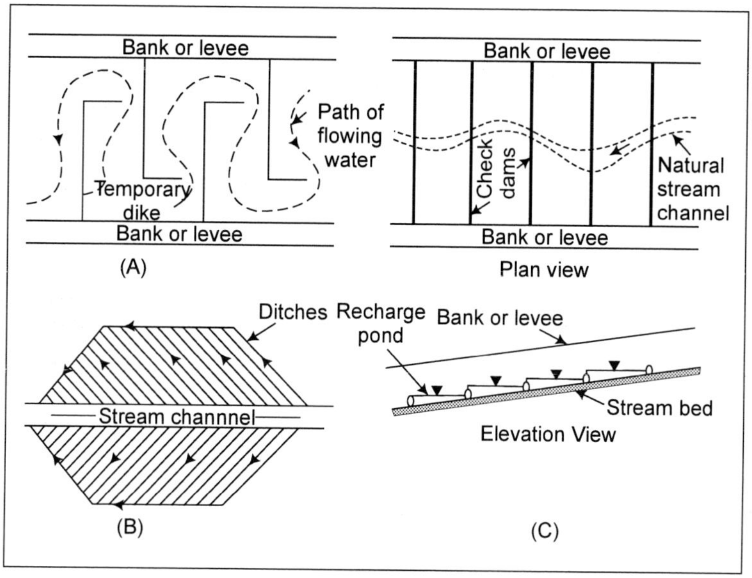

Fig. 6 Stream channel modification.

upstream reservoirs can be constructed to dispense water in accordance with stream bed infiltration capacity. Dredging flowing channels also increase the infiltration efficiency and widening, levelling, increase the wetted area as well as infiltration rate. Most stream channel modification structures are temporary and are designed to increase recharge only seasonally. Many are destroyed by floods. Nevertheless stream channel modification is effective wherever suitable because construction costs are relatively low, maintenance is inexpensive, and the procedure rarely conflicts with other land uses.

Stream Flow Augmentation

Stream flow augmentation involves application of recharge to a stream channel near the head of its drainage area to reestablish or increase infiltration through the stream bed. Stream flow augmentation is considered an alternative to artificial recharge in areas where streams fed by groundwater have ceased to flow or have become dry in their upper reaches because of lowered groundwater level.

This method is generally less efficient than other techniques because stream velocities exceed infiltration rates, and economical sources of recharge water are not always available.

Overirrigation

When irrigation water is applied during ongrowing seasons, the groundwater is artificially recharged. Methods of irrigation are similar to those of artificial recharge and include overland flow, ditch and furrow systems, sub-irrigation, flooding, and spray systems. Because overirrigation is usually created by excess surface water, one of the first four techniques described is usually employed. Implementation costs are minimal because the water distribution works have already been constructed.

Direct surface recharge method is suitable to recharge phreatic aquifers and semi- confined aquifers, which are in hydraulic connection with phreatic aquifers. Recharge to be more effective, the materials in zone of aeration should have good vertical permeability to aid percolation and the underlying aquifer should have good transmissivity to transport the recharged water away from the spreading area.

3.2 Direct Subsurface Recharge

Direct subsurface recharge is achieved when water is conveyed and emplaced directly into an aquifer. Direct subsurface recharge generally is used when a semi-permeable confining stratum separates the source of recharge water from the aquifer requiring replenishment. Techniques of direct-subsurface recharge includes injection of water into natural openings in the aquifer, pits or shafts, wells and drainage pipe networks.

In all methods of subsurface recharge the quality of the recharge water is of primary concern because the water enters the aquifers without the filtration or oxidation that occurs when water percolates through the unsaturated zone. All direct-subsurface methods of recharge use structures that occupy much less land than is required for surface methods.

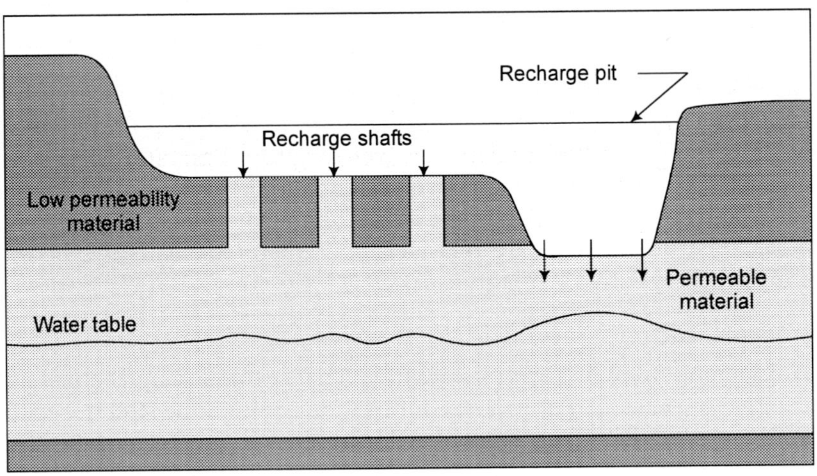

Fig. 7 Recharge pits and shafts.

Natural Openings

Natural openings, such as those caused by fracturing or solution in cavernous limestones or other soluble rock, can act as a drain beneath an impounded body of water or as the extension of a pipeline delivering water to it. Depending on the source of water and size, configuration and location of the openings, maintenance, protection and improvement may not be necessary. Although this type of recharge system is relatively inexpensive, unfavourable terrain and geologic condition may preclude its use.

Pits and Shafts

Where a semi-pervious confining layer is at or near land surface, the aquifer below must be recharged through deep pits or shafts penetrating that layer. Pits are variable in dimension; their depth generally is dependent on the thickness of the confining unit. The steep sides of pits minimize the opportunity for clogging, which occurs mainly at the pit bottom. Cost of construction and maintenance are high compared to surface techniques, especially if comparatively small quantities of water are to be applied. In some areas these costs can be avoided by use of abandoned gravel pit or quarries.

Shafts are deeper than pits and smaller in diameter. They are either constructed by hand with draglines or are drilled or bored. They can be filled with coarse material or simply lined. As with all sub-surface techniques, shaft walls and fill materials are susceptible to clogging by suspended solids or biologic activity. Shafts are particularly difficult to maintain. Although similar in operation to large-diameter wells, shafts that terminate above a static water table cannot be redeveloped by pumping. Rest periods and chemical treatment are generally the only means of restoring infiltration capacity and often only partly successful. Badly clogged shaft fill material must be removed and replaced with clean fill. Because of high construction and maintenance costs, pits and shafts have limited application.

Reverse drainage

Reverse drainage, similar to sub-irrigation, is used to avoid interference with surface land use. In this method water is piped into underground networks of perforated drainage conduit, from which water infiltrates into the soil. The primary advantage of reverse drainage is its negligible effects on the surface land use. Conduits may be emplaced directly in the soil or may be encased with coarser material to increase effective size and infiltration area. Size of pipe, depth and type of installation and spacing of conduits depend on the quantity of water to be applied, depth to the water table, and hydraulic properties of the medium. Reverse drainage techniques are desirable where land is relatively expensive.

Wells

Recharge wells, commonly called injection wells, are generally used to replenish groundwater when an aquifer is deep and separated from the surface by materials of low permeability. If injection wells are installed in unconsolidated material, the upper section of the well is cased the screen is directly placed in the aquifer or surrounded by an artificial gravel pack (Fig. 8). Typically a concrete seal is constructed at the point where the well passes into the aquifer to prevent injection pressures from forcing water upward around the casing. For a consolidated aquifer overlain by impermeable consolidated deposits, casing and screens may not be required. Wells are also used to recharge unconfined aquifers where available land is limited. Recharge wells are not limited to replenishing just one aquifer nor does the recharge water need to be derived from a surface source. Recharge wells can be constructed to supply water to two or more aquifers simultaneously (Fig. 9) and where hydraulic conditions permit, can be used as passive connectors between adjacent aquifers separated by impermeable material.

In addition to the primary purpose of replenishing of potential aquifers, recharge wells have also been used to recharge groundwater used for air-conditioning and to add freshwater to coastal aquifers experiencing salt water intrusion. Injection well design depends on the recharge purpose, the amount of water to be injected and the acceptance rate of the aquifer. The

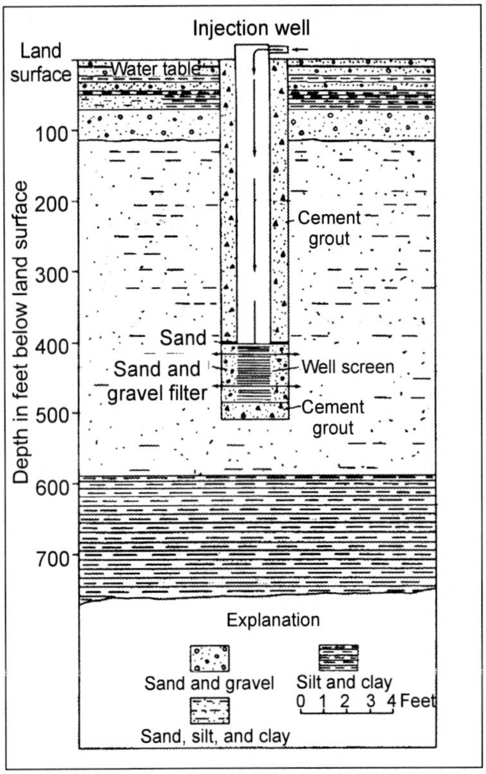

Fig. 8 Recharge well.

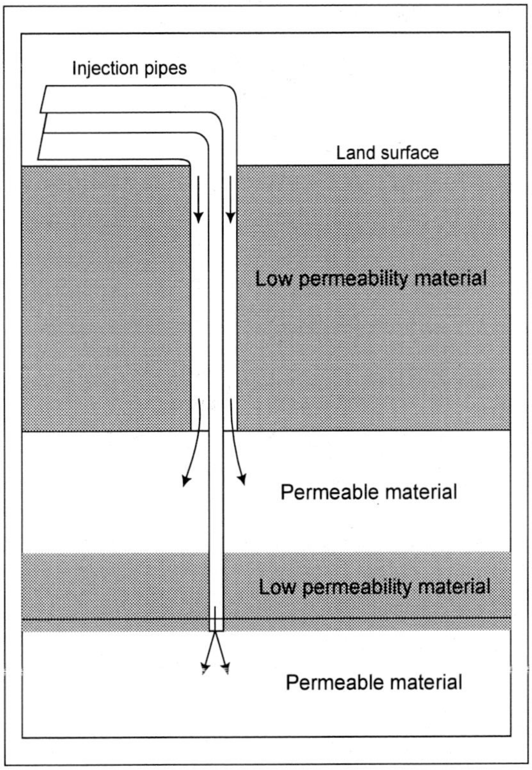

Fig. 9 Multi-aquifer Recharge well.

latter is a function of hydraulic gradient, aquifer permeability and length and type of screen. Because the contact area between well screen and aquifer is small, gravel packs surrounding the screen are generally used to increase the effective well diameter.

Recharge well performance can be severely hindered by the accumulation of the suspended solids and biologic and chemical impurities, as well as by dissolved air and gases and entrained air from turbulence. Most clogging effects can be avoided by proper treatment of the recharge water beforehand, but to correct once it has occurred requires various well redevelopment procedures, which include:

1. Pumping and surging the well to remove inorganic material and loosely attached organic material.
2. Adding biocides and oxidizing agents to eliminate organic matter stemming from bacteria and their waste products.
3. Using specific chemical treatments to remove encrustation caused by chemical precipitation.

Aquifer storage recovery (ASR) wells

One method of artificial recharge that is growing in popularity is Aquifer Storage Recovery (ASR). ASR is the storage of water in a well during times when water is available, and recovery of the water from the same well during times when it is needed.

Water is stored deep underground in water bearing geologic formations, or "aquifers," that may be in sand, clayey sand, sandstone, gravel, limestone, dolomite, glacial drift, basalt and other types of geologic settings. Stored water displaces the water naturally present in the aquifer, creating a very large bubble around the well. The bubble is usually confined by overlying and underlying geologic formations that do not produce water.

The hydro-geologic characteristics of a successful storage zone include moderate permeability, confinement above and below by low permeability sediments, and water quality as fresh as possible to minimize mixing. ASR provides a cost-effective solution to many of the world's water management needs, storing water during times of flood or when water quality is good, and recovering it later during emergencies or times of water shortage, or when water quality from the source may be poor.

ASR reduces or eliminates the need to construct large and expensive surface reservoirs. In many cases, the storage zones are aquifers that have experienced long-term declines in water levels due to heavy pumping to meet increasing urban and agricultural water needs. Groundwater levels can then be restored if adequate water is recharged. Other benefits of ASR are that there is little or no loss to evaporation or seepage, and there are minimal land requirements.

Soil Aquifer Treatment (SAT)

Municipal and industrial wastewater can be reclaimed through various methods of artificial recharge depending on local conditions. The process of purifying and reclaiming wastewater by allowing it to pass through soil and aquifer is referred as Soil Aquifer Treatment (SAT) system. In SAT systems wastewater or other low quality water is purified by physical, chemical and biological process as it moves through soil and aquifer. The purified water is collected from aquifer with drains or wells in such a way that the movement of the renovated water is confined to certain section of the aquifer.

SAT systems should be designed to provide sufficient underground detention time and travel distance for the wastewater, so that the renovated water will be of required quality. Care should be taken to avoid undue rise of water table below infiltration systems and pollution of groundwater outside the SAT systems. This system is more effective in alluvial areas as granular nature of materials has good filtration characteristics. Consolidated rocks where groundwater occurs in fractures are not suitable for SAT systems. For SAT systems to be more effective, aquifer should have good infiltration properties. To achieve maximum possible quality improvement it is advisable to keep underground detention time at least one month and 100 m undergroundwater travel distance. The coarser the material, the longer the detention time and travel distance is required.

3.3 Combination of Surface and Subsurface Recharge

Several direct-surface and sub surface recharge techniques can be used in conjunction with one another to meet specific recharge needs. Among the advantages of surface application are ease of maintenance, large infiltration areas, and the ability to store water; advantages of subsurface methods include access to deeper aquifers and minimal land requirements.

Subsurface Drainage Collectors With Wells

The recharge potential of recharge basins underlain by a thick zone of low permeability can be greatly increased by subsurface drainage collectors used in conjunction with wells. In such systems recharge water within the basin percolates into the ground where it is collected by horizontal drains between the basin and the zone of low permeability. The drains carry the water to wells that penetrate the zone of low permeability, to replenish an aquifer at depth.

Basins with Pits, Shafts, or Wells

Systems that include a combination of basins with pits, shafts or wells rely on the basins to store water and on the shafts, pits, or wells to dispense water rapidly to more permeable zones beneath the recharge site. Such systems may incorporate large diameter pre-cast perforated concrete cylinders installed below a less permeable zone.

3.4 Indirect Recharge

Indirect recharge techniques of artificial recharge involve: pumping aquifers to induce recharge from hydraulically connected surface waters, modifying aquifers or constructing new aquifers to enhance or create groundwater reservoirs.

Induced surface water recharge

Although induced recharge from surface sources is not always considered to be artificial recharge because it does not normally increase the amount of groundwater in storage, it does allow an increased rate of groundwater withdrawal from an aquifer. In this type of recharge, pumping facilities or infiltration galleries are installed near a gaining lake or stream and the increased pumping and consequent lowering of groundwater levels induces infiltration from the adjacent surface water body.

Induced surface water recharge systems are typically installed near perennial streams that are hydraulically connected to an aquifer through the permeable, unconsolidated deposits that form the stream channel. The quantity of surface water that can be induced to recharge the aquifer varies with amount and proximity of surface water, the hydraulic conductivity and transmissivity of the aquifer, the area and permeability of the stream bed (or lake bottom), and the hydraulic gradient created by pumping. Silt deposition, the primary cause of decreased stream bed permeability, can be avoided by placing pumping facilities near stream reaches having adequate velocity to prevent deposition, such as outer edge of the bend in the stream.

Where aquifer material beneath a stream is thin, collector wells or infiltration galleries can be installed at the base of the aquifer to allow greater drawdowns and a more efficient use of the aquifer than would otherwise be possible. The groundwater pumped from the aquifer contains a mixture of surface water and native groundwater. The surface water is filtered by percolation through earth material, which removes suspended solids, organic material, and bacteria and mineralised groundwater may be improved by dilution with the surface recharge.

Aquifer Modification

Aquifers can be modified by structures that impede outflow or by techniques that creates additional storage capacity. Groundwater barriers have been built in many places to obstruct and detain groundwater flow. Structures are built across the channels of intermittent rivers beneath river deposits. A trench is dug across the riverbed and back filled with relatively impermeable material obtained locally and therefore inexpensively. The barrier extends only partly through the alluvial deposits above the bedrock to create a regulated "back water" effect, which sustains elevated groundwater levels resulting from monsoon rains. Some times construction of small-scale aquifers is good for artificial recharge scheme. Lining a natural drainage channel, filling it with a clean, uniform sand, and covering it with a gravel mulch to protect against evaporation can create an artificial recharge that will supply filtered water under gravity flow with minimum maintenance. For construction of artificial aquifer governing factors are suitable size of drainage catchments, design for proper storage capacity, and reservoir geometry that minimise clogging.

4 SOME CASE STUDIES OF ARTIFICIAL RECHARGE IN INDIA

Many organizations in the country have attempted artificial recharge schemes with assistance of UNDP, UNESCO, Central Groundwater Board Government of India, NGRI, NEERI and other organizations. Some of them are as mentioned below.

4.1 Moti Rayan Project

Moti Rayan Project has been initiated in the state of Gujarat and the area is underlain by the Manchar sediments consisting of reddish conglomerates, sandstones, calcareous grit and clays. Large part of cultivated land of the village has become hard pan which is difficult to plough and has low permeability. Irrigation water, when applied remains on the surface. This is due to salinity and alkalinity problems. Before the monsoon in 1989, 18 check dams, 1 percolation pond and 2 recharge wells were constructed. Recharge through the water harvesting structures resulted in water level rises varying from 1.2 m to 3.65 m. The amount of water recharged was of the order of 66,500 cubic meter. During 1990, 2 more percolation ponds with recharge tube wells and 1 subsurface dam with 4 recharge tubewells were constructed. Although there was less rainfall in village that year, five check dams; two of the percolation ponds and one subsurface dyke did receive some water from storms higher in the catchments.

4.2 Tatiana Project

The work undertaken at Tatiana in the state of Haryana is of induced recharge type. The Tatiana site is situated on the left bank of the Ghaghhar river in the Kurukshetra district of Haryana state. One test well and four observation wells tapping unconfined aquifer of fine to medium sand with some kankar and gravel down to a depth of 24.5 m were constructed. Two observation wells were constructed parallel to the river (20 m from the test well) while two observation wells were constructed perpendicular to the river (50 m from test well). It was estimated that if a well field along a river at a distance of 100 m is developed with spacing of 200 m between two wells and if each well is pumped at a rate of 1728 cum per day, about 97% of recharge is contributed by the river discharge to the aquifer by means of induced recharge and it takes 94 days for river water to reach the aquifer after the commencement of the pumping. It is further calculated that if the clogging from the river bed is removed by scrapping the surface clay/silt layer from the river bed and assuming that the resistance of the river bed is nil, there can be 35% increase in the well discharge without any increase in the drawdown and the river water took only 40 days to reach the aquifer.

4.3 Jeur Sub Basin

The work at this site is an example of percolation tanks and is situated at Jeur in the Ahmednagar District of the state of Maharashtra. More than 7000 percolation tanks were constructed in the state of Maharashtra to store excess run-off for recharging groundwater. A case study by the Central Ground Water Board on the effectiveness of the 12 percolation tanks in Jeur sub basin, Ahmednagar district showed that around 10% of stored water was lost by evaporation while 50% recharged the groundwater reservoir. The study shows that if site of percolation tank is designed properly, the groundwater recharge through tanks can go upto 70%.

4.4 Peint Project

The technique used in this project was rock fracturing. In the tribal area of the western Maharashtra, average rainfall is between 2000 to 3000 mm, yet parts of Nasik and Thane districts and Akola taluka of Ahmednagar district face problem for drinking water. The area is occupied by Deccan Trap. The terrain is mostly undulating with slopes and valleys, which are unfavourable for storage of groundwater and its onward movement. Open wells go dry before March and bore well-drilled upto depths of 60 to 100 m are also not found successful which indicates that there is no recharge in the deeper zone. To find solution to this problem rock fracturing experiment was tried to create artificial joints and fractures below surface for storage of groundwater. The site selected for experiment was in the heart of tribal area of Nasik district near village Saraste, Taluka Peint. The fracturing in the Deccan trap rock at a depth of 5 to 6 m was made by bore hole, blasting in a grid of 10 m × 10 m in 8260 m^2 area. The normal rainfall was spread over the area to effect natural infiltration. The result revealed that rock fracturing is effective, resulting in augmenting storage of groundwater.

4.5 Sabarmati River Bed

A pilot scheme of Soil Aquifer Treatment (SAT) near Ahmedabad in Sabarmati riverbed was taken up jointly by NEERI and PRL. The pilot scheme having a capacity of 5000 cubic meter per day comprises a settling basin, a central recovery well and four piezometers. Sewage is pumped from the nearby river channel for settling of the suspended material and it is allowed to percolate through the soil below. The treated wastewater is then pumped out. The findings indicate that there is a substantial improvement in chemical and biological quality and is technically feasible for treatment of wastewater. Based on the above findings, it is proposed to install a SAT system in a 1.2 km stretch of the Sabarmati river downstream of Vasna barrage and it is expected to produce renovated wastewater to the tune of 55 million cubic meters annually.

5 CONCLUSION

Artificial recharge is an excellent water resources management tool and provides numerous benefits. By adopting artificial recharge techniques a substantial amount of water lost through evaporation in reservoir storages could be saved. A part of flood flow in the rivers could be saved from wastage by storing them in the aquifer to be retrieved when necessary.

The choice of recharge method depends on hydrogeologic characteristics, topography and stream flow, water supply characteristics, legal aspects, availability of land, land use in adjacent areas and public areas. The major constraint for artificial recharge is availability of fresh water and cost involved in construction of structures and their maintenance.

The need of the day is to conserve every drop of water and recharge the depleted aquifers. People's participation is a must to make the scheme of artificial recharge a success. Mere slogan shouting will not carry us far.

References

Herman Bouwer 2000. Integrated water management; emerging issues and challenges. *Agricultural Water Management.* (45): 217–228.

Herman Bouwer 2002. Artificial recharge of groundwater: hydrogeology and engineering. *Hydrogeology Journal.* (10): 121–142.

Huisman, L & Olsthoorn. 1983. *Artificial Groundwater Recharge.* London: Pitman Books limited.

Mishra, G. C. 2000. Methods of Artificial recharge. *Short term course on Groundwater Recharge Assessment and Quality.* Department of Hydrology, I. I. T. Roorkee. (I): AR/GCM 1-15.

National Water Policy 2002. Government of India, Ministry of Water resources 1-9.

Pliakas, F. and et al. 2005. Modelling of groundwater artificial recharge by reactivating an old streambed. *Water Resources Management.* (29): 279–294.

Raju, K. C. B. 1998. Importance of recharging depleted aquifers: state of the art of artificial recharge In India. *Journal of Geological Society of India.* (51): 429–454.

Shroff V. N. and *et al.* 1993 Recharging underground water through dugwells .*Journal of Indian water works association* (4) 241-243.

Singhal, D. C. 2000. Investigation techniques for artificial recharge. *Short term course on Groundwater Recharge Assessment and Quality.* Department of Hydrology, I.I.T. Roorkee. (I):AR/DCS 1-19.

Takashi Asano (ed). 1985. *Artificial recharge of groundwater.* Boston: Butterworth Publishers.

Todd, D. K. 1959. *Groundwater Hydrology.* New York: John Wiley & Sons.

19

Hydrogeological Investigations in Pathri Rao Watershed, Uttarakhand, India Using an Integrated Isotopic Technique

Laxmi Kant Kachhwal[1], D.C. Singhal[2], Bhishm Kumar[3], Vinay K. Sharma[4], S.K. Verma[5] and M. Israil[6]

[1,2]Department of Hydrology, Indian Institute of Technology Roorkee, India

[3]National Institute of Hydrology, Roorkee, India

[4]Department of Earth Sciences, Indian Institute of Technology Roorkee, India

[5]National Institute of Hydrology, Roorkee, India

[6]Department of Earth Science, Indian Institute of Technology Roorkee, India. Phone: 91-1332-285078; Fax: 91-1332-273560; E-mail: mohdfes@iitr.ernet.in

ABSTRACT

Tritium tagging technique and environmental tritium measurements have been integrated to determine groundwater recharge due to precipitation, recharge zone and possible groundwater flow direction in Pathri Rao watershed of Himalayan foothills region, Uttarakhand, India. On the basis of soil type four sites were selected for the recharge estimation due to monsoon precipitation using tritium tagging technique. Groundwater recharge due to precipitation is estimated by monitoring the vertical movement of injected tritium. The method is based on the assumption that the soil water in the unsaturated zone moves vertically downward as discrete layers. Water added on the surface either as precipitation or irrigation will move downwards by pushing the older water beneath and this in turn will push the still older water further below, thereby the water from the unsaturated zone is added to the groundwater reservoir. This flow mechanism is known as piston flow. The estimated recharge varies from 9% to 29% in the area. A linear relationship is obtained between the sand content and estimated recharge at these sites. Environmental tritium dating of six groundwater samples collected, from the shallow and intermediate aquifers, has indicated the relative age of shallow and intermediate aquifer in the area and possible recharge zone. The results have been used to define the interconnections of shallow and intermediate aquifers, and their possible recharge zones. The study is useful in understanding the aquifer system, their possible interaction and groundwater flow direction.

1 INTRODUCTION

Hydrogeological investigation plays an important role in groundwater resource evaluation, conservation and water balance estimation. Groundwater recharges have two components, first is vertical component of recharge, which includes recharge from precipitation, and from irrigation return flow and second is horizontal component of recharge through subsurface flow due to natural hydraulic gradient.

The conventional method for recharge estimation is based on the water level fluctuation monitored between pre-monsoon and post-monsoon period. The Central Ground Water Board (CGWB), India, based on data from some representative basins, has suggested specific yield values for common water bearing formations, which can be used in areas for which pumping test data are not available (CGWB, 1997). However, the estimation of specific yield is very critical in computing groundwater recharge from water level fluctuation data. Usually, the specific yield value of the aquifer is estimated from pumping test data. In order to have realistic values of specific yield in unconfined aquifers, long duration pumping tests are necessary. Alternatively groundwater recharge can be estimated by monitoring the vertical movement of artificially injected radioactive tracer tritium. This is known as tritium tagging technique. This method is based on the assumption that the soil water in unsaturated zone moves vertically downward as discrete layers. This flow mechanism is known as the piston flow (Zimmerman *et al.*, 1967 a, b).

Tritium is the most commonly employed radioisotope used to determine the recent age (less than 50 years) of groundwater. "Modern" groundwater is recharged within the past few decades and is thus, part of an active hydrological cycle.

In regions where direct and rapid infiltration of rain water occurs, the composition of the groundwater will have same isotope character as that of precipitation. The analysis of stable isotopic (oxygen) content, together with that of dating data gives a reliable 'fingerprint' enabling identification of each group of water of different origins.

Present paper discusses the amount of recharge due to precipitation using tritium tagging technique. Relative age of groundwater has been determined using environmental tritium dating of water samples. The results have been integrated to define the aquifer system and their possible interconnection in the study area.

2 STUDY AREA

The study area is located between latitudes 29° 55′ 48″ to 30° 02′ 24″ North and longitudes 78° 00′ 00″ to 78° 06′ 00″ East falling in Pathri Rao watershed of District Haridwar of Uttarakhand state, covering an area of approximately 51 km^2 shown in Fig. 1. The river Pathri Rao is a seasonal river, originates from Siwalik foothills, and flows towards south-west direction.

Geologically the area comprised Siwalik rocks and alluvial deposits. Towards north, the Siwalik sedimentary rocks are composed of indurated to compacted clastic sediments exposed

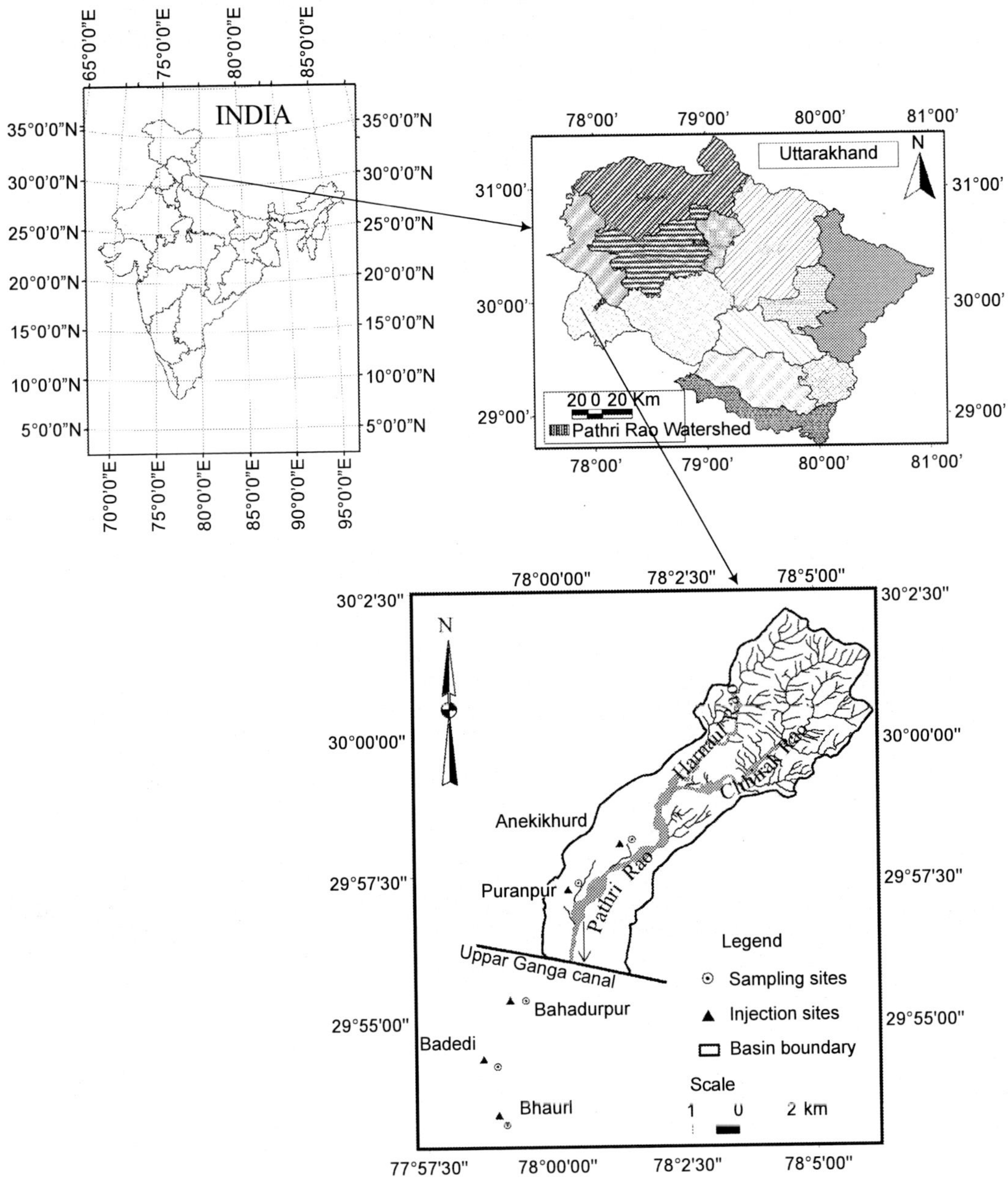

Fig. 1 Location map the study area showing tritium injection sites and groundwater sampling sites along with basin boundary.

in Haridwar district and represent a over 6000 m thick sequence, consisting of interbedded mudstones, sandstones, conglomerates and subordinate marls (Kumar *et al.*, 1989).

The formations occurring to south of the Siwalik hills are alluvial fan deposits of recent age. The alluvial fan deposits (known as piedmont zone) are made up of assorted sands and gravel associated with occasional clays.

The depth of water table monitored from the observation wells in and near the study area ranges from 3.3 m to 32 m below ground level. The occurrence of shallow water table near Lower piedmont zone can also be attributed to the presence of clay lenses at shallow depth, which act as confining layer. Reason for occurrence of deep water levels (15 m to 32 m (bgl)) in the Bhabhar zone seems to be due to high permeability of assorted and unconsolidated deposits due to general absence of confining clay layers at shallow depths.

3 METHODOLOGY

Three isotopic techniques, viz. tritium tagging technique, environmental tritium dating technique, and stable isotope of oxygen (O^{18}) analysis have been used to estimate vertical recharge due to precipitation, age of groundwater and possible recharge zones in the study area. The data of each technique is discussed below.

3.1 Tritium Tagging Technique

Tritium tagging studies have been carried out at four sites (Fig. 1), selected as representative of the area on the basis of soil type, and regional topography. The groundwater recharge has been estimated by monitoring the vertical movement of injected tritium. The position of the injected tracer is indicated by a peak (maximum) in tritium activity versus depth plot. The radioactive tritium obtained from Bhabha Atomic Research Centre (BARC), India with specific activity of 40 micro Curie/cc were injected in five holes placed in a circular geometry at each site on 23 July 2004. The soil samples were collected at the time of injecting the tritium (pre-monsoon) and after four months in post-monsoon in December, 2004. The soil samples at each 10 cm level up to a depth of 2.5 meters were collected from each site. Volumetric moisture content of each soil sample was estimated in the laboratory at National Institute of Hydrology, Roorkee. The liquid scintillation counter (LSC) has been used to measure the tritium activity of soil moisture extracted from each soil sample mixed with 10 ml of cocktail (whose background count was already measured with LSC) in the scintillation vials. The vials containing 1 ml of soil water and 10 ml of cocktail were used in LSC for each sample and count rate (counts per minute) for each sample was recorded. These count rates were corrected for background counts in order to get net tritium counts per minute. The counting rates so obtained are plotted with depth for each site. The shift in tritium peak is estimated by using the depth of injection and centre of gravity of peak position. The shift in tritium peak and volumetric moisture content in pre-monsoon and post-monsoon season is shown in Fig. 2. The amount of recharge during the time interval of tritium injection (pre-monsoon) and sampling (post-monsoon) can be estimated by multiplying the tritium peak shift and average volumetric moisture content in the tritium peak shift region (Zimmerman *et al.*, 1967 a, b). Mathematically the equation for the estimation of percentage of recharge to groundwater can be written as:

$$R = \theta_v d(100/p)$$

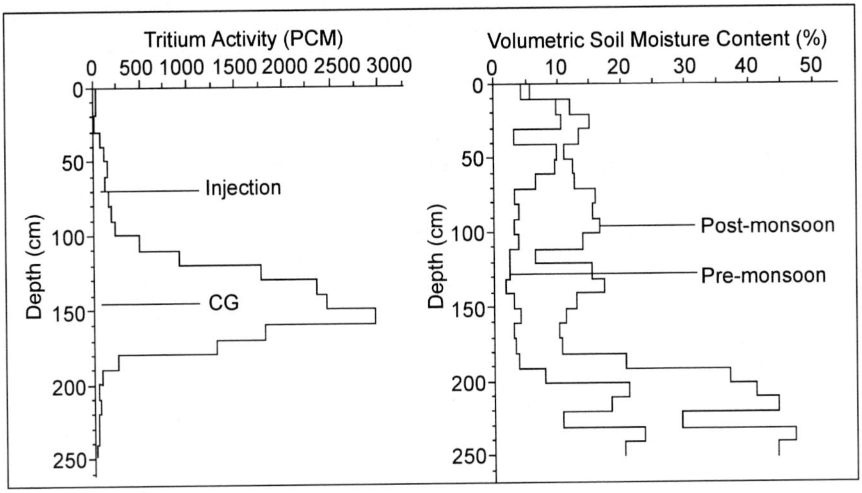

Fig. 2 Movement of Tritium peak and soil moisture at Anekikhurd site (TIS-4).

Here, R = the percentage of recharge to groundwater, θ_v = the effective average volumetric moisture content in tritium peak shift region, d = the shift of tritium peak in cm, and p = precipitation/irrigation in cm.

The particle size distribution analysis has also been done of the soil samples collected from each site. Linear relationship between the average sand content and recharge per cent shown in Fig. 3, is obtained in the area.

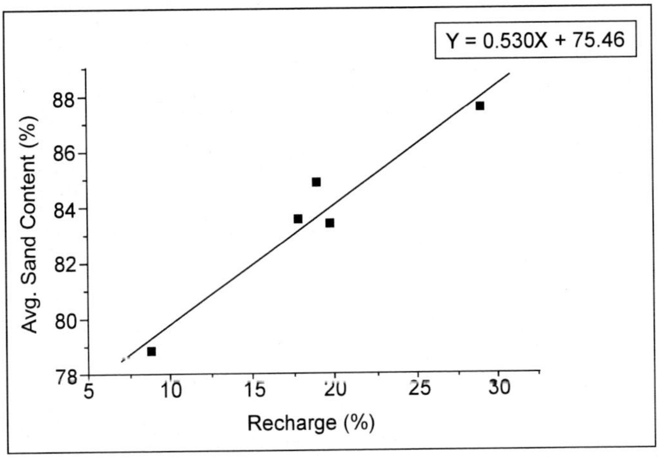

Fig. 3 Relationship between average sand content and percentage recharge rates measured by tritium tagging technique.

3.2 Environmental Tritium Dating Technique

Six groundwater samples were collected from shallow and deep aquifers during pre-monsoon period of 2004, for environmental tritium dating. The locations of these sampling sites are shown in Fig.1. Tritium is a short-lived isotope of hydrogen with a half life of 12.32 years. It exists in water molecule as $^1H^3HO$ and so is the only isotope that actually dates the groundwater. Kaufman and Libby (1954) first recognized the potential for dating groundwater with cosmogenic tritium. At present natural concentration in atmosphere is about 10-12 Tritium Unit (TU). Dating of groundwater by decay of tritium is based on the assumption that the tritium input into a groundwater is known and that the residual tritium measured in a groundwater is the result of decay alone. The decay equation can be written as:

$$C_i = C_i \exp(-\lambda t) \quad ...(1)$$

Here, C_i is the initial tritium concentration (expressed in TU), C_t is the residual activity remaining after decay over a time t, λ is the decay constant, $\lambda = 0.693/t_{1/2}$ (= 0.05617 for Tritium) and $t_{1/2}$ is the half life (12.32 years).

In order to measure the activity of environmental tritium in the groundwater samples, 13 gm Canberra-Packard Pico-flour LLT (Low Level Tritium) cocktail was added in 8 ml of each enriched sample, along with these samples two vials containing STD-B and STD-C, three vials containing tritium free water were also prepared for counting too. These standards are used to calculate the tritium activity in the groundwater samples while the activity recorded in the tritium free water is treated as background and subtracted from the measured activity in each sample. All these vials are placed in counter tray according to the preset protocol. The measurement is done by measuring its beta activity through the beta counting using Ultra Low Level Liquid Scintillation Spectrometer (ULL LSC, 1220 Quantulus Model), a facility available in Nuclear Hydrology Laboratory at the National Institute of Hydrology (NIH), Roorkee. The standards and T-free vials were put in a spatially distributed manner. The whole counting was done for 10 cycles and each vial is counted for 40 minutes in each cycle. The cycling helps in discarding any particular data, which is not statistically acceptable.

The count rate (cpm = counts per minute) so obtained along with associated statistical error was recorded and further used to calculate the tritium value in Tritium Units (TU). The tritium values so obtained in TU were used to calculate uncorrected age of groundwater samples using simple exponential decay equation as given in Eq. (1).

The maximum tritium activity of 9.01 was recorded at one site at some deeper depth, therefore, the present-day atmospheric activity, i.e. 10 TU was assumed as an initial activity of tritium in precipitation.

3.3 Stable Isotope O^{18} Analysis

Oxygen and hydrogen isotopes are commonly used isotopes to determine the elevation of recharge zone for a particular groundwater. In regions where direct and rapid infiltration of rain occurs, the composition of the groundwater will have same isotope character as that of precipitation. Three samples which were used for tritium dating also used to find out their δO^{18} values. Out of these one sample is of shallow aquifer and other two are of deeper aquifer. These samples were analyzed through the "Iso Prime Dual Inlet Isotope Ratio Mass Spectrometer (GV)" installed at Nuclear Hydrology laboratory, the National Institute of Hydrology, Roorkee. Table 1, shows the estimated values of tritium in TU and percent of δO^{18} from groundwater samples collected from the study area.

Table 1 Tritium concentration and corresponding tentative age, electrical conductivity, and δO^{18} values at different sites.

Sample No.	*Sampling site/ Type of aquifer*	*Tritium Value (TU)*	*Relative age (years)*	*Electrical Conductivity (µS)*	*δO^{18} values (‰)*
WS-5	Anekikhurd/Shallow	7.37	5	450	−6.9
WS-6	Anekikhurd/Intermediate	1.28	40	3000	−6.7
WS-4	Puranpur/Shallow	4.60	14	800	
WS-3	Bahadurpur/Intermediate	6.38	8	350	
Sample	Sampling	4.62	14	600	
WS-1	Bahuri/Intermediate	9.01	2	650	−6.6

4 DISCUSSION AND CONCLUSIONS

The results obtained from these techniques gives a view about possible recharge zones for shallow and deeper aquifers at different sites in the study area. Age of groundwater means time taken by water to travel the path from the recharge zone of the aquifer of the sampling location. The water sample, with high tritium concentration or smaller age, is recharged recently from nearby area, whereas the water that has low tritium concentration or old age is recharged from a location far away from the sampling location.

The schematic diagram shown in Fig. 4, provides a pictorial representation of the flow regime in the study area deduced from the present analysis.

The tritium activity at site Anekikhurd in shallow aquifer was recorded 7.37 ± 0.33 TU (Age ≈ 5 years) indicating the slow rate of recharge due to comparatively deeper depths of aquifer. However, as the site located near river Pathri Rao, some recharge is expected from the riverbed. It has little depleted values of δO^{18} (–6.9‰) than the local precipitation (–6.6‰). The lower tritium value, i.e. 1.28 ± 0.27 TU (Age ≈ 40 years) at the deeper depth (deeper aquifer) at the site Anekikhurd clearly indicates that the second aquifer at this site is not at all connected with the shallow aquifer. This is also supported by the electrical conductivity (EC) value (3000 µS), observed in the deeper aquifer. It appears that recharge zone for this aquifer is located quite away (18-20 km) but not having higher altitude, as the δO^{18} value is –6.7%.

At Puranpur village tritium value is measured only for shallow aquifer. The observed tritium value is 4.6 TU (~14 years) and estimated vertical recharge by tritium tagging technique is 8.94%. In this area shallow aquifer has high clay (21.2%) content too. So this aquifer is not getting rapid vertical recharge of water and it is also possible that this is getting fair recharge from the water which is recharging the shallow aquifer at Anekikhurd village and flowing down towards Puranpur village because of altitude difference. During the movement from Anekikhurd to Puranpur in shallow aquifer some local salinity may dissolved and its EC is modified from 450 µS to 800 µS.

On the other hand, at the Bahadurpur site, measured environmental tritium value in intermediate aquifer is 6.38 TU (~8 years) and estimated vertical recharge rate in shallow aquifer by tritium tagging technique is very high 29.01%. The water, which is recharging the shallow aquifer, is also connected with deeper aquifer and the vertical flow rate 260 cm/year estimated from tritium tagging technique. The water from shallow aquifer will take around 9 years to join deeper aquifer while the tritium value indicates the age around 8 years.

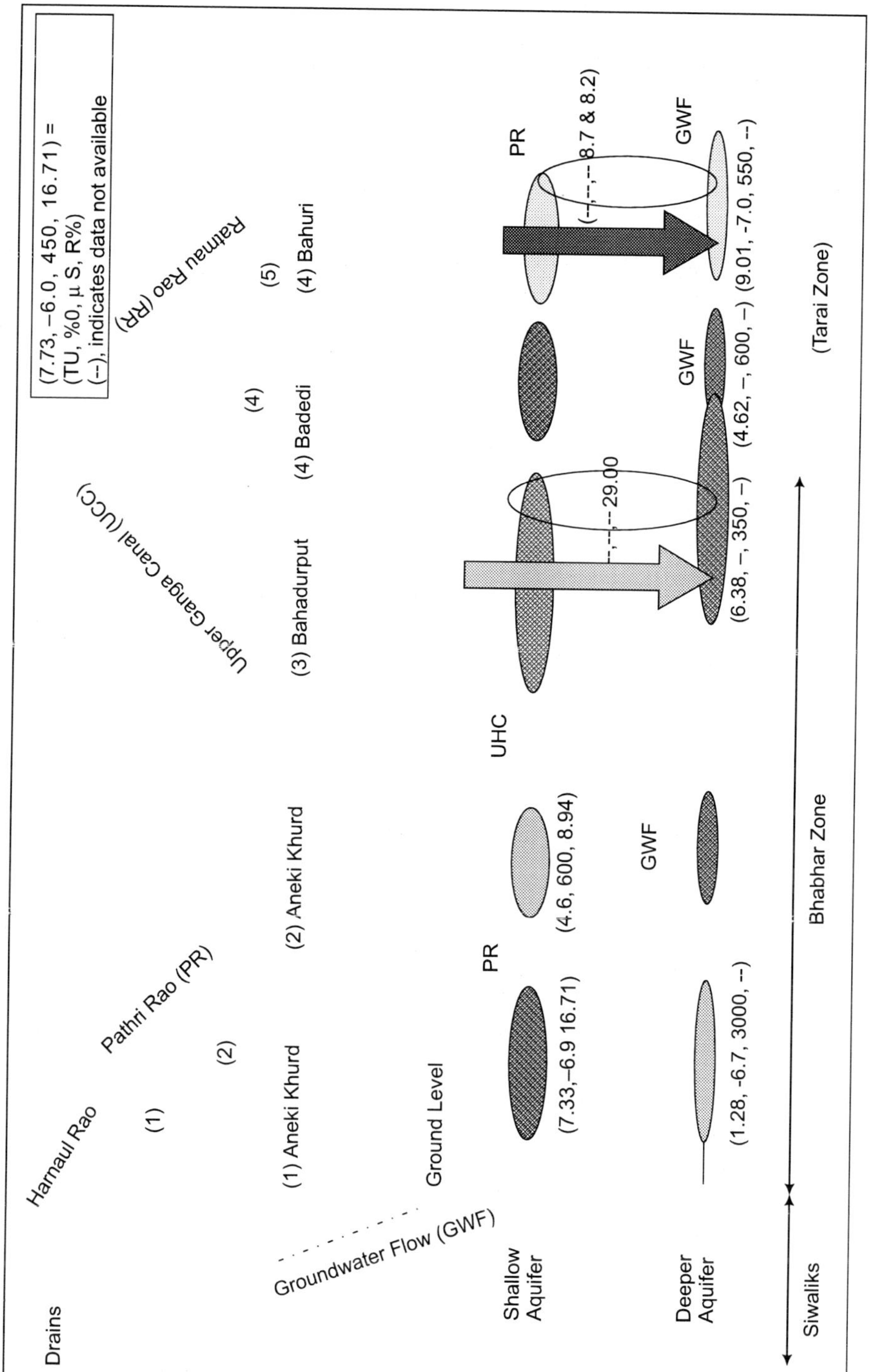

Fig. 4 Schematic section representing groundwater flow direction, recharge sources and aquifer interconnections

At Bahuri site, the tritium value is comparatively high 9.0 TU (2 years) in deeper aquifer which is very near to natural value in precipitation (10 TU). It indicates that it is rapidly recharged from shallow aquifer. This is quite possible as there is very shallow water table (<2 m) and the shallow aquifer is well connected with the deeper aquifer as confirmed by the electrical resistivity sounding data. The δO^{18} value is –6.6‰ which is same as of local water.

References

CGWB 1997. Groundwater resources estimation methodolog, report of the ground water resources estimation committee, Ministry of Water Resources, India.

Kumar, R. & Nanda, A.C. 1989. Sedimentlogy of the middle Siwalik subgroup of Mohand area, Dehradun valley, India, *Jour. Geol. Society of India*, pp. 597–616.

Zimmerman, U., Ehhalt, D. & Munnich, K.O. 1967a. Soil water movement and evapotranspiration: changes in isotopic components of soil water. *In: Isotopes in Hydrology Proc. Symp. Vienna*, pp. 567–585.

Zimmerman, U., Munnich, K.O. & Roether, W. 1967b. Downward movement of soil moisture traced by means of hydrogen isotopes. In: Glenn ES (Ed.) Isotope techniques in the hydrologic cycle, *American Geophysics Union, Geophysics Monograph*, Vol. 11, pp. 28–36.

Sustainable Lei-Drainage Groundwater of Rawalpindi-Islamabad (North Pakistan) and the Potentials of Recharge

20

A.H. Malik

Department of Environmental Science, COMSATS Institute of Information Technology, CIIT Abbottabad, Pakistan

ABSTRACT

The Lei-Drainage-System of Rawalpindi-Islamabad in North Pakistan is about 210 km^2 area. Aquifer depletion, pollution, floods, droughts, urbanization, variability of rainfall, evaporation losses are the main challenges. Monsoon yields about 60% of the total annual. Annual average is about 1,200 mm/a, 30% flows off, 56% evaporates and 14% recharges the groundwater. The Basin is recharged with 36 million m^3/a and about 35 mm^3/a of groundwater flows into it from the neighboring Kurang basin. Water table drop is 1.40 m/a and more than 44 mm^3/a of groundwater is being over-mined annually. In case of Islamabad, new environmental techniques can still be applied to the areas under construction planning. The naturally sealed areas can be used for the construction of houses and buildings, etc. but not the infiltration areas. Since evaporation losses in dry period would be high, methods of recharge would be restricted to involve little surface exposure to sun/wind. Surface water spreading is fit for direct recharge in the periods of low evaporation. Small check or diversion dams can also contribute towards groundwater recharge. The remaining old ancestral village irrigation areas should be retained for groundwater recharge. Recharge system could integrate dried-lined-dug-wells. Broadly conceived measures would improve groundwater both qualitatively and quantitatively.

1 INTRODUCTION

The Lei catchment area lies in the cities and suburbs of Islamabad and Rawalpindi between latitude 33° 00′ and 34° 00′ north and longitude 72° 45′ and 73° 30′ east. The stream Lei and its tributaries rise in the Margalla Hills bordering Islamabad to the north. It is a fairly typical mountain river, or hill torrent producing dendritic patterns of steep gradient in limestone in its northern parts and meandering designs of gentle gradient in its southern alluvial parts in the heart of Rawalpindi (Fig. 1) after being joined by its main tributaries. The Lei-River-Basin catchment is about 210 km^2 having annual average precipitation value of about 1200 mm, of which 30% (358 mm) flow off, 56% (674 mm) evaporate and 14% (168 mm) recharge the local aquifer (Malik 1999a, b). The population of both the cities of Islamabad and Rawalpindi is now about 3 to 4 million, out of which about half of the urban population resides in the poor to

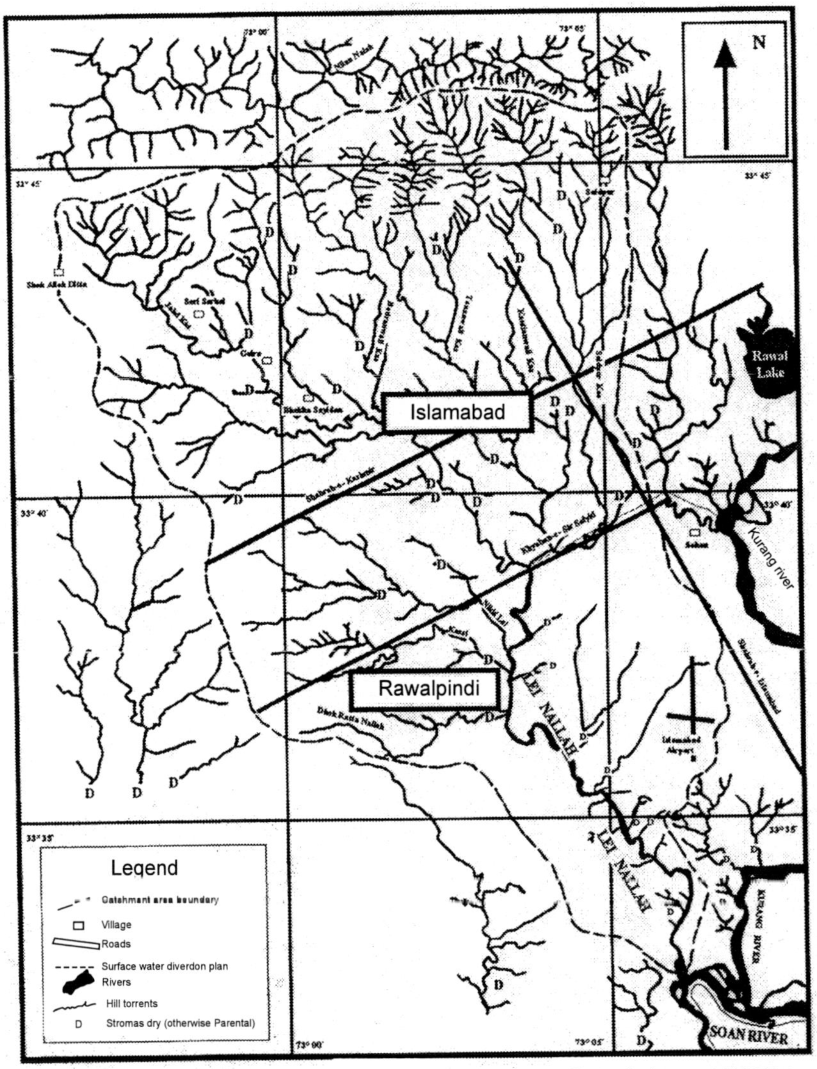

Redrawn: Malik, A.H. (1995) 0 1 2 km Data: Lei-report (1987) Islamabad-Rawalpindi guide map (1981)

Fig. 1 The Lei Drainage System (Islamabad-Rawalpindi).

middle class family areas of Rawalpindi (Malik, 2002). The stream of Lei begins in the Margalla Hills at a height of about 1,000 m (above NN) to the north of Islamabad and drains into the Soan river at a height of about 415 m. Floods, droughts and water pollution in the Lei River Basin of both the cities are mainly caused by inadequate management of water, settlement and urbanization patterns, institutional and regulatory arrangements (Malik 1998).

Four streams – Saidpur Kas (also known as Upper Lei), Kanitanwali Kas, Tenawali Kas and Bedranwali Kas – rise in the Margalla Hills, cross Islamabad, come together upstream of Khyaban-I-Sir Syed and constitute the main Lei (Fig. 1). Another tributary known as Nikki Lei joins it from the west side about 3.2 km downstream of the above mentioned confluence. About 20 other small streams draining the city join it from both sides. Finally, it falls into the "The river Soan" after crossing the G.T. Road, where it approaches Rawalpindi from the Lahore side near the building of the Fauji Foundation. The Lei-Basin on the whole is pear-shaped (Fig. 1), but the whole catchment along the Northern Margalla Hills is narrow and steep and leads to flashy run-off and rapid drainage.

2 LEI-DRAINAGE SYSTEM AND DRAINAGE BASIN CHARACTERISTICS

Total length of the Lei is about 43 km. The catchment area is bounded to the north by the Margalla Hills, to the east by the Kurang-Basin and to the west by a low ridge bearing the Pakistan railway track. Lei catchment drains an area of about 210 km^2, of which 55% lies in Islamabad i.e., up to the Khyaban-i-Sir Syed and the remaining 45% lies in Rawalpindi city and Cantonment (Malik, 2002). The Lei-Basin is connected with the neighbouring basin of the river Kurang underground. It is being recharged with groundwater internally. The surface drainage-divide in the north and north-east affects only surface water, whereas the divide in the west affects groundwater as well.

The original Lei river bed and the neighboring meandering banks and flood plains are being disturbed due to illegal encroachments specially along the floodplains. Only about 35% of the total solid waste is being collected properly, the rest ends up in sewerage pipes and streams. New settlement patterns (urbanization) since 1961 of Islamabad and Rawalpindi have reduced the groundwater infiltration areas, thus disturbing the groundwater quality and reducing its quantity (Malik 2002). There is overexploitation of groundwater resources and overuse of drinking water in the posh areas in Islamabad even to run the flush system, for gardening, laundry, fire extinguishing purposes, while the people of the low earning groups living in the un-regulated areas are not getting enough water even for drinking purposes. The same is the case if it is flooding, inundation is more in Rawalpindi areas (Malik 2002). Out of the 160 million liters of waste water produced in Rawalpindi and 140 million liters per day in Islamabad, only about 25 million liters are being treated daily, that too also partially and mechanically (Malik 2002). This treated/untreated wastewater is being thrown into the natural streams thus causing increase of water contamination for the downstream areas. The industrial units lying according to Master Plan of 1960, in the southern part of Islamabad are also contributing towards flood peaks and polluting the local aquifers especially for Rawalpindi (Malik 2002). Artificial erosion and mining of limestone and other rock extraction is also contributing negatively.

The original Lei bed and the neighboring meandering banks, flood plains should be kept free from settlements and to establish the proper system of solid-waste collection and management of the area. The regional city planning (Master Plan) which unfortunately is leading to sealing of more and more infiltration areas or pollution of the naturally available local aquifers should be reformed according to the latest environmental techniques available. The industrial zones lying between both the cities should be transferred to other geological suitable sites to save the local aquifers.

The watershed management should commence from upper upstream reaches and should progressively proceed towards downstream areas integrating also the pollution free groundwater recharging zones for local tube wells and for other confined aquifers. The remaining village irrigation settlements in Islamabad ought to be protected as a contribution towards groundwater recharge and its quality, thus also consuming the floodwater load for groundwater recharge and irrigation (Malik 1997a). Application of world wide accepted rain and floodwater harvesting and water conservation measures, especially along the still un-urbanized Margalla Hills and other areas should be applied (Prinz and Malik 2002). The stored rain and floodwater (even from rooftops) can be used for laundry, sanitation and gardening or be brought to the open unlined natural streams for groundwater recharge (Malik 1977a, b). It is high time to consider even official-subsidiary for the rain and flood water harvesting and water conservation models for households, industries, mosques, churches, schools and other installations to introduce and demonstrate models for multiplications (Prinz and Malik 2002).

The complex jurisdiction of the Lei-Basin (Federal Government for Islamabad, Punjab Government for Rawalpindi) due to administration of 6 or more different sub-authorities for different sub-sectors of the same catchment, the development work of one authority cannot be environmentally sustainable for the downstream or upstream environment. Governance is characterized by a lack of transparency, lack of inclusion of the local community, NGOs and quite often, there is a poor coherence between government system, legal system and civil society activities (Malik 2002). There is a need of creation of a new Lei-Basin Development Authority (LBDA). This body should be responsible for the finalization of projects forwarded from different concerned sub-authorities like Capital Development Authority (CDA), etc., for project implementation and to look after the long term water resources and other environmental interests of the whole comprehensive integrated Lei-River-Basin sustainable development (Malik 2000).

3 MAIN WATER RESOURCES PROBLEMS OF THE STUDY AREA

The main water resources concerning problems of the study area are floods, droughts, groundwater over-mining and water pollution and these are mainly caused by (Malik 2005):

- inadequate water resources management and long-term sustainable environmental development (Malik 1998, 2000),
- inadequate settlement and urbanization patterns,
- inadequate institutional and regulatory arrangements,
- inadequate involvement of NGOs, mosques, churches, schools, etc.
- illegal encroachments along the floodplains and groundwater recharging zones,
- inadequate solid waste and waste water management,
- over exploitation of groundwater resources,
- use of drinking water to flush the toilets, gardening, laundry, etc.,

- locations of industrial unit lying between the southern part of Islamabad and northern part of Rawalpindi thus polluting the Rawalpindi aquifers (Malik 2002),
- artificial erosion and mining of limestone and other rocks in the catchment area,
- inadequate quality and quantity control of private groundwater exploitation,
- inadequate official incentive for rain/flood water harvesting and water conservation measures (Prinz and Malik 2002),
- inadequate forest management and its destruction through fire and cutting of trees,
- complex jurisdiction of the Lei-Basin (Federal Government for Islamabad, Punjab Government for Rawalpindi).

The original Lei river bed and the neighbouring meandering banks and floodplains are being disturbed due to illegal encroachments. Only about 35% of the total solid waste is being collected properly, the rest ends up in sewerage pipes and streams. New settlement patterns (urbanization) since 1960 in Islamabad and Rawalpindi have reduced the groundwater infiltration areas, thus disturbing the groundwater quality and reducing its quantity (Malik 2002). There is overexploitation of groundwater resources and overuse of drinking water in the mostly posh areas in Islamabad even to run the flush system, for gardening, laundry, fire extinguishing purposes, while the people of the low earning groups living in the unregulated areas are not getting enough water even for drinking purposes. The same is the case if it is flooding; inundation is more in Rawalpindi areas (Malik 2002). Out of the 160 million liters of wastewater produced in Rawalpindi and 140 million liters per day in Islamabad, only about 25 million liters are treated daily, and that is also only mechanically (Malik 2002). This treated/untreated wastewater is being thrown into the natural streams thus causing increase of water contamination for the downstreamers. According to Master Plan of 1960, the industrial units located in the southern part of Islamabad are also contributing towards flood peaks and polluting the local aquifers especially for area around Rawalpindi (Malik 2002). Artificial erosion and mining of limestone and other rock extraction is also contributing negatively. There is an increasing encroachment on the waterways, especially in Rawalpindi area (Malik 2002).

4 WATER RESOURCES CHALLENGES OF THE STUDY AREA

4.1 Shape of the Lei-Catchment System

The Lei-Catchment System is pear-shaped, individual small sub-catchments along the limestone hills are narrow and steep and lead to flash run-off, flood and rapid drainage (Fig. 1).

4.2 Seasonal Precipitation Variability

The Lei-River-Basin catchment is about 210 km^2 having annual average precipitation value of about 1200 mm, of which 30% (358 mm) flow off, 56% (674 mm) evaporate and 14% (168 mm) recharge the local groundwater (Malik 1997a, b, 1999a). About 60% of the precipitation takes place in two to three monsoon months (Fig. 2). Contribution from the perennial streams, presence of perennial streams along the eastern side as compared to runoff flow after only rainy seasons in the western side.

4.3 Geology and Topography

Soils comprised loess, widespread in the catchment, are not very good for infiltration. The Margalla hill slopes of the streams are steep and generate quick and maximum run-off and the limestone aquifer can retain the water. Margall hill limestone rocks can contribute a lot to store rain and flood water underground to compensate the quality and quantity of ground and surface water of the whole catchment area.

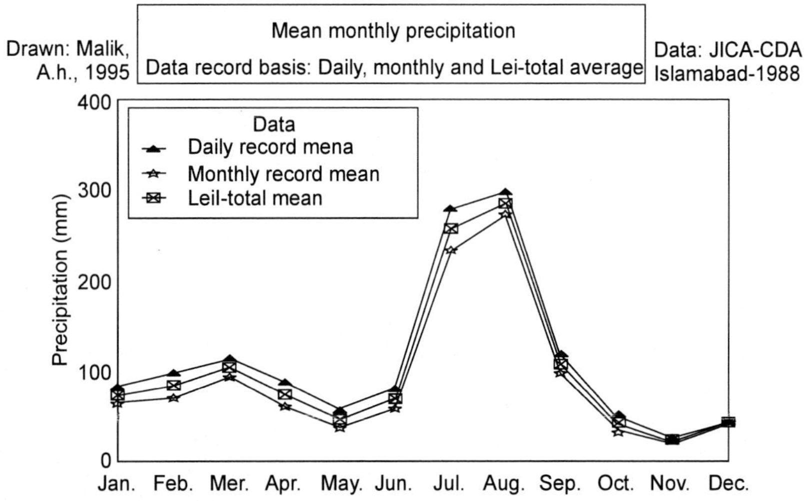

Fig. 2 Mean monthly precipitation in the Lei-drainage basin.

5 MAJOR RECOMMENDATIONS

The water resources concerning problems of the study area, i.e. floods, droughts, over-mining of groundwater, and water pollution problems can be solved with the following measures (Malik 2005):

- sustainable long-term water resources development and management of the area (Malik 1998, 2000),
- environmentally sustainable settlement and urbanization patterns planning (Malik, 1998, 2000),
- institutional and regulatory arrangements,
- involvement of NGOs, mosques, churches, schools, etc.
- floodplains and groundwater recharging zones should be kept free from encroachments and pollution,
- provision of proper solid waste and waste water management,
- check on overexploitation of groundwater resources,
- use of harvested water for flush system, gardening, laundry, etc.
- to establish a Lei-Basin Development Authority (LBDA),
- to dislocate industrial units to other geologically impermeable sites,

- to control artificial erosion and mining of limestone and other rocks,
- to control quality and quantity control of private groundwater exploitation,
- to introduce official incentive for rain/flood water harvesting & water conservation measures (Prinz & Malik 2002) for new buildings (construction by laws),
- to do forest management and to check fire and cutting of trees in the catchment,
- to reserve groundwater recharge zones (Malik 1977a, b) according to the geological suitability (Malik 1999a, b),
- remaining village irrigation settlements ought to be protected as a contribution towards groundwater recharge (Malik 1977a, b) and its quality (Malik 1997a, b, 1999a, b),
- curb the straightening of courses of natural streams and to learn from experience of other countries,
- extra drainage facility for extra portion of water being consumed from outer catchment systems (Simly + Rawal + Khanpur) in the catchment of the Lei Basin,
- to construct more dams in the catchment, specially along the Margalla Hills,

The faulty regional city planning (Master Plan) which unfortunately is leading to sealing or pollution of the naturally available local aquifers should be reformed according to the latest environmental techniques available. The original Lei bed and the neighboring meandering banks, floodplains should be kept free from settlements and to establish the proper system of solid-waste collection and management of the area. The regional city planning (Master Plan) which unfortunately is leading to sealing of more and more infiltration areas or pollution of the naturally available local aquifers should be modified in accordance with the latest environmental techniques available. The stored rain and floodwater (even from rooftops) can be used for laundry, sanitation and gardening or be brought to the open unlined natural streams and artificial canals for groundwater recharge (Malik 1977a, b, 1999a, b). It would also save huge amounts of drinking water. The industrial zones lying between both the cities should be transferred to other geologically impermeable suitable sites to save the local aquifers from further pollution.

The watershed management should commence from upper upstream reaches and should progressively proceed towards downstream areas integrating also the pollution free groundwater recharging zones for local tube wells and for other confined aquifers. The remaining village irrigation settlements in Islamabad ought to be protected as a contribution towards groundwater recharge and its quality (Malik 1999a, b), thus also consuming the flood water load for groundwater recharge and irrigation (Malik 1977a, b). Application of world-wide accepted rain and flood water harvesting and water conservation measures (Prinz and Malik 2002), specially along the still un-urbanized Margalla Hills and other still not constructed areas should be applied. It is high time to consider even official-subsidiary for the rain and flood water harvesting and water conservation (Prinz and Malik 2002) for households, industries, mosques, churches, schools and other installations to introduce and demonstrate models for multiplications. There is a need of more dams specially along the Margalla Hills, as seen from the example of Rawal dam. Three dams have already been proposed by the Punjab Small Dams Organisation as the anti-flood measures and these will also compensate the declining water tables of the area, and improve the ground and surface water quality. Know-how and manpower can be provided locally.

Groundwater recharge in the study area must be based on (Malik, 2005), water surface covered (only in the period of high evaporation rate) methods and water surface un-covered methods (in the period of low evaporation), not only to compensate flood peaks and droughts, but also to compensate declining water table and to improve the water quality of the area. To select groundwater recharging areas, the methodology can be further confirmed by integrating the following steps:

- Watershed analysis, land cover and land use analysis,
- Identification of natural depressions (basins), and
- Identification of permeable and impermeable surfaces.

Potential Recharge locations of the catchment can be classified as: (Malik, 1997a, b, 2005)

- Margalla Hills
- Other permeable areas of the Lei-Basin except few bedrock impermeable pockets.

6 GENERAL IMPORTANCE OF GROUNDWATER RECHARGE AND ECONOMY

According to Fox (1999), the three commonly used methods for aquifer recharge are important. Recharge basins are the most common low technology method that require large amounts of land. Direct injection wells allow for injection into confined aquifers and require high technology water pre-treatment. Vadose zone injection wells are an emerging technology that provides some of the advantages of both recharge basins and direct injection wells unconfined aquifer with sufficient storage capacity. Direct injection wells can inject water directly into unconfined aquifers or confined aquifers. When unconfined aquifers are unavailable, direct injection wells are the only alternative for groundwater recharge and are capable of simultaneously injecting water into several aquifers. However, direct injection wells are expensive, require advanced pretreatment technology and advanced technology for maintenance. Therefore, direct injection is not a viable option when low technology solutions are desired. The major costs associated with recharge basins are the required land and the conveyance system to deliver water to the recharge basins. Therefore, it is often desirable to locate recharge basins near water conveyance systems where land that is located in floodplains might be available.

Groundwater recharge should be considered an integral part of every water resource planning (Fox, 1999). The advantages of groundwater recharge far outweigh the potential disadvantages of groundwater recharge. Groundwater recharge provides greater flexibility in water resources planning and groundwater banking systems provide economic incentives for groundwater recharge projects. Prevention of land subsidence is difficult even when systems for groundwater banking have been developed. As with most water resources projects, the development of infrastructure for the conveyance of water for groundwater recharge projects can be prohibitively expensive. Consideration of groundwater recharge in the initial stages of water resources planning will reduce large expenditures by optimizing the location of projects. According to Asano (1999), the purposes of artificial recharge of groundwater have been practiced in order to reduce, stop, or even reverse declines of groundwater levels, protect groundwater in coastal aquifers against saltwater intrusion from the ocean, and to store surface water, including flood or other surplus water, and reclaimed municipal wastewater for future use.

According to Dillon (1999), an undergroundwater banking technique known as Aquifer Storage and Recovery (ASR) has emerged as a means of expanding urban water resources by harvesting waters that would otherwise be foregone. Where peri-urban areas are endowed with suitable aquifers, even those where groundwater may be saline, these present opportunities for generating new water resources. Injecting storm-water or treated effluent may restore saline aquifers so that they become underground reservoirs for irrigation supplies. With appropriate pretreatment these supplies can even be made potable. In South Australia, five years of experience with storm-water recycling has led to the development of an ASR trial for reclaimed municipal effluent for recovery as an irrigation supply in the dry season. In Bandung, Indonesia, CGS and local partners have commenced preliminary research on the potential for reversing groundwater declines by injection of roof top harvested water run-off (Dillon 1999). There are numerous other opportunities for ASR to prevent land subsidence and saline intrusion for cities located on deltas and overusing groundwater from confined aquifers. The main limitation in Asia is the quality of the water available for injection, but by being selective with sources of water, such as in Bandung, and with the type of aquifers into which to inject, or to exclude drinking as a potential use of recovered and affected water, it is possible to manage ASR so that substantial benefits can occur safely and in a sustainable manner.

References

Asano, T. 1999. Groundwater recharge with reclaimed municipal wastewater – regulatory perspectives, Session 4, Augmentation of Groundwater Resources through Aquifer Recharge, Proceedings of the International Symposium on "Efficient Water Use in Urban Areas - Innovative Ways of Finding Water for Cities" June 8–10, 1999 at the WHO Kobe, Japan.

Dillon, P. 1999. Aquifer Storage and Recovery in Urban Areas - Technology, Risks, and Implementation Issues Session 4, Augmentation of Groundwater Resources through Aquifer Recharge, Proceedings of the International Symposium on "Efficient Water Use in Urban Areas - Innovative Ways of Finding Water for Cities" June 8–10, 1999 at the WHO Kobe, Japan.

Fox, P. 1999. Advantages of aquifer recharge for a sustainable water supply, Session 4, Augmentation of Groundwater Resources through Aquifer Recharge, Proceedings of the International Symposium on "Efficient Water Use in Urban Areas - Innovative Ways of Finding Water for Cities" June 8–10, 1999 at the WHO Kobe, Japan.

Malik, A.H. 1997a. Sustainable Groundwater Development of the Lei-Basin (Islamabad-Rawalpindi, North Pakistan), as an Example of Groundwater Recharge Prospects for Pakistan and other Developing Countries, Volume of Abstracts: Pakistan Academy of Geological Sciences, Colloquium on Geology and the Human Life, Lahore, Pakistan, 4-6 Jan., 19–22, 1997.

Malik, A.H. 1997b. Hydrological Investigation of the Lei Drainage System (Islamabad-Rawalpindi, North Pakistan) with Special Reference to the Possibilities of Natural and Artificial Groundwater Recharge, Doctoral Thesis, Technical University of Berlin, Germany 1–281, unpublished, 1997.

Malik, A.H. 1998. Sustainable Groundwater Exploitation of the Lei-Basin, Proceedings of the 24th WEDC Conference, Islamabad, Pakistan, Sanitation and Water for All, John Pickford, Ed., 1998.

Malik, A.H. 1999a. Lei-Drainage-Basin (Islamabad-Rawalpindi): A Model of Groundwater Recharge, Proceedings of Regional Workshop on Artificial Groundwater Recharge, June 10–14, Quetta, Pakistan.

Malik, A.H. 1999b. Lei Drainage Basin (ISLAMABAD - RAWALPINDI), an Example of Sustainable Groundwater Exploitation, Quarterly SCIENCE VISION, 5(4), April-June,27–38,1999.

Malik, A.H. 2000. Integrated Urban Lei River (North Pakistan) Water Resources Management, International Journal of Water Resources Development, Special Issue: Urban Water Management for Developing Countries, A. K. Biswas, Ed., 16, (1), 97–117, 2000.

Malik, A.H. 2002. Leh-River-Basin (Islamabad-Rawalpindi, North Pakistan) A Model Case Study of Urban Dynamic Liveable Cities and Water Resources? 12th Stockholm Water Symposium, Abstracts: Workshop 8, Urban Dynamics, Livable Cities and Water, 2002.

Malik, A.H. 2005. Groundwater recharge system of the Lei-Basin, (Rawalpindi-Islamabad, North Pakistan), International Conference on Hydrological Perspectives for Sustainable Development (HYPESD-2005) 23-25, Feb. 2005, IIT Roorkee, India, 901–912.

Prinz, D. & Malik, A.H. 2002. More Yield with Less Water. How Efficient Can be Water Conservation in Agriculture?. Keynote Address, in Tsakiris (Ed.), Proceedings, 5th International Conference on Water Resources Management in the Era of Transition, Athens, Greece, 4-8 September, 2002, European Water Resources Association, Athens 18–35.

Groundwater Modeling for an Aquifer Storage Recovery Well under Cotton Wheat Cultivation

21

R.S. Malik and B.S. Jhorar[1], J. Ingwerson and T. Streck[2]

[1]Department of Soil Sciences, CCSHAU, Hisar – 125 004, India.

[2]Institute for Soil Science and Land Evaluation, Section Biogeophysics, University of Hohenheim, 70593 Germany

ABSTRACT

Groundwater situation was projected on 81.7 ha farm having brackish underground water growing cotton wheat with and without an Aquifer Storage Recovery (ASR) well operation for 1 year using Hydrus 2D model. Farm having ASR well used 0.375 annual canal, 0.3815 m annual rainfall and 0.4265 m groundwater irrigation, however; farm without ASR well did not use groundwater irrigation. The ASR well had recovery rate of 15 L/s and recharge rate of 20 L/s. Actual wheat cotton daily field measured evapotranspiration ETa values using SWAP model and remote sensing technique, served as surface variable atmospheric boundary condition of the model. Higher ETa values (1.183 m/annum) from well-irrigated field were used for farm with ASR well and the lower ETa values (0.648 m/annum) were used for farm without ASR well. Water table depth increased by 2.54 m at ASR well farm and it decreased by 2.12 m at without ASR farm in one year. Net groundwater abstraction was 0.2963 m in farm with ASR well farm and there was nil groundwater abstraction in farm without ASR well. The study has shown that artificial groundwater recharge through an ASR well having brackish native groundwater could be a viable technology for lowering the water table and increasing crop productivity at a large farm over the years.

1 INTRODUCTION

Aquifer Storage Recovery ASR well is an ideal means for storage of excess surface water and can provide viable solution to the uneven spatio-temporal distribution of surface water supplies. Studies have shown that if excess good quality water is directly injected into the tube wells in saline regions, subsequent withdrawl of groundwater will be of better quality than the native groundwater (Pyne, 1995; Gerges *et al.*, 2002a, b; Malik *et al.*, 2002; Pelvic *et al.*, 2002a, b) the rising groundwater levels in the central and western part of Haryana State (India) underlain by marginal to saline groundwater forms a direct threat to the sustainability of agriculture with the potential problems of waterlogging and secondary soil salinization.

Modeling long-term water pressures around the ASR well would not only quantify this temporal and spatial rise or fall in water levels but also be helpful in assessing the environmental impacts on long-term basis for the planners and researchers. Temporal and spatial water pressure responses near injection/pumping/ASR strainer wells have been modeled using analytical (Anonymous, 1993; Simpson *et al.*, 2003) in saturated flow conditions in homogeneous areas and numerical two/three-dimensional approaches (Boochs *et al.*, 1981; Chiang and Kinzelbach, 1998; Williams, 2000; Kohfahl *et al.*, 2002; Jorgensen and Helleberg, 2002; Sahrawat et.al. 2005) under steady and transient saturated flow conditions for confined and unconfined aquifers in heterogeneous areas with fair degree of success. To our knowledge, a few numerical modeling studies have been reported for cavity type ASR wells. A scientifically documented and evaluated (Diodato, 2000), HYDRUS-2D software package (Diodato, David M. 2000) having extensive interface capabilities for simulating saturated and unsaturated water, solute and heat flow under bare and cropped condition would simulate the water table situation at a farm operating a cavity type ASR well on short-term and long-term basis under the varied field heterogeneities.

A study was, therefore, initiated with the objectives: to develop scenario for quantifying change in water pressures with and without operating the ASR well for one year under cotton wheat in brackish groundwater zones.

2 MATERIALS AND METHODS

2.1 Field Site ASR Facilities for Calibration and Validation

Calibration and validation was done in short-term experiment of recovery-recharge experiment at farmer's field at village Badhyan Rangran about 35 km south of Hisar where the cavity type ASR was fitted with centrifugal pump. The Hydrus-2D model was used to project the effect of of recharging and recovery from an ASR well in piezometer and observation well installed at 7 and 50 m radial distance from ASR well in the field on water pressure h as a function of time t. The ASR well has been working here since 1995 under Indo-German collaborative Volks-Wagen funded project. In order to make the injection possible, it was necessary to remove the valve from the foot valve assembly below the T connection (Fig. 1). The rate of injection was 20 L/s, measured by a pipe water meter specifically installed in a separate piece of pipe. The rate of recovery was kept controlled at a lower rate 15 L/s by the farmer to avoid collapsing of the thin clay layer above the cavity. The calibration and validation results have been presented elsewhere (Malik *et al.*, 2004).

2.2 Parameter Estimation

Soil samples were collected up to the active aquifer while installing piezometers and observation wells. The litholog consisted of 5 distinct layers of loamy sandy loam and hard clay

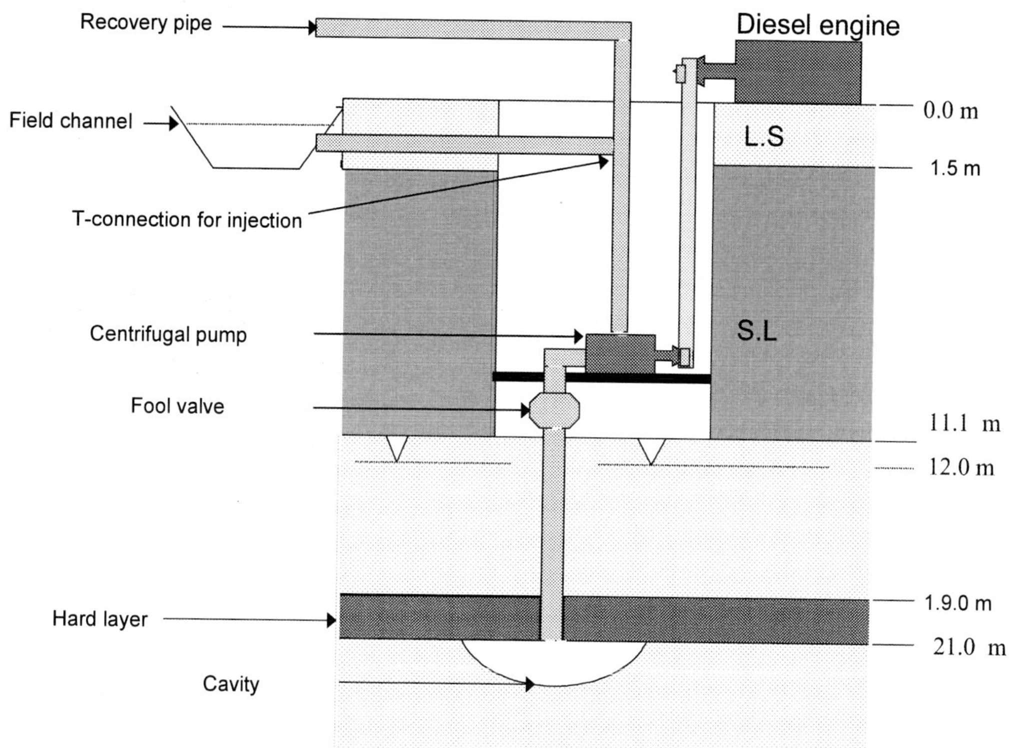

Fig. 1 Schematic diagram of cavity type ASR tube well at Badyan Ranghram.

layer as depicted in Fig. 1. Soil water retention functions were derived from pressure head and water content data measured on pressure plate apparatus using Van Genuchten-Mualem (1980) equation. The short-term calibration numerical experiments data of piezometric and observation well pressure heads, time (hi, ti) pairs were used to optimize effective saturated hydraulic conductivity Ks by inverse method. The hydraulic parameter values used are depicted in Table 1.

Table 1 Hydraulic parameters used in numerical experiments.

Parameters	θr (m^3m^{-3})	θs (m^3m^{-3})	α (m^{-1})	η (–)	*Ks* (md^{-1})	*k*	*Ak**
Loamy sand	0.005	0.43	5	2.9	10	0.5	15
Clay	0.07	0.54	1.1	1.001	0.7	0.5	5
Sandy loam	0.03	0.47	1.7	2.2	2.8	0.5	2

2.3 Numerical Experiments

Water table projection was done on 2 farms of 81.7 ha each with and without ASR well growing wheat cotton in a Bhakra canal command area of canal water supply of 0.375 m and annual rainfall of 0.3815 m. Daily rainfall values were adopted from 1992 Hisar rainfall data. The windows based HYDROUS-2D package solved the Richard equation for variably saturated flow numerically, assuming applicability of the Van Genuchten-Mualem soil hydraulic functions. To run HYDROUS-2D, we implemented a relatively coarse numerical mesh involving nodes depicting the geometry of the flow region, 27 m depth with radius of 510 m and an exocentric elliptical cavity with 2 m horizontal radius and 1m vertical radius. Recovery and recharge fluxes were applied as variable flux boundary condition at axisymmtric elliptical cavity surface (Table 2 and Fig. 2). These fluxes may also be seen as variable fluxes in Fig. 2.

Table 2 Recovery and recharge schedule for one year starting from 1st December.

Recovery		Storage		Recharge	
t (d)	Rate (m^3/d)	t (d)	Rate (m^3/d)	t (d)	Rate (m^3/d)
0 to 180 and 273 to 365	1255.7	243 to 272	0	181 to 242	1727.0

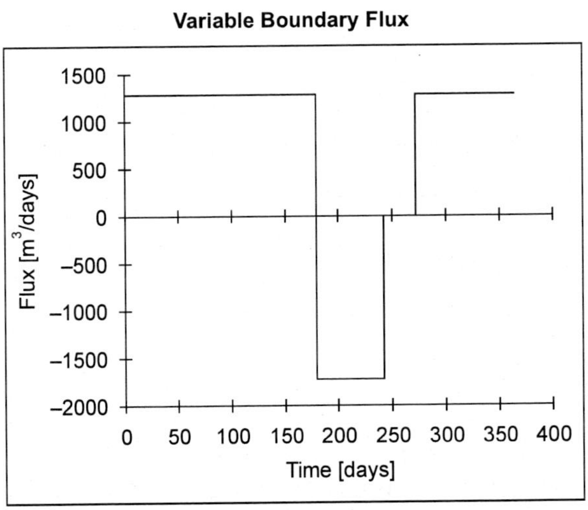

Fig. 2 Variable recovery flux (+) and recharge fluxes (–) applied to the operating ASR well for one year.

An atmospheric variable boundary condition accounting for rainfall including irrigation and evaporation including actual transpiration on daily basis was imposed at the soil surface and may be seen in Fig. 3 for ASR well and without ASR well. Other outer boundaries were closed for flow. Observation nodes numbering 1, 2, 3, 4, and 5 for piezometers and 6, 7, 8, 9, 10 for observation wells were inserted at x, z positions 6, 2; 100, 2; 187, 2; 291, 2; 423, 2 for piezometers; and at 418, 6; 293, 9; 187, 8; 105, 8; 7, 8 m for observation wells to observe the water pressure fluctuations. Initial condition was provided by pressure head in equilibrium with the bottom pressure head of 14 meter. Graphic presentation of boundary condition as appear in the model may be seen in Fig. 3.

The year started from 1st Dec as t = 1 day from the sowing of wheat. The actual daily evapotranspiration ET_a consisted of actual evaporation from soil surface Ea and crop

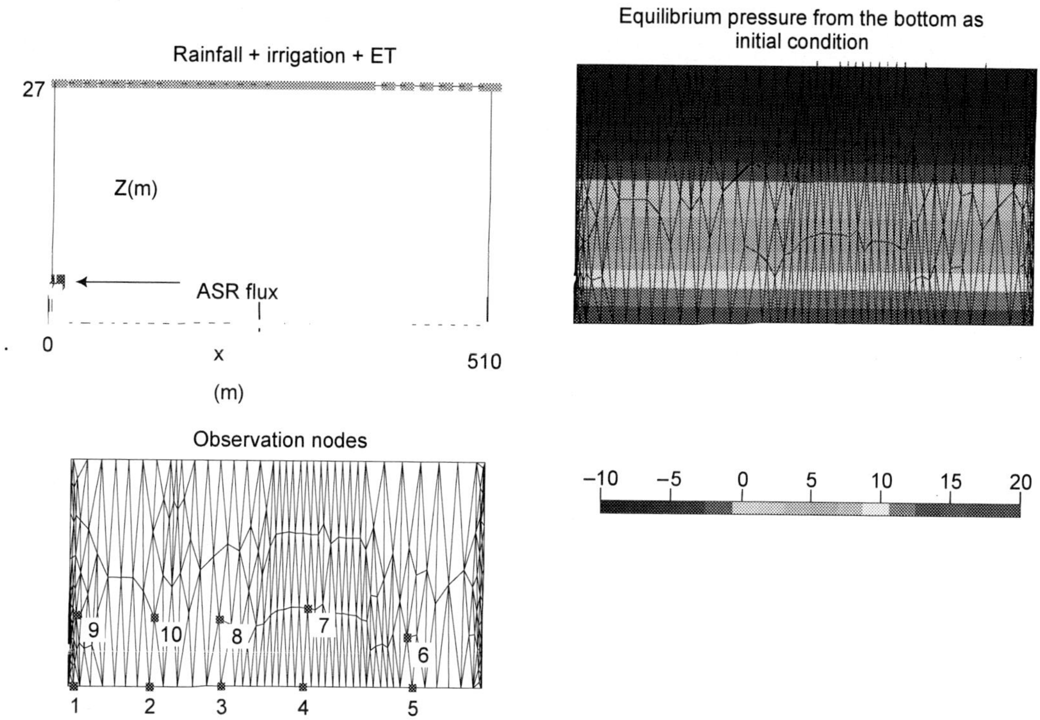

Fig. 3 Atmospheric variable boundary condition (R+I+ETa) imposed at the soil surface of with and without ASR well under cotton wheat.

transpiration Ta. These were estimated using field measured soil moisture irrigation amounts and remote sensed actual evapotranspiration data in SWAP/WOFOST model as given in Table 3 in summary form on annual basis were taken from Van dam and Malik (2003) for cotton and wheat. Higher ETa values (1.183 m/annum) of well irrigated field were used for farm with ASR well and the lower ETa values (0.648 m/annum) of waterlogged-salt stressed fields were used for farm without ASR well. The absolute values of the minimum allowed pressure head at

Table 3 Annualized variable atmospheric boundary conditions for farm with and without ASR well.

Parameter	*With ASR well*	*Without ASR well*
Ea (m)	0.253	0.183
Ta (m)	0.93	0.465
ETa (m)	1.183	0.648
R (m)	0.3815	0.3815
Ci (m)	0.3750	0.3750
Ui (m)	0.4265	0
R+Ci+Ui (m)	1.183	0.7565
Net groundwater abstraction (m)	0.2965	0

the soil surface was kept at 100 m on all days so as to keep the given ETa values equal to potential evaporation values on all days. Daily rainfall R, canal irrigation Ci and undergroundwater irrigation from ASR well Ui were given as rainfall and ETa as potential evaporation in variable atmospheric boundary condition (Fig. 3). All the above boundary condition values used in the model are presented in annualized summarized form in Table 3 for with and without ASR well facilities.

Long term scenario in this paper was aimed at solving the rising water levels problem of the state in the western zone having canal water supply and brackish under groundwater quality.

3 RESULTS AND DISCUSSION

Fig. 4 showed that the cumulative positive rainfall, irrigation and actual evapotranspiration ETa (R+I+ETa) applied as atmospheric fluxes were much larger and positive in farms with ASR well showing a net withdrawal of groundwater. Cumulative recharge and recovery ASR fluxes are shown in Fig. 5. Positive sign indicated the outgoing and negative sign indicated the incoming fluxes because *z* was kept as positive upward in this model. The recovery was more than the recharge. There was no groundwater abstraction where ASR well was not installed. Net groundwater abstraction was more (0.2965 m/annum) in farm with ASR well than that in farm without ASR well (nil, see Table 3). This would further aggravate the problem of rising water table problem under cotton wheat where no ASR well facility was provided.

The net result of these negative and positive fluxes can be seen on pressure head at different time of the year at different observation nodes no. 2, 6 and 7 in Fig. 6 and on the piezometric and observation well observation nodes at different times of the year in Fig. 7. These graphs of pressure head show that groundwater level responds to rainfall and irrigation. Water table depth increased by 2.54 m at ASR well farm and it decreased by 2.12 m at without ASR farm. The rising water table in farms without ASR well would imply that continued canal irrigation in farms having brackish underground water would cause water-logging and surface salinization with poor crops and lower ETa.

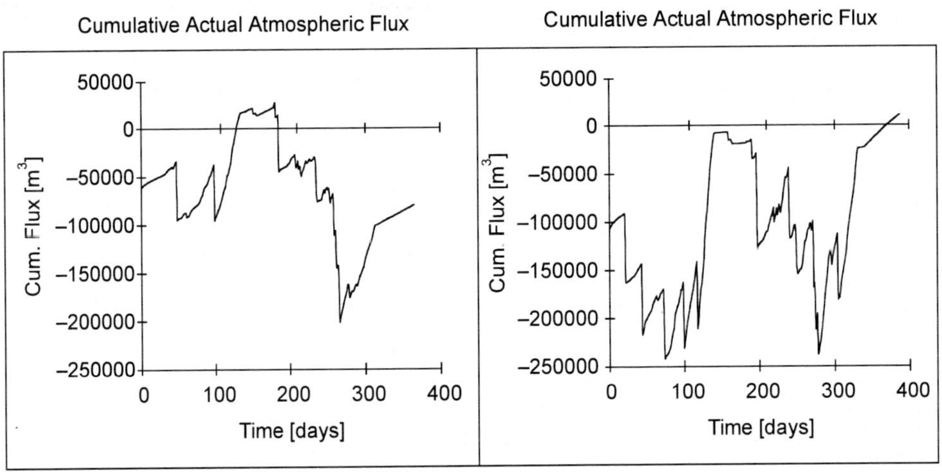

Fig. 4 Cumulative rainfall, irrigation and actual evapotranspiration under cotton wheat in indicated numerical experiments.

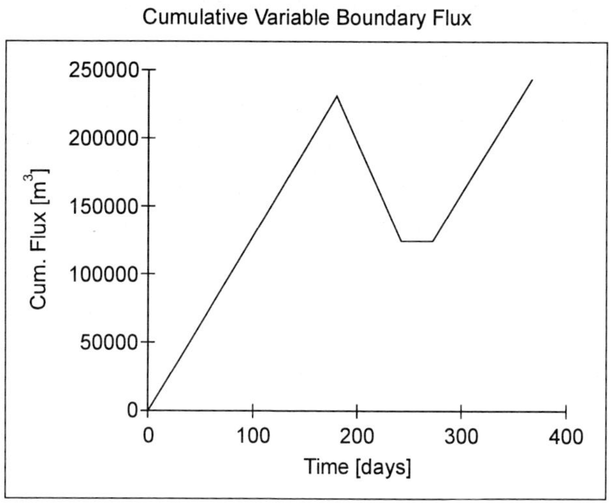

Fig. 5 Cumulative groundwater recovery and recharge in ASR well experiment.

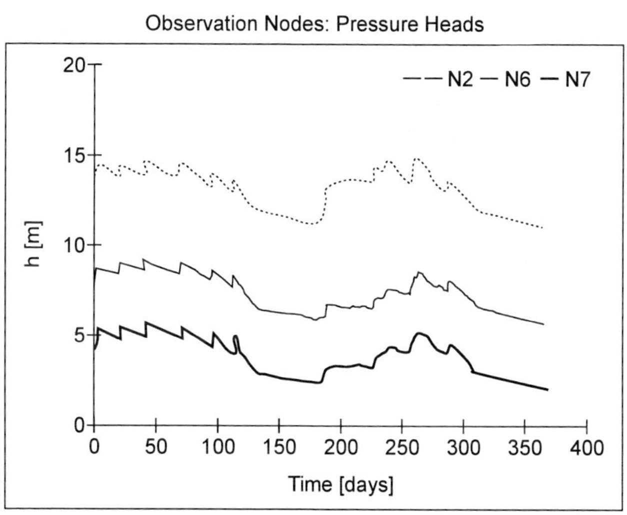

Fig. 6 Pressure head variation with time in indicated observation node numbers.

4 CONCLUSION AND PRACTICAL UTILITY

Operation of ASR well would increase the groundwater abstraction by improving groundwater quality through freshwater injection. Water table will be lowered down and the increased irrigation facility would increase the ETa and crop yields. Practical utility of this study is in canal irrigated area with brackish groundwater to encourage conjunctive use of fresh rainwater and groundwater. For improving the quality of groundwater recharge and recovery volume ratio can be adjusted depending upon the availability of surface excess water of good quality and on the salinity level of the groundwater and the salt resistance of the crop grown. It

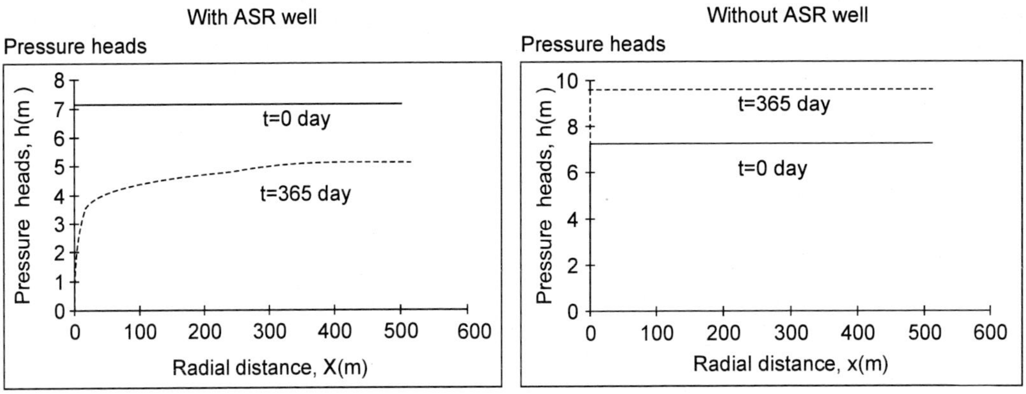

Fig. 7 Pressure head changes at indicated times at z = 31.7 m.

may easily be estimated that 3:1 recovery recharge ratio would dilute the salinity to 1/3 if the fresh water injected water bubble does not move far away because of regional hydraulic gradients and leakage in confining layers. In fact, the decrease in salinity will be governed by many hydrological and operational factors which are not the subject of this paper. This study further showed that Hydrus 2D model can be used successfully for groundwater studies around ASR wells.

Acknowledgements

The authors wish to acknowledge with thanks and gratitude Alexander von Humboldt Foundation for financial supporting the revisit to Germany during which the modeling work for this paper was done; and Volkswagen Foundation, Germany for providing basic facilities for collecting the data under Indo German collaborative project between Chaudhary Charan Singh Haryana Agricultural University, Hisar-125 004, India and University of Hohenheim, Institute for Soil Science and Land Evaluation (Biogeophysics Section), 70593 Stuttgart, Germany.

References

Anonymous 1993. Progress report of ICAR coordinated research schemes on optimization of ground water utilization through wells and pumps. *Department of soil and water Engineering college of Agriculture Engineering*. PAU, Ludhiana: 4–26.

Boochs, P. W. & Barovic, G. 1981. Numerical Model Describing Groundwater Treatment by Recharge of Oxygenated Water, *Water Resour. Res.* 17: 49–56.

Chiang, W.H. & Kinzelbach, W. 1998. Processing Modflow. 3D-Groundwater Modeling with PMWIN: *A simulation system for modeling ground water flow and pollution.* ISBN: 3540677445: 3rd edition, Vpringer-Verlag Berlin.

Diodato, David M. 2000. Review: Hydous-2D. *Ground Water,* 38(1): 10–11.

Gerges, N.Z; Dillon, P.J. Sibenaler, X. P; Martin, R. R; Pavelic, P; Howles, S. R. & Dennis, K. 2002a. South Australian experience in aquifer storage and recovery. In P.J. Dillon (Ed.) *Management of aquifer recharge for Sustainability; Proc. of Intl. Symposium on Artificial Recharge Adelaide 22-26 September. 2002.* Tokyo: Balkema : 453-458.

Gerges, N.Z; Howles, S.R. & Dillon, P.J. 2002b. Town water supply purification using aquifer storage and recovery in a saline aquifer. In P.J. Dillon (Ed.) *Management of aquifer recharge for Sustainability; Proc. of Intl. Symposium on Artificial Recharge Adelaide 22-26 September. 2002*. Tokyo: Balkema: 453-458.

Jorgensen, N.O., and B. B. Helleberg 2002. Stable isotopes (2H and 18O) and chloride as environmental tracers in a study of artificial recharge in Denmark. In P.J. Dillon (Ed.) *Management of aquifer recharge for Sustainability; Proc. of Intl. Symposium on Artificial Recharge* Adelaide *22-26 September. 2002*. Tokyo: Balkema: 245 – 250.

Kohfahl, C., E. Hamann, A. Pekdger 2002. Modeling of artificial water oscillations in a flooded lignite mine. . In P.J. Dillon (Ed.) *Management of aquifer recharge for Sustainability; Proc. of Intl. Symposium on Artificial Recharge* Adelaide *22-26 September. 2002*. Tokyo: Balkema: 261-264.

Malik, R.S., Jhorar, B. S; Jhorar, R.K; Streck, T. & Richter, J. 2002. Long-term successful operation of existing brackish cavity wells for ASR to improve quality for irrigation by Indian farmers. In P.J. Dillon (Ed.) *Management of aquifer recharge for Sustainability; Proc. of Intl. Symposium on Artificial Recharge Adelaide 22-26 September. 2002*. Tokyo: Balkema : 465- 268.

Malik, R.S; Jhorar, B. S; Jhorar, R. K. & Streck, T. 2004. Modeling groundwater recharge from an ASR well. In R. S. Malik; B. S. Jhorar; R. K. Jhorar & T. Streck.(Eds).ground water recharge and management through ASR techniques. Proc. Workshop on groundwater recharge. CCSHAU, Hisar. 15-17 April. 2004. Soil Sci. Department.: 65-75.

Pavelic, P., Dillon, P. J. & Simmons, C.T. 2002a. Lumped parameter estimation of initial recovery efficiency during aquifer storage and recovery. In P.J. Dillon (Ed.) *Management of aquifer recharge for Sustainability; Proc. of Intl. Symposium on Artificial Recharge* Adelaide *22-26 September. 2002*. Tokyo: Balkema: 285-290.

Pavelic, P., Dillon, P. J; Barber, C. & Yin Foo, D.A. 2002b Water banking in the Australian tropics: results from a trial on south Goulburn Island, northern territory. In P.J. Dillon (Ed.) *Management of aquifer recharge for Sustainability; Proc. of Intl. Symposium on Artificial Recharge* Adelaide *22-26 September. 2002*. Tokyo: Balkema : 441-446.

Pyne, R.D.G. 1995. Ground water recharge and wells- *A guide to aquifer storage recovery*. CRC press Inc 2000 Corporate Blvd, N.W. Boca Raton, Florida 33431. USA Lewis Publisher: 6-320.

Saharawat, Y.S; Malik, R. S; Jhorar, B. S. & Goyal, V. 2005. Simulating effect of successive cycles in Aquifer Storage and recovery well in India. In 5th International symposium on Aquifer storage. Berlin 11-16 June, 2005. In press.

Simpson, Mathew J., Prabhakar, C. T. & Francis E. Yeomans 2003. Analytical model for computing residence times near a pumping well. *Ground Water*, 41(3): 351-354.

Van Dam, J.C. & Malik,R. S. 2003. Water productivity of irrigated crops in Sirsa district, India. Integration of remote sensing, crop and soil models and geographical information system. *WATPRO final report. ISBN 90-6464-864-6*. Wageningen, UR.

Van Genuchten, M. Th. 1980. A closed form equation for predicting the hydraulic conductivity of unsaturated soil. *Soil Sci. Soc. Am. Proc.* J. 44: 692-698.

Williams, A. T. 2000. Using an aquifer storage and recovery (ASR) trial as a large scale tracer test. *In Tracers and modeling in Hydrogeology. Proc. of the TraM' 2000 Conference, IAHS Publ.No 262. 2000.*

Delineation of Groundwater Zone by Magnetics in Basement Complex, West of Cuddapah Basin, India

22

C. Rama Rao, M.R.K. Prabhakara Rao, Sitarama Sinha and G.D.J. Sivakumar Sinha
National Geophysical Research Institute, Uppal Road, Hyderabad–500 007, India

ABSTRACT

Total intensity aeromagnetic maps of hard rock terrains provide useful information in isolating fault/shear/fracture patterns, lithological variations, shallow and deep heterogeneities within the crustal rocks. Magnetic data collected over a part of exposed granite-gneissic terrain west of Cuddapah basin has been analyzed in conjunction with the available well data and drainage network to locate potential ground water zones. Total intensity ground magnetic data has been collected along a north-south profile of length 2500 m at a station interval of 20 m to bring out the lithological variations and structural controls for the delineation of groundwater accumulations. Analysis of the data has yielded a shallow subsurface dyke, a shear zone and a subsurface ridge-like structure formed due to the intrusions of dykes. This subsurface structure appears to have been helped for the formation of a good aquifer zone towards south.

1 INTRODUCTION

Aeromagnetic surveys have been found successful in delineating, indirectly, fault/fracture and dyke controlled aquifer systems in crystalline and metamorphic regions of Western Africa and other parts of the world (Astier and Paterson, 1987; Paterson and Bosschart, 1987). Astier and Paterson (1987) have shown in some parts of crystalline basement complex of Western Africa a direct correlation between the yield of bore holes and wells and their proximity to faults and dykes inferred from aeromagnetic data and subsequent ground checks. Astier (1969) based on photo geological and other data sets took up ground magnetic follow up surveys for confirming the presence of fault lineaments before pinpointing sites for resistivity surveys.

Chandra and Reddy (1987) carried out ground geophysical surveys comprising magnetic, resistivity and EM over a suspected shear zone in the Singhbhum metamorphic belt in Eastern India. Their surveys have resulted in finding out artesian flows in the shear zone. Rama Rao *et al.* (2002) have made a correlative study of aeromagnetic data with drainage network and distribution pattern of yielding wells in a part of crystalline basement complex west of the proterozoic Cuddapah basin. Their study revealed a one to one correspondence with inferred magnetic lineaments and drainage network and well distribution.

Here we present a case study related to the delineation of a shallow subsurface ridge-like structure and its implication on the surface and groundwater regimes in the exposed granite-gneissic basement terrain located towards the south western part of the Cuddapah basin.

2 STUDY AREA

2.1 Aeromagnetic Map of the Study Area

Figure 1 is the total intensity aeromagnetic map of the study area. The NW-SE trending positive anomaly feature along Ghantapuram-Musthuru, just south of Buttapalle, is inferred to be an intense shear zone which has been intruded by dolerite dykes. This shear zone divides the area into two groundwater regimes with different orientations of drainage patterns. The southern part of the shear zone is associated with N-S oriented drainage channels and the northern part with E-W channels (Fig. 2).

A ground magnetic profile of length 2500 m along N-S direction with a station spacing of 20 m has been laid along Battulapalle-Venkatagaripalle (V. Palle) road in the northern part of the shear zone (Fig. 1). This ground magnetic survey has been taken up to delineate zone of small scale lithological/structural heterogeneities within the granite-gneissic terrain which might act as controls of groundwater movement.

2.2 Drainage Network

The distribution of yielding wells, drainage channels and elevation of the area are shown in Fig. 2. Major E-W drainage channels are prominently seen in the northern part whereas N-S trending channels are observed in the south. The region marked 'A' between V. Palle-Narsampalle seems to be devoid of significant yielding wells. The area located west of the road is reasonably covered with wells and is occupied by a tank fed by E-W and N-S trending drainage channels. The drainage channel just south of V. Palle traverses along east west direction and takes an abrupt north-south turn at 1500 m mark and takes a road parallel track. The area between the E-W trending channels on either side of V.Palle is a region of scarce groundwater potential ('A'). The north-south trending elevated zone (marked as B) on the eastern side of the road divides the area into two parts : the eastern part with more number of yielding wells and the western part with less wells.

Ground magnetic data at a station interval of 20 m has been collected in a north-south direction starting from Battulapalle in the south up to V. Palle in north totalling a distance of 2500 m.

Fig. 1 Total intensity aeromagnetic map of the study area. This area is located towards west of the SW corner of the Cuddapah basin.

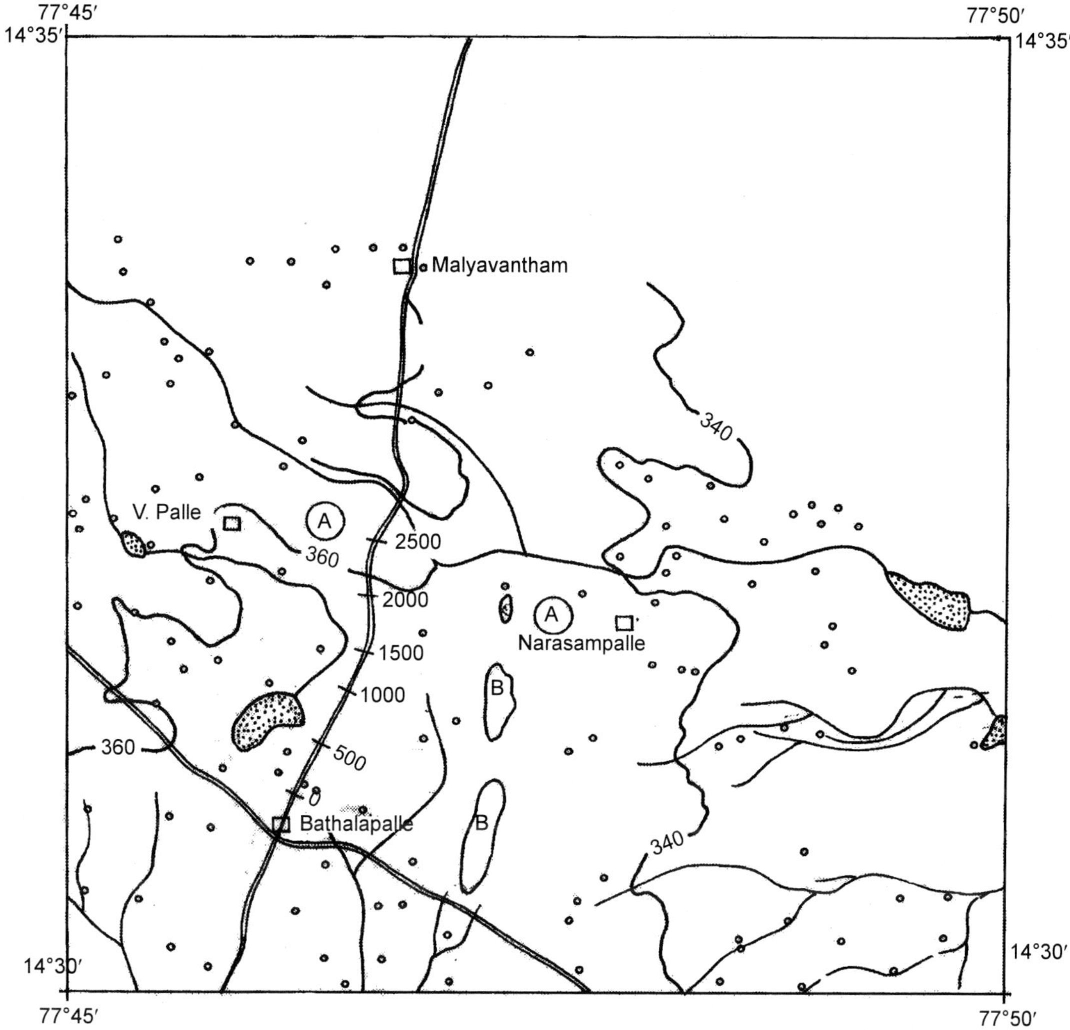

Fig. 2 Area of study showing the drainage channels, water tank, elevation contours, yielding wells and the magnetic profile along the road (0-2500 m).

3 RESULTS AND INTERPRETATION

3.1 Interpretations of Ground Magnetic Data

Aeromagnetic data along the length of the present ground magnetic profile appears as a smoothly increasing gradient in the total field component and is not amenable for any meaningful interpretation. Finer and subtle features are naturally suppressed in the aeromagnetic data at this flight height of 150 m. However, the ground magnetic profile has brought out finer lithological variations all along the profile.

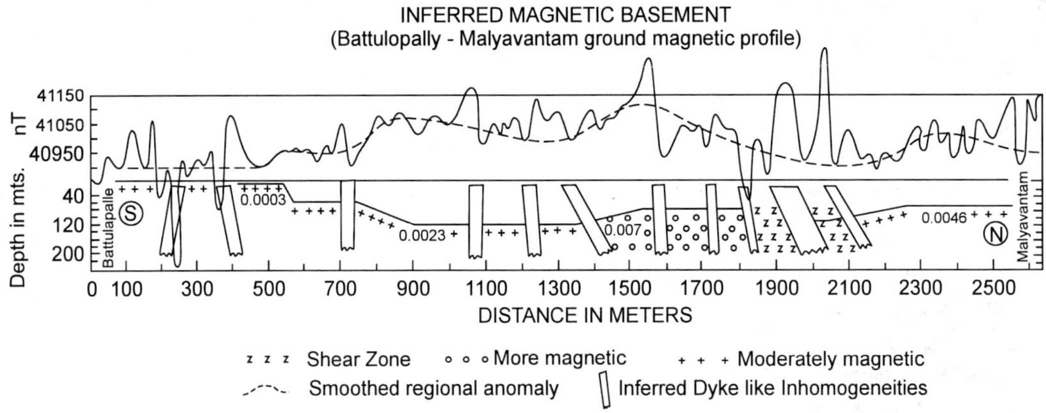

Fig. 3 The ground magnetic profile along with the regional component with dashed line on the top and the interpreted section below the profile.

A reasonable polynomial fit representing the regional/long wavelength component of the observed magnetic profile is computed and is shown as dashed line (Fig. 3). The residual anomalies are superposed with varying amplitudes on this smoothly varying field and these have been interpreted using the MAGMOD software of Paterson, Grant & Watson Ltd. Canada (1981). The interpreted section is presented below the magnetic profile and the results are discussed in the following section.

The region between 0–400 m with high amplitude anomalies is inferred to be associated with emplacement of dykes at shallow depths. The region between 400–1400 m shows a gradual stepwise increase in the regional component. Anomalies with moderate amplitudes within this section are inferred to be associated with variations in magnetic mineral content within the granite-gneissic terrain.

However, the magnetic anomalies between 1500–2100 m are associated with high amplitudes and differ significantly from those anomalies between the section 500–1400 m. This can be inferred to be associated with a considerable variation in lithology in this section.

3.2 Amplitude Spectrum

Amplitude spectra of these two parts of the magnetic profile have been computed and are presented in Fig. 4. Analysis of the spectral characters of these two parts clearly indicates that the lithological variations are more prominent at shallow levels and there are no significant variations at deeper levels with in the granite-gneissic country. The slightly higher values associated with amplitude spectrum between the section 1500–2500 m denotes a region of more magnetic content with in the gneisses compared to that of the section between 400–1400 m. The average depth computed from the spectrum for the section between 400–1400 m is 65 m where as the depth obtained for the part between 1500–2500 m is 85 m. These average depths can be inferred to be associated with the thickness of weathered/fractured horizons within the gneissic terrain overlying the consolidated basement beneath.

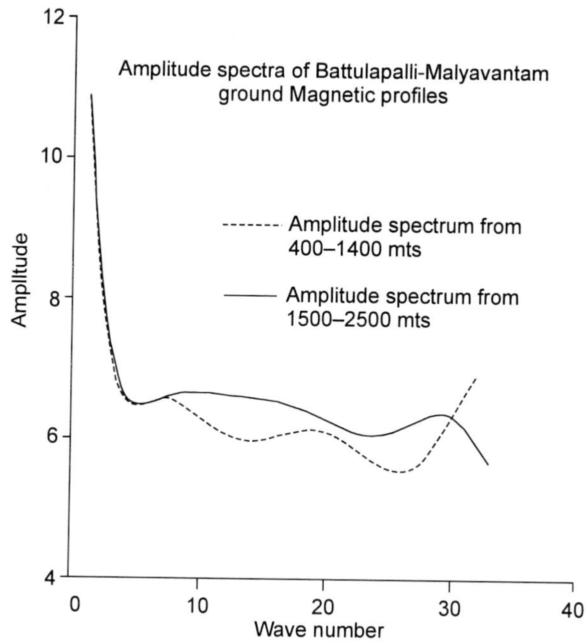

Fig. 4 The amplitude spectra computed for the southern part (400–1400 m section) and the northern part (from 1500–2500m). Notice the variation in the energy on the higher frequency part of the profiles which indicate that the variations are confined to shallower levels.

3.3 Magnetic Basement Configuration

Modelling of the long wavelength positive magnetic anomaly between 600–1200 m has yielded a maximum depth of 100–120 m with a susceptibility range of 0.0003– 0.0023 cgs units representing the granite-gneiss basement. The shallow overlying layer above the inferred basement can be viewed as weathered/fractured granite-gneiss terrain (with an average thickness of 65 m as obtained from spectral studies) intruded at places by dykes. This overlying layer may form as a possible zone for groundwater accumulation and percolation. The region between 1400–2200 m is associated with slightly higher magnetic susceptibility (0.007 cgs units) indicating the presence of more mafic minerals in this part of the granite-gneissic terrain. Also the cumulative effect of the five inferred dyke like intrusions between 1600–2200 m section seems to have been elevated the terrain forming a ridge-like structure. This inferred subsurface ridge like structure at shallow depth might be acting as a barrier to both the surface water flow and the groundwater movement. The higher susceptibility in this region can be attributed to consolidated rock matrix having more mafic content. The inferred dyke at 1600 m traversing in E-W direction seems to be controlling the surface channel south of V. Palle and making it change its course to north south.

3.4 Inferred Shear Zone

A shear zone is inferred between 500–1100 m as shown in Figure 5 and this area is occupied on the west by a tank. A reversely magnetised dyke has been delineated at 400 m at a shallow

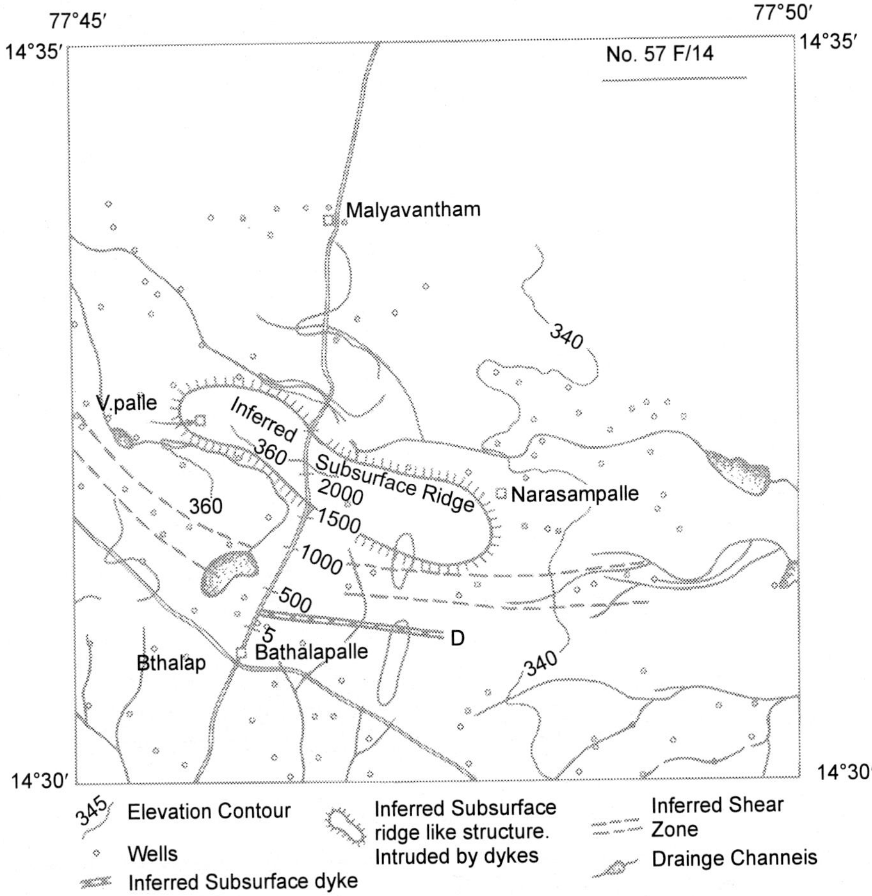

Fig. 5 The study area indicating the inferred subsurface ridge-like structures, viz. dyke, and the shear zone.

depth of 6–8 m south of the shear zone. This dyke seems to be separating the groundwater regimes into two parts one towards north and the other towards south whose effect is evident by the distribution of wells on either side of the dyke and change of surface channel courses.

4 CONCLUSIONS

Total intensity airborne and ground magnetic data is most useful in identifying lithological variations, delineating shear/fault/fracture lineaments and detecting the subsurface dykes that are responsible for accumulation of groundwater resources mainly in crystalline basement complexes.

Ground magnetic data along a profile of length 2500 m has been analysed in a granite-gneissic terrain in the vicinity of Cuddapah basin to bring out structural features that aid in isolating zones of groundwater. Spectral analysis and modeling of the magnetic data has resulted in delineating:

(i) a shallow subsurface reversely magnetised dyke at a depth of 6-8 m in the southern part of the profile which is separating the aquifer into two parts in this area.

(ii) a shear zone housing a big tank acting probably as surface water source and also helping the replenishment of groundwater.

(iii) a shallow subsurface ridge-like structure formed due to the intrusions of moderately high magnetic dykes intruded into the granite-gneissic terrain in the northern part of the profile. This structure appears to be responsible for the change of course of surface channels and acting as a subsurface barrier for groundwater movement.

(iv) Two magnetically different granite-gneiss rock suits: the southern part of the profile with less magnetic minerals and the northern part containing consolidated rock matrix with high mafic content.

References

Astier, J.L 1969. Etude des Eaux souterain au Dahoney:interpretation des measures Geophysiques F.A.O et. Service de I hydrauliaqe.

Astier, J.L & Paterson N.R. 1987. Hydrogeological interest of aeromagnetic maps in crystalline and metamorphic areas. *Exploration'87 Proceedings.*

Chandra, P.C & Reddy, P.H.P. 1987. Geophyscxal surveys in hard rock and sedimentary terrains – status, findings and recommendations. *Govt. of India ministry of Water Resources, CGWB report no.29, project in ground water studies in Kasi and Subarnareka river basins.*

Paterson, N.R. & Bosschart, A.R. 1987. Air Borne geophysical exploration for ground water, Ground Water , *Vol. 25, No.1.*

Rama Rao, Ch.. Prasanta Laxmi M. & Rambabu, H.V. 2002. Delineation os structural controls using aeromagnetics and their relation with the occurrence of ground water on the exposed basement complex in a part of south Indian shield. Proceedings of IGC- 2002., In : *On sustainable development and management of ground water resources in semi arid region with special reference to hard rocks.* pp. 49–56.

Investigation of Aquifer Remediation Strategies by Coupled Finite Element Model for Groundwater Flow and Contaminant Transport

23

S.M.V. Sharief, T.I. Eldho and A.K. Rastogi

Department of Civil Engineering, Indian Institute of Technology, Bombay, India.

Corresponding author: SMV.Sharief, E-mail: smvsharief@iitb.ac.in Fax. 022-25767302

ABSTRACT

Groundwater is one of the important sources of drinking water for many countries in the world. Long-term withdrawal of groundwater resources has created a need to predict its effect on future water quality and sustained availability. With the increasing water demands, growing agricultural and industrial activities, the protection of groundwater resources has become a major challenge. Remediation of contaminated groundwater is complicated, time-consuming and an expensive practice. Identifying the source of groundwater pollution is imperative before a specific groundwater pollution remediation strategy can be applied. This paper describes two strategies, i.e. self-cleaning (no-action) and combination of flushing and pumping for remediation of contaminated aquifers using a coupled model for groundwater flow and contaminant transport by finite element method. The study investigates a problem of cleaning up the aquifer, which has been contaminated due to seepage from a polluted pond and from a recharge well entering to the flow domain. A simulation period of 10,000 days is considered involving withdrawals from three potential wells. From the remediation results, it has been observed that the initial pollutant level in the aquifer is reduced to drinking water standards (≈400 ppm) within 1.3 years by self cleaning alternative for three different scenarios, while the same was achieved in less than 1.3 years in the second alternative. However the second option demands more financial resources and their suitability depends upon the water necessity in a particular region.

Keywords: Aquifer remediation, Self-cleaning, Flushing and Pumping, Groundwater pollution, Finite element method.

1 INTRODUCTION

Groundwater is a major source of drinking water in both urban and rural areas, also for the agricultural and industrial sectors in India. It is an integral part of the hydrological cycle and its availability depends on the rainfall and recharge conditions. The demand for water has increased over the years and this has led to water scarcity in many parts of the country. The situation is aggravated by the problem of groundwater contamination. The quality of groundwater is getting severely affected because of the widespread pollution of surface water, discharge of untreated waste water through bores and leachate entering from poorly designed landfills. When a pollutant finds access to the groundwater flowing through an aquifer, it can travel larger distances with the flowing water. The main processes involved in solute transport are, convection or advection, mechanical dispersion and molecular diffusion. Due to advection, the non-reactive solutes are carried at an average rate equal to the average linear velocity of the water. As a result of the convection and dispersion, the solute is transported by the groundwater with varying degrees of concentration, throughout the domain of flow. The basic governing equations of flow and contaminant transport through a porous media are non-linear, coupled second order partial differential equations. The exact solutions are extremely complex and hence numerical techniques are being increasingly used for the contaminant transport simulation. Among the various numerical techniques available, finite difference and finite element techniques have been found to be effective by many researchers. Early application of numerical simulation to solute transport problems was studied by Shamir and Harlman (1967). One of the first attempts at numerical simulation of solute transport in a regional scale was carried out by Bredehoeft and Pinder (1973), who investigated salt water contamination of the groundwater system. Marino (1974) developed one and two-dimensional analytical solutions for contaminant transport in porous media. Hunt (1978) studied the dispersive phenomena of contaminants in porous media. Sykes *et al.* (1982) examined the leachate migration from the landfill using a finite-difference model. Liji *et al.* (1991) presented several analytical solutions for three-dimensional solute transport in semi-infinite porous media with simplified boundary conditions. Fang and Singh (1995) modified the Galerkin's finite element formulation for solute transport. Before the remediation strategies for a confined aquifer is initiated, the flow and transport phenomena in the system is to be studied in detail to understand the movement of the plume. In this paper aquifer remediation strategies are investigated using a coupled finite element model for flow and contaminant transport.

2 GOVERNING EQUATIONS FOR GROUNDWATER FLOW AND CONTAMINANT TRANSPORT

The governing equation describing the flow in a two-dimensional inhomogeneous confined aquifer is given by (Bear, 1979)

$$\frac{\partial}{\partial x}\left[T_x\frac{\partial h}{\partial x}\right]+\frac{\partial}{\partial y}\left[T_y\frac{\partial h}{\partial y}\right]=S\frac{\partial h}{\partial y}+Q_w\delta(x-x_1)(y-y_1)-q \qquad ...(1)$$

Subject to the initial condition

$$h(x,y,0)=h_0(x,y) \qquad x,y\in\Omega$$

and the boundary conditions

$$h(x, y, t) = H(x, y, t) \qquad x, y \in \partial\Omega_1$$

$$T\frac{\partial h}{\partial n} = q(x, y, t) \qquad x, y \in \partial\Omega_2$$

where $h(x, y, t)$ = piezometric head (m); $T(x, y)$ = transmissivity (m^2/d); S = storage coefficient; x, y = horizontal space variables (m); Q_w = source or sink function ($-Q_w$ source, Q_w = Sink) ($m^3/d/m^2$); t = time in days; Ω = the flow region; $\partial\Omega$ = boundary region ($\partial\Omega_1 \cup \partial\Omega_2 = \partial\Omega$); $\frac{\partial}{\partial n}$ = normal derivative; $h_0(x, y)$ = initial head in the flow domain (m); $H(x, y, t)$ = known head value of the boundary head (m) and $q(x, y, t)$ = known inflow rate ($m^3/d/m$).

The governing partial differential equation describing the physical processes of convection and hydrodynamic dispersion is basically obtained from the principle of conservation of mass (Freeze and Cherry, 1979). The derived partial differential equation for transport of a single chemical constituent in groundwater, considering advection dispersion, fluid sources/sinks can be given (Wang and Anderson, 1982; Sun, 1996) as follows:

$$h(x, y, 0) = h_0(x, y) \qquad x, y \in \Omega$$

$$R\frac{\partial C}{\partial t} = \frac{\partial}{\partial x}\left(D_{xx}\frac{\partial C}{\partial x}\right) + \frac{\partial}{\partial x}\left(D_{xy}\frac{\partial C}{\partial y}\right) + \frac{\partial}{\partial y}\left(D_{yx}\frac{\partial C}{\partial x}\right) + \frac{\partial}{\partial x}\left(D_{yy}\frac{\partial C}{\partial y}\right) - \frac{\partial}{\partial x}(V_x C)$$

$$-\frac{\partial}{\partial y}(V_x C) - \frac{\partial w}{nb} + \frac{W}{\theta}C_0 - \frac{Q}{\theta}C \qquad ...(2)$$

where V_x and V_y = seepage velocity in x and y direction; D_{xx}, D_{yy}, D_{xy}, D_{yx} = Components of dispersion coefficient tensor [L^2T^{-1}]; C = Dissolved concentration [ML^{-3}]; θ = Porosity (dimensionless); w = elemental recharge rate with solute concentration c; n = elemental porosity; b = aquifer thickness under the element; R = Retardation factor: C_0 = water with concentration injected into an aquifer; W= water injected per unit time per unit aquifer volume (T^{-1}); Q = volume of water extracted away from per unit aquifer volume per unit time.

The initial condition for the problem is

$$C(x, y, 0) = f \qquad x, y \in \Omega \qquad ...(3)$$

and the boundary conditions are

$$C(x, y, t) = g_1 \qquad (x, y) \in \Gamma_1$$

$$\frac{\partial C}{\partial n}(x, y, t) = g_2 \qquad ...(4)$$

where Γ_1 = boundary sections of Ω; f = a given function in Ω; g_1 = given function along Γ_1, which is known solute concentration from the present study. g_2 = concentration gradient normal to the boundary Γ_2; n = unit normal vector.

3 FINITE ELEMENT FORMULATION FOR GROUNDWATER FLOW AND TRANSPORT

3.1 Finite Element Formulation for Groundwater Flow

For numerical analysis, the finite element method is selected because of its relative flexibility to discretize a prototype system which represents boundaries accurately. The finite element method (FEM) approximates the governing partial differential equation by integral approach. In the present study, two-dimensional linear triangular elements are used for spatial discretization and Galerkin approach (Reddy 1993) is used for finite element approximation. In the FEM, the head variable in Eq. (1) is initially approximated as

$$\hat{h}(x, y, t) = \sum_{L=1}^{NP} h_L(t) N_L(x, y) \qquad ...(5)$$

where h_L = unknown head; N_L = known basis function at node L; $\hat{h}(x, y, t)$ = trial solution; NP = total number of nodes in the problem domain. A set of simultaneous equations is obtained when residuals weighed by each of the basis function are forced to be zero when integrated over the entire domain. Applying the finite element formulation for Eq. (1) gives

$$\iint_{\Omega} \left[\frac{\partial}{\partial x}\left(T_x \frac{\partial h}{\partial x}\right) + \frac{\partial}{\partial y}\left(T_y \frac{\partial h}{\partial y}\right) - Q_w + q - S\frac{\partial h}{\partial t} \right] N_L(x, y) dxdy = 0 \qquad ...(6)$$

Equation (6) can further be written as the summation of individual elements as

$$\sum_e \iint \left(T_x^e \frac{\partial \hat{h}^e}{\partial x} \left\{ \frac{\partial N_L^e}{\partial x} \right\} + T_y^e \frac{\partial \hat{h}^e}{\partial y} \left\{ \frac{\partial N_L^e}{\partial y} \right\} \right) dxdy + \sum_e \iint \left(S \frac{\partial \hat{h}^e}{\partial} \right) \left\{ N_L^e \right\} dxdy$$

$$= \sum_e \iint (Q_w) \left\{ N_L^e \right\} dxdy - \sum_e \iint (q) \left\{ N_L^e \right\} dxdy \qquad ...(7)$$

where, $\left\{N_L^e\right\} = \begin{Bmatrix} N_i \\ N_f \\ N_k \end{Bmatrix}$ and the trial solution $\hat{h}(x, y, t)$ can be defined as

$$\hat{h}^e(x, y, t) = N_i^e(x, y)h_i(t) + N_j^e(x, y)h_j(t) + N_k^e(x, y)h_k(t) \qquad ...(8)$$

In equation (8), i, j, k are the element node numbers in the counterclockwise order and the interpolation function can be given as

$$N_1^e(x, y) = \frac{1}{2A^e}[a_1 + b_1 x + e_1 y] \qquad ...(9)$$

where $1 = i, j, k$; $b_i = y_j - y_k$; $c_i = x_k - x_j$; $b_j = y_k - y_i$; $c_j = x_i - x_k$; $b_k = y_i - y_j$; $c_k = x_j - x_i$, (x_i, y_i), (x_j, y_j) and (x_k, y_k) are the coordinates of nodes i, j, k respectively. A^e = area of triangular element, and i, j, k

expressed in terms of its coordinates at the corners. Eq. (7) can be written in matrix form for an element as

$$[G^e]\left\{h_i^e\right\} + [P^e]\left\{\frac{\partial h_i^e}{\delta t}\right\} = \{F^e\}$$

The element matrices for the linear triangular element are given as follows (Reddy, 1993)

$$[G] = \left[\frac{T_x^e}{4A^e}\begin{bmatrix} b_i^2 & b_ib_j & b_ib_k \\ b_jb_i & b_j^2 & b_jb_k \\ b_kb_i & b_kb_j & b_k^2 \end{bmatrix} + \frac{T_y^e}{4A^e}\begin{bmatrix} c_i^e & e_ie_j & e_ie_k \\ c_jc_i & c_j^2 & c_jc_k \\ c_kc_i & c_kc_j & c_k^2 \end{bmatrix}\right];$$

$$[P] = \left[\frac{S^e.A^e}{12}\begin{bmatrix} 2 & 1 & 1 \\ 1 & 2 & 1 \\ 1 & 1 & 2 \end{bmatrix}\right]; \{F\} = \begin{Bmatrix} \frac{qA}{3} \\ \frac{qA}{3} \\ \frac{qA}{3} \end{Bmatrix} \pm \begin{Bmatrix} Q_{wi} & N_i \\ Q_{wj} & N_j \\ Q_{wk} & N_k \end{Bmatrix}$$

The summation for all the elements gives the global matrix as

$$[G]\{h_1\} + [P]\left\{\frac{\partial h_i}{\partial t}\right\} = \{F\} \quad \text{...(10)}$$

Applying the finite difference scheme for the $\frac{\delta h}{\partial t}$ term in time domain for Eq. (10) gives

$$[G]\{h_1\}_{t+\Delta t} + [P]\left\{\frac{h_{i+\Delta t} - h_t}{\Delta t}\right\} = \{F\} \quad \text{...(11)}$$

The subscripts t and $t + \Delta t$ represent the groundwater head values at previous and present time steps. By rearranging the terms of Eq. (11), the final form of the equation reduce to (Istok, 1989)

$$[[P] + \omega\Delta t\,[G]]\{\mathrm{h}\}_{t+\Delta t} = [[P] - (1-\omega)\,\Delta t\,[G]]\{h\}_t + \Delta t\,(1-\omega)\,\{F\}_t + \omega\{F\}_{t+\Delta t} \quad \text{...(12)}$$

where $[G]$ = conductance matrix containing transimisivity terms; $[P]$ = storage matrix containing storage coefficient terms; Δt = time step size, $\{h\}$ = groundwater head, {F} = nodal recharge or discharge vector. $h_{t+\Delta t}$ = unknown head vector; h_t = known head vector at time t. All these are global matrices. ω = Relaxation factor which depends on the type of finite difference scheme used. In the present study Crank-Nicholson scheme with $\omega = 0.5$ is used. These global matrices can be constructed using the element matrices of different shapes based on discretization of the domain. The linear simultaneous equations derived above are solved to obtain the head distribution at nodal points using Choleski's method for the given initial and boundary conditions, recharge, pumping, transmissivity, and storage coefficient values. After getting the nodal head values the time step is incremented and the solution proceeds in the same manner by updating the time matrices and recomputing the nodal head values. From the obtained nodal head values, the velocity vectors in x and y directions can be calculated using Darcy's law.

3.2 Finite Element Formulation for Contaminant Transport

Applying Gelarkin's FEM to solve partial differential Eq. (2), similar to the groundwater flow equation, final form of equation in implicit form that is used in the contaminant transport model is given as follows.

$$[[S] + \omega\Delta t[D]]\{C\}_{t+\Delta t} = [[S] - (1-\omega)\Delta t[D]]\{C\}_t + \Delta t(1-\omega)\{F\}_t + \omega\{F\}_{t+\Delta t} \quad ...(13)$$

where $[S]$ = element sorption matrix; $[D]$ = element advection–dispersion matrix; $[F]$ = flux matrix; $[C]$ = nodal concentration column matrix; the subscripts t and Δt indicates the concentration at previous time step and current time step. As in the case of groundwater flow, Eq. (13) is applied to all elements and assembled to form system of equations. The system of equations are solved, similar to the groundwater flow, after the application of initial and boundary conditions to get the unknown concentration distribution in the aquifer region.

4 SIMULATION OF TWO-DIMENSIONAL COUPLED FLOW TRANSPORT MODEL

Based on the Gelarkin's finite element formulation, one-dimensional and two-dimensional flow and transport models have been developed using C programing. The code is applied for steady and transient groundwater flow and solute transport problems in confined aquifer including discharge term. The model uses linear triangular elements. The aquifer parameters can vary from element to element, but assumed constant over an element.

The developed model for groundwater flow and contaminant transport in confined aquifer is checked and verified with analytical solutions and also compared with the mass balance equations. The coupled model consists of groundwater flow and solute transport model. The head distribution in the aquifer domain is calculated from the flow model and these head values are used to calculate flow velocities. The resulting velocity vectors used for the transport model to find the concentration distribution in the aquifer region.

In the present study a hypothetical problem of two-dimensional coupled flow and solute transport model was investigated. The aquifer (Fig. 1) is assumed to be homogeneous and anisotropic and confined. The boundary conditions for the flow model are constant head on east (70 m) and west sides (75 m), and no flow on the north and south sides. For transport model, the boundary conditions are no-mass flux on all the sides. The aquifer is discretised into finite element mesh with a grid spacing of 200 m × 200 m. Three production wells extracting water at the rate of 500, 1000, 250 m^3/d, respectively are considered in the flow region. These wells are located at nodes 23, 33, and 35 respectively. A recharge well with inflow rate of 750 m^3/d is located at node 21.

The problem involves seepage from a polluted pond into the underlying aquifer. The bottom of the pond is assumed to be sufficiently close to the groundwater surface so that the flow in the unsaturated region is neglected. The recharge rate from the pond controls the amount of solute introduced into the system and contributes the defining velocity in the field. The aquifer is assumed initially uncontaminated. In this problem, the contaminant is entering into the flow field through seepage from the polluted pond and a polluted recharge well having concentration 1000 ppm (total dissolved solids). Present study examines the coupled flow and solute transport problem for three different scenarios. For the first and second scenarios, pumping was not involved during the simulation period. In the third scenario the system was affected along with the two input sources by pumping wells for the existing simulation period of 10,000 days. The following data is used for finding the head and concentration distribution for steady and unsteady state conditions for coupled flow and contaminant transport problem. Thickness of the confined aquifer = 25 m; size of the confined aquifer = 1800 m × 1000 m; storativity = 0.0004; aquifer is recharged through some aquitard in the area of 0.6 km^2 (between node numbers 19 and 42) at the rate of 0.00012 m/d; porosity 0.2; longitudinal dispersivity = 200 m; transverse dispersivity = 20 m; rate of seepage from the pond = 0.05 m/d; T_{xx} = 1250 m^2/d, and T_{yy} = 900 m^2/d. Finite element discritization of the problem is shown in Fig. 1.

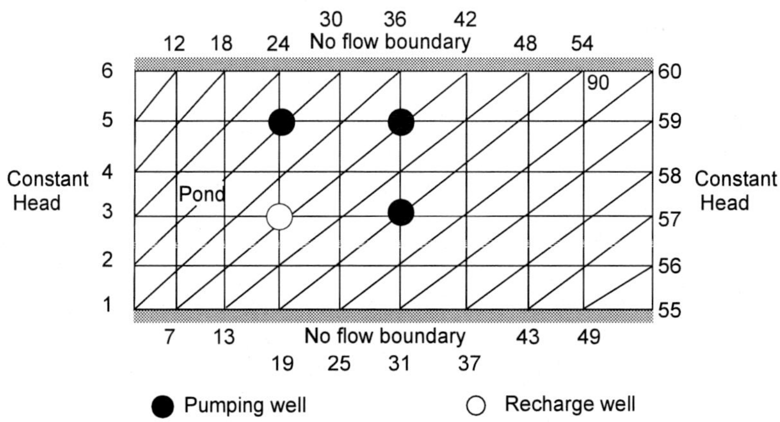

Fig. 1 Finite element discretization.

5 STRATEGIES OF AQUIFER REMEDIATION

In this study, the decontamination process in an aquifer is analyzed by two different strategies that are self cleaning and combination of flushing and pumping. Three scenarios have been considered for each remediation strategy, i.e. case 1 (a, b and c) and case 2 (a, b and c). In an aquifer contaminated by pollutants, if the sources of the pollutants are removed or the entry of the pollutants to the aquifer is stopped by some means, then the pollutants move away and the decontamination process takes place in the aquifer. It is assumed that from the beginning to a specified period, the contamination process is going on. At a particular time, the contamination source is removed and the resulting concentration movement is analyzed. For case 1 (a), initially the system was affected by seepage at the rate of 0.05 m/d, from the polluted pond with solute concentration of 1000 ppm. For case 1 (b), the input sources are seepage from the polluted pond and a contaminated recharge well. Considering these as input sources, pumping wells in the flow domain are treated as inactive, and the system was simulated for a period of 10,000 days for case 1 (a) and 1(b). For the case 1 (c) apart from these input sources, pumping wells are active during the simulation period. The obtained simulated plume is considered initial conditions for remediation strategies. The no-action strategy is the self-cleaning option, in which the aquifer is left to natural cleaning process. In this remediation strategy the input sources are cutoff, and no pumping is involved during the remediation period. The movement of the plume is due to the existing hydraulic gradient in the system and higher and lower head boundaries.

In the second strategy, the aquifer is flushed from the seepage of the polluted pond having concentration of 50 ppm at the same rate (0.05 m^3/d), also three pumping wells are extracting water at the rate of 500, 1000 and 250 m^3/d respectively from the system. Before the remediation process starts the input sources are arrested, and hence the flow in the system is due to constant head boundary conditions and abstraction wells.

6 RESULTS AND DISCUSSION ON REMEDIATION BY SELF-CLEANING

Case 1 (a)

A 400 ppm concentration contour is considered for this analysis to study the movement and the influencing area of this contour for a particular period. The behavior of this contour is analyzed for a period of 500 days and 1000 days after the decontamination begins. Initially this contour is distributed for a length of 1700 m, influencing an area of 0.85 km^2 (1700 m × 500 m). Once the decontamination is initiated, the reduction in length of this contour is very significant. The length of this plume is reduced to 750 m, and area is reduced to 0.3 km^2 (750 m × 400 m) after 500 days. The effect of this contour diminishes in the system after 1000 days. The progress of the aquifer decontamination for this alternative is shown in Fig. 2 (a, b and c).

Case 1 (b)

The area under 400 ppm contour is 1.02 km^2 (1700 m × 600 m), which is larger, compared to case 1 (a) as the system was affected by a polluted recharge well in addition to the seepage from the contaminated pond. Once the decontamination process starts, the influence of this contour is reduced to a length of 900 m and the area is reduced to 0.54 km^2 (900 m × 600 m) after a period of 500 days. Further, the concentration for the chosen contour is insignificant after 1000 days. The concentration distribution for this scenario is shown in Fig. 3 (a, b and c). Therefore after this period the aquifer can be considered completely cleaned-up for practical purpose.

Case 1 (c)

The same contour (400 ppm) is studied for the case 1 (c) in this decontamination alternative. For this scenario the system was affected by seepage from the contaminated pond, recharge well, and additionally three pumping wells were also in operation during the simulation period. Before the remediation, the concentration distribution for the 400 ppm contour is 1700 × 400 m which is influencing an area of 0.68 km^2. The length of this contour is reduced to 700 m after 500 days; the reduction in the area of this contour is (0.28 km^2) (700 m × 400 m). In this decontamination alternative, there is no effect of this contour in the system after 1000 days. Concentration distribution for different times have been shown in Fig. 4 (a, b and c).

From the above three cases for this remediation strategy, it has been found that the effectiveness of this strategy mainly depends upon severity of the initial concentration conditions in the system. Results have shown that when the pollution from the source is ceased,

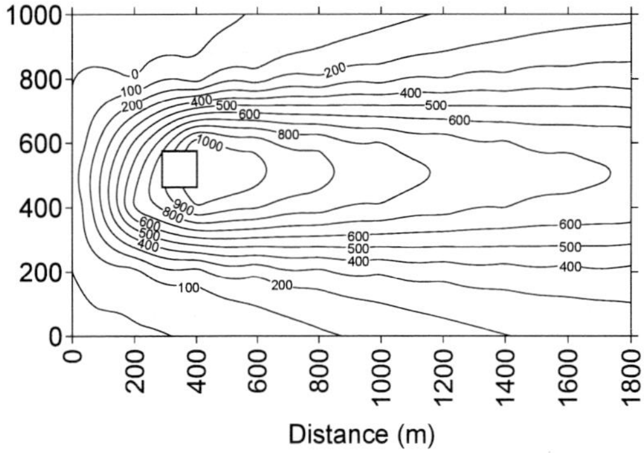

Fig. 2(a) Concentration distribution after 10,000 days of simulation for case 1 (a).

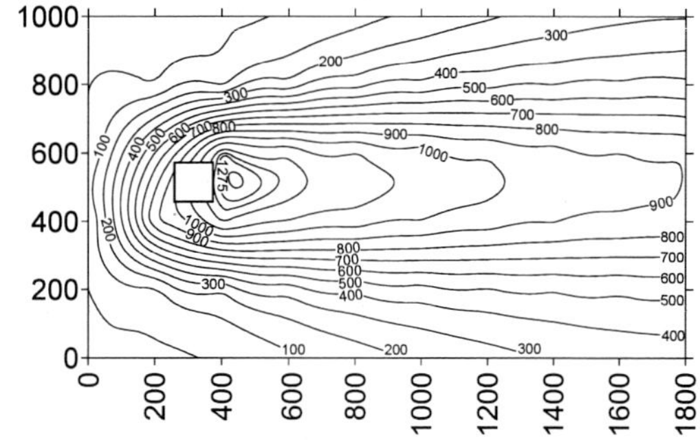

Fig. 3(a) Concentration distribution after 10,000 days of simulation for case 1 (b).

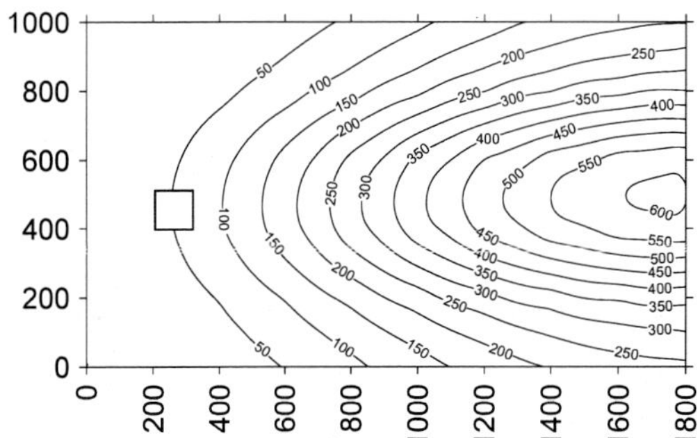

Fig. 2(b) Remediation by self-cleaning after 500 days for case 1 (a).

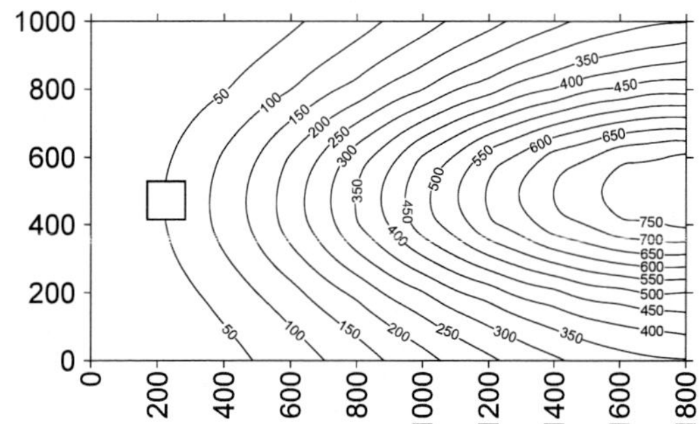

Fig. 3(b) Remediation by self-cleaning after 500 days for case 1 (b).

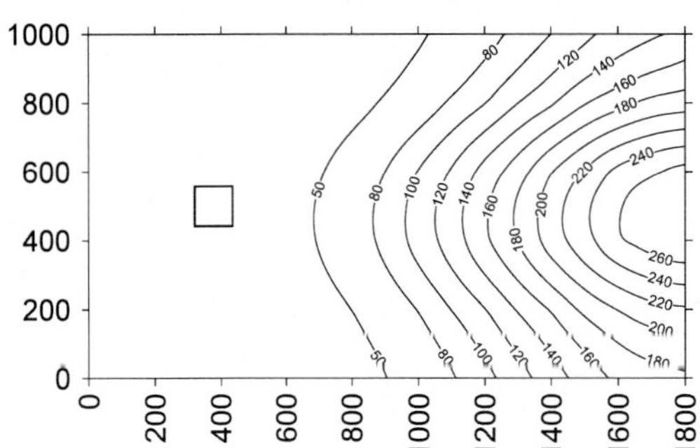

Fig. 2(c) Remediation by self-cleaning after 10,000 days for case 1 (a).

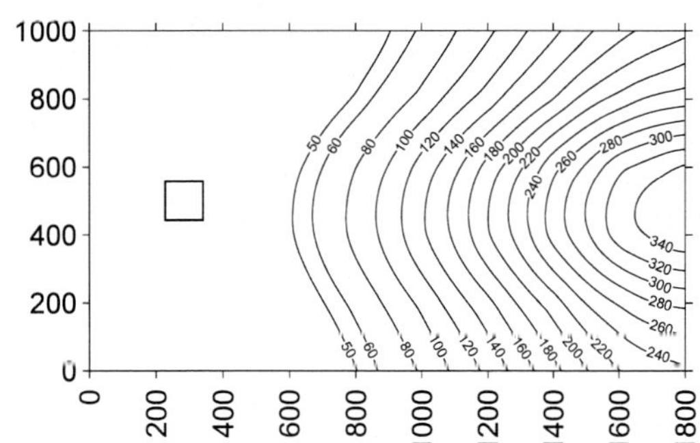

Fig. 3(c) Remediation by self-cleaning after 1000 days for case 1 (b).

the remediation by self-cleaning takes more time. This is due to the aquifer is left to natural cleaning process. As expected the cleaning process is expedited, when pumping wells are also operated.

7 RESULTS AND DISCUSSION ON REMEDIATION BY FLUSHING AND PUMPING

Case 2 (a)

The initial conditions for this remediation strategy are same as for case 1 (a). Once the decontamination is initiated, the length of this plume is reduced to 300 m and the area under this contour is reduced to 0.195 km^2 (650 m × 300 m) after a period of 500 days. Further, the effect of this contour in the system is not found after 1000 days. The progress of the aquifer decontamination for this alternative is shown Fig. 5 (a, b and c).

Case 2 (b)

The initial conditions for this strategy are same as case 1 (b). After the decontamination process starts, the influence of this contour is reduced to a length of 500 m and area is reduced to 0.4 km^2 (800 m × 500 m) after 500 days. The results have shown that, the system was not influenced by the chosen contour (400 ppm) after 1000 days. The Fig. 6 (a, b and c) illustrates, the state of contamination of the aquifer after 500 and 1000 days in this strategy.

Case 2 (c)

The initial conditions for this strategy are similar to case1 (c). After 500 days, the spreading of this contour is reduced to 600 m and the area is reduced to 0.15 km^2 (600 m × 250 m) in this decontamination process. The effect of this contour on the system is insignificant after 1000 days. The distribution of concentration contours are presented in Fig. 7 (a, b and c).

Based on the results obtained from the second remediation strategy, it could be concluded that with the increase of remediation time, the influence of the chosen contour is reduced on the system. This happens because of the abstraction wells are extracting the contaminated water and simultaneously seepage from the pond pushes forward the concentration plume. The clean-up time for the contaminated aquifer can be reduced further by increasing the number of pumping wells located near the maximum polluted regions and also by increasing the rate of freshnwater recharge from the pond.

8 CONCLUSIONS

Contamination of groundwater is a widespread problem and remediation process is very complex. In this study the contaminated aquifer remediation is investigated using a coupled flow and transport modeling with two different strategies, i.e. self-cleaning, combination of flushing and pumping. From the results it is observed that the chosen initial concentration contour in the aquifer is reduced to a reasonable extent within 500 days and further reduced by 1000 days in the self-cleaning alternative. Hence, it is understood that when the pollution from the source is ceased, remediation by self-cleaning takes more time. Since it depends upon small natural groundwater flow velocities. In case of second decontamination strategy, the clean-up time is less as compared to the first strategy. This is because pumping wells extract the contaminated water and at the same time favorable hydraulic gradients are produced due to

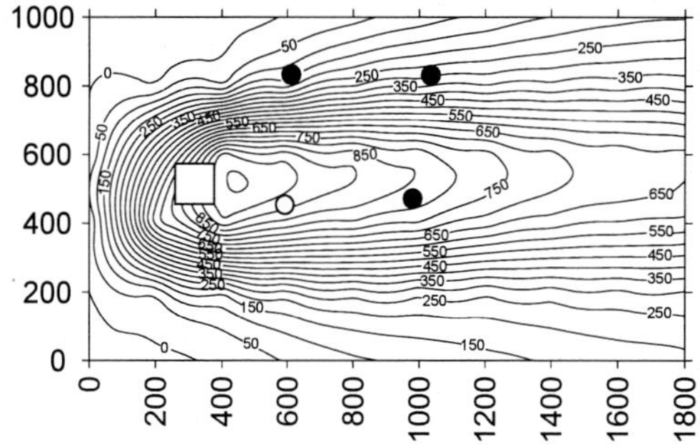

Fig. 4(a) Concentration distribution after 10,000 days of simulation for case 1 (e).

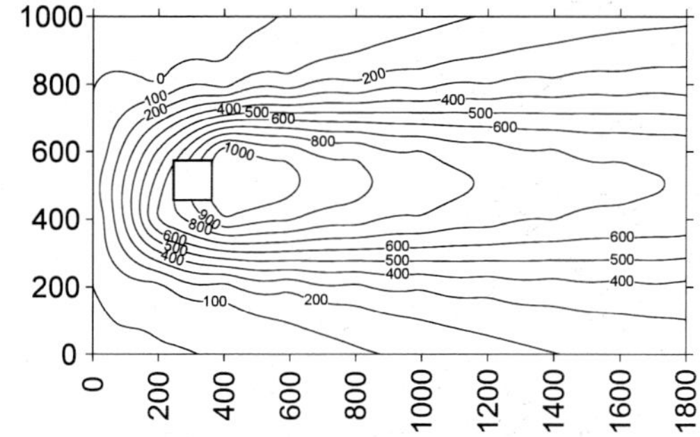

Fig. 5(a) Concentration distribution after 10,000 days of simulation for case 2 (a).

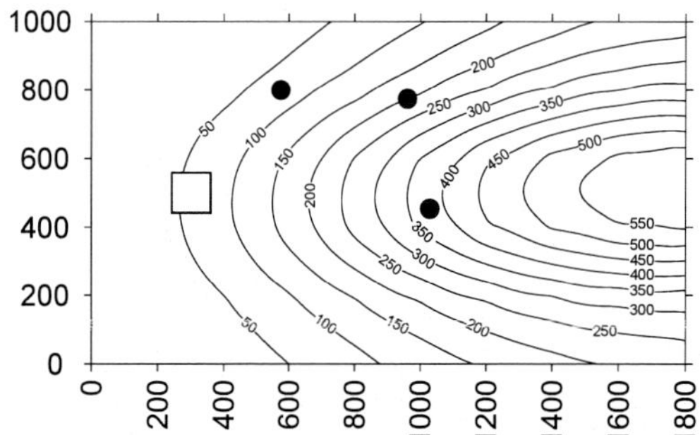

Fig. 4(b) Remediation by self-cleaning after 500 days for case 1 (e).

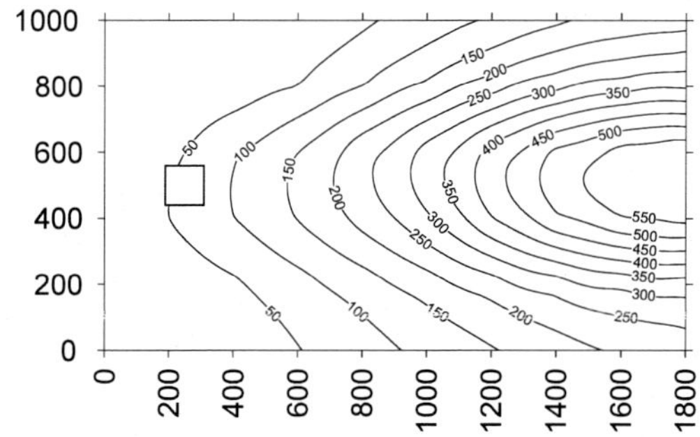

Fig. 5(b) Remediation by self-cleaning after 500 days for case 2 (a).

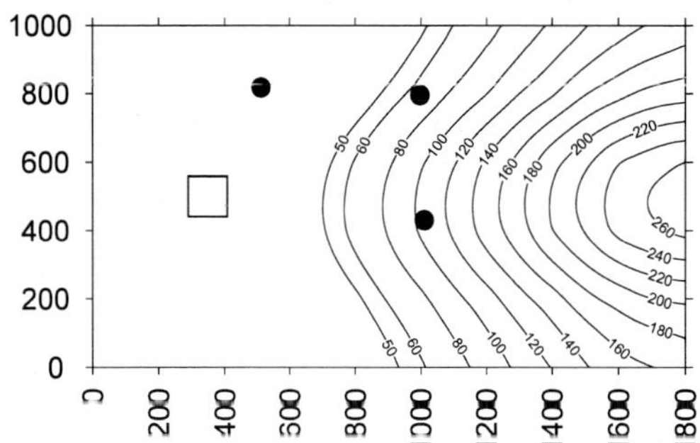

Fig. 4(c) Remediation by self-cleaning after 1000 days for case 1 (c).

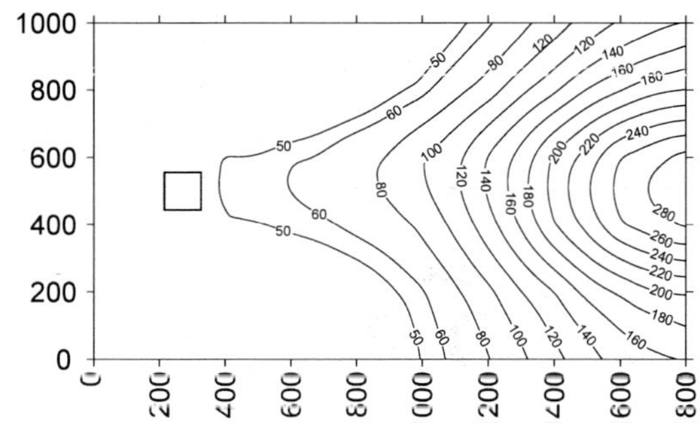

Fig. 5(c) Remediation by self-cleaning after 1000 days for case 2 (a).

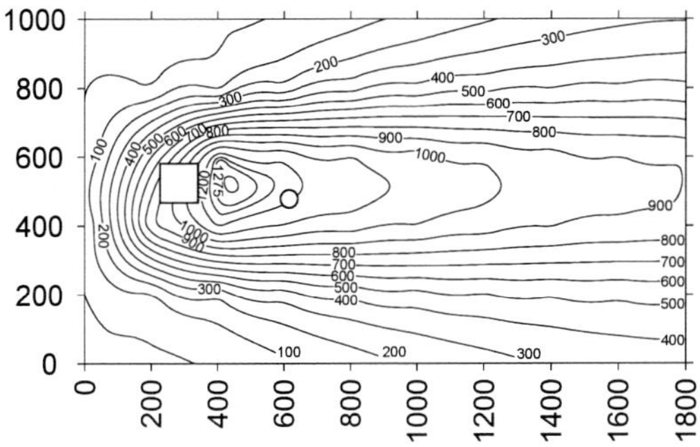

Fig. 6(a) Concentration distribution after 10,000 days of simulation for case 2 (b).

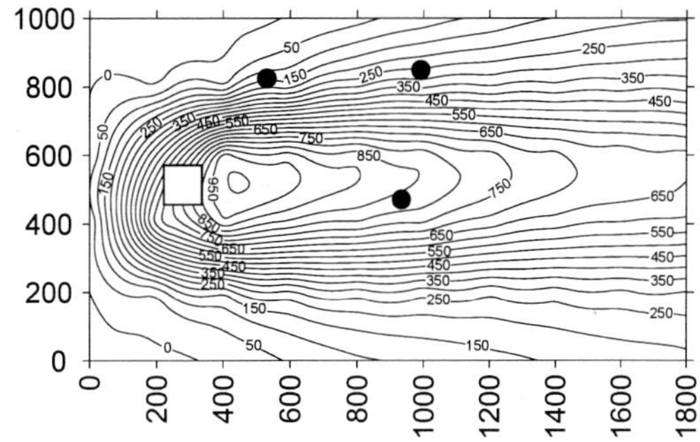

Fig. 7(a) Concentration distribution after 10,000 days of simulation for case 2 (c).

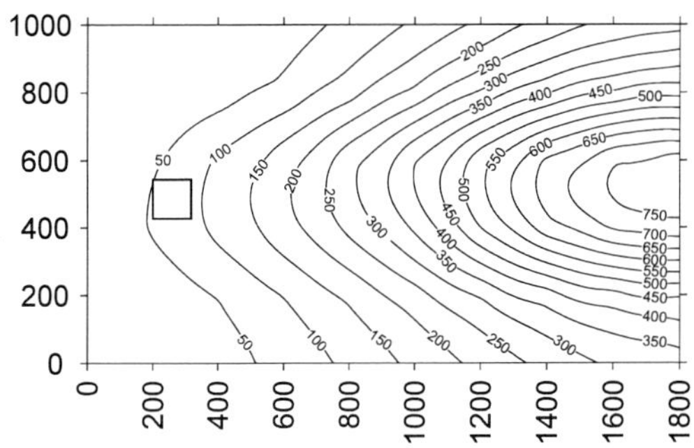

Fig. 6(b) Remediation by self-cleaning after 500 days for case 2 (b).

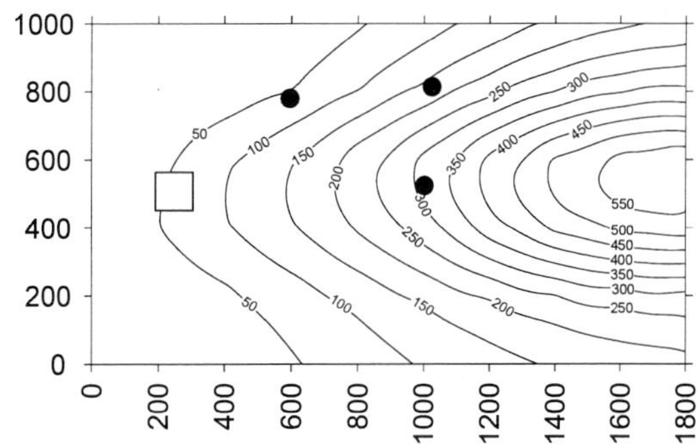

Fig. 7(b) Remediation by self-cleaning after 500 days for case 2 (c).

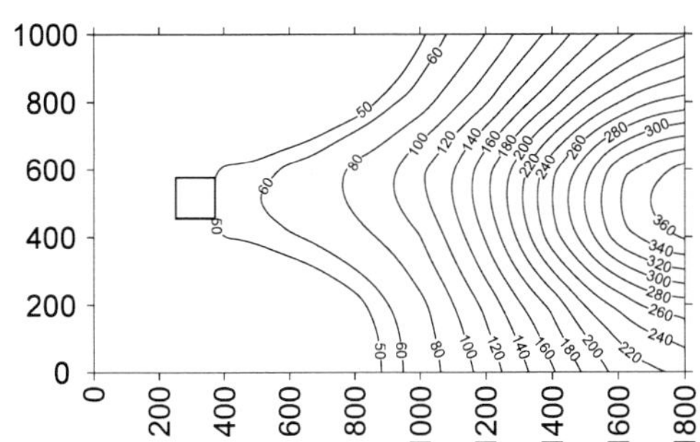

Fig. 6(c) Remediation by self-cleaning after 1000 days for case 2 (b).

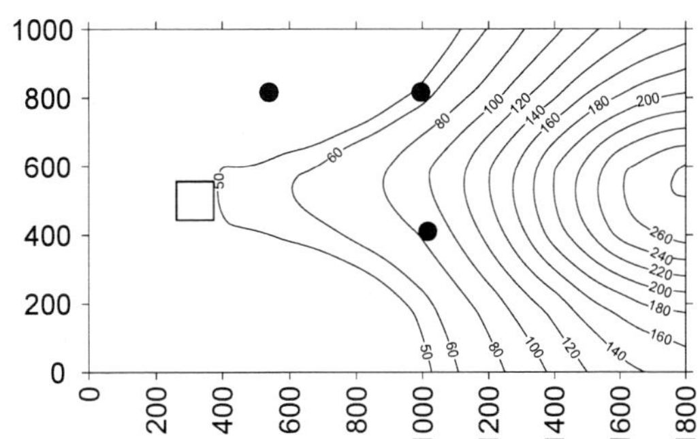

Fig. 7(c) Remediation by self-cleaning after 1000 days for case 2 (c).

freshwater flushing from the seepage pond causing contaminated plume move away from the pond. The area of the plume is significantly reduced in this strategy as the contaminant mass is extracted out of the aquifer through pumping wells. The clean-up time in this alternative mainly depends on the rate of pumping and also the number of pumping wells in the contaminated area. Many such strategies therefore can be examined by the coupled FEM model which can eventually help to pick up the most economically viable method (or least time taking) from practical consideration.

References

Bear, J. 1979. *Hydraulics of groundwater.* New York: McGrw-Hill Publishing.

Bredehoeft, J. D. & Pinder, G. F. 1973. Mass transport in flowing groundwater. Water Resource. Res., 9(1): 194–210.

Fang Xin Yu. & Singh, V. P. 1995. Improved finite element method for solute transport,. *Journal of Hydraulic Engineering.* 121(2), 145–158.

Freeze, R.A., & Cherry. J. A.1979. *Groundwater.* New York: Prentice-Hall: Englewood Cliffs.

Hunt, B. 1978. Dispersive sources in uniform ground water flow, *Journal of Hydraulics division.* 104 (HY1), 75–85.

Istok, J. 1989. *Groundwater modeling by the finite element method* .Washington: American Geophysical union.

Liji, F.J. Skaggs, T. H. & Genuchten, V. 1991. Analytical solutions for solute transport in three-dimensional semi infinite porous media., *Water Resour. Res.,* 27 (10), 2719–2733.

Marino, M. A. 1974. Longitudinal Dispersion in saturated porous media. *Journal of Hydraulics division.* 100 (HY1), 151–157.

Reddy, J.N. 1993. *An Introduction to the Finite Element Method.* New York: McGraw-Hill Publishing.

Sharmir, Y.U. & Harleman, D.R.F. 1967. Numerical solutions for dispersion in porous mediums. *Water Resour. Res.,* 3(2): 557–58.

Sun, N. 1996. *Mathematical modeling of groundwater pollution.* New York Springer Publishing House.

Sykes, J. F. Pahwa, S. B. Lantz, R. B. & Ward, D.S. 1982. Numerical simulation of flow and contaminant migration at an extensively monitored landfill. *Water Resour. Res.,* 18 (6), 1687–1704.

Wang, H. & Anderson, M.P. 1982. *Introduction to groundwater modeling finite difference and finite element methods.* New York : W. H. Freeman and Company.

Infiltration Model for Safe Fly-ash Disposal in the Compound of IB Thermal Power Plant, Orissa, India

24

Ram Karan Singh[1], S.K. Singh[2] and Satyajeet Nanda[3]

[1]Department of Natural Resources, TERI-SAS, Deemed University, Habitat Place, Darbari Seth Block, New Delhi-110003, India.

[2]Department of Civil Engineering, Delhi College of Engineering, Bawana Road, Delhi, India

[3]Department of Civil Engineering, KIIT-Deemed University, Bhubaneswar, Orissa, India

ABSTRACT

In this study, we have developed an infiltration model for safe fly-ash disposal in the compound of IB thermal power plant. This study is based on detailed fieldwork carried out in the study area sample collection and laboratory investigation. We have also carried out detailed literature review on the subject and proposed most appropriate model for the investigated sites as Explicit Green-Ampt Model. The major findings of this research are: At a constant geological and topographic condition infiltration rate increases with increased in ponding height. Hydraulic conductivity or permeability of soil at saturated condition shows linear relationship with infiltration rate i.e. the more the permeability, the greater will be the infiltration rate. It has been observed that for coarse grain soils, having coefficient of curvature varying between 0.13 and 1.15, infiltration rate increases with increase in coefficient of curvature. Assessing infiltration based on soil types will not give correct result.

1 INTRODUCTION

Protection of land, air and water from human activity are the most difficult problems in present scenarios. The threat to nature primarily coming from existence of industries as well as the wastes produced from industries. The waste from thermal power plant is one of them. Fly-ashes are finely dived residue resulting from combustion of powdered coal at about 2500

Fahrenheit. The primary sources of fly-ash are thermal power plant, steel industry, metal industry etc. Fly-ashes generally finer than cement and consist mainly of glassy-spherical particles as well as residues of hematite and magnetite, char and some crystalline phases found during cooling. Fly-ashes are no more treated as waste because of their extensive use in concrete industry but immediate disposal of fly-ash became major problem for factory management since ashes are small in size and lighter in weight, so it is quite possible that ashes can mix with air and water. For disposal of fly ashes several methodologies have been developed and which are based on ecologically safe and economically viable. In the present study, we have carried out field investigation along with laboratory test to develop an appropriate model for safe fly-ash disposal site in the compound of IB thermal power plant in Orissa, India mainly keeping in mind the leaching effect and protect the groundwater in the vicinity of the plant.

IB thermal power plant, which is situated on western part of state Orissa annually, produces of fly-ashes. The power plant adopted fly-ashes slurry disposal in ash pond. Technical problems always persist with the ash pond and the system associated with disposal of fly-ashes, sometimes problem is so acute that it leads to closer of plant. Alternative to ash pond, it may be considered to disposal of fly-ashes in low land area with proper technology. When fly-ash slurry disposal in low land area is studied, the feasibility or the economic aspect largely depends on the time required to safely converts ash slurry to dry form. Looking at the pollution control board regulation, the disposed slurry convert to dry form only by utilizing natural agency, i.e. evaporation and infiltration. The rate of evaporation is expected to be constant in a region because it largely depends upon the site temperature and wind. The rate of infiltration is not considered to be constant as it depends on soil type, topography and groundwater table, which are expected to widely vary in a region.

Infiltration is defined as the process of water movement into vadose zone through the soil surface, indirectly it may be defined as a term generally used in soil physics and hydrology to define the maximum rate at which rain or irrigation water can be absorbed by the soil under given condition. Movement of water into and through the vadose zone is very important as it has potential threat to pollute the existing groundwater.

The movement of water through soil is very dynamic, changing dramatically over time and space. Several researchers have developed various mathematical models to predict infiltration but the use of model is very specific with respect to site condition and the input parameter given to it. When ash slurry is deposited in lowland area it produces a ponding case and to predict infiltration rate in ponding case Explicit Green-Ampt Model is the best model. This paper includes prediction of infiltration rate by using Explicit Green-Ampt Model and analysis of model result with respect to different input parameter. The input parameters are collected from different locations of IB thermal power plant.

2 LITTERATURE REVIEW

Soil water infiltration is controlled by the rate and duration of water application, soil physical properties, slope, vegetation and surface roughness. At the beginning the rate of infiltration is generally very high thereafter decreasing rapidly and then slowly until it approaches a constant rate. Fig. 1 represents distribution of water into five zones during infiltration process. Saturation zone is extended into few millimeters and it assumes that all the pores are filled with water. The transition zone is extended into few centimeters and rapid decrease in water content

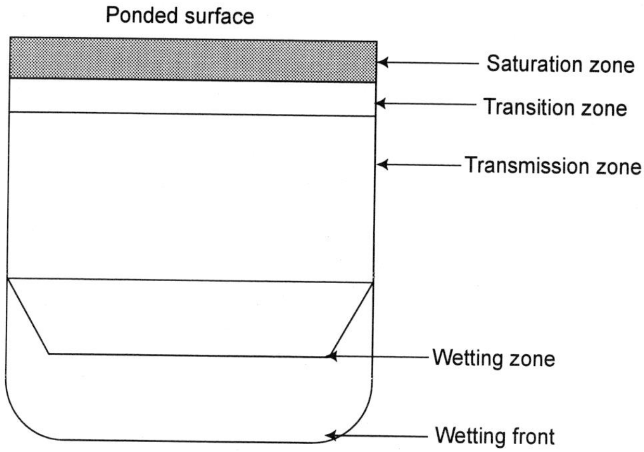

Fig. 1 Distribution of water into five zones during infiltration process.

occurs. The transmission zone is characterized by a small change in water content with depth. In general, the transmission zone is a lengthening unsaturated zone with uniform higher water content. Gravitational forces primarily drive the hydraulic gradient in this zone. In wetting zone the water content sharply decreases with depth from the water content of the transmission zone to near the initial water content of the soil. Wetting front is characterized by a steep hydraulic gradient and forms a sharp boundary between the wet and dry soil. Beyond the wetting front, there is no visible penetration of water as found by Philip (1969) and Hillel (1982).

Infiltration models broadly classified into six types, i.e. Semi-Empirical Model, Homogeneous Model, Non-Homogeneous Model, Infiltration Model for Ponding Condition, and Infiltration Model for Non-Ponding Condition and Wetting and Drying Model. Semi-Empirical Model is based on data obtained from field with very little physical basis. The commonly used semi-empirical infiltration models in the field of soil mechanics and hydrology are Kostiakov's model, Horton's model, and the Soil Conservation Service model (Hillel, 1982). Homogeneous Models are developed based on well-defined physically based theories of infiltration. Infiltration models commonly used for homogeneous soil profiles include the Green-Ampt infiltration model, the Philip infiltration model, the Burger infiltration model, and the Parlange infiltration model. Non-Homogeneous Model considers non-uniformity of soil profile with depth. The application of the Green-Ampt equation to calculate cumulative infiltration into no uniform soils has been studied by Bouwer (1969), Childs and Bybordi, (1969), Fok (1970), Moore (1981), and Flerchinger *et al.* (1988). Infiltration Model for ponding condition will be applicable when the application rate of water to the soil surface exceeds the art of infiltration, free water tends to accumulate over the soil surface. Several infiltration models for ponding conditions have been developed, including those by Parlange *et al.* (1985), Haverkamp *et al.* (1990, 1994). Infiltration Model for non-ponding condition is applicable when the rate of water supply to the soil surface does not exceed the soil infiltrability, all water can percolate into the soil and no surface ponding of water occurs. Infiltration models for non-ponding conditions have been developed by several researchers (Philip, 1957; Childs and Bybordi, 1969; Hillel and Gardner, 1970). Wetting and Drying Model based on alternate infiltration and exhilaration of water at the soil surface will result in an unsteady diffusion of water into the soil. Infiltration models for such conditions have been developed by Eagleson (1978) and Corradini *et al.* (1994).

Explicit Green-Ampt Model is a physically base equation, which described infiltration of water into soil. This is the best model for predicting infiltration rate in ponding condition because its mathamecal equation takes ponding height as one parameter. Mathematical formulations for this model are as follows:

$$\frac{1}{Ks} = \left(1-\sqrt{2}/3\right)t + \frac{\sqrt{2}}{3}\sqrt{Xt+t^2} + \left(\frac{\sqrt{2}-1}{3}\right) \times X[\ln(t+X) - \ln X]$$

$$+\frac{\sqrt{2}}{3} \times \left[\ln\left(t+\frac{X}{2}+\sqrt{Xt+t^2} - In\left(\frac{X}{2}\right)\right)\right]$$

$$\frac{q}{K_s} = \frac{\sqrt{2}}{2}\tau^{-1/2} + \frac{2}{3} - \frac{\sqrt{2}}{6}\tau^{1/2} + \frac{1-\sqrt{2}}{3}\tau$$

$$X = \frac{(h_s - h_f)(\theta_s - \theta_0)}{K_s}$$

$$\tau = \frac{t}{t+X}$$

$$\eta = (2+3\lambda)$$

$$h_f = \frac{\eta h_o}{\eta - 1}$$

where,

- q is infiltration rate (cm/h),
- K is saturated hydraulic conductivity (cm/h),
- t is time (h),
- Hs is ponding depth or capillary pressure head at the surface (cm),
- hf is capillary pressure head at the wetting front (cm),
- θs is saturated volumetric water content (cm/cm) and $\theta 0$ is initial volumetric water content (cm/cm).
- he air exit head in cm
- λ is the exponent of the Brooks-Corey water retention model

3 RESULT AND DISCUSSION

Soil sample collected from different locations of IB thermal plant mainly coarse grain soil of SP, SW, GP. The coefficient of curvature varies between 0.13 and 1.15 and coefficient of uniformity varies between 2.1 and 22.5. Fig. 2 represents effect of ponding height and rate of infiltration,

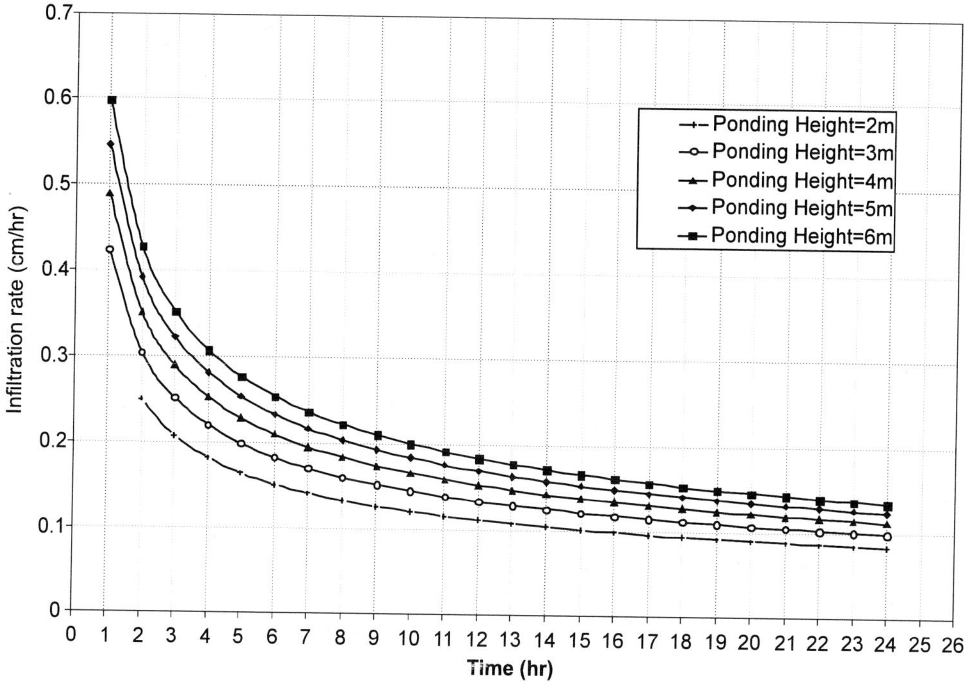

Fig. 2 Variation of infiltration rate with ponding height at site 1.

which are indicated with increase in ponding height rate of infiltration. It is observed from Fig. 2 that the effect of ponding height persist with time and also when infiltration reaches to a constant rate.

Variation of constant infiltration rate and the coefficient of curvature (C_c = D302/D10 D60) is plotted in a semi-log plot (Fig. 4). The plot shows that at lower value of C_c infiltration rate increase rapidly and then rate of increase in infiltration rate gradually decreases.

4 CONCLUSIONS

On the basis of Explicit Green-Ampt Model results and analysis of test result with soil properties and hydrological condition, following conclusions are arrived.

- At a constant geological and topographic condition infiltration rate increases with increase in ponding height.
- Hydraulic conductivity or permeability of soil at saturated condition shows linear relationship with infiltration rate, i.e. the more the permeability, the greater will be the infiltration rate.
- It has been observed that for coarse grain soils, having coefficient of curvature varying between 0.13 and 1.15, infiltration rate increases with increase in coefficient of curvature.
- Assessing infiltration based on soil types may not give correct results.

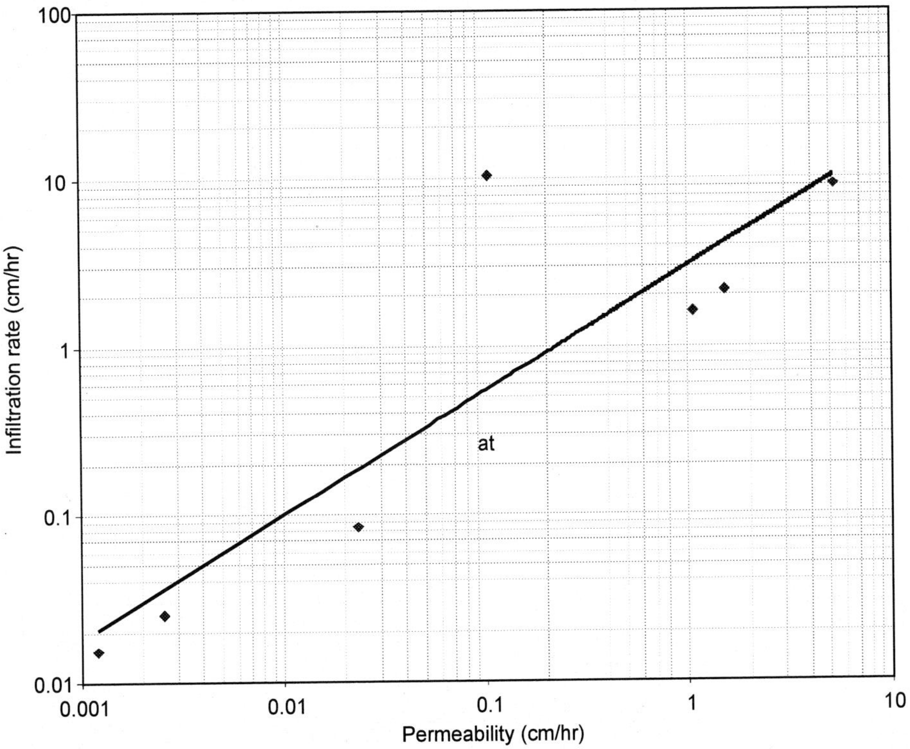

Fig. 3 Variation of infiltration rate at 24 hr at constant with ponding height of 2 m.

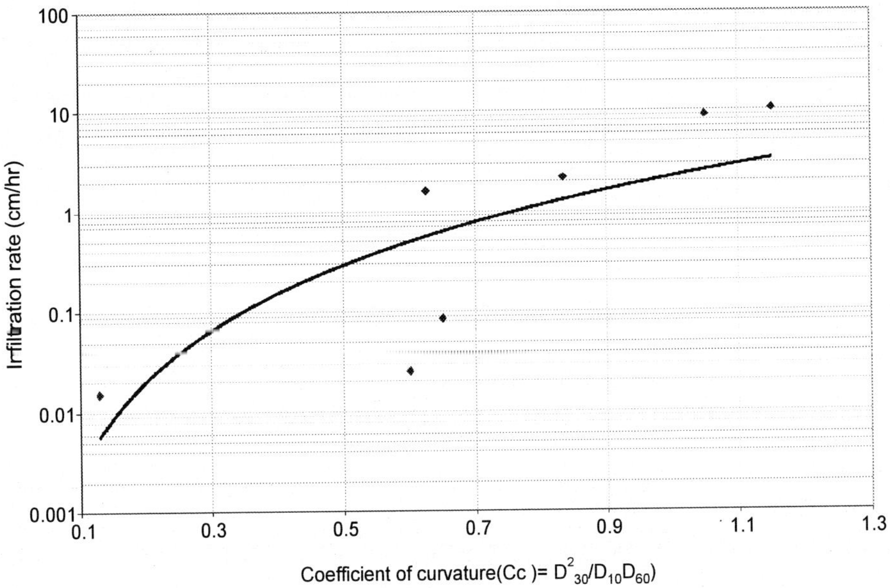

Fig. 4 Variation of coefficient of curvature with the infiltration rate at 24 hr and constant ponding height.

References

Bouwer, H. 1969. Infiltration of water into nonuniform soil, Irrigation and Drainage Division of ASCE, 95: 451–462.

Childs, E.C. and Bybordi. M. 1969. The vertical movement of water in stratified porous materials infiltration, Water Resour. Res., 5: 446–459.

Corradini, C., F. Melone, and R.E. Smith. 1994. Modeling infiltration during complex rainfall sequences, Water Resour. Res., 30: 2777–2784.

Eagleson, P. 1978. Climate, soil, and vegetation. 3. A simplified model of soil moisture movement in the liquid phase, Water Resour. Res., 14: 722–730.

Flerchinger, G.N., F.J. Waltts, and G.L. Bloomsburg. 1988. Explicit solution to Green-Ampt equation for nonuniform soils, Journal of Irrig. and Drain. Div., ASCE, 114: 561–565.

Fok, Y. S. 1970. One-dimensional infiltration into layered soils, Journal of Irrig. and Drain. Div., ASCE, 90: 121–129.

Haverkamp, R., J.Y. Parlange, J.L. Starr, Schmitz, and C. Fuentes. 1990. Infiltration under ponded conditions: A predictive equation based on physical parameters. Soil Sci., 149: 292–300.

Haverkamp, R., P.J. Ross, and J.Y. Parlange. 1994. Three-dimensional analysis of infiltration from the disc infiltrometer, Water Resour. Res. 30: 2931–2935.

Hillel, D. 1982. Introduction to Soil Physics. Academic Press. New York.

Hillel, D., and W.R. Gardner. 1970. Measurement of unsaturated conductivity diffusivity by infiltration, Soil Sci., 109: 149.

Moore, I.D. 1981. Infiltration equation modified for subsurface effects. Journal of Irrigation and Drainage Division, ASCE 107: 71–86.

Parlange, J. Y., R. Haverkamp, and J. Touma. 1985. Infiltration under ponded conditions optimal analytical solution and comparison with experimental observations, Soil Sci., 139: 305–311.

Philip, J.R. 1957. The theory of infiltration sorptivity and algebraic infiltration equations. Soil Sci., 84: 257–264, Academic Press, New York, USA.

Philip, J.R. 1969. Theory of infiltration. Advances in Hydroscience. 5: 215-296. Academic Press, New York, USA.

25

Stochastic Algorithm for Simulation of NAPL Dissolution and Degradation Reactions in Porous Media

P. Jegathambal[1] and S. Mohan[2]

[1]Department of Civil Engineering, Indian Institute of Technology, Chennai.

[2]Department of Civil Engineering, Indian Institute of Technology, Chennai.

ABSTRACT

The kinetic approach is suitable for solving complex reactive transport processes in groundwater. If the transport scale is faster than the reaction time scale, the kinetic formulation of reaction modeling can be studied by two different approaches. The time behaviour of the homogenous chemical reaction system can be described either by the deterministic approach or by the stochastic approach. The deterministic approach considers the time evolution as continuous, which is governed by a set of coupled Ordinary Differential Equations (ODEs). The stochastic approach considers the time evolution similar to a random walk process. It is governed by a single differential equation, which is called the master equation. This paper deals with the conventional way of modeling geochemical reaction systems using a deterministic approach with its limitations and clarifies the necessity and application of stochastic algorithms (Gillespie's Algorithms) to PCE_{NAPL} dissolution degradation reaction. The simulation of PCE_{NAPL} dissolution degradation reaction was done using Gillespie's Direct Method, a stochastic. The next reaction to be occurred and time of occurrence of reaction are obtained through generation of two random numbers. The following figure represents concentration profiles of PCE_{NAPL} dissolution degradation simulation using ODE and Stochastic Algorithm. From the above results, it is concluded that Stochastic Algorithm can be a better approach for reaction involving low concentration species.

1 INTRODUCTION

1.1 Context

Reactive transport models couple mass transfer processes, abiotic transformations and microbial activities with the transport processes such as advection and dispersion. The quantitative description of all these subsurface processes limits the practical application of models, for example, uncertainty in hydraulic conductivity limits the model accuracy.

1.2 Stochastic Approach for Reaction Modeling

In the deterministic approach, the species population is described by a continuous state although a chemical reaction involves random collisions between individual species. Also a predictable system is assumed for the reaction rate or velocity and time evolution. These two assumptions are not appropriate for a system with low concentration of species where high relative fluctuations due to stochastic effect occur. Stochasticity in the state change occurs as intrinsic and extrinsic stochasticity. The intrinsic stochasticity is inherent to the system and mainly arises due to low concentration of species and extrinsic stochasticity arises due to the random variation of environmental factors.

1.3 Gillespie's Direct Method (GDM)

Stochastic Algorithms, the Direct Method and the First Reaction Method to solve the chemical master equation, based on the assumption that the reaction system is well mixed and homogenous. The method provided exact stochastic realization of master equation (Gillespie, 1977a) and the results can be obtained without solving master equation numerically. The algorithm is executed in the following manner.

- The rate of each reaction a_μ is determined by multiplying the reactant partners (here only single species is involved in degradation) with the specified rate constant. The total reactivity a_0 of the system is calculated by summing up reaction rate of each reaction.
- The probability value of occurrence of each reaction is calculated using a_μ / a_0.
- A reaction is chosen according to a uniform random number in $(0, ..., a_0)$.
- The time step is generated based on an exponentially distributed random number with a mean of $1/a_0$.
- Based on the stoichiometric values, the particles of reactant species are removed from the system and the particles of product species are added on to the system. Thus, the state of the system is updated.

2 PREDICTION OF BIODEGRADATION COUPLED WITH NAPL DISSOLUTION GETTING STARTED

2.1 Introduction

The origin of the groundwater plumes is mostly due to spills, leaks of immiscible liquids which are referred to as NAPL (Non Aqueous Phase Liquids). The leachate travels though

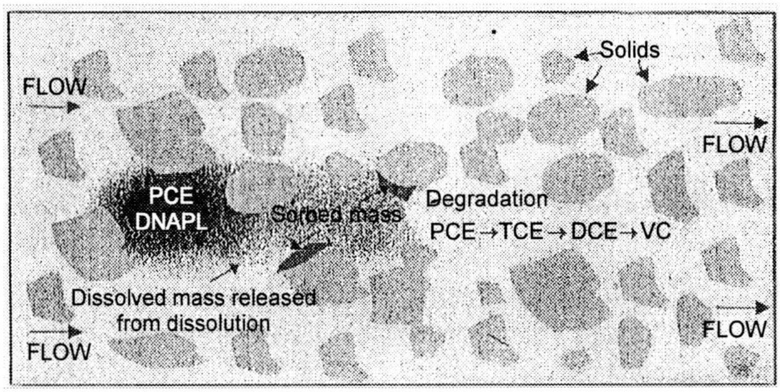

Fig. 1 Conceptual representation of rate limited reactive transport processes (Clement, 1998).

unsaturated media and reaches saturated porous media. During its transport they may either spread in lateral direction or get trapped on to the surface space of pores as given in Fig. 1. Miller and Benson (1983) have described the transfer from solid phase to liquid phase is described by first-order mass-transfer model. The fate and transport of such contaminant is described by the following equation Clement (2001).

$$\frac{\partial C}{\partial t} = \frac{\partial}{\partial x_t}\left[Dy\frac{\partial C}{\partial x_j} - \frac{\partial}{\partial x_t}(v_t C)\right] + \frac{qx}{\phi}C_s + k_L a_0 (C - C) \qquad ...(1)$$

$$\frac{\partial C_{NAPL}}{\partial t} = -\,k_L a_0 (C - C) \qquad ...(2)$$

where C is the mobile phase concentration in mg/L, C* is equilibrium concentration in mobile phase in mg/L, kL is the mass transfer coefficient mg/L, a_0 is the interfacial area between NAPL and mobile phase in 1/L. Since NAPL is defined as the concentration per unit volume of aqueous phase, the porosity and bulk density need not be considered in the transport equation.

2.2 Sorption and Desorption Process

When the dissolved contaminants travel faster in more permeable layers and more slowly in the clay lenses of aquifer then Local Equilibrium Approach (LEA) is not valid. Even when the plume has crossed certain point, the slow kinetic desorption starts to transport the contaminant from clayey lenses to the permeable area. Due to this, even after pump and treatment process, the slow desorption continuous to feed the contaminant over long period. These conditions require a mass-transfer description for the sorption reactions which is dictated by the value of mass-transfer coefficient. During this situation, it is indispensable to track the contaminant concentration in mobile and immobile species where the rate is represented in mathematical form as given in the following reactions:

$$\frac{dC}{dt} = -\varepsilon\left(C - \frac{S}{\lambda}\right) \qquad ...(3)$$

$$\frac{dS}{dt} = -\frac{\phi\varepsilon}{\rho}\left(C - \frac{S}{\lambda}\right) \quad ...(4)$$

where C is the contaminant concentration in mobile phase in mg/L, S is the contaminant concentration in immobile phase in mg/L; ε is the mass transfer coefficient.

3 SIMULATION OF NAPL DISSOLUTON AND DEGRADATION USING GILLESPIE'S DIRECT METHOD

3.1 Simulation Parameters

In this problem, the simulation of PCE_{NAPL} dissolution and biodegradation is done where PCE and TCE are assumed to degrade only in aqueous phase. The system parameters are given in Table 1.

Table 1 System parameters for NAPL dissolution and degradation.

S. No.	Parameter	Values
1.	Bulk density	1.6 kg/L
2.	PCE solubility	200 mg/L
3.	NAPL-Phase PCE	1000 mg/L
4.	Porosity	0.3
5.	Mass transfer rate	1
6.	PCE decay rate	0.2
7.	TCE decay rate	0
8.	Retardation factor for PCE	2
9.	Retardation factor for TCE	1

3.2 Simulation using Gillespie's Direct Method

In this problem, PCE_{NAPL} dissolution and biodegradation is simulated using Gillespie's Direct Method. The deterministic rate equation for concentration of PCE species are calculated based on the following equations, where the Eq. 5 represents the rate reaction for the PCE concentration in mobile- phase, Eq. 6 represents the concentration of the formation of TCE in mobile phase and the rate of dissolution of PCE_{NAPL} is given in Eq. 7.

$$k_{LA}([PCE]^* - [PCE]) = \frac{k_{PCE}[PCE]}{R_{PCE}} \quad ...(5)$$

$$\frac{d[TCE]}{dt} = \frac{Y_{TCE/PCE}\,K_{PCE}[PCE]}{R_{TCE}} - \frac{K_{TCE}[TCE]}{R_{TCE}} \quad ...(6)$$

$$\frac{d[PCE_{NAPL}]}{dt} = -k_{LA}([PCE]^* - [PCE]) \quad ...(7)$$

For the simulation using GDM, a scale of 1:1 is used to transfigure the concentration into number of particles. In this problem, three reactions are assumed in which the deterministic rate is taken for NAPL immobile phase (for simulation of dissolution reaction) and the linear rate constant is assumed for other two reactions based on the decay rate constants of PCE and TCE and their yield values in mobile phase (biodegradation process). The algorithm generates two random numbers to choose the reaction and its time of occurrence based on the reaction rates to update the state of the system. The results obtained were compared with the results using ODE solver in MATLAB as given in Fig. 2. The PCE profile in the mobile phase reaches the saturation concentration and remains stable, whereas TCE profile in mobile phase continues to increase due to degradation of PCE to TCE.

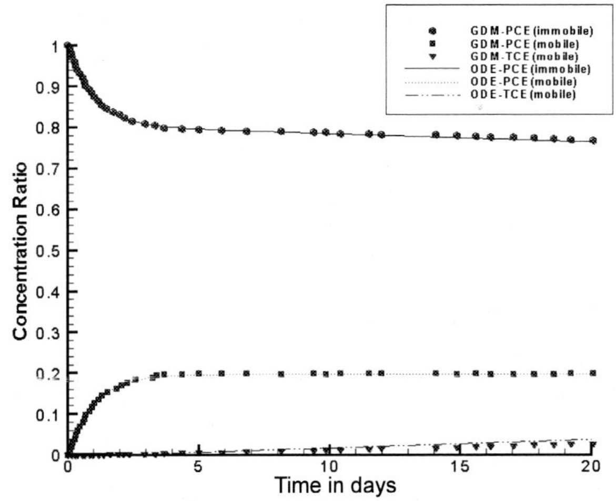

Fig. 2 Simulation results using GDM and ODE solver for 20 days.

For the study of NAPL-PCE dissolution and degradation, the rate of the dissolution reaction is calculated based on the deterministic rate equation. So when the concentration to particle ratio is 1:1, a good agreement of results in concentration profiles of NAPL-PCE, PCE (mobile) and TCE (mobile) is obtained. The simulation is carried out to determine the time period of degradation of all particles of NAPL-PCE to TCE. The time of complete degradation to TCE is found to be 400 days. For this simulation period, the model was run with concentration to particle ratio of 1:4, which has no significant change in the concentration profiles. The simulation results are depicted in Figs. 3, 4, 5 and 6.

4 CONCLUSIONS

Reactions where a high degree of coupling exists between various species can be easily modeled using stochastic algorithms. When different types of reactions such as sorption, precipitation, degradation, etc. occur in the subsurface medium simultaneously, the realistic behaviour of the low contaminant concentration species can be modeled using Gillespie's algorithms. The difficulty in formulation of Ordinary Differential Equation (ODE) for reaction simulation can be overcome using Gillespie's algorithms where the reactions can be

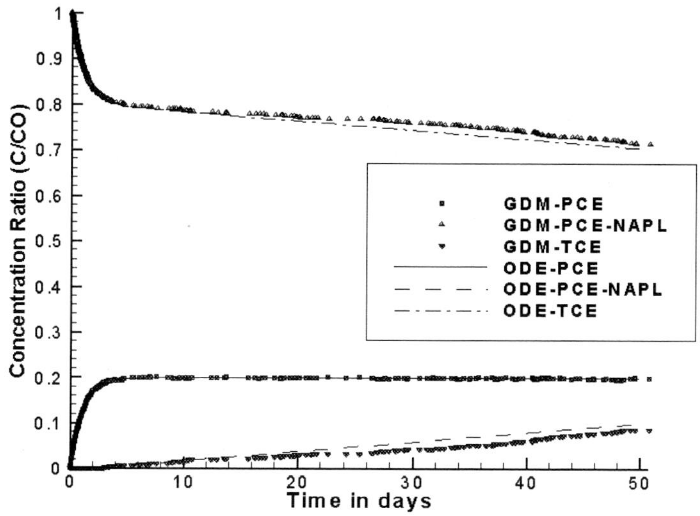

Fig. 3 Simulation results using GDM and ODE solver for 50 days.

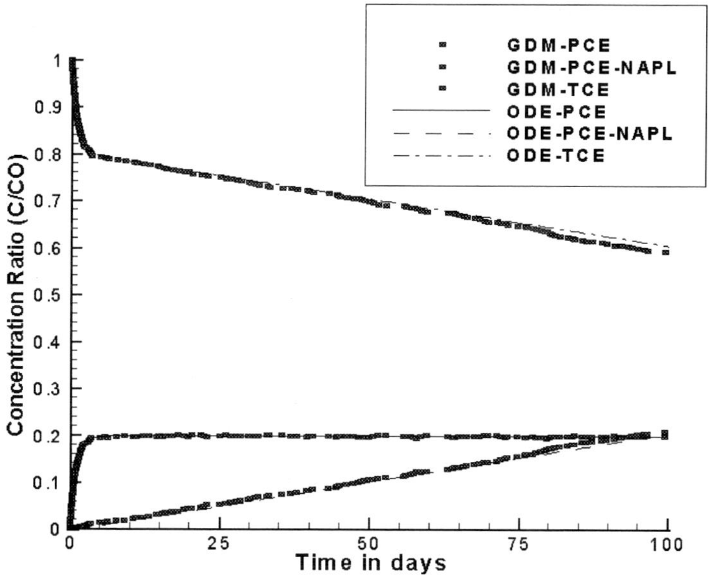

Fig. 4 Simulation results using GDM and ODE solver for 100 days.

represented in the form of matrix form which involves the stoichiometric constants and the reaction rate constants. GDM is suitable for modeling first order homogenous and heterogeneous reactions. For heterogeneous types of reactions simulation as in sorption reactions using a Langmuir isotherm, GDM is sensitive to the particle to concentration ratio (1:1) since the sorption capacity is taken into account.

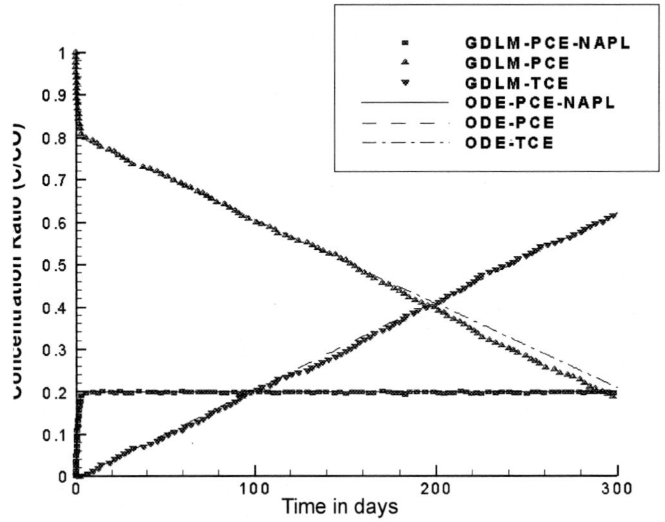

Fig. 5 Simulation results using GDM and ODE solver for 300 days.

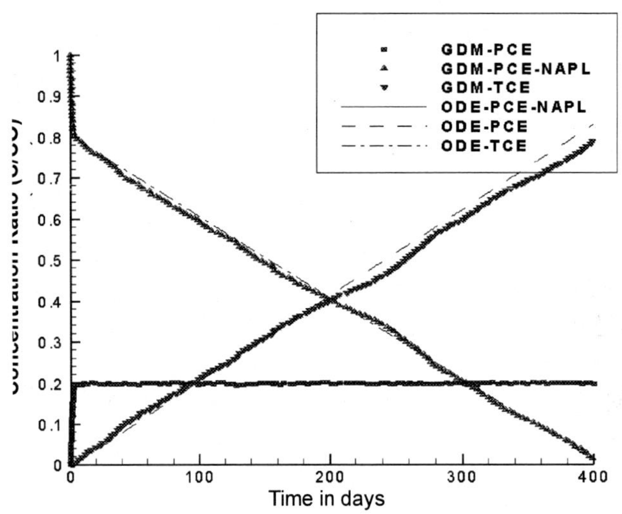

Fig. 6 Simulation results using GDM and ODE solver for 400 days.

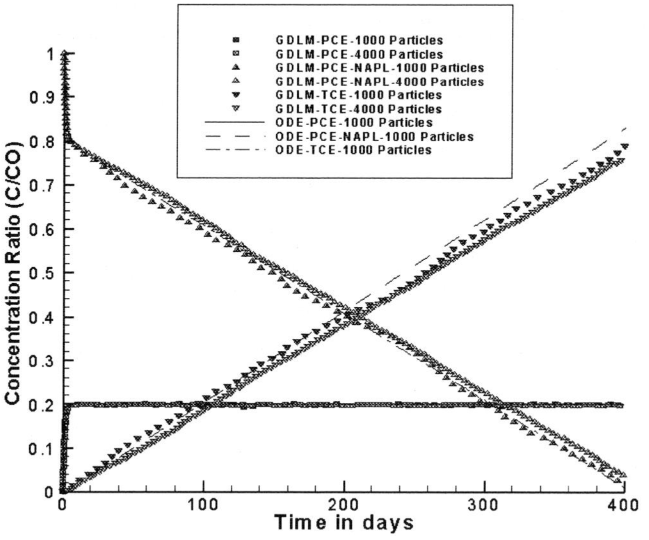

Fig. 7 Simulation results using GDM and ODE solver for 400 days with concentration to particle ratio of 1 : 1 and 1 : 4.

References

Becker, T. (2003). Stochastic simulation algorithm for coupled reactive transport processes. Proc. XXX IAHR Congress, Student Paper Competition, AUTh, Thessaloniki, Greece, 49–56.

Clement, T. (1997). A modular computer code for simulating reactive multispecies trans- port in 3-dimensional groundwater system, RT3D manual.

Clement, T. (1998). Rt3d reaction module for modeling bio degradation coupled with NAPL dissolution processes. Draft-NAPL dissolution module.

Gillespie, D. (1977a). Concerning the validity of the stochastic approach to chemical kinetics. Journal of Statistical Physics, Vol. 16, 311–318.

Gillespie, D. (1977b). Exact stochastic simulation of coupled chemical reactions. The Journal of Physical Chemistry, Vol. 81, No. 25, 2340–2361.

Gillespie, D. (1978). Monte carlo simulation of random walks with residence time dependent transition probability rates. Journal of Computational Physics, 28, 395–407.

Gillespie, D. (2001). Approximate accelerated stochastic simulation of chemically reacting systems. Journal of Chemical Physics, Vol. 115, No. 4, 1716–1733.

Miller, C. & L. Benson (1983). Simulation of solute transport in a chemically reactive heterogeneous system: Model development and application. Water Resources Research, Vol. 19, No. 2, 381–391.

Modeling of Riverbed Aquifer with Radial Collector Well and Stream Recharge Using MODFLOW and Analytic Element Method: A Case Study

26

H. M. Patel[1], T. I. Eldho and A. K. Rastogi[2]

[1]Civil Engg. Deptt. M.S.University of Baroda,Vadodara— 390001 India. Email: haresh@iitb.ac.in

[2]Dept. of Civil Engineering., Indian Institute of Technology Bombay,Mumbai—400 076 India.

ABSTRACT

An effective and cost efficient development of limited water resources need application of modeling techniques in designing water works. In alluvial river, a Radial Collector Well (R.C. Well) is appropriate solution to satisfy large demand under low-flow condition. Due to asymmetry of radials and position of recharge channel, no exact analytical solution exists to provide discharge drawdown relationship. A MODFLOW model and Analytic Element Method (AEM) based model were used to simulate stream aquifer interaction process for a given radial position under steady state. Results of AEM are comparable with MODFLOW model. AEM is found to be very simple to construct the model for design purpose. Initial trials for radials layout can be done using AEM and final design can be checked by MODFLOW model. This approach saves considerable time during design process.

1 INTRODUCTION

As urban, industrial and agricultural water use grows, the effective and cost efficient development and management of limited water resources is quickly increasing in importance. The firm supplies of non-perennial rivers are limited in absence of storage facilities such as reservoir. In such situations, a careful design of intake structure is required to ensure supply of water for specified needs. In alluvial river, a radial collector well (R.C. Well) is appropriate solution to satisfy large demand under low-flow condition. Another advantage of providing radial collector well is that it provides naturally filtered water and saves the treatment cost.

As the radial collector well comprises horizontally placed screen pipes in riverbed aquifer, the complete structure behaves like well with large effective radius. However, due to asymmetry of radials and position of recharge channel, no analytical solution exists to exactly provide discharge drawdown relationship. The discharge and design of radials are governed by saturated thickness of aquifer, position of river channel with reference to radial collector well, hydraulic conductivity of riverbed aquifer, gradation of aquifer material and depth of scour. An appropriate methodology to design radial collector well is to decide radials orientation as per site condition and then to verify the design discharge by constructing riverbed aquifer flow simulation model. In this paper, a MODFLOW model and AEM based model were used to simulate radial collector well and stream recharge. The methodology is demonstrated through a case study.

2 RIVERBED AQUIFER MODELING USING MODFLOW

In alluvial rivers, the riverbeds generally comprise a layer of highly pervious sand underlain by impervious clay of rocky strata. In case of non-perennial rivers, during winter-flow, the surface channel carries very low discharge but the saturated riverbed aquifer contains large amount of clean water. This riverbed reservoir receives induced recharge from the surface channel when the aquifer is pumped. Due to shallow depth, pumping of the aquifer is generally carried out by radial collector wells or infiltration galleries. Under constant pumping and constant river inflows during most of the time, the flow condition can be analyzed as steady state. The governing equation for groundwater flow is solved by applying appropriate boundary conditions and by using numerical techniques. Various modeling tools are available to construct aquifer flow model. One of the most popular modeling tools is MODFLOW (Mc Donald and Harbaug, 1988) which uses finite difference scheme. For riverbed aquifer system, RIVER module of MODFLOW simulates river channel. The banks of the rivers can be simulated as *no-flow* boundary or *flux* boundary as the case may be. In case of radial collector well, the horizontal wells are difficult to simulate using *well* boundary which considers constant discharge or *drain* boundary which considers atmospheric pressure at drain level. The flow into radials of collector well is governed by the difference between head in aquifer outside the screen pipe and water surface elevation in central collector well. The screen resistance also affects the discharge. In this case *head dependant* boundary of MODFLOW is appropriate.

3 ANALYTIC ELEMENT METHOD

Due to spatial variability of alignment of river channel, and radials, grid type construction of MODFLOW make the creation of model difficult and unwieldy. An Analytic Element Method (AEM) was evolved to simulate the aquifers flow problems with accuracy similar to analytic solution and ease of representing hydro-geologic elements similar to finite elements or finite difference technique. Strack (1989) gave detailed mathematical descriptions of the analytic elements and their numerical implementation. By adopting the Dupuit–Forchheimer assumption, shallow confined flow and unconfined flow can be modeled as two-dimensional flow. Strack introduced a discharge potential, *F*, which can always be expressed in terms of the piezometric head f. The relation between head and potential may be either linear (e.g. for the cases of semi-confined and confined flow) or non-linear (e.g. for the case of unconfined flow) as shown in Eq. (1).

$$\Phi = kH\phi - \frac{1}{2}kH^2 \qquad (\Phi \geq H)$$

$$\Phi = \frac{1}{2}k\phi^2 \qquad (\Phi \leq H) \qquad ...(1)$$

where, k represents the hydraulic conductivity, H the thickness of the confined aquifer and Φ the piezometric head. The discharge potential defined above satisfies the continuity equation and gives Laplace equation. (Haitjema, 1995) It is possible to represent features of the aquifer by analytic function which is known as analytic element. Most elements are considered in the form of complex potential, $\Omega = \Phi + i\Psi$, whose real part is the potential function $\Phi = \Re(\Omega)$, and the imaginary part is the stream function, $\Psi = \Im(\Omega)$. All the potentials represent one particular feature of the aquifer. To obtain a full description, these particular features are superimposed. Due to the linearity of the Laplace-equation it means a simple sum of the individual potentials. Once the unknowns of various potential functions are worked out, sum of all potential functions gives discharge potential at any point in the aquifer. The value of discharge potential is then converted to head. The AEM model has been developed using a tool Visual Blue Bird (Craig and Matott, 2005) which provides a library of various elements such as well, lake, river/stream, inhomogeneity to represent no-flow boundaries, uniform flow, etc. Both the MODFLOW and AEM based Visual Blue Bird are used in present case study.

4 CASE STUDY

A radial collector well was proposed to be constructed at confluence of river Kanhan and Jam near Borgaon of Madhya Pradesh (Fig. 1), India for supplying 22.5 million liter per day water to an industrial area of M.P. Audyogic Kendra Vikas Nigam. Necessary field data were collected and pumping tests were carried out.

The design discharge and details of radials were decided by the following parameters:

1. Saturated thickness of aquifers around the collector well.
2. Position of river channel with reference to the collector well.
3. Hydraulic conductivity of the aquifer material.
4. Gradation of aquifer material.
5. Depth of scour.

A saturated thickness of 6.9 to 8.6 m was found except on right side of radial collector well. Radials cannot be therefore placed in some portion on right side. River channel position with reference to collector well largely affects the yield of collector well. As river channel position is likely to change with season, following two critical cases are considered in this study to find out minimum discharge available from well during worst conditions.

Case A: River channel 50 m on left side of the collector well.

Case B: No river channel, i.e. surface channel dries while water table just at riverbed level.

Gradation curves of aquifer samples between 4.0 m and 9.0 m depth, i.e. effective zone of radials and within 25 m radius of well were analyzed to determine slot size of radials. It was proposed to provide 265 mm diameter radial screen pipes with 150 mm slot with 4 mm outer and 6 mm inner width. Design flood discharge was taken as 3505 m^3/s for 50 years return

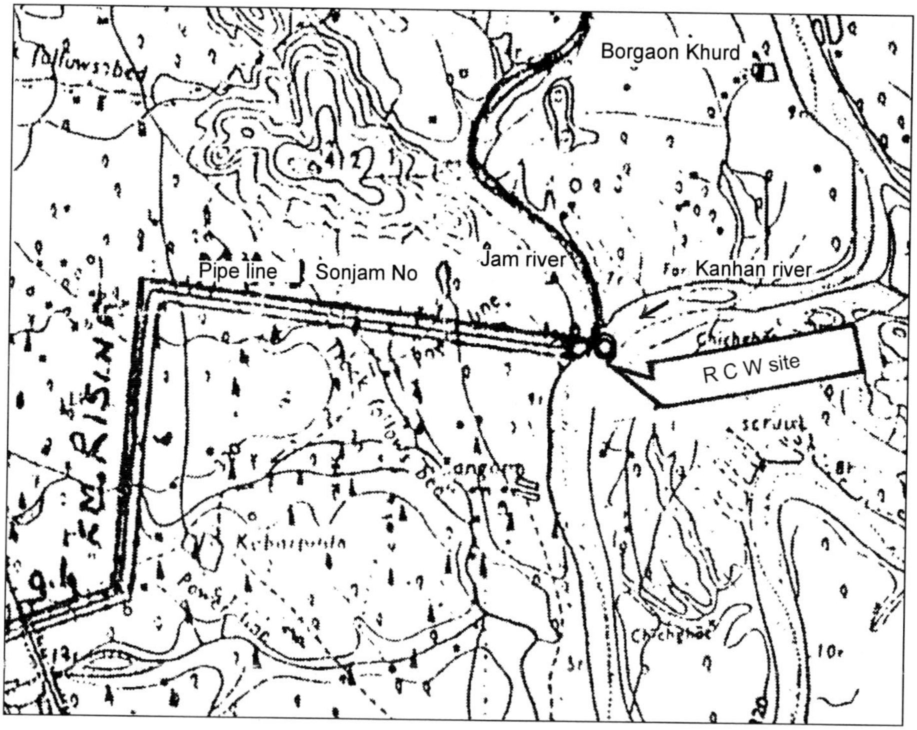

Fig. 1 Index map of site of radial collector well.

flood. The corresponding HFL was worked out as R.L. 100.0 m. Using silt factor of 1.75, the scour depth was worked out as 9.4 m. Hence, the radials were required to be located below RL 81.6 m.

The diameter of central collector well is 10 m, which is large enough to accommodate total 27 potholes in two tiers. It was decided to provide total 14 radials, six radials in tier 1 and eight radials in tier 2. As impervious layer encountered at 25 m distance on right side, radials of only 20 m of length are provided. Total radial length of 380 m was provided. Position of all radials with reference to center of bridge and their lengths are shown in Table 1 below.

4.1 MODFLOW Model

Using MODFLOW, aquifer was divided into seven layers as shown in Fig. 2. Area measuring 301 m × 301 m was discretized into 113 cells in rows and columns. Hydraulic conductivity of 0.3 cm/sec was chosen for design purpose from analysis of pumping test data. Radial locations were modeled by head dependent boundary cells. Constant head boundary was considered at 150 m from center of collector well. Along the river flow direction, the uniform flow was introduced by varying constant head boundary values. Recharge channel was introduced using constant head boundary cells. A steady state simulation was carried out for different positions of recharge channel and under different drawdowns. The summary of results is given in Table 2.

Table 1 Details of radials.

Tier No.	*Radial*	*Invert R.L. (m)*	*Angle from bridge center (degree)*	*Length (m)*
1	R1	78.5	90	20
1	R2	78.5	130	25
1	R3	78.5	170	35
1	R4	78.5	210	35
1	R5	78.5	240	25
1	R6	78.5	280	20
2	R7	79.5	40	20
2	R8	79.5	80	25
2	R9	79.5	100	25
2	R10	79.5	140	35
2	R11	79.5	185	35
2	R12	79.5	230	35
2	R13	79.5	270	25
2	R14	79.5	310	20

Table 2 Discharge of RCW for different river channel positions using MODFLOW.

River channel position		*Discharge in million liters per day at drawdown of*		
		3 m	*4 m*	*5 m*
Case A:	River channel 50 m on left	22.7	28.5	33.2
Case B:	No river channel	12.2	15.1	17.0

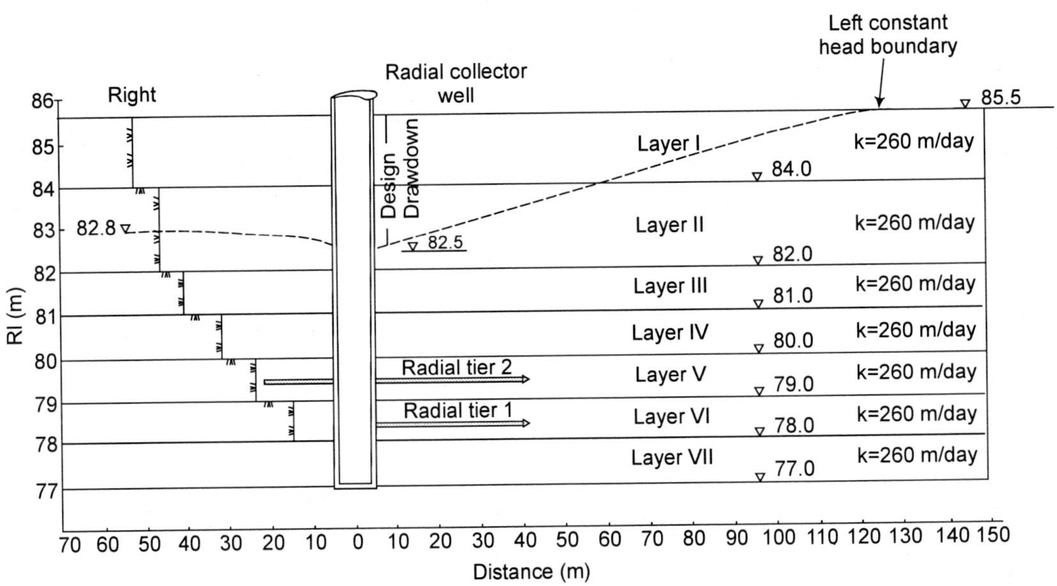

Fig. 2 Arrangement of radials in collector well.

Simulation results indicated that minimum 17 million liter per day of water can be drawn from R.C. well even in the absence of surface channel in summer. The actual discharge may increase considering the depletion of storage in the riverbed aquifer.

4.2 AEM Model Using the Visual Blue Bird

AEM model was also constructed for riverbed aquifer considering 8.5 m thickness and 77.0 m base elevation. Hydraulic conductivity of 260 m/day and uniform flow of 8 m^2/day were assigned in river flow direction. Average water level was assumed at 85.5 m at 150 m away from centre of R.C. Well. Fourteen radials of well were created in model using line sink element i.e. river element with bed resistance. The central collector well which creates obstruction to aquifer flow was created using circular inhomogeneity. The right side of the river bank at 30 m from well, which is an impervious boundary, was simulated by constructing thin inhomogeneity with very low permeability. The river channel was constructed for Case A at 50 m on left of well using river element with constant water level of 85.5 m. The elements of the AEM model are shown in Fig. 3.

To perform simulation under drawdown of 2 m, 3 m and 4 m, the appropriate heads were assigned to each radial. The total discharge of all the radials was computed. The water level

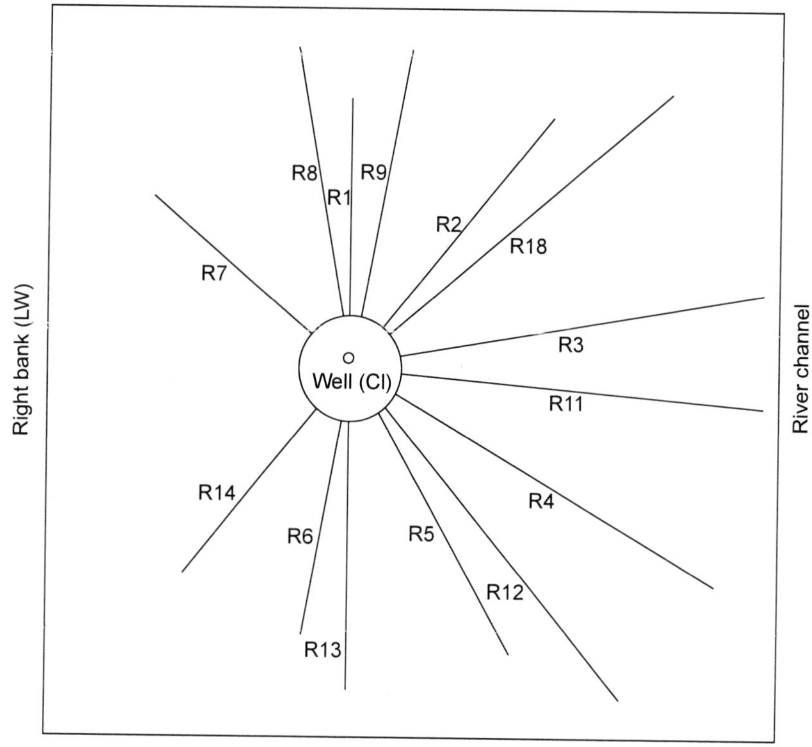

Fig. 3 Layout of analytic elements for radial collector well.

contours for the Case A with 4 m drawdown are shown in Fig. 4. The comparison of AEM results and MODFLOW results are shown in Table 3.

Table 3 Comparison of discharges obtained using AEM and MODFLOW.

Drawdown (m)	*Discharge in m³/day under different drawdown*			
	Case A		*Case B*	
	AEM	*MODFLOW*	*AEM*	*MODFLOW*
3.0	22130	22692	13727	12213
4.0	28635	28509	17375	15105
5.0	33971	33212	20525	16993

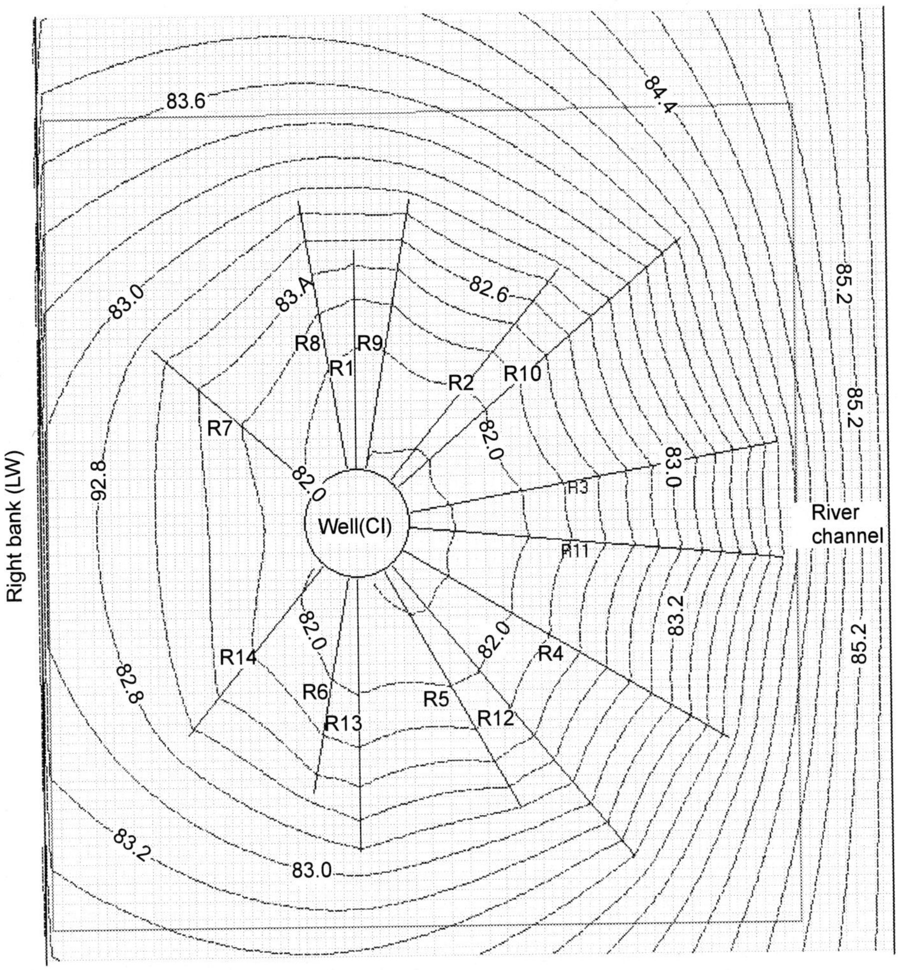

Fig. 4 Contours of water levels for Case B with 4 m drawdown.

5 DISCUSSIONS AND CONCLUSION

The minimum discharges found out using both MODFLOW and AEM shows good agreement in Case A while 12% to 20% deviation is observed in discharge obtained by AEM in Case B. Actual field data indicated that river channel remains dry only for about 45 days. During this dry period, the aquifer storage may be depleted by pumping to maintain required discharge. Thus, assured design discharge of 22.5 million liter per day was finalized considering Case A.

In the absence of analytic solution that represents site condition of R.C. Well, numerical modeling was resorted to form design which led to economical viability and provide safe yield. Finite difference based three-dimensional model, MODFLOW and Analytic element method based two-dimensional model were used in this study. Results of Analytic Element Method are comparable with MODFLOW model. AEM is found more suitable when the area under study is influenced by surface water features such as river channels. AEM is found to be very simple to construct the model for design purpose as it requires minimum field data. Initial trials can be done using AEM by varying radial lengths, numbers and orientation with different surface water channel location. A final design can be checked by three-dimensional MODFLOW model. This approach saves considerable time during design process and provides ease to the designer.

References

Craig , J. & Matott, S. 2005. *Visual Bluebird Users Manual, Version 2.0.*

Haitjema, H.M. 1995. *Analytic Element Modeling of Groundwater Flow*. California: Academic Press.

McDonald, M.G. & Harbaugh, A.W. 1988. A modular three-dimensional finite-difference ground-water flow model, *U.S. Geol. Surv. Techn. Water-Resour. Inv.*, Book 6, Chapt. A1.

Mitchell-Bruker, S. & Haitjema, H.M. 1996. Modeling Steady State Conjunctive Groundwater and Surface Water Flow with Analytic Elements, *Water Resource. Res.*, Vol. 32, No. 9, pp 2725–2732.

Strack, O.D.L. 1989. *Groundwater Mechanics*, New Jersey: Prentice Hall.

Strack, O.D.L. 1999. Principles of the analytic element method, *J. Hydrology*, 226, 128–138.

27

Role of Subsurface Colloids in Groundwater Contamination

T.K. Sen
Department of Chemical Engineering, National Institute of Technology, Rourkela—769 008, Orissa, India

ABSTRACT

Until some two decades ago, it was believed that only the soil liquid and gaseous phases were mobile and could facilitate the transport of chemicals and nutrients in subsurface flow. It is now generally accepted that also part of the soil solid phase is mobile under different geochemical conditions and that mobile organic and inorganic soil particles, colloidal in nature may facilitate or retard the contaminant transport. The mobility of contaminants in subsurface zone is dependent on their distribution between the immobile solid phase and the mobile aqueous phase. Many contaminants readily sorbs to immobile aquifer media and therefore considered to be virtually immobile in the subsurface, thus presenting little danger to groundwater supplies. Therefore, most predictions of contaminant transport were based on two-phase equilibrium adsorption models (a dissolved phase and a sorbet immobile phase). The greater the extent to which a contaminant partitions onto the immobile phase, the slower is its average transport velocity in the groundwater. This is the case of contaminant transport in the absence of mobile colloidal fines in subsurface environment. The unexpected appearance of low-solubility contaminants some distance from known source or sooner than would be expected from their solubility, led to examination of the possible involvement of nonaqueous, mobile colloids in contaminant transport. The presence of mobile colloidal fines can explain such observations and led to the thinking of three phases in contaminant transport models, i.e. mobile liquid phase, mobile colloidal phase and the immobile solid phase. It is a complex phenomenon in porous media involving several basic processes such as adsorption, colloidal fines release, migration and fines entrapment/plugging at the pore constrictions. The effects of these basic processes on the contaminant transport are presented in this work.

1 INTRODUCTION

1.1 Contaminant Transport in Groundwater Flows

The mobility of contaminants in saturated subsurface zone is dependent on their distribution between the immobile solid phase and the mobile aqueous phase. Many contaminants readily sorbs to immobile aquifer media and therefore considered to be virtually immobile in the subsurface, thus presenting little danger to groundwater supplies. For example, many metals and radionuclides bind strongly to mineral components and particulate organic matter of the solid phase, while many non-polar organic contaminants have a high affinity for binding to particulate organic matter (Sen and Khilar 2005; Kretzschmar *et al.*, 1999). Therefore, most predictions of contaminant transport were based on two-phase equilibrium adsorption models (a dissolved phase and a sorbed immobile phase). The greater the extent to which a contaminant partitions onto the immobile phase, the slower is its average transport velocity in the groundwater shown in Fig. 1. This is the case of contaminant transport in the absence of mobile colloidal fines in subsurface environment.

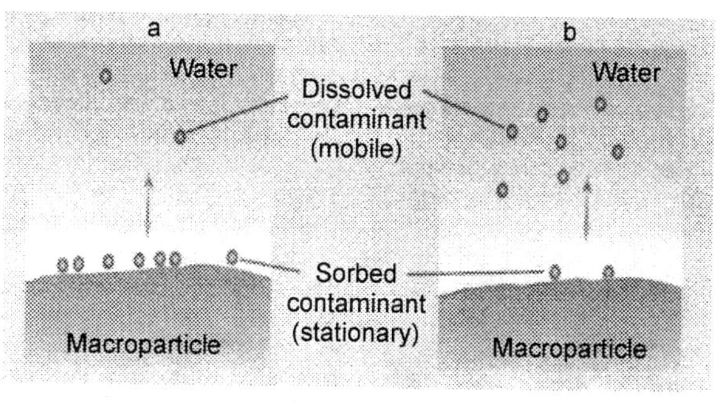

Fig. 1 Contaminant transport in a simple two-phase groundwater system. (a) High sorption and therefore low contaminant solubility; (b) low sorption, high contaminant solubility. Macroparticles are the stationary components of a groundwater aquifer and include clays, metal oxides (Honeyman 1999).

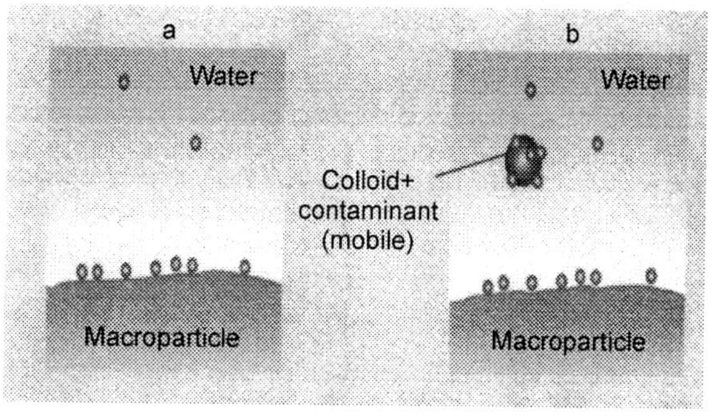

Fig. 2 Comparison of generalized two and three-phase groundwater system. (a) Two-phase; (b) three-phase. The third phase in *b* is a colloidal fines or microparticle, shown here with contaminant molecules sorbed to it, thus making them mobile (Honeyman 1999).

The unexpected appearance of low-solubility contaminants some distance from known source or sooner than would be expected from their solubility, led to examination of the possible involvement of nonaqueous, mobile colloids in contaminant transport (Sen and Khilar, 2005; McCarthy and Zachara, 1989; Kersting *et al.*, 1999; Ryan and Elimelech, 1996). The presence of mobile colloidal fines can explain such observations and led to the thinking of three phases in contaminant transport models, i.e. mobile liquid phase, mobile colloidal phase and the immobile solid phase (Fig. 2). Figure 2 compares two and three-phase contaminant transport models. Figure 2a shows low solubility contaminants distributed between an aqueous phase and immobile aquifer solids called the macroparticles. Colloidal fine materials can increase the apparent solubility of low solubility contaminants (Fig. 2b) if those contaminant strongly associate with the colloids and the colloids minimally interact with the stationary macroparticle phase. This contaminant transport has been known as "Colloid-facilitated contaminant transport" in literature. Based on research progress towards the role of colloidal fines on the transport of contaminants, which is not limited to facilitation only, we shall term this phenomenon as "Colloid-associated contaminant transport" (Sen and Khilar 2005; Sen *et al.*, 2002).

2 EXPERIMENTAL STUDIES ON THE ROLE OF COLLOIDAL FINES ON CONTAMINANT TRANSPORT IN SUBSURFACE FLOW

2.1 Laboratory Column Experiments

Several recent experimental studies have found that colloidal fines in addition to radionuclides can often carry contaminants such as metal ions and hydrophobic organic thus, significantly accelerate the transport of contaminants through porous media (Grolimund & Borkovec 2005; Sen *et al.*, 2005; De Jong *et al.*, 2005; Grolimund & Borkovec, 1996; Kretzschmar *et al.*, 1999; Puls & Powell, 1992; Roy & Dzombak, 1997; Sen *et al.*, 2002). Puls and Powell (1992) have performed experiments on the transport of As^{2+} in the presence of colloidal ferric oxide particles and showed that colloid associated arsenate transport is 21 times faster than that of dissolved arsenate.

Experimental studies have been conducted for systems where the porous media contain the mobilizable colloidal fine particles, thus representing the underground soil mass (Grolimund & Borkovec, 2005; Sen *et al.*, 2005; De Jong *et al.*, 2004; Grolimund & Borkovec, 1996; Kretzschmar *et al.*, 1999; Puls & Powell, 1992; Roy & Dzombak, 1997; Sen *et al.*, 2002).

Roy and Dzombak (1997) conducted a systematic laboratory investigation on Ni^{2+} contaminant transport in presence of *in situ* mobilized colloidal particles, through sand packed column. Significant enhancement of transport by colloids was observed. Without colloids, Ni^{2+} concentrations approached inlet concentration after about 15-pore volumes, whereas with colloids it reached only after 3-pore volumes. These experiments were conducted at a Ni^{2+} concentration of 10^{-6} M. When the experiments were performed at a higher concentration of Ni^{2+}, 10^{-4} M, the trends were quite different: the deposition efficiency of the colloids increased substantially and almost no colloids were measured in the effluent. The concentrations in the effluent were substantially lower than that for the low concentration Ni^{2+} experiment. In the experiment performed to observe colloid transport without Ni^{2+}, it was observed that the effluent colloid concentration was essentially identical to those for the 10^{-6} M Ni^{2+} case.

The potential of natural *in situ* mobilized colloids for facilitating the transport of strongly sorbing contaminants in natural porous media was demonstrated by Grolimund *et al.* (2005, 1996). In column packed with a non-calcareous soil material, they observed significant colloid release and mobilization of Pb^{2+} when feed solution was switched from 50 mM NaCl to 0.15 mM $CaCl_2$.

Sen *et al.* (2002) conducted laboratory column study on the effect of mobilization and migration of colloidal fines, kaolin on the transport of Ni^{2+} through sand beds containing 6 wt% kaolin particles under non-plugging condition (Fig. 3). From Fig. 3, breakthrough curves (BTC) for Ni^{2+} gives significant differences as compared to those for the BTC with sand bed. Breakthrough plots are presented as plots of Cc/Cc_0 against pore volume (PV), where Cc is the aqueous phase effluent Ni^{2+} concentration, Cc_0 is the influent Ni^{2+} concentration. SR represents the bead to particle size ratio.

First, the breakthrough has occurred earlier at around 6 pore volumes. Secondly, the dimensionless outlet of Ni^{2+} concentration has increased beyond 1.0 implying that colloidal kaolin particles with Ni^{2+} ions adsorbed on their surface have migrated out of the bed. Both these observed effects clearly show that the presence of colloidal kaolin particles in the bed has facilitated the Ni^{2+} transport process.

The fact that colloidal particles were migrating out of the bed can be confirmed by the measurement shown in Fig. 3b. One observes from the Fig. 3(b) that particles migrated out of the bed for about 25 pore volumes during which period, the dimensionless Ni^{2+} concen-tration was above 1.0. Such measurements carried out by Sen *et al.* (2002) which clearly document the well-established phenomenon of colloid-facilitated contaminant transport. Further, it documents that the facilitation occurs as long as the fine particles are migrating out of the bed.

2.2 Colloid-associated Retardation of Inorganic Ions Due to Plugging Phenomena

The literature available so far has primarily focused on the facilitation of the contaminant transport due to presence of colloidal fine particles. The focus of most of these studies has been on documenting this important phenomenon which has been reviewed first by McCarthy & Zachara (1989) and subsequently by Ryan & Elimelech (1996) and Kretzschmar *et al.* (1999) respectively.

Migration of colloidal fines in porous media is a reasonably well-understood topic. It has also been found out that under certain conditions which have been discussed earlier section in details, entrapment of the migratory colloidal particles can occur at the pore constrictions, which could eventually lead to plugging of porous media (Sen *et al.*, 2002; Sen & Khilar, 2005).

This plugging results in drastic reduction in flow rate and thereby, there exists a possibility of retardation of contaminant transport and subsequent containment technique in soil remediation. No reported study has considered the possibility of occurrence of plugging of colloidal fine particles. Therefore, it is necessary to consider this plugging phenomenon on colloid-associated contaminant transport in porous media.

Sen *et al.* (2002) presented their results on Ni^{2+} transport under plugging condition (Fig. 4). Figure 4 presents the measured BTCs for different packed beds which indicates that the Ni^{2+} breakthrough gets delayed as the content of kaolin in the bed is increased. One further observes

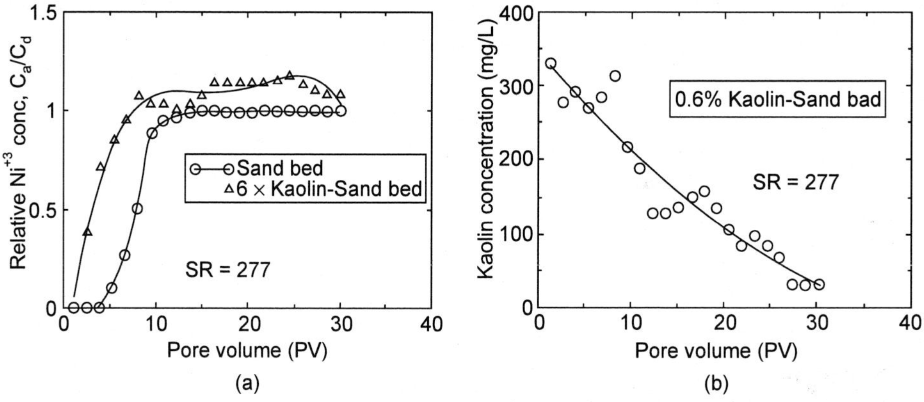

Fig. 3 Role of mobilizing kaolin particles on the facilitation of Ni metal ions transport during flow through kaolin-sand bed under non-plugging condition. (a) Comparisons of experimental BTCs for Ni metal ions during upward flow through the sand bed and the 6% kaolin-sand bed; (b) Condition of effluent kaolin concentrations during flow through same bed. Initial Ni metal ion concentration = 11.0 ppm, length of bed (L) = 0.28 m, Porosity (€) = 0.32, Superficial velocity = 6.69×10^{-4} m/s and SR = 277 (non-plugging).

that the dimensionless effluent Ni^{2+} concentration becomes close to unity after passing approximately 12 pore volumes in the case of 1% and 18 pore volumes in the case of 2% kaolin bed where little plugging occur. Plugging, however, occur for bed containing 4% and 6% kaolin. At these conditions, the dimensionless effluent Ni^{2+} concentration is below 0.65 after passing 18 pore volumes through the beds. At the same plugging conditions, the dimensionless effluent Ni^{2+} concentration is almost one after passing approximately 44 pore volumes for 4% bed and 54 pore volumes for 6% bed.

The BTCs in the plugging condition, exhibit a long, slow increase in effluent Ni^{2+} concentration with variations in the slope of the curve (Fig. 4). Therefore, at plugging condition retardation of Ni^{2+} transport occurs and it is due to increase in adsorption. Such an increase can be attributed to the increase in accessibility to adsorption sites arising out of more mobilization and higher sweeping at higher pressure drops and at higher concentration of kaolin (Sen *et al.*, 2002).

3 CONCLUSIONS

Both field and laboratory research have clearly demonstrated that colloidal particles can be transported through subsurface zone under certain hydrogeochemical conditions. If present in sufficiently large concentrations, existing mobile colloidal particles can provide potentially relevant transport pathways for strongly sorbing contaminants. These colloidal particles act as highly mobile contaminant carrier and thereby enhance the spreading of sorbing pollutants in groundwater flows. Here, the contaminants may be inorganic or organic, must bind strongly and essentially irreversible to the colloids for colloid-facilitated transport to occur. This present work which is not limited to facilitation only but also presented the compilation of retardation of contaminant transport under certain conditions due to entrapment/plugging conditions. Therefore, depending on geoenvironmental conditions, colloidal particles not only facilitate the contaminant transport but also retard in transport in subsurface flows.

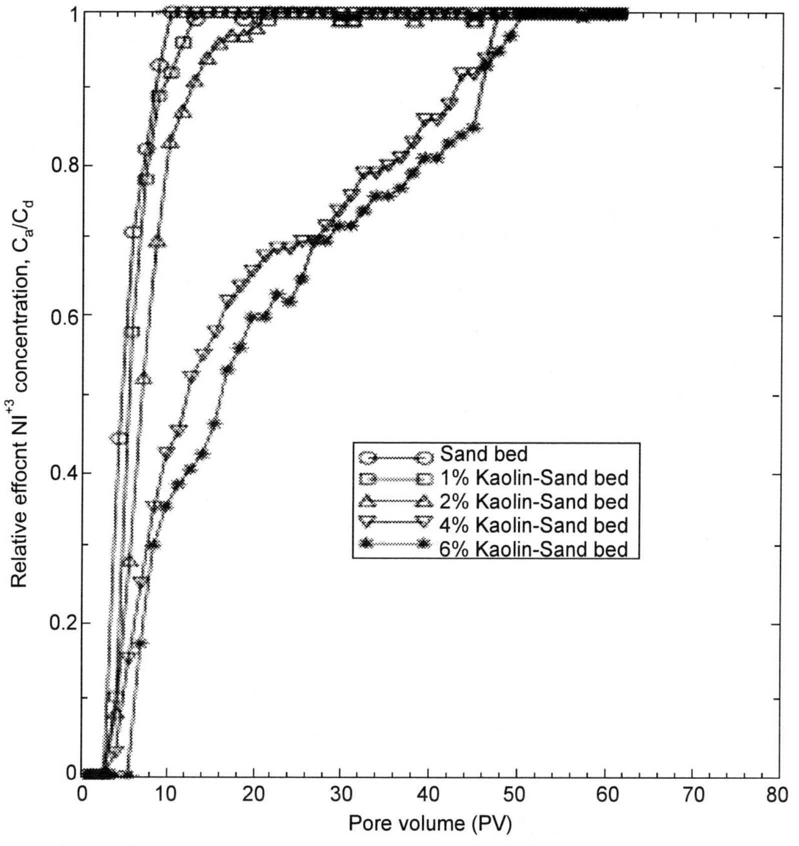

Fig. 4 Experimental BTCs for Ni metal ions during upward flow through different composite kaolin-sand beds under plugging condition. Initial Ni metal ion concentration = 20.5 ppm, Length of bed (L) = 0.28 m, SR = 50, porosity (€) = 0.29 and superficial velocity = 3.39×10^{-4} m/s.

Acknowledgements

Author thanks Prof. K. C. Khilar of I.I.T Bombay for his suggestions on this work. The financial support received from MHRD, New Delhi under R&D Project No: F-26/1/2004.T.S.V. dated 31.3.05 is acknowledged.

References

Bekhit H. M. & Hassan, A.E. 2005. Stochastic modeling of colloid-contaminant transport in physically and geochemically heterogeneous porous media. *Water Resour. Res.* 41(2): 1–18.

De Jong, L. W., Moldrup, P., Rubaek G. H., Schelde K. & Djurhuus, J. 2004. Particle leaching and particle-facilitated transport of phosphorus at field scale. *Vadose Zone Journal*, (Soil Sci. Soc. Am. J.) 3: 462–470.

Grolimund, G. & Borkovec, M. 2005. Colloid-facilitated transport of strongly sorbing contaminants in natural porous media: mathematical modeling and laboratory column experiments", *Environ.Sci.Technol.* 39(17): 6378–6386.

Grolimund, D., Borkovec, M., Barmettler, K. & Sticher, H. 1996. Colloid-facilitated transport of strongly sorbing contaminants in natural porous media: A Laboratory column study. *Environ. Sci. Technol.*, 30, 3118–3123.

Honeyman, B. D, 1999. Colloidal culprits in contamination. *Nature*, 397: 23–24.

Kersting, A. B., Efurd, D.W., Finnegan, D.L., Rokop, D.J., Smith D.K. & Thompson, J.L. 1999. Migration of plutonium in groundwater at the Nevada test sites. *Nature*, 397: 56–59.

Kretzschmar, R., Borkove, M. Grolimund, D. & Elimelech, M. 1999. Mobile subsurface colloids and their role in contaminant transport. *Advances in Agronomy*, 66:121–193.

McCarthy, J. & Zachara, J. 1989. Subsurface transport of contaminants. *Environ. Sci. Technol.* 23: 496–502.

Puls, W.R. & Powell, R.M. 1992. Transport of inorganic colloids through natural aquifer media: Implications for contaminant transport. *Environ. Sci. Technol.*, 26: 614–621.

Roy, S.B. & Dzombak, D.A. 1997. Chemical factors influencing colloid-facilitated transport of contaminants in porous media. *Environ. Sci. Technol.*, 31: 656–664.

Ryan, J.N. & Elimelech, M.1996. Colloid mobilization and transport in groundwater. *Colloids and Surfaces A*, 107: 1–56.

Sen, T.K. & Khilar, K.C.2005. Review on subsurface colloids and colloid-associated contaminant transport in saturated porous media. *Advances in Colloids and Interface Science* (in press).

Sen, T.K., Mohajan, S.P. & Khilar, K.C. 2002. Colloid-associated contaminant transport in porous media: 1. Experimental study, *American Institute of Chemical Engineers Journal*, 48(10): 2366–2375.

28

GIS Based Water Balance Model of Rift Valley Lakes, Ethiopia

Ram Karan Singh, Ph.D. and Post Doc. (NIRE, Japan)
Associate Professor and Head, Department of Natural Resources
TERI-SAS, Deemed University, Habitat Place, Darbari Seth Block, New Delhi-110003, India.
Getachew, A.; Lecturer, Department of Hydraulic Engineering, AWTI, Ethiopia

ABSTRACT

Information availability is limited regarding the water resources systems of Abaya and Chamo lakes, which are found in Rift Valley Region of Ethiopia. In this study, modelling of the water balance parameters and their impact on water use, sediment transport and its deposition are carried out. For developing the water balance model, the watersheds of the lakes system are simulated under GIS environment. The watershed characteristics are developed. Undertaking bathymetry survey carries out the optometry of the Lakes. Various databases have been integrated to model the water balance of the Lakes, which simulates the water level in the lakes. Based on the model results, it is found that sediment inflow and deposition has threatened the future existence of these Lakes.

Keywords: GIS, Rift Valley, Sediment transport and Depositions, Water Balance

1 BACKGROUND

In the Southern Rift Valley area of Ethiopia there are two important Lakes, Abaya and Chamo. The watershed catchments of the Lakes are around 18,000 km^2 with surrounding population of around 3 million people. These Lakes are important shallow Lakes and among others importance, to inhabit a number of aquatic animals and to modify the climate of the area. Yet there are evidences that they are heavily influenced by human activities and endangered Lakes unless proper precautionary measures are undertaken in the Lake's watersheds. In order to investigate the available water resources quantity in the lakes as well as impacts and influences of natural and man-induced factors a water balance model of the two Lakes have been developed.

The knowledge of water balance of lakes and reservoirs is an essential component of water management. Water management decision as far as possible should be based on a thorough quantitative understanding of the hydrologic cycle of the lakes/reservoirs in the basin. In developing country like Ethiopia, the quantitative understanding of individual components of the hydrological system themselves is difficult. This paper focuses on developing a water balance model, and describes shortly methods employed to obtain the physical characteristics of the watersheds and the lakes as well as the components of hydro-meteorological elements of the water balance components under limited data situation. The developed information is used for the simulation of lake water level. Furthermore, impacts of various influences such as sediment and water use are described and expected life of the Lakes under sediment load are discussed and measures to reverse the deteriorating situation are briefly discussed.

2 METHODOLOGY

The input and output components of the water balance of a lake or reservoir depend not only on the physical dimension of the water body, but also on the climatic, hydrological and geological factors affecting the water body and its surrounding areas. The water balance equation can be written, from continuity equation at any time, which is governed by the conditions that the water volume remains constant. The continuity equation inturn is governed by conservation of matter, which described by equilibrium between added water volume or depth, lost water volume or depth and change in volume or depth as:

$$V_i - V_o + P - E \pm \phi = \Delta S$$

Where V_i is surface and subsurface inflow; V_o is surface and subsurface outflow; P is precipitation volume; E is evaporation volume; ΔS is change in storage and ϕ is error function.

In the above equations, parameters can be distributed as follows:

$$V_i = \sum_{i=1}^{ngw} V_g + \sum_{i=1}^{nuw} V_{ug} + V_{ssi}$$

$$V_0 = V_{so} + V_{sso}$$

Where V_{si} and V_{so} are sums of surface inflow and outflow; V_{ssi} and V_{sso} are subsurface inflow and outflow; V_g and V_{ug} are gauged and ungauged inflows; *ngw* and *nuw* are number of gauged and ungauged watersheds. The error term ϕ is treated part wise.

The water balance equations written above have been solved using depth-based simulation. The program developed to compute water balance in lake consists of selecting the model parameters and input, computation and results.

2.1 Data of Water Balance Parameters

The components of the water balance, such as runoff, rainfall and evaporation, Lake morphometry, etc., did not exist and these components have been developed in the wider scope of a research associated to this paper. The data can be stated as watershed data, lakes morphometric data, meteorological data, runoff data and peculiarities in hydrology of the watersheds. The composed data for hydrometeorology totals 27 years of Abaya and 22 years of Chamo Lake on monthly basis. The watersheds of the two lakes drainage basin have been modelled combining Arc View GIS (developed by ESRI) and Watershed Modelling System (WMS) hydrological model. While the former has been employed for Digital Elevation Modelling (DEM) the latter has been used for Digital Terrain Modelling (DTM) in derivation of extensive hydrological parameters, such as slope, basin area, perimeter, etc. Figure 1 shows the derived main sub-watersheds of the two Lakes Basin. After refining the sub-watersheds capturing the locations of interest such as gauge stations and lake outlets, the entire basin was subdivided in to 50 sub-watersheds. The watershed data under the GIS environment masks the Lake morphometry due to data unavailability. In order to close this gap, bathymetry surveys of the two Lakes have been undertaken combining Global Positioning System (GPS) and Echo sounder. As a result the digital information systems of the two Lakes water body, physical morphometric characteristics such as area, perimeter, length, etc., as well as capacity curves have been developed. Figure 2 (a) and (b) provide a derived elevation contour map of the Lakes [ref 1-4].

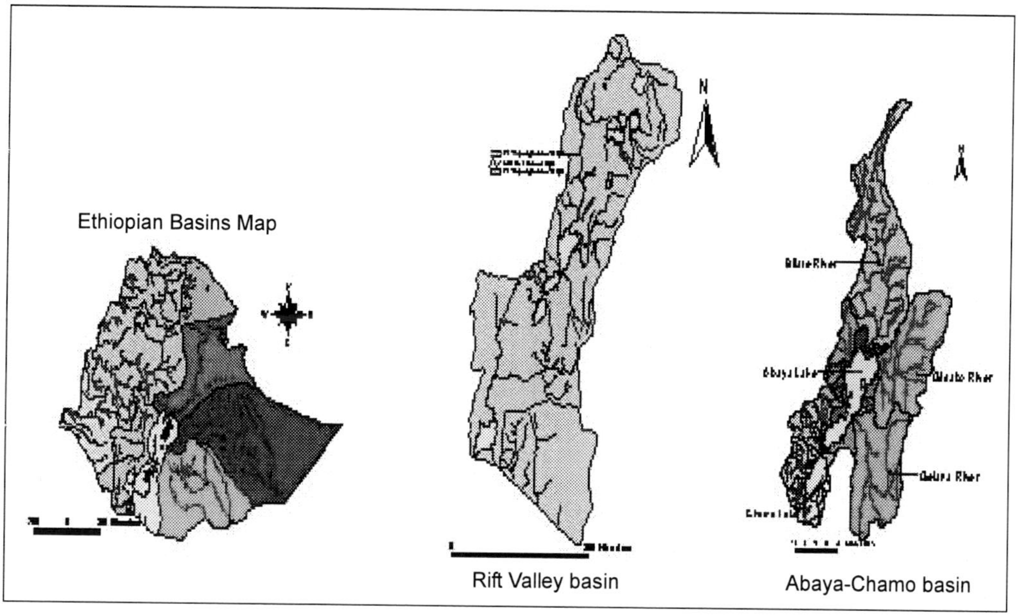

Fig. 1 Ethiopian Basin Map, Rift Valley Basin and Abaya and Chamo Lake.

Meteorological data such as rainfall and evaporation on the Lakes and the identified watersheds are important elements to compute the water balance model and to generate runoff using rainfall-runoff model where inflow data are not available. In order to estimate the watershed and lake meteorological monthly data, a modified Thiessen polygon has been developed to estimate rainfall; temperature-elevation relationship has been established to estimate temperature and thereby evapotranspiration estimation is made possible using Thornthwaite's procedure [reference 3]. Most of the area in the two Lakes basin is ungauged

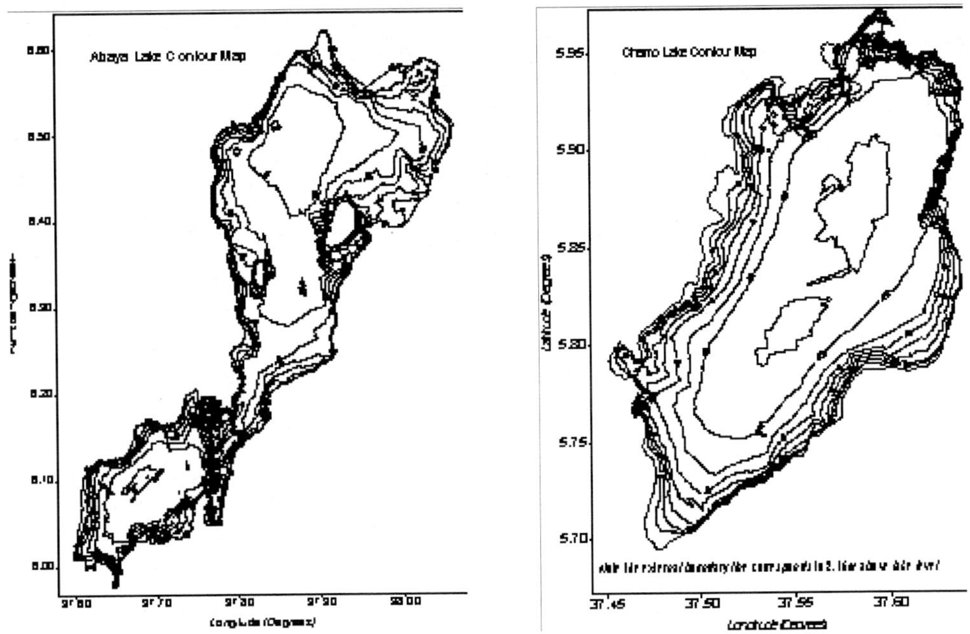

Fig. 2 (a) Abaya Lake Depth Contour Map (b) Chamo Lake Depth Contour Map.

(only 40.5% of land area is gauged). While the data of gauged areas have been used, a rainfall-runoff model based on monthly water balance concept was developed, to estimate ungauged portion of the watersheds as well as the missing elements of data in the simulation period. Components of the subsurface flows are accounted via surface modelling and direct accounts are disregarded. Certain peculiar characteristics regarding the hydrology of the watersheds such as intensive evaporation due to wetland and inundation areas, hot springs, water uses in irrigation are accounted as these elements affect the magnitude of inflow components. Furthermore, surface outflows are accounted in the simulation model from the gauged information or by trial and error where such information is unavailable. The existing Lake level records constitute four gauge stations on Abaya and two stations on Chamo Lake. While the Abaya Lake data could be filtered, comparing the stations fills corrected and missing values, in the case of Chamo Lake data certain years of monthly uninterrupted records at the beginning and end of simulation periods could be used, because the intermediate values are associated with shift of stations and change of recording mechanisms.

2.2 Calibration of Parameters

The *EF* and *PF* parameters are used as calibration parameters. Optimum values of these parameters can be selected by minimising the error terms between measured and simulated Lake water level using ordinary least square procedures or by visual inspection of plotted results combined with regression equations. The optimum values of the calibration parameters can be set for various scenarios of investigating impacts on the lakes.

2.3 Simulation Assuming Outflow

The simulations considering outflows are converged with *EF* parameters of 1.85 and 1.70 and *PF* parameters of 0.85 and 0.90 respectively for Abaya and Chamo Lakes. The higher evaporation factor for Abaya Lake is caused by higher heat absorption capacity of the Lake, underestimation of the method of evapotranspiration as well as uncertainty of the data derived from Lake Periphery. The lower rainfall factor is due to possible pluviometric depression effect of the rainfall falling on the Lakes. The coefficient of determination of the regression of measured and gauged values for Abaya and Chamo Lakes under these conditions is 0.87 and 0.99 respectively. Analysis of error shows, the simulation result is highly sensitive to local error, and as such if an error due to data occurs at a particular point, the error propagates in the subsequent simulation periods and shows apparent errors.

2.4 Simulation with Sediment Deposit

The effect of deposited sediment on the lake is clear and is detrimental to the water carrying capacity of the Lakes. The sediment delivered in to the Lakes has no outlets and fully deposited. Sediment load entering the Lakes particularly that of Abaya is high in magnitude. The causes of the large quantity of sediment are associated to intensify erosion in the catchments area due to deforestation, overgrazing and poor farming practice. In terms of Lake level simulation, the effect of inclusion of sediment deposition is to loose the bottom capacity of the Lakes and thereby an increased water level. In order to quantify the effects of sediment, unfortunately data of neither monitoring of sediment transported and delivered in to the Lakes nor adequate sediment measurement at some points in the rivers systems is available. In order to assess impact of sediment on the Lakes, two pairs of three various cases of sediment inputs were evaluated. These are yield or concentration based under tolerable sediment, moderate and high sediment rates. These rates are based on watershed yield based to be 0.15, 2.5 and 26 t/ha per year respectively and the concentration based to be 0.03, 5 and 26 kg/m^3 respectively. Based on the various scenarios of sediment input the water level simulation and the life expectancy of the Lakes were evaluated. The results of the simulation under concentration based high sediment consideration for both Lakes are included. Under this case (i.e. worst scenario with sediment input of 26 kg/m^3), the existence to complete disappearance of Abaya Lake is estimated at 200 years and that of Chamo is estimated at 460 years. The latter would be dramatically reduced once Abaya is filled with sediment and the existence of Chamo would be then few decades after Abaya's disappearance. Investigation of research catchments located outside the watershed of the Lakes and other rivers in Ethiopia which have similar watersheds and climatic characteristics as well as evidences of the sediment load (short period sampled data) in the tributary rivers of the Lake show the sediment input in the two Lakes is rated medium to high rate. As a result the two shallow Lakes are threatened by the large deposition of sediment.

3 CONCLUSIONS

As per results achieved from simulation in Abaya Lake rainfall contributes 27.5%, runoff contributes 72.5% of the total volume. During no outflow condition, evaporation on the lake reduces 97.60%, existing water use reduces only 0.20% and other factors reduce 2.2% of the available water of the Lakes. Surface outflow occurred during the simulation period during the period of 1998 to 2002. The water balance simulation enables assessment of impact of water use as well as modification of input parameters. While existing water uses is insignificant, the impact of sediment input into the lakes, which directly caused by inappropriate use of resources such as forest and land in the watershed have threatened the existence of these two lakes. Added to the above problem there is a growing need to expand water utilisation such as in drinking water supply, irrigation projects, etc. The water resource in the study area is not abundant and well distributed in time and space. The resource is scarce and the Lakes are highly vulnerable to disappearance if proper corrective measures couldn't be undertaken. On the other hand, there is a pressing need to utilise the land and water resources potential of the basins due to growing population. If the available water resource in the system is properly managed, it could be effectively utilised for regional development and preservation of the entire natural environment.

Acknowledgements

Authors wish to thank the Librarian and Dean (Academics) Mr. Bogale G. Mariam of Arbaminch Water Technology Institute for extending all kinds of help to carry out this study. We also thank GTZ for their academic and financial support to the Institute, which directly and indirectly helped to carry out this study.

References

Legates, D.R., and C.J. Willmott (1990), "Mean Seasonal and Spatial Variability in Gauge-Corrected Global Precipitation", International Journal of Climatology, 10, 111–127.

Reed S. (1996), "Soil-Water Budget. Part I: Methodology and Results", http://www.ce.utexas.edu/prof/maidment/GISHydro/seann/explsoil/method.htm.

Thornthwaite, C.W. (1948), "An Approach toward a Rational Classification of Climate", Geographical Review, 38, 55–94.

United Nations - Food and Agriculture Organization (1994), "Digitized Soil map of the World", Version 3.0.

Willmott, C.J. (1977), "WATBUG: AFORTRAN IV Algorithm for Calculating the Climatic Water Budget", Publications in Climatology, 30, 2.

Groundwater Resources Development in the Western Ghats, India

29

Pradeep K. Naik[1] **and Arun K. Awasthi**[2]

[1]Central Ground Water Board, Central Region, NS Bldg., Civil Lines, Nagpur – 440001, India.
Ph. +91-712-2534415/2565314. Email: pradeep.naik@water.net.in

[2]Department of Earth Sciences, Indian Institute of Technology, Roorkee–247 667, India

ABSTRACT

The Western Ghats (hills) region of the Indian Peninsula in western India is bestowed with heavy precipitation (4000-6000 mm/yr), but the headwater basins coalescing runoff from these hills retain very small quantity of groundwater due to poor surface slope conditions. The narrow valleys in these hills offer ample scope for surface water development, and several medium to major irrigation projects have already been constructed in these areas with well-defined canal network. These developments have truly boosted agricultural productivity in the region, but at the same time they have also brought an economic disparity between the command (areas irrigated by these canals) and non-command areas. Waterlogging problems are also increasingly seen in low-lying areas. These problems are mainly due to poor groundwater management strategies in the region, and call for implementation of suitable measures for a uniform groundwater development. As a first step in this direction, several measures have been suggested as a case study for the groundwater resources development in the Koyna River basin, a headwater basin on the east of the main ridge of the Western Ghats. Conjunctive use of surface water and groundwater has been suggested in the command areas of the lift-irrigation schemes. Artificial recharge measures have been suggested for the water-scarcity areas in non-command areas, and also for the development of springs at higher elevations. Suggestions have also been given for the development of existing dugwells and construction of new dugwells and borewells in both command and non-command areas.

Keywords: Western Ghats, Deccan terrain, conjunctive use, artificial recharge, groundwater resources development.

1 INTRODUCTION

The Western Ghats hill ranges are one of the major physiographic provinces of the Indian sub-continent running for a distance of 1600 km parallel to the Indian West Coast. These hill ranges have an average elevation of 1000-1300 m above mean sea level (amsl) (maximum 2700 amsl). These hills are bestowed with heavy precipitation (4000-6000 mm/yr), but the headwater basins coalescing runoff from these hills retain very small quantity of groundwater due to poor surface slope conditions. The narrow valleys in these hills offer ample scope for surface water development, and several medium to major irrigation projects have already been constructed in these areas with well-defined canal network. These developments have truly boosted agricultural productivity in the region, but at the same time they have also brought an economic disparity between the command and non-command areas. Waterlogging problems are also increasingly seen in low-lying command areas, while in non-command areas aquifers are gradually drying up due to over-exploitation of groundwater. These problems are mainly due to poor groundwater management strategies in the region, and call for implementation of suitable measures for a uniform groundwater development. Several measures have been suggested as a case study for the groundwater resources development in the Koyna River basin, a headwater basin on the east of the main ridge of the Western Ghats, that characteristically represents the physiographic set-up of other river basins in the Western Ghats region. This river basin generally trends north-south and covers an area of 2,036 km^2 in the Deccan terrain of the district of Satara, Maharashtra State. Naik *et al.* (2001) gave a detailed account of this river basin.

2 HYDROGEOLOGY

Basalt is the primary aquifer in the Western Ghats region, while laterites, alluvium and talus deposits are shallow aquifers of secondary importance. In a typical weathered zone at shallow levels, the basaltic lava flows consist of four water-bearing horizons: (1) black cotton soil with yellow or reddish clay (0.5-8 m); (2) highly weathered and highly jointed basalt (3-10 m); (3) poorly weathered and highly jointed basalt (1-2 m); and (4) poorly weathered and poorly jointed basalt, which overlies the hard massive basalt (about 1-2 m). The second horizon consists of both weathered vesicular and non-vesicular (massive) basalt, in which the joints in the upper portion are completely filled with clay. Gradually the clay content decreases downward and the joints are more open because of the decreasing effects of complete weathering. The lower part of the horizon 2 and the entire thickness of horizon 3 have the highest permeability. With greater depth, the joints are closed (horizon 4), and the underlying massive basalt yields little groundwater. Deeper groundwater occurs at the basaltic flow contacts, within the vesicular basalts, and in the fractured and jointed sections in a basaltic flow. Groundwater occurs under phreatic conditions in shallow aquifers (<30 m), while in deeper aquifers (>30 m), it generally occurs under semi-confined to confined conditions. Transfer of groundwater does take place between the shallow and deeper aquifers, and the fractures/lineaments act as conduits for such groundwater movements.

Dugwells tapping shallow aquifers are commonly used for drinking and irrigation purposes, while borewells tapping deeper aquifers are used only for drinking purposes. Dugwells are mostly 2-14 deep, and rarely are they 25 m deep. Diameters range from 2-10 m; 50% of them have a diameter of 2-4 m. The borewells, mostly 150 mm in diameter, are generally

30-90 m deep; most of them are 50-70 m deep. Depth of water levels (DTW) in shallow aquifers vary between 0.50 to 11.50 m below ground level (mbgl) during premonsoon period and 0.05 to 4.80 mbgl during post-monsoon period. Seasonal fluctuations (pre-monsoon to post-monsoon) vary between 0.05 to 7.60 mbgl. DTW is generally above 1 mbgl in the central valley portions. Lower reaches in the irrigation command areas already show signs of waterlogging conditions.

3 GROUNDWATER RESOURCES DEVELOPMENT

3.1 Conjunctive Use of Water Resources

The Koyna River is a perennial river due to the release of water two times in a month from the Koyna dam (Fig. 1). About 61 MCM of surface water is lifted annually from this river by farmers through several cooperative lift-irrigation schemes (Naik and Awasthi 2003). Farmers usually practice flooding method of irrigation while watering their lands. Because they have to wait for a long period, often more than a month, for their turn to come for irrigation, whenever their turn comes they apply as much water as their lands can hold. This practice causes unnecessary wastage of surface water. Assuming that 30% of the applied irrigation water augments the groundwater table (CGWB, 1998a), an amount of 18 MCM recharges annually to the Koyna River basin for an application of 61 MCM of surface water. But unfortunately, since minimal quantity of groundwater is actually being utilized, such an accrual of groundwater makes the water table shallower thus leading to waterlogging to near waterlogging conditions. Salt encrustations are already seen in the lower reaches, and productivity of crops in these areas has significantly declined. Strong measures, therefore, need to be taken immediately before things aggravate and the lands become permanently unclamable.

Conjunctive use of water resources seems to be an easy remedy for the waterlogging problems. The use of surface water resources is not so far regulated. To initiate this process it would possibly be proper to start metering the quantum of surface water supplied to an individual farmer, or a group of farmers. But this is something not very easy to practice at the present set-up. Therefore, metering at the pumping stations of the lift-irrigation schemes (LIS) and charging accordingly to the cooperative society handling the particular LIS would put pressure on the society itself to regulate its water use. At the moment, individual farmers are charged as per the acreage of land they irrigate, not as per the quantum of water they actually consume. This is a major flaw in the LIS, and once tackled, part of the water logging problem could possibly be solved.

One solution to motivate farmers to use appropriate quantity of surface water would be to allow night time irrigation. This will shorten the length of the irrigation cycle. However, night time irrigation might give rise to more uncontrolled irrigation. But, at the moment when the day time irrigation is not well controlled, the issue of controlling the night time irrigation does not arise at all. Therefore, until the time the farmers are not made conscious of the consequences of the irregulated use of water, night time irrigation is an option the authorities might possibly consider to reduce the water use problems.

Elsewhere in the Western Ghats region, waterlogging problems are already seen in many areas, viz., in the canal command areas of the Vir Irrigation Project in the Nira River basin, the low-lying areas of the Dhom Irrigation Project in the Krishna River basin, etc. In the Koregaon Taluka of Satara District which receives canal water from the Dhom Irrigation Project,

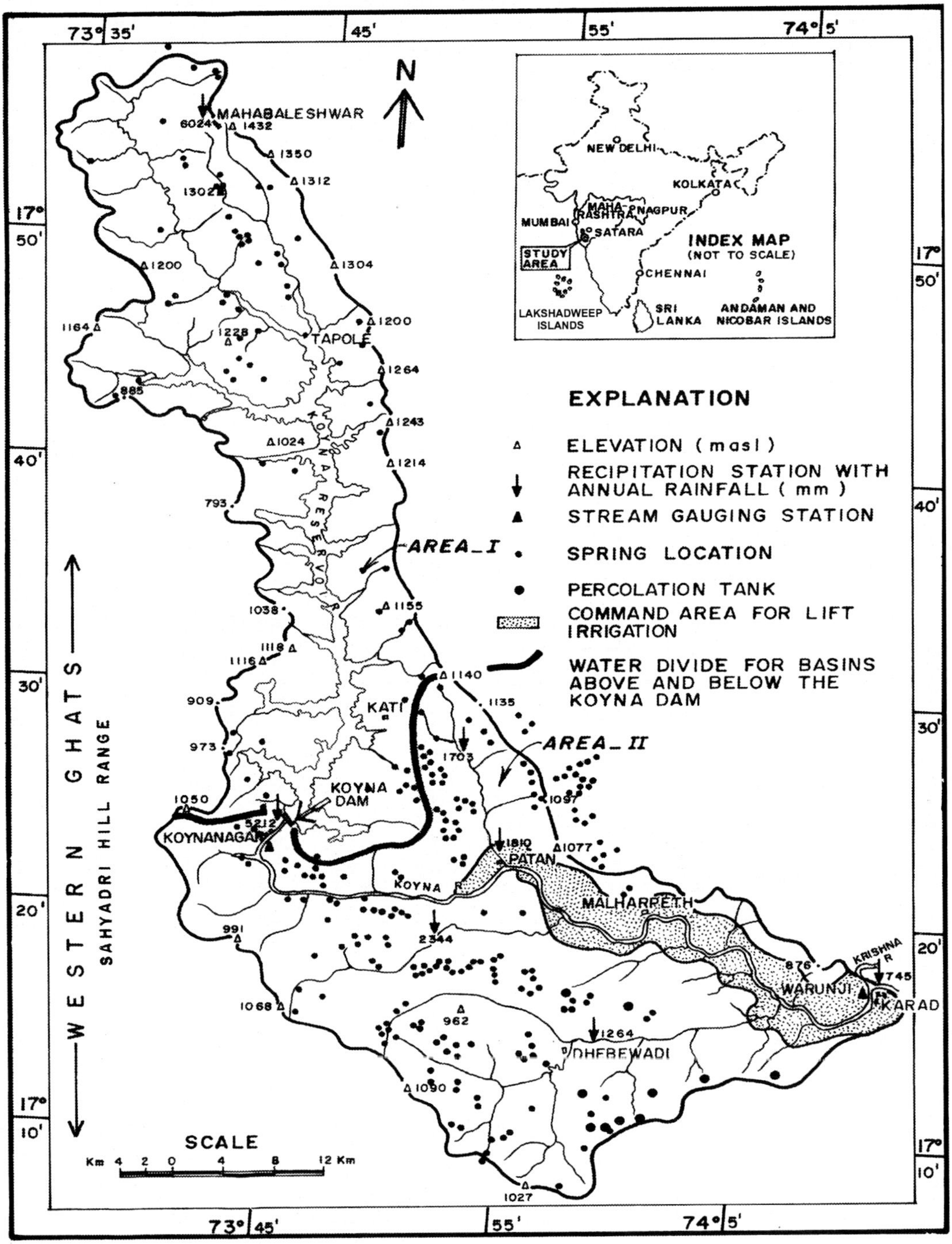

Fig. 1 Location of the Koyna River basin, India.

groundwater irrigation decreased 40% from 5443 ha (1985-86) to 3240 ha (1991-92), and surface water irrigation increased 440% from 1578 ha (1985-86) to 8523 (1991-92) in a span of just five years. These statistics show that conjunctive use practices need to be initiated in the Dhom canal command areas from now itself before things aggravate and it becomes another case study for waterlogging problems.

Government of Maharashtra (GOM) has issued a resolution in July 2001 notifying that irrigation water is to be supplied only to farmers' Water User Associations (WUAs) in the canal command areas. Under the WUA-managed irrigation systems, distribution of water is usually rotated with farmers at the farthest reaches of the command area getting the first share. Since the management is local, the system is transparent. But the GOM's desire has received limited acceptance among the farmers so far. Mitigation of waterlogging problems would really get a new boost with the formation of such WUAs in the irrigation command areas.

3.2 Artificial Recharge of Aquifers

Artificial recharge of aquifers is a solution to the declining trends of groundwater levels in most parts of the non-command areas in the Western Ghats region due to over-exploitation of the groundwater resources. The common artificial recharge structures recommended for a basaltic terrain are percolation tanks, and small-sized check dams in the form of Kolhapur type weirs (KT weirs), cement plugs, gabions, etc. (Naik, 1990; CGWB, 1998b). Other similar structures such as brush wood dams and loose boulder structures are designed especially for soil conservation measures (Naik, 1997). Of all these structures, percolation tanks are by far the most popular and useful artificial recharge structures. These are comparatively larger structures with greater storage capacity and greater beneficial area. These structures are usually constructed in first/second order streams. However, the percolation tanks constructed in the Western Ghats region in the hilly tracts are not very successful due to a variety of reasons. These structures should actually sit on a relatively planer ground, and tap weathered/ fractured rocks, and should have adequate command areas downstream. But, most of the percolation tanks meant for artificial recharge through downward percolation of stored water lie on poorly weathered and poorly jointed massive basalts in foothill zones and do not produce desired results. Poor site selection of these structures possibly due to several field constraints is the main reason for their poor performance. In the future, selection of these sites should be based on thorough examination of the local geomorphological and hydrogeological situations. Care must be taken that the percolation tanks lie on moderately or highly weathered basaltic horizons. In some places these structures need additional deepening to expose these horizons. Creation of artificial fractures through use of explosives on the tank beds, in case they sit on poorly weathered and poorly jointed massive basalts, may help in the downward percolation of surface water.

Western Ghats tributaries are very narrow and deep, and small check dams (cement plugs/KT-weirs) constructed across these tributaries can store considerable amount of surface water with backwater extending for long distances upstream. Dugwells in this region, especially in the sloping grounds, are generally constructed beside these smaller tributaries. Therefore, backwater in these streams will cause induced-recharge in these wells thus enhancing their life for a longer period of time. Such smaller structures are yet to be introduced in the Western Ghats region. Construction of these structures in every 1 km through several employment generation schemes will store huge quantity of surface water in the region, and will benefit thousands of dugwells. Farmers can even lift water from these structures, and if necessary irrigate their lands for rabi crops. One problem that could be encountered while constructing these smaller structures is the higher velocity of flow in these streams during the rainy season. These structures, therefore, have to be designed in such a way that they sustain these higher velocity flows.

3.3 Augmentation of Dugwells

Redbole horizons in the Western Ghats region invariably indicate that below them lie the weathered vesicular basalts and below the vesicular basalts lie the highly weathered and fractured massive basalts (horizon 2 – discussed in the section on hydrogeology). Therefore, if a well ends in redboles, it should be deepened further to tap the entire thickness of the weathered/fractured zones (horizon 2 and horizon 3) until it encounters hard massive basalt (horizon 4) below.

Many of the dugwells in the area go very deep in hard massive basalt (horizon 4) with really little success of getting enough water. In such cases, it would be appropriate to drill a shallow vertical borehole of about 4 inches diameter to decipher the depth of the next basaltic flow that begins with a redbole. Depending on the depth of the redbole bed, if it becomes economical, it is better to dig the dugwell further downward to the zones below the redbole bed. Otherwise, only the borehole may be extended downward until it taps the entire thickness of the potential horizons (horizon 2 and horizon 3) below the redbole bed. In such a case, the dugwell will be a dug-cum-borewell.

The yields of the fully penetrating wells (wells ending in horizon 4) can also be enhanced by constructing one or more horizontal or gently inclined boreholes at suitable points at depth in them. The typical diameter of these boreholes should be about 4 inches. These boreholes may be drilled towards the upslope direction to make passage for the groundwater to flow towards the concerned dugwells in the central valley. These boreholes should be located in horizon 3 just above the horizon 4.

3.4 Popularization of Borewells

The deeper aquifers receive about 1 MCM of recharge annually during the non-monsoon period in the Koyna River basin (Naik and Awasthi, 2003). The recharge may be more during the monsoon months. However, borewells that tap deeper aquifers are not very popular in the Western Ghats region for irrigation purposes because of the high risk involved in the drilling of a successful borewell. About 50% of the borewells drilled either give poor yields or go dry (Naik, 1993). Only about 20% of the borewells are high-yielding (3000 L/h) which sustain pumping for irrigation purposes, while others are good enough for drinking and domestic purposes with the use of hand pumps or 1-2 HP electric motors. To make the borewells popular to tap the static groundwater reserve, the Government probably needs to come forward and help farmers in locating high-yielding borewells through use of the latest technologies, such as remote sensing and geographic information system (GIS). Government may possibly construct high-yielding borewells at its own and provide them to the farmers at a price on a cooperative basis. Water being a State subject, the State Government has to take new initiatives in this direction.

3.5 Selection of Well Sites

Following is an analysis of the high-yielding borewells and dugwells in terms of their occurrence with respect to lineaments and geomorphic features in the Koyna River basin. Since hydrogeological conditions are not different in other river basins of the Western Ghats region, these analyses hold good for these basins too.

1. Borewells along or near lineaments have substantially higher yields than those not associated with lineaments. Conversely, in shallow dugwells, lineaments have little influence on well yields.
2. Where two or three lineaments intersect, yields of borewells at those sites are greater than the yields of wells away from lineament intersections.
3. Borewell yields are higher along the NE-trending lineaments than along lineaments in other directions.
4. Borewells intersecting NE-trending lineaments are more high-yielding in deeply buried pediments than those intersecting NE-trending lineaments on the dissected plateau. The borewells in such areas have yields as high as 39,000 L/h.
5. Borewells along the NW-trending lineaments are high-yielding if a stream is flowing along these lineaments and the borewells are very close to it. The borewells along lineaments in this setting have yields as high as 6,000 L/h.
6. Dugwell yields are related to the type of geomorphic features. Yields are higher in wells dug in deeply buried pediments than those in shallow buried pediments and dissected plateaus.

3.6 Harnessing of Springs

The Western Ghats region is dotted with numerous cold water springs (Fig. 1). The high rate of precipitation and the terrace-like topographic features in the hills facilitate recharge of the aquifers through weathered/fractured/jointed rocks. These aquifers, wherever dissected by fractures, joints, and hillsides, are drained by springs. They generally occur at an elevation range of 600–1350 masl. The mean discharge of the individual springs is about 46 m^3/d (range, 2.06–432 m^3/d) in winter and 28 m^3/d in the summer (range, 0.94–216 m^3/d).

People at higher elevations depend exclusively on these springs. On many occasions, these springs are thoughtlessly diverted causing heavy ecological damage and public outcry. Government has tapped many of these springs for community water supply under a 'gravity water supply scheme'. A small tank is generally constructed at the origin of a spring and water is allowed to flow by gravity through a pipeline to a distribution tank located higher than the village to be served. Depending on the quantity of water stored, water is released either periodically or continually by pipelines to the villages for their use. Among the 30 water supply schemes investigated, 23 (~80%) were found ineffective (Naik *et al.*, 2002) due to several reasons. Following are few suggestions for effective development of these springs in the Western Ghats region.

1. Springs should be tapped after a thorough examination of their discharges in different seasons including the summer. The water requirements of the villages to be supplied with the spring waters must be satisfied.
2. Whenever necessary two or more springs should be tapped together to cater to the demand of a relatively larger population.
3. After installation of the water supply schemes, they should be monitored periodically.
4. In case of excessive water after its use, the surplus water may be stored in a tank to practice horticulture or to irrigate some acres of land.
5. Rainwater harvesting structures may be constructed at suitable locations on the hilltops for enhancement of the life of springs and to increase their discharges. The life of these springs can be enhanced through construction of several rainwater harvesting structures, such as large ponds, on hilltops. In case these hilltops are hard enough not to allow the downward percolation of rainwater, shallow borewells may be constructed at the base of these ponds to connect them to the source aquifers of the springs below. This way not only the rainwater will have direct passage towards the source aquifers to recharge them from time to time whenever rainfall takes place, but also many new springs may develop that can further be used by the local inhabitants.
6. The pipelines connecting the spring origins and the storage tanks may be aligned underground in about a metre depth so as to protect them from detachments due to rolling down of rock boulders or interference of wild animals.
7. A barbed wire fence may be constructed around the spring origins under use to protect them from human/wildlife interference.

4 CONCLUSIONS

The Western Ghats region is characterized by both scarcity and abundance of groundwater – scarcity in the dissected plateaus and abundance in the command areas of the irrigation schemes, especially in the central portions of the river valleys. The issues involved in the optimal groundwater development in the region are (1) water scarcity in the dissected plateau is a direct function of lack of saturation of the aquifers, owing to the rocks being well drained, and (2) abundance in the central valley is because of its situation in the discharge area of the basin and also in a region of maximum irrigation both of which cause high water levels. Countering these challenges is a very difficult task and needs systematic integrated water resources development and management.

Water scarcity in the dissected plateaus can be addressed through construction of several artificial recharge structures at suitable locations for augmentation of the available groundwater resources. Percolation tank is by far the most suitable structure, but selection of sites for these structures should be based on thorough examination of the local geomorphological and hydrogeological conditions.

In the water-abundant areas in central valleys, conjunctive use of water resources is highly recommended to tackle the impending danger of waterlogging and soil salinity. To check over-application of water, it is suggested that the surface water supplied to the fields be metered, and farmers be charged as per the quantum of water they actually use rather than the area they irrigate. Night time irrigation could also be an option to reduce the timing of the irrigation cycle. Formation of Water User Associations (WUAs) is highly recommended for timely and equitable distribution of water in the irrigation command areas.

Existing dugwells need to be revitalized. The partially penetrating wells may be deepened to tap the full sections of the weathered and fractured basalts. They can further be revitalized by constructing horizontal trenches in them or converting them as dug-cum-borewells.

Borewells are not very popular in the region. A failure rate of 50% discourages farmers to venture for borewell construction. While these structures are essential to tap the static groundwater reserves of the region, the government will have to initiate some confidence building measures, such as introduction of subsidies for borewell construction and installation of pumps, popular loan schemes, free technical guidance etc., to encourage farmers to go for deeper borewells. Areas having NE-trending lineaments can be expected to result in higher yields of borewells, and dugwells constructed in deep buried pediments are likely to have high groundwater potential.

Springs may be developed at higher elevations. With no any Government scheme for their protection, they are gradually dwindling, and therefore need special attention. Springs may be thoroughly examined before they are tapped for water supply. If needed, two/three springs may be tapped together to cater to the demand of a larger population. Rainwater harvesting structures may be constructed at suitable locations on hilltops for enhancement of the life of springs and also their discharges.

Acknowledgements

The authors thank the people of the Koyna River basin for their cooperation in the field studies. Thanks are also due to Central Ground Water Board, Ministry of Water Resources, Government of India, for providing necessary field support and facilities. Views expressed in the paper are the authors' own and do not necessarily reflect the opinions of their affiliated departments.

References

CGWB (Central Ground Water Board) (1998a). Detailed guidelines for implementing the ground water estimation methodology - 1997. Central Ground Water Board, Ministry of Water Resources, Government of India, New Delhi, 219 pp.

CGWB (Central Ground Water Board) (1998b). Artificial recharge to groundwater in orange growing areas, watershed WR-2, Taluka Warud, District Amravati, Maharashtra, Central Ground Water Board, Central Region, Nagpur, 190 pp.

Naik, PK (1990), Possibilities of surface water development and artificial recharge of aquifers in parts of Western Ghats, Maharashtra, Proc., All India Seminar on 'Appropriate Methods of Artificial Recharge of Groundwater in Hard Rock Areas', Hyderabad, 1.7.1–1.7.4.

Naik, PK (1993), Systematic hydrogeological studies in parts of Satara, Sangli, Kolhapur and Ratnagiri Districts, Krishna Basin and Coastal Tracts, Maharashtra, India, Central Ground Water Board, Central Region, Nagpur India, 147 pp.

Naik, PK (1997). Impact of the National Watershed Development Program in Sukli watershed, District Nagpur, Maharashtra, India. Central Ground Water Board, Central Region, Nagpur, 17 pp.

Naik PK, Awasthi AK, Anand AVSS, Mohan PC (2001) Hydrogeologic framework of the Deccan terrain of the Koyna River basin, India. Hydrogeology Journal 9(3): 143–264.

Naik PK, Awasthi AK, Mohan PC (2002) Springs in a headwater basin the Deccan Trap Country of the Western Ghats, India. Hydrogeology Journal 10(5):553–565.

Naik PK, Awasthi AK (2003) Groundwater Resources Estimation of the Koyna River basin, India. Hydrogeology Journal 11(5):582–594.

Groundwater and Water Scarcity–Legislation and Litigation: A Tamil Nadu Experience

30

S. Ganapathy Venkatasubramanian[1] and T.V. Lakshmanan[2]

[1]Assistant Professor, Centre for Environmental Studies, Anna University, Chennai–600 025
E-mail: sgvsmanian@hotmail.com

[2]Advocate, Hon High Court of Madras, Chennai–600 002
E-mail: lexit69@gmail. com

ABSTRACT

The sharing of river water relates to irrigation and is linked to food security. The state of Tamil Nadu has an agreement with the state of Andhra Pradesh for bringing Krishna water to the city of Chennai, as a solution to its drinking water scarcity. Even within States, the sub-regionalism prevails. These aspects should be seriously addressed by the water managers for the purpose of allocation of water and the existing legal framework has to be attuned to tackle legal disputes surcharged with emotional content.

Chennai, the capital of Tamil Nadu, is prone to water scarcity. Drinking water is a constant issue to be addressed. A solution to it has evaded successive governments. One of the objective of the project is to solve the water problem by bringing water from Veeranam Lake to Chennai. This has been a non-starter for a long time.

1 DIFFERENT RIGHTS TO WATER

The Right to Water has been seen as an environmental issue for a long time. The Supreme Court of India recognised pollution-free water as one of the components of environmental rights. Thus, even in early decisions the Supreme Court also included water as forming part of environmental rights. Environmental rights found their place in the Constitution of India in Article 21, which guaranteed right to life. In Rural Litigation and Entitlement Kendra vs. State of U.P.[1], dealing with limestone quarries the Supreme Court observed that "protecting and

[1]1985-(002)-SCC-0431-SC

safeguarding the rights of the people to live in healthy environment with minimal disturbance of ecological balance and without avoidable hazard to them and to their cattle, homes and agricultural land and undue affectation of air, water and environment."

The multifarious beneficial use of water is well recognised by policymakers. Its scarcity makes different uses competing with each other for primacy. Water is essential for development and industrial use, as for domestic purpose and sanitation as also for agriculture.

In Virender Gaur vs. State of Haryana[2] the Court held "Art. 21 protects right to life as a fundamental right. Enjoyment of life and its attainment including their right to life with human dignity encompasses within its ambit, the protection and preservation of environment, ecological balance free from pollution of air and water, sanitation without which life cannot be enjoyed. Any contra act or action would cause environmental pollution.

Again the Supreme Court held "The Constitutional and statutory provisions protect a persons right to fresh air, clean water and pollution-free environment, but the source of the right is the inalienable common law right of clean environment."[3]

In Narmada Bachao Andolan vs. Union of India[4], the Supreme Court held that since people have a Right to Water and need water for sustenance, that dam should be allowed to be constructed and Right to Water would have the overriding effect over environmental rights.

2 WATER DISPUTES

Different water rights have created potential conflicts of different dimensions as the demand for water increases.

In a writ petition[5] before the Supreme Court by a resident of Delhi against Haryana for release of water of Yamuna to Delhi, an argument was advanced that among different uses of water right to drinking water would prevail over other uses. The argument was attractive to the court also. But, because of a memorandum of understanding among the riparian states the legal question was not dealt with. Again in Sardar Sarovar case[6], the Supreme Court specifically adverted to the beneficial effects of the dam that it would alleviate drinking water scarcity in parts of Rajasthan. These dimensions add to the complexity of the problem.

The sharing of river water relates to irrigation and linked to food security. The state of Tamil Nadu has an agreement with the state of Andhra Pradesh for bringing Krishna water to the city of Chennai, as a solution to its drinking water scarcity. Even within states, the sub-regionalism prevails.

It is said that in future, wars would be fought over water. Already at the national level we see how much water disputes can strain relationship among states and how much it can be politicised. Cauvery water dispute is only one such instant. Even within the state such a

[2]1995-(002)-SCC-0577-SC

[3]Vellore Citizens Welfare Forum vs. Union Of India 1996-(083)-AIR-2715-SC

[4](2000) 10 SCC 664

[5]Delhi Water Supply & Sewage Disposal Undertaking and Another vs. State of Haryana and Others, 1996-(2)-SCC -572

[6]Narmada Bachao Andolan, vs. Union of India and Others. 2000-(10)-SCC-664

dispute can have an aggravating effect. A recent litigation[7] has highlighted the gravity a water dispute can attain including emotional aspect which the court had to grapple with.

3 NEW VEERANAM PROJECT AND ITS EXTENSION

To tackle the frequent drinking water problem first Veeranam project was conceived in the year 1969, to transport water through cement pipes from Veeranam Lake to Chennai City. Again in 1991 another project was formulated which was approved in 1996. Earlier there was an agreement between Tamil Nadu and Andhra Pradesh for supply of water to Chennai known as Telugu Ganga Project. However, the project has not been fulfilled. In this background the government in the year, 2002, commenced New Veeranam Project. Water was being drawn from Veeranam Lake at the rate of 180 MLD and supplied to Chennai. Due to continuous drought for three years, recently water table went down in Chennai and its surroundings, while Kollidam received copious rainwater. An Expert Committee[8] was constituted. It conducted detailed investigation, and officials heard farmers in the four districts. It recommended extending New Veeranam Project and drawing subsurface water in Kollidam riverbed by constructing six collector wells.

4 THE DISPUTE

The agriculturists objected to the proposal on the ground that the project gave "preferential treatment and ignored the hardship being caused to the agriculturists in the aforementioned four districts. "They also argued that 2224 towns and villages were benefiting and meeting their agricultural and drinking needs from existing 30 wells and additional withdrawal through collector wells would deplete existing wells and when there is no flow in Cauvery there will not be any water in Kollidam and Veeranam, and excessive tapping of water would expose the 30 wells to seepage of non-potable water into them.

While the petitioners also cited another alternative of Desalination project, more importantly, they relied on Tamil Nadu Ground Water (Development and Management) Act, 2003 and contended that the State Government should constitute the committee in terms of the Act and ought to have consulted Central Ground Water Authority. More importantly, they sought the ground level report of the expert committee which is not privileged under section 123 0[9] of the Evidence Act.

The government objected on the grounds of locus standi, objected to the supply of the report on the ground of "possible misuse and abuse" and that the decision was taken as a public policy and no mala fide was alleged, the court could not interfere.

[7] S. Ramamirtham vs. The State of Tamil Nadu & Others 2005 Writ L.R 451

[8] The Expert Committee, which is a study team consisted of (1) Chief (SG&SWRDC)PWD Chennai, Chairman (2) Chief Engineer, WRO, Chennai Region, Chennai, Member (3) Engineering Director, TWAD Board, Chennai, Member, and (4) Chief Engineer (CWSAP) CMWSSB, Chennai, Member, all were bureaucrats.

[9] **Evidence as to affairs of State:** No one shall be permitted to give any evidence derived from unpublished official records relating to any affairs of State, except with the permission of the Officer at the Head of the Department concerned, who shall give or withhold such permission as he thinks fit.

5 THE DECISION[10]

Regarding locus standi the court overruled the objection of the government, as the dispute related to "administrative decisions related to development ... harmful to the environment and jeopardize people's right to natural resources such as air or water". The court considered various decisions of the Supreme Court and High Courts and relied on the decision in Balco Employees' Union vs. Union of India[11]. Further the Court also took into account the facts that the representations were made by the petitioners and agitations against the extension were organised and supporting affidavits of villages were annexed as documents.

However, regarding claim of privilege to the report, the court accepted the government's plea on the ground "When such an apprehension is being felt by the Secretary to Government through the affidavit, in the absence of any material to show that the said apprehension is misconceived" and that there was no reason to hold that such apprehension was not genuine. After considering the report the court concluded that there was illegality in the report. Incidentally, the court held that as Ground Water Act had not come into force it was not mandatory for the State to consult the Central Ground Water Authority, as it was satisfied with the report of its own officials and did not accept the contention of the petitioners. From the decision it does not appear that any argument objecting to the composition of the Expert Committee, as containing exclusively government officials, was taken.

6 THE MENTAL ATTITUDE OF THE PETITIONERS

More than the legal angle, it is the emotional aspect of the litigation that has to be taken note of. Because a litigation on basic need like water cannot be resolved by governmental decision, however fair, or even by decisions of the court. At the national level, the complexity is well known. At the intra-state level it is more complex as there is no single representative body to speak for the rival claimants. A government may be accused of tilting in favour of one region over another. In the above case, as noted in the decision itself, even as the litigation was going on and arguments were heard at length, a flood of letters numbering about 1000, addressed to the judges by names were sent. In some of the letters the signatures of the villagers were in blood. The letters sought for suitable orders and not to permit the extension project. In spite of court asking the counsel to advise the villagers not to send such letters in future and instructions given by the counsel, this continued. In fact, some of the letters intimated that if water was drawn pursuant to the order of the court, the agriculturists would end their lives.

The court had to merely lament and express its sadness. Though the action in other context could amount to contempt, considering the anxiety of the agriculturist the court had to sympathetically deal with it. Even so, the court held, it would go by law in administering justice. In the absence of any contradictory material the court upheld the policy of the government. The court gave direction to the government to send special delegates and explain to the agriculturists in arranged meetings and make sincere efforts to allay their apprehensions.

---

[10] The decision was given by Justice M. Karpagavinayagam on behalf of the bench consisting of Justices M. Karpagavinayagam and C. Nagappan

[11] 2002 (2) SCC 333

7 CONCLUSION

The important pointer in the whole litigation is the portent it holds for similar litigation in the future. Avoiding water disputes is one way. More important is to make legal disputes strictly confined to the legal bounds. The apprehension that the authority would have taken a decision unmindful of interest of one section of the public is the reason for such uncontrollable expression of emotion. Participatory management in environmental issues is a well-known principle. If such participation had not been limited to consultative level but made at the decision-making level, then perhaps the litigation may not have been strained.

Briefing: As title indicates, the paper is more about laws related to the use of water and its sharing. A case study of Veeranam Lake Project, Chennai is discussed.

Groundwater Quality Assessment in Tsunami Affected Tamil Nadu Coast, India: A Case Study along Coleron to Karaikal

31

J.F. Lawrence, M. Karthikeyan and R. Ranjit Kumar
Department of Geology, Presidency College (Autonomous), Chennai – 600 005, India

ABSTRACT

On 26 Dec. 2004, Tsunami was triggered due to the earthquake that rocked Sumatra Island. This tsunami has affected Indonesia, Sri Lanka, and the entire south-eastern coast of India. East coast of Tamil Nadu had experienced severe environmental devastation such as erosion and deposition of sediments, disturbances in the environment of microbes and water quality deterioration both in surface and groundwater resources. This study is aimed to evaluate both surface and groundwater quality changes due to Tsunami. The study area is bounded by the river Coleron in north, southern boundary of Karaikal (Union Territory of Pondicherry) in south, Bay of Bengal in east and 5 km stretch perpendicular distance from the coast in west. Water sample collection and conduction of Vertical Electrical Sounding (VES) have been carried out from the pre-determined 24 locations that are equally distributed in the study area. These water samples were analysed for major cations and anions and the data are processed statistically. Various groundwater quality maps have been prepared in GIS environment. These output maps indicate the changes in the water quality along the coast due to tsunami. The VES field data have been processed both manually and digitally. With the geophysical output, resistivity maps for different layers and isoapparent resistivity maps for different depths have been prepared. Hydrogeochemical output matches with the findings of the geophysical investigation.

1 INTRODUCTION

The earthquake that rocked Sumatra Island has generated a high rising tsunami in Indian Ocean on 26th December 2004. This tsunami has affected many parts of the world. The worst

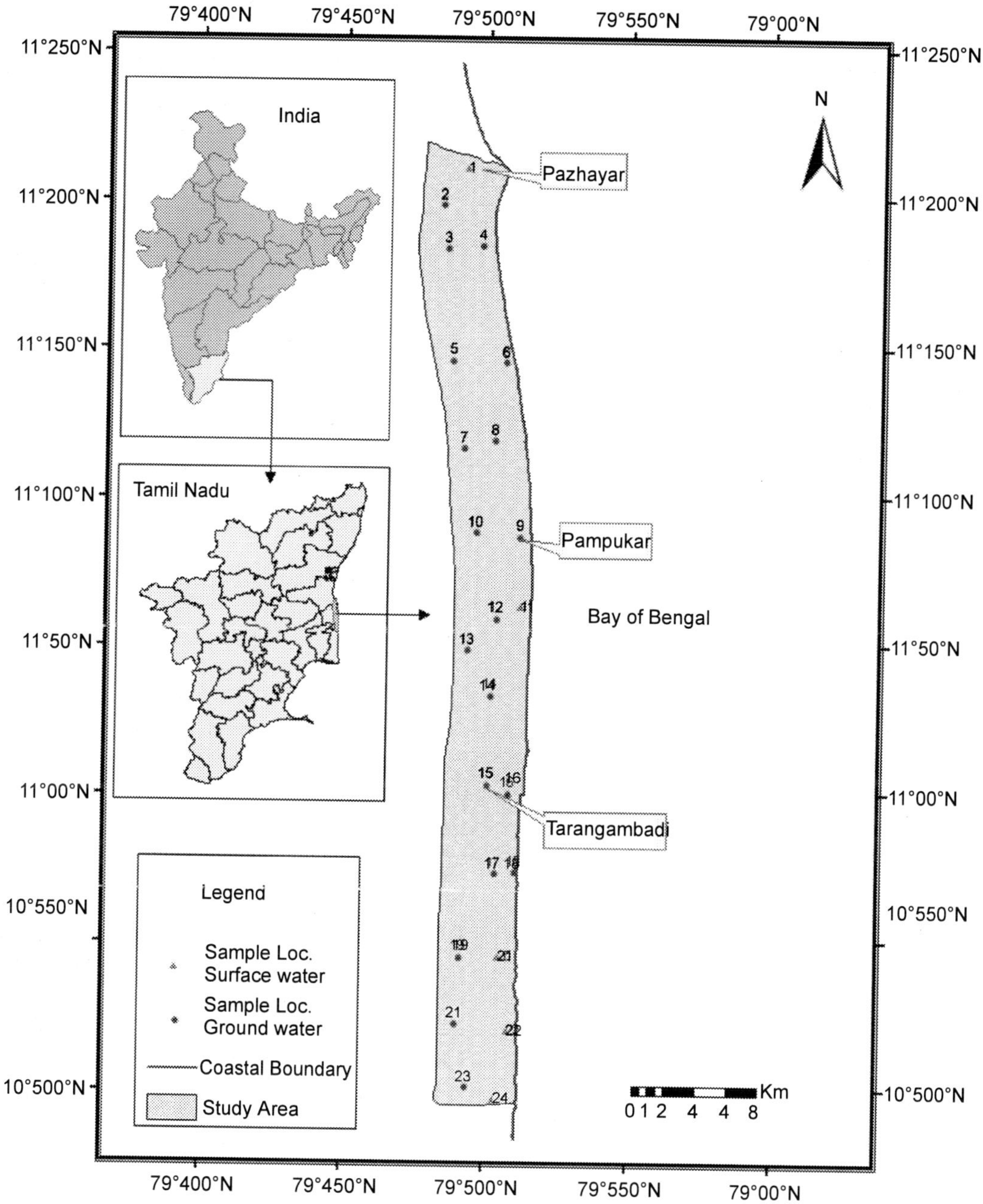

Fig. 1 Location map.

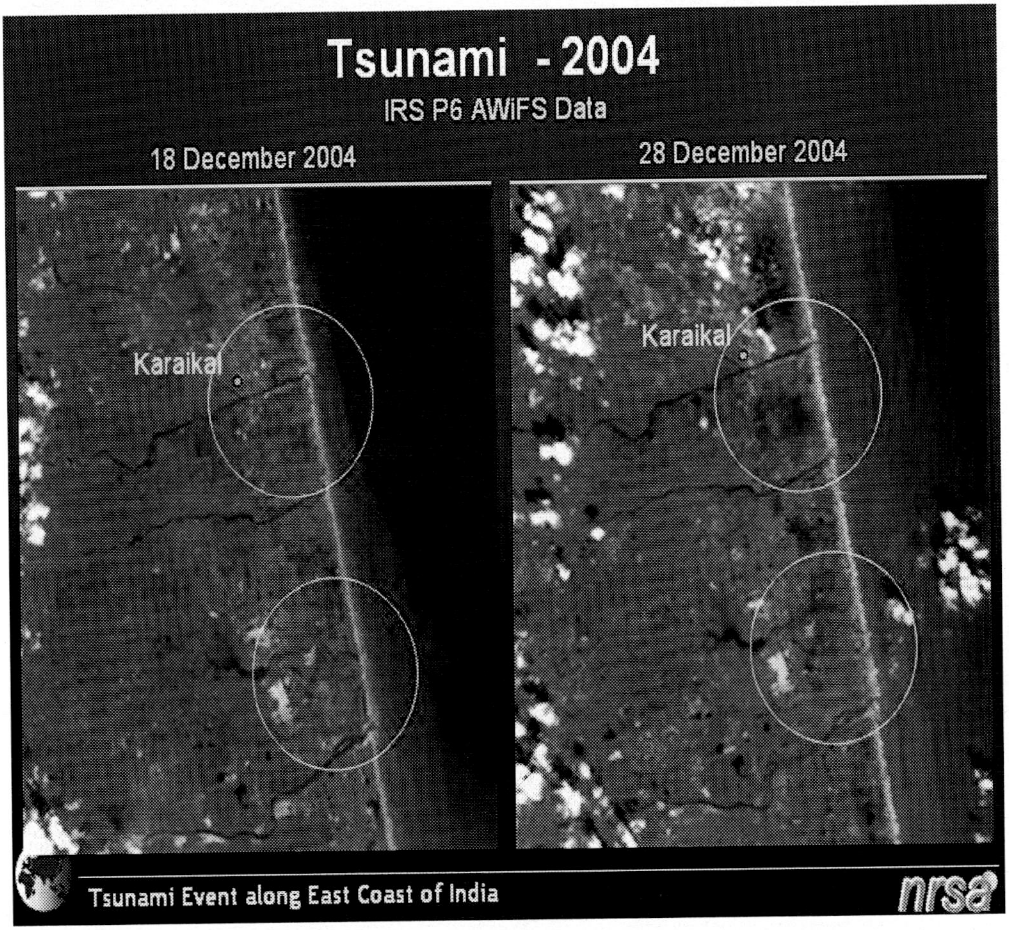

Fig. 2 Satellite imagery of the study area.

affected areas are Indonesia, Sumatra, Andaman, Nicobar, Sri Lanka entire eastern coast India up to Orissa. The entire Tamil Nadu coast was inundated especially Cuddalore, Nagapattinam and Kanyakumari Districts suffered maximum loss. This study was aimed to assess impact of tsunami on ground and surface waters of part of Nagapattinam District and Karaikal region.

1.1 Study Area

The River Coleron forms the northern boundary; Bay of Bengal in the east and the southern boundary of Karaikal territory of Pondicherry forms the southern boundary of the study area. The study area is in the shape of long rectangular stretch. The western boundary has been marked 5 km from the coast. The northern part of the study area falls in revenue boundary of Nagapattinam District and the southern stretch forms a part of Karaikal.

The base map has been prepared through on screen digitization on scale of 1: 50,000 Survey of India toposheets 58 M/15, 58 M/16 and 58 N/13 in ARC GIS environment.

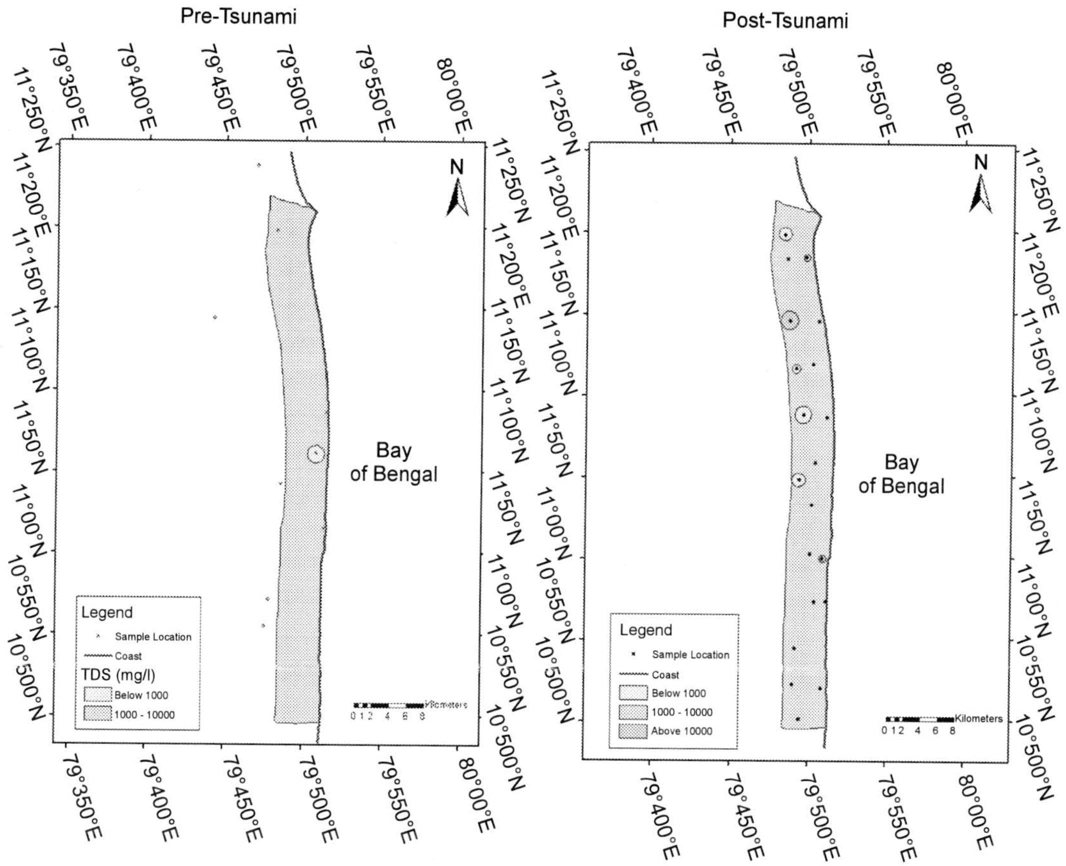

Fig. 3 Total dissolved solids in groundwater.

Ground/surface water samples have been collected in early March 2005 from 24 pre-determined geocoded locations. One stretch of sampling location is situated at a distance of 0.5 km from coast another stretch 2 km from coast (Fig. 1).

Aim and objective

- To prepare water quality thematic maps from the water samples collected for different field seasons.
- Micro level resistivity survey to assess salinity and existing fresh water horizon
- Generation of micro level primary data base in tsunami affected area

In order to assess the post tsunami (ground and surface) water quality of the project area the following geochemical parameters have evaluated

- Total Dissolved Solids
- pH
- Major cations and anions
- Total Hardness

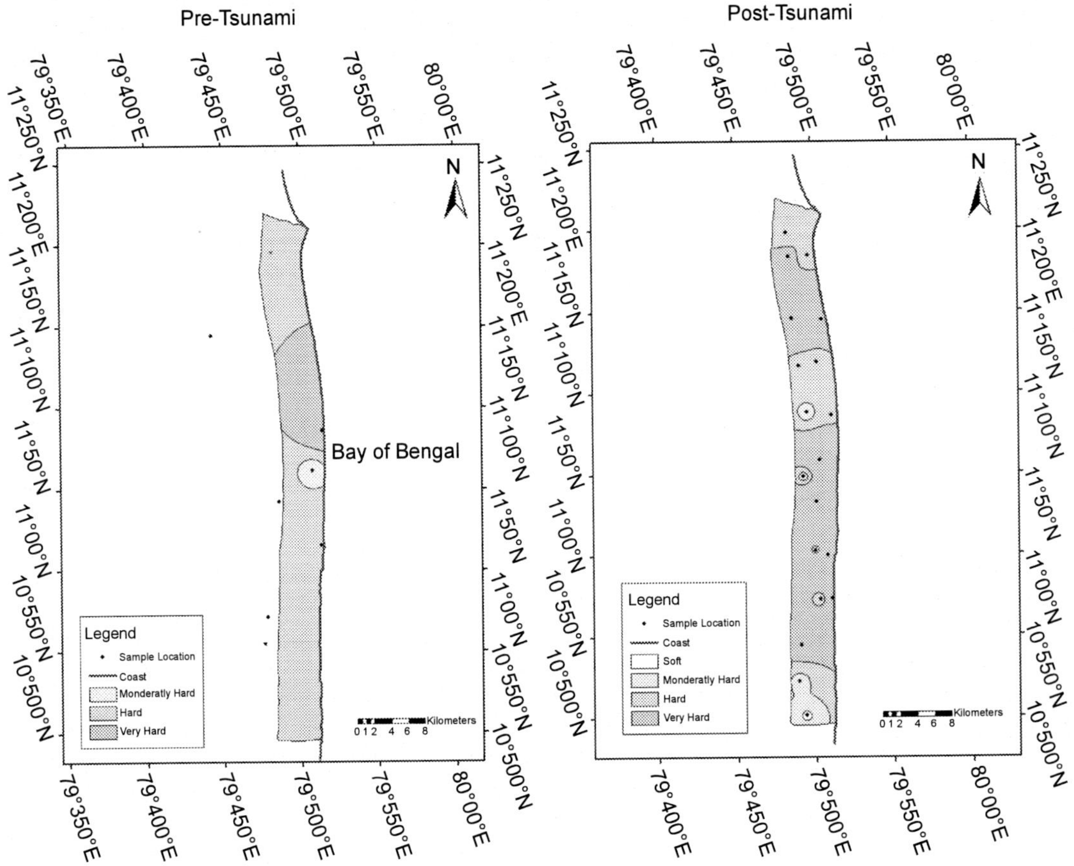

Fig. 4 Total hardness in groundwater.

- $Cl^-/HCO_3^- + CO_3^{2-}$ Ratio (Sea water Intrusion Studies)
- Corrosivity Ratio
- Classifications of ground and surface water
- Water Type
- Determination of Major Cations and Anions (Ca^{2+}, Mg^{2+}, Na^+, K^+ and CO_3^{2-}, HCO_3^-, Cl^-, SO_4^{2-}) using APHA (1993) method.
- Conducting geoelectrical soundings in the same locations where the water samples were colletted in order to assess aquifer nature and its vertical extent using Aquameter CRM 500.

1.2 Geology

Nagapattinam and Karaikal coast a simple geological configuration. The project area– the coastal plain mainly comprises recent to sub recent alluvium and beach sands. Older Tertiary and Cretaceous sediments encountered only at deeper levels.

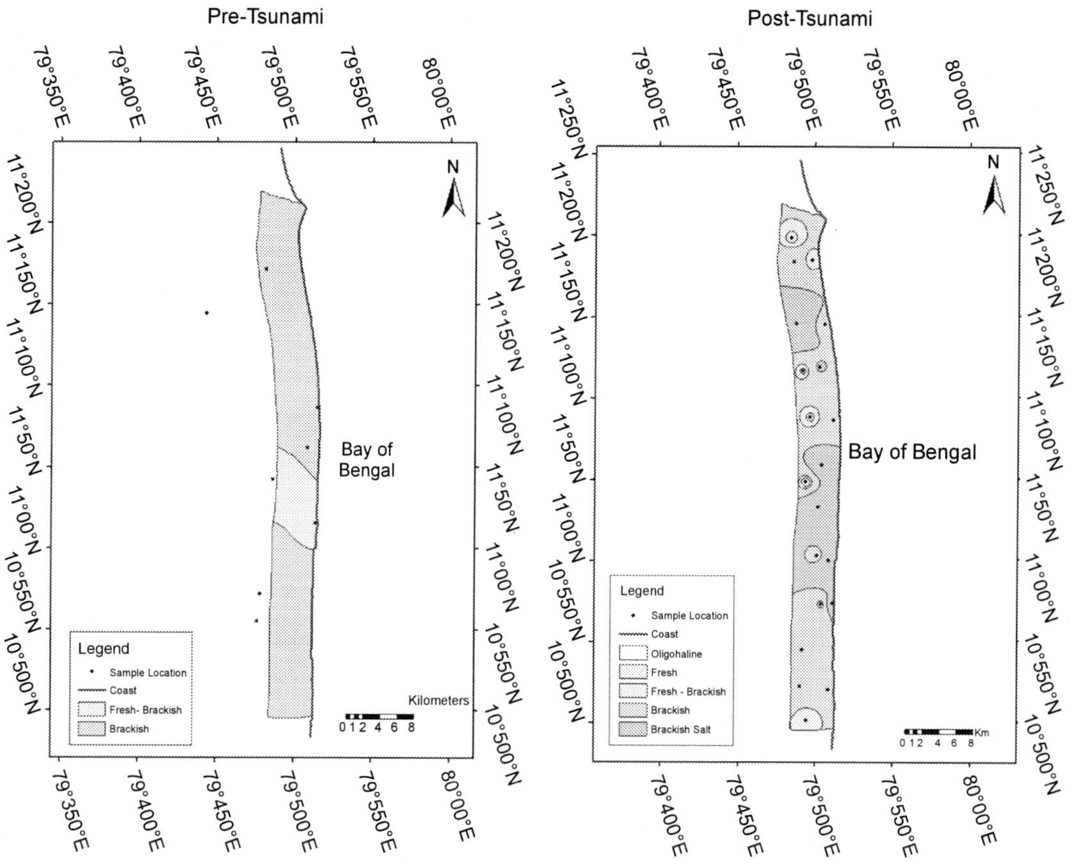

Fig. 5 Groundwater classification based on chlorinity.

1.3 Geochemistry

1.3.1 Pre-Tsunami Scenario

The groundwaters of part of Nagapattinam and Karaikal coastal belt had range of EC values from 840 to 20500 μ s/cm and the highest at Achalapuram in Sirkali Taluk. Depth to water level varied from 2.9 to 8.1 m bgl and the deepest was at Semmangudi of Sirkali Taluk.

1.3.2 Post-Tsunami Scenario

The collected water samples have been analysed for major cations and anions viz, pH, EC, TDS, Ca^{2+}, Mg^{2+}, Na^{+}, K^{+}, CO_3^{2-}, HCO_3^{-}, Cl^{-}, SO_4^{2-} using APHA (1993) standard procedure. From the results the ground/surface waters have been classified adopting standard water classifications.

The total dissolved solids (TDS) of groundwater ranges from 200 (Loc No. 13-Pillai perumal nallur) to 24550 (Melvanjiyur- Loc. No. 24) mg/L and the surface water ranges from 9410 (Loc.No. 22- Mariamman pettai) to 36350 (Loc.No.1- Pazhayar) mg/l. As per Freeze and Cheery (1979) classifications water ranging from fresh to saline water exist in the study area (Fig. 3).

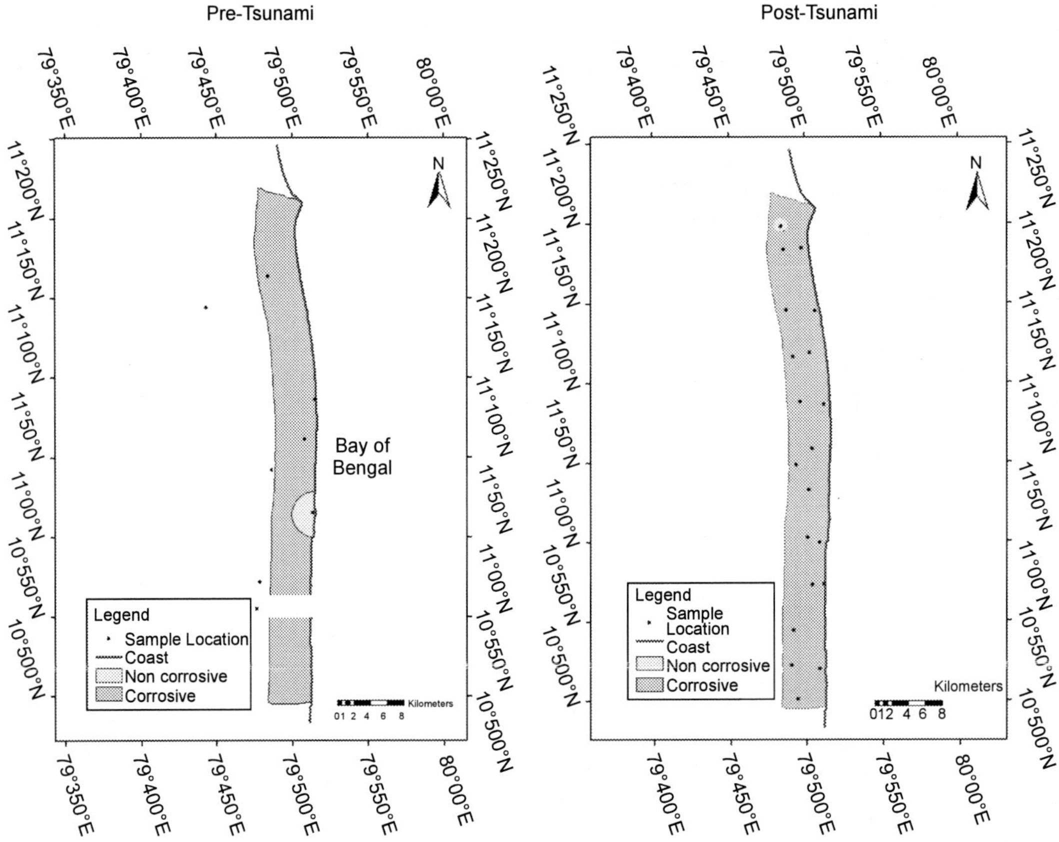

Fig. 6 Corrosivity in groundwater.

The Total Hardness (TH) of groundwater ranges from 69.7 (Loc.No.23 – Sannamangalam) to 2861 mg/L (Loc. No.24 - Melvanjiyur). As per Sawyer and Mc Carty (1987) groundwater of soft to very hard water exist. The total hardness of surface water ranges from 1740 mg/L (Loc. No. 22 – Mariamman Pettai) to 4262 mg/L (Loc. No. 1 – Pazhayar) (Fig. 4).

The chlorinity of the groundwater ranges from 26 mg/L (Loc. No. 13 - Pillai perumal nallur) to 13799 mg/L (Loc.No.24 - Melvanjiyur) and the surface water ranges from 3678 mg/L (Loc. No. 22 – Mariamman Pettai) to 40768 mg/l (Loc. No. 1 – Pazhayar). As per Stuyfzand's (1989) (Fig. 5) Classification, groundwater of Oligohaline to brackish salt exists and the surface water falls under the category of brackish salt to salt water.

Most samples are corrosive in nature except the Pudupattinam sample (Loc. No. 2). Chloride/Bicarbonate ratio ranges from 0.3 (Loc. No. 23 – Sannamangalam) to 68.5 (Loc. No. 24 – Mel Vanjiyur) (Fig. 6). The higher Cl/HCO_3 ratio indicates saline intrusion (Revelle, 1941). From the above study it could be seen that there is water quality deterioration due to tsunami. Freshwater exists in a few locations.

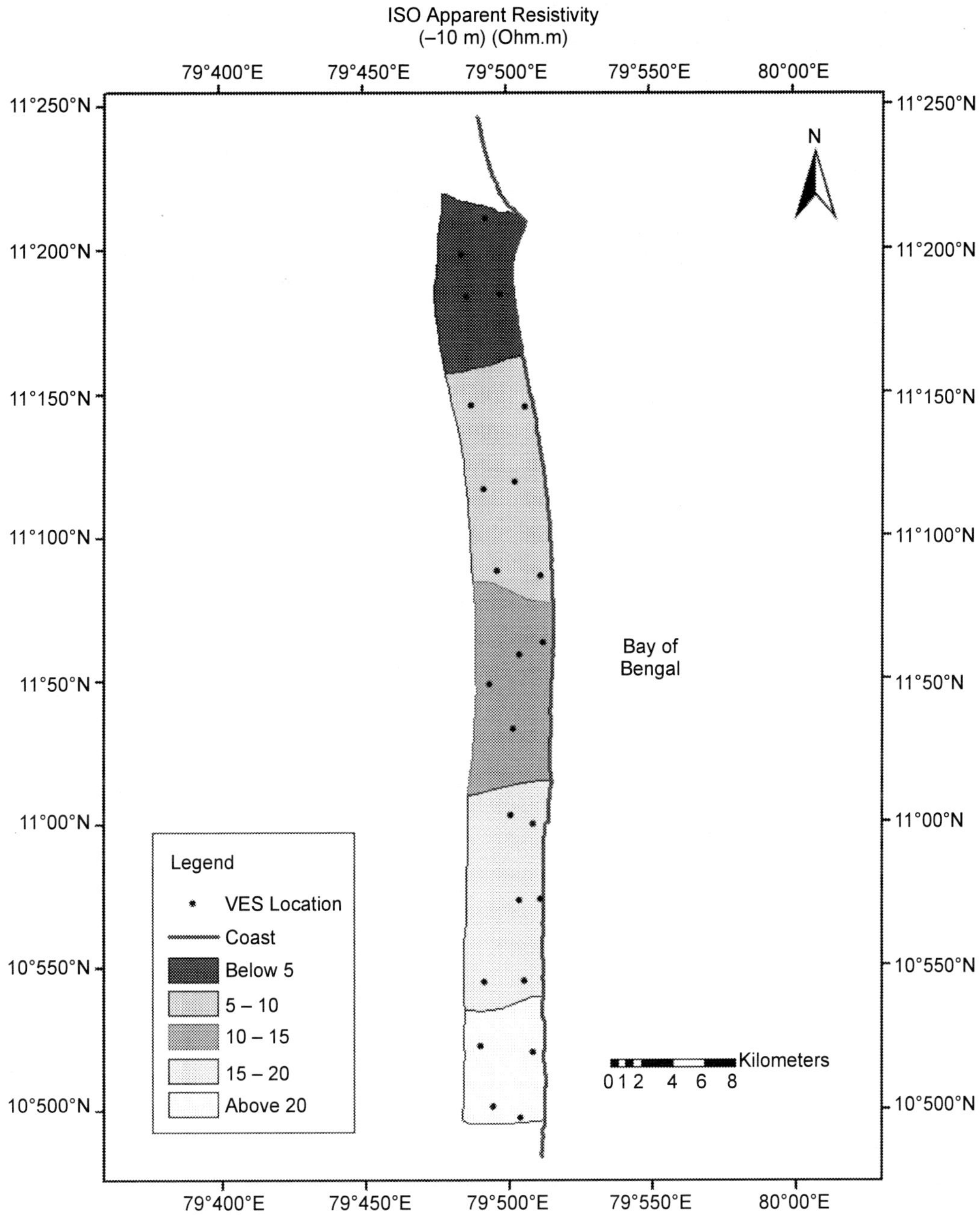

Fig. 7 Isoapparant resistivity map of study area.

1.4 Geophysics

The VES sounding field data have been interpreted both manually and computer inversion method. With the output thematic maps have been prepared (Fig. 7). The geophysical findings also confirm the results of geochemical investigation.

2 CONCLUSION

The tsunami water has come inside the land, stayed for some time and receded. Little quantity got infiltered. Quantum of water depends on area under inundation. Northern part of the study area and region south of Karaikal tsunami water has entered nearly 1 km from the coast (Fig. 6). The successive field work results indicate the decrease of ionic concentration with time.

References

APHA - AWWA and WPCF (1984) Standard methods for the examination of water and wastewater, 16th Ed., Washington D.C, APHA, 1268 p.

Balasubramanian. A, Subramanian. S and Sastri. J.C.V (1991b) HYCH - Basic Computer Program for Hydrochemical Studies, Proc. Vol. Nat. Sem. on Water. Govt. of Kerala, Trivandrum.

Carroll. D (1962) Rainwater as a chemical agent of geologic processes–A review, U.S Geological Survey Water - Supply Paper 520-f, pp. 97–104.

C.G.W.B (2005) Hydrogeological conditions of Tamil Nadu.

Freeze, R.A and Cherry, J.A(1979) Groundwater, Prentice- Hall, Inc. Englewood Cliffs, New Jersey, 604p.

Ryzner, J.W (1944) A new index for determining amount of calcium carbonate scale formed by a water, J. Amer. W.W. Assn., 36, pp 472–486.

Sawyer, C.N., and P.L., McCarty (1987) Chemistry for sanitary engineers, 2nd ed., McGraw-Hill, New York, 518 pp.

Stuyfzand, P.J. (1989) A new hydrochemical classification of water types with examples of application, IAHS, 184, pp 89–98.

Evaluation of Groundwater Quality for Irrigation and Drinking Purposes in Selected Taluks of Dakshina Kannada District in Karnataka, India

32

Oinam Jayalakshmi Devi[*1], S.L. Belagali[2]., A. Balasubramanian[3], Sutapa Bose[1], Senthilkumar G.[1], S. N. Ramaswamy[2], AL. Ramanathan[1]

[1]School of Environmental Sciences, Jawaharlal Nehru University, New Delhi-110067, India

[2]DOS in Environmental Science, University of Mysore, Manasagangotri, Mysore-570006, Karnataka, India

[3]DOS in Geology, University of Mysore, Manasagangotri, Mysore-570006, India

ABSTRACT

An attempt has been made in this work to evaluate the environmental chemistry of groundwater in two taluks of Dakshina Kannada district, Karnataka, India. The results show that, the ground water of this district is well within the permissible limits. The data are subjected to various hydrochemical interpretations. The entire samples corrosivity ratio was less than 1 epm except two samples. For one of the samples (sample no. 1) from Bantwal Taluk, the corrosivity ratio was 7.176 epm. According to USSL Classification, the ground water falls under mostly two types as C_2S_1 and C_3S_1. The values of SAR in study area were in excellent type (<10 epm). According to Handa's classification, the study area is characterized by water having both temporary and permanent hardness. However, most of the ground water samples have shown temporary hardness (75 % of the samples). The ground water salinity and sodium Hazards (Handa) which are much useful for the agriculturists. According to Gibb's plot, all samples in the study area are under the category of rock weathering Dominance zone except one in under evaporation. The data

Corresponding Author Postal Address:

Dr. Oinam Jayalakshmi Devi, School of Environmental Sciences, Jawaharlal Nehru University, New Delhi-110067, India. E-mail: jayalakshmioinam@gamil.com, Phone: +91-9811618836, Fax No. +91-11-26106501 were subjected R-mode factor analysis. A systematic calculation was made to determine the correlation coefficient 'r' amongst the parameters.

Key Words: Groundwater, Dakshina Kannada, Hydrochemical Facies, Corrosivity Ratio, Factor analysis.

1 INTRODUCTION

Ground water is one of the earth's renewable resources which occur as a part of hydrological cycle. Ground water serves as a major source for drinking, irrigation and industrial purpose. Ground water is generally considered to be cleaner than the surface water. Never the less, several factors, like discharge of industrial, agricultural and domestic wastes, land use practices, geological formation, rainfall patterns and infiltration rate are reported to affect the quality of ground water in an area and once contamination of ground water in aquifers occurs, it persists for hundreds of years because of very slow movement of water in them.

Many researchers in the world studied the groundwater quality in different aspects of metals. (Bhattacharjee *et al.* 2005) and groundwater contamination by chlorinated naphthalenes (Gero Vinzelberg *et al.*, 2005). Bilgehan Nas et al. (2006) has studied the groundwater contamination by nitrates in the city of Konya (Turkey): A GIS perspective. They concluded that the distribution of nitrate concentrations is not correlated with well depths within the study area. Andres Navarro et al. (2007) was characterized using isoconcentration maps, hydrogeochemical diagrams (Piper, Schoeller–Berkaloff) and by analyzing hydrogeochemical changes along a flow-path that crosses an urban and peripheral industrial area in the main alluvial aquifer (the La Llagosta unit) in the groundwater hydrogeochemistry of the La Llagosta aquifer.

Balasubramanian (1986) has investigated the Hydrogeological aspects of Tambraparani River Basin, Tamil Nadu and has pointed out that the ratios are positive in the recharge areas and negative in the discharge areas. Venkateswaran, *et al.* (1993) has investigated the hydrogeochemical aspects of Cumbam Valley, Madurai District, Tamil Nadu.

In Karnataka, water quality studies are not adequate- Jayanthi (1993), Ayed, (2002) Mahadev, *et al.* (2004) have investigated the water quality of some places. Meenakumari *et al.* (2003) have observed that Hardness, TDS, Calcium and Iron were very high in the ground waters in Mysore city. Thirumala *et al.* (2005) have reported that water quality in Ayyankere Tank of Chikkmangalore District was good and could be used for drinking irrigation and agriculture. However, a detailed environmental assessment of the water resources have not been attempted by the earlier researchers. Hence, the present study has been undertaken to evaluate the groundwater quality in two Taluks State with reference to drinking and irrigation purposes.

2 MATERIALS AND METHODS

2.1 Characteristic of the Study Area

It lies between $12^0\ 4^/\ 15^{//}$ and $13^0\ 58^/\ 30^{//}$ north latitude and $74^0\ 43^/\ 26^{//}$ and $75^0\ 44^/\ 31^{//}$ east longitude. The latitude of Mangalore, the chief town of South Canara, is one of the two most important points in India. It is at the extreme end of the arc of parallel, which crosses the peninsule of India in latitude 13^0 from Madras to Mangalore, passing through Bangalore Midway. An attempt has been made in this work on the results of water quality in Dakshina Kannada of Karnataka state, India.

Except for the coastal region, where there are formations to be closed as recent and sub-recent deposits and the lateritic formations, the rock formations in the rest of the area belong to the earliest period of the earth's history- the Archaen epoch. The main rock types occurring in the district are arranged in the following sequence:- *Recent and Sub-Recent, Intrusive and Dharwars.*

2.2 Groundwater Sampling and Analysis

15 water samples were collected for analysis from 15 locations in Bantwal and Mangalore Taluks of Dakshina Kannada District, Karnataka (Fig 1). The Physico-Chemical parameters were determined by employing Standard Methods (APHA, 1995; Trivedi & Goel, 1986).

The major ion concentration data has been processed using the HYCY- basic Computer program developed by Balasubramanian et al., (1991). In the program, hydrochemical facies classification is attempted using the criteria of Piper (1944), USSL (1955) schemes, Handa (1964) and Stuyfzand (1989). Aquachem was used for plotting the Piper, Durov and Tenery diagrams.

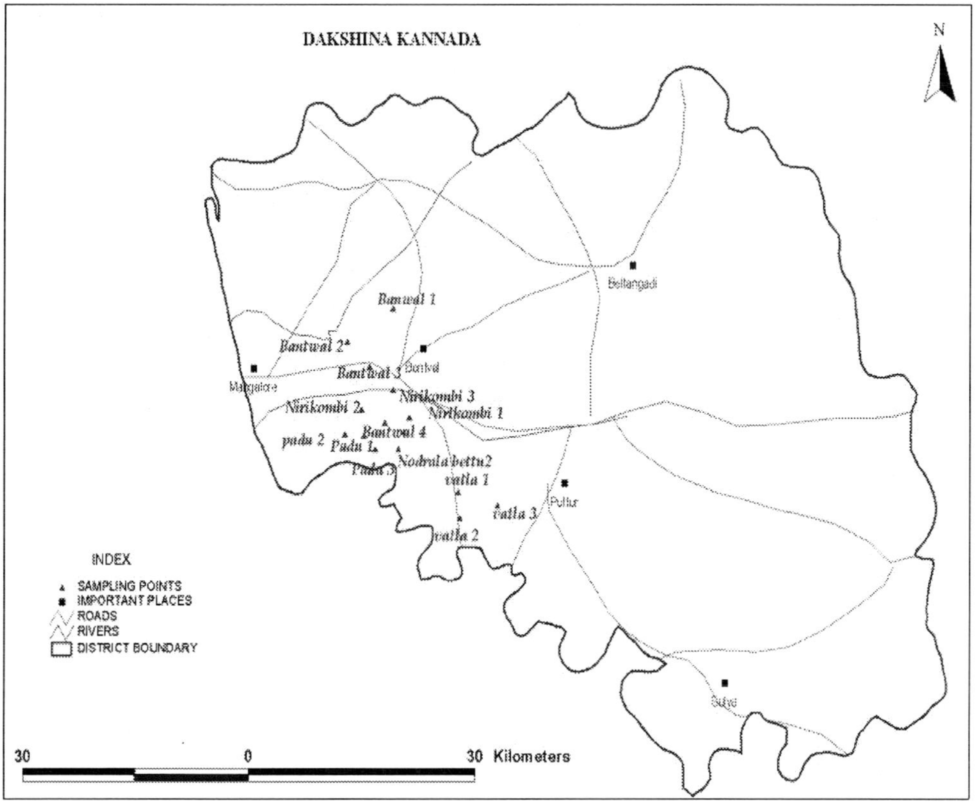

Fig. 1 Showing the sampling sites of the study area

Statistical package for Social Sciences (SPSS) version 15.0 was used to perform factor and correlation analysis.

The output contains information pertaining to the ionic strength, Index of Base Exchange (IBE), total Hardness, water types of Schoeller and permeability Index of Doneen, Piper Hydrochemical Facies, USSL classes of water quality and the mechanisms controlling water Chemistry. A comprehensive picture of the water quality criterion is obtained using this software.

3 RESULTS AND DISCUSSION

3.1 Anion and Cations in Groundwater

Water samples of Dakshina Kannada district were analysed for Physico-Chemical and Hydrogeochemical characteristics from 15 sites. The results were compared with Indian Council of Medical Research (ICMR, 1975), Bureau of Indian Standards (BIS, 1991) and World Health Organization (WHO, 2004) standards. The average, minimum and maximum values for each water quality parameter analyzed for ground water samples are given in Table 1. Regarding to the Physico-Chemical parameters of water samples in study area, only Fluoride was found to be high (1.6 mg L^{-1}). All the other parameters were well within the permissible limits in all the samples.

3.2 Hydrogeochemical Aspects of Groundwater

The hydrogeochemical analysis shows that almost all water samples are useful for human consumption as well as for the irrigation purpose (Table 2).

Table 1 Summary of chemical constituents in groundwater samples. (All values are given in mg L^{-1} except pH and EC, EC value has given in μS cm^{-1})

Parameters	*Maximum*	*Minimum*	*Average*
Caicium	93	5	31.117647
Magnesium	15	1	6.1176471
Na^{+}+K^{+}	53	1	27.058824
HCO_3^{-}	428	10.9	142.31765
CO_3^{-2}	21.6	0	5.1882353
Chloride	27.9	3.9	14.582353
Nitrate	1.48	0.3	1.0458824
Sulphate	70	0	23.532353
TDS	808	150	369
EC	1263	234	576.82353
pH	8.3	6.1	7.5705882
Fluoride	1.6	0.9	1.3058824
Phosphate	0.09	0	0.0188235
Nitrite	1.15	0	0.2323529
DO	5.9	0	3.8117647
COD	24.4	0	7.5823529
T H	315	42	109.94118

Table 2 Hydrogeochemical analysis for Dakkshina Kannada District

	HANDA	Hydrogeochemical		Schoeller's			IBE				
Sl.no	*PI*		*Cation*	*Anion*	*USSL*	*Water Type*	*CR*	*RSC*	*SAR*		
1.	105.553	A3C1S1	Na+K,Ca+Mg	$Cl+SO_4$	C2S1	II	7.176	–0.1531	3.202	–1.9643	–0.6548
2.	75.3570	B1C2S1	Ca+Mg,Na+K	$HCO_3+CO_3,Cl+SO_4$	C2S1	II	0.235	1.01041	1.636	–7.2327	–0.3605
3.	107.590	B2C1S1	Na+K,Ca+Mg	$HCO3+CO_3,Cl+SO_4$	C3S1	I	0.658	0.20501	1.200	–1.7317	–0.4083
4.	103.979	B2C2S1	Na+K,Ca+Mg	$HCO3+CO_3,Cl+SO_4$	C2S1	I	0.706	0.77401	2.674	–1.9931	–0.5413
5.	67.1361	B1C2S1	Ca+Mg,Na+K	$HCO_3+CO_3,Cl+SO_4$	C3S1	II	0.378	0.0810	1.342	–4.0597	–0.2761
6.	107.077	B2C2S1	Na+K,Ca+Mg	$HCO3+CO_3,Cl+SO_4$	C2S1	III	0.533	0.78911	2.606	–2.0432	–0.5147
7.	108.140	B1C1S1	Ca+Mg,Na+K	$HCO_3+CO_3,Cl+SO_4$	C2S1	II	0.227	0.37873	0.895	–2.4002	–0.2757
8.	126.128	B1C1S1	Ca+Mg,Na+K	HCO_3+CO_3	C2S1	II	0.133	0.49522	0.985	–1.9562	–0.3155
9.	102.603	B1C1S1	Ca+Mg,Na+K	HCO_3+CO_3	C2S1	II	0.161	0.39442	0.777	–4.5334	–0.2701
10.	100.777	B3C2S1	Na+k,Ca+Mg	$Cl+SO_4,HCO_3+CO_3$	C2S2	II	1.030	0.37427	2.920	–1.9282	–0.5246
11.	62.7155	B1C3S1	Ca+Mg,Na+K	HCO_3+CO_3	C3S1	II	0.069	1.55050	1.262	–10.7443	–0.2565
12.	119.021	B2C*1S1	Na+K,Ca+Mg	$HCO3+CO_3,Cl+SO_4$	C2S1	I	0.381	0.62559	1.909	–0.8135	–0.3737
13.	56.0776	A1C2S1	Ca+Mg	HCO_3+CO_3	C2S1	I	0.141	–0.489	0.274	0.2526	0.0368
14.	57.9854	A1C2S1	Ca+Mg	HCO_3+CO_3	C2S1	I	0.174	–0.2776	0.177	0.6259	0.1323
15.	65.6610	A1C*1S1	Ca+Mg	HCO_3+CO_3	C1S1	I	0.172	–0.2491	4.2E-02	0.8875	0.1859

PI= *Permeability Index*; USSL=US *Salinity Laboratory*; SAR= *Sodium Adsorption Ratio*; CR= *Corrosivity Ratio*; IBE=*Indices of Base Exchange*; RSC= *Residual Sodium Carbonate*

3.2.1 Corrosivity Ratio

Corrosion is an electrolytic process that takes place on the surface of the metals, which severely attacks and corrodes away the metal surfaces. Most of the corrosion problems are associated with salinity and encrustation problems associated with alkalinity.

Badrinath et al. (1984) used the Corrossivity Ratio to evaluate the Corrosive tendencies of river waters. The ratio is of the form:

$$\frac{\dfrac{Cl^{-} mg L^{1-} (mg\ L^{-1})}{35.5} + \dfrac{SO_4^{2-}\ (mg\ L^{-1})}{48}}{\dfrac{(HCO_3^{-} + CO_3^{2-}) * (mg\ L^{-1})}{50}}$$

Water having corrosivity ratio less than 1 is considered to be non-corrosive, while the value above 1 is corrosive.

All the samples corrosivity ratio was less than 1 epm except two samples. For one of the samples (sample no. 1) from Bantwal Taluk, the corrosivity ratio was 7.176 epm (Table 2).

3.2.2 Schoeller's water type

Schoeller (1965, 1967) suggested two chloro alkaline indices CAI 1 and CAI 2 to indicate the exchange of ions between the groundwater and its host environment. The exchange of ions Na^{+} and K^{+} from the water with the Mg^{2+} and Ca^{2+} in the rock is designated as direct exchange. If the exchange is reversing, it is known as reverse exchange, this may be represented as a reversible chemical reaction.

$$Na^{+} + K^{+}\ (\text{Direct}) \rightarrow Ca^{2+} + Mg^{2+}$$
$$(\text{Reverse}) \leftarrow$$

The chloro alkaline indices used in the evaluation of Base Exchange are expressed as

$CAI\ 1 = [Cl^{-} - (Na^{+} + K^{+})] / Cl^{-}$

$CAI\ 2 = [Cl^{-} (Na^{+} + K^{+})] / Cl^{-}[SO_4^{-} + HCO_3^{-} + CO_3^{-} + NO_3^{-}]$

If the values of both these indices are positive, it denotes a direct exchange while in the case of reverse exchange the indices are negative.

From the study area, 80% of the samples have shown negative indices (Table 2).

Schoeller pointed out that the first and foremost water are those in which

$rCO_3^{-2} > rSO_4^{-2}$ ...Type I

As the total concentration increases the above relation changes to

$rSO_4^{-2} > rCl^{-}$ ...Type II

Still at concentration the water may change to

$RCl^{-} > rCO_3^{-2}$ ...Type III

And in the final stages

$rCl^{-} > rSO_4^{-2} > r\ rCO_3^{-2}$ and $Na^{+} > Mg^{+2} > Ca^{2+}$ ...Type IV

Where r = epm content of the ions.

Based on Schoellers (1965, 1967) water type classifications, all the samples fall under type I and II except one sample in type III (Table 2).

3.2.3 Sodium Adsorption Ratio (SAR)

The expression of the equilibrium between exchangeable positive ions (Cations) in the soil and cations in the irrigation water is known as Sodium Adsorption Ratio. It gives a measure of suitability of water for irrigation with respect to the sodium (alkali) hazards. High SAR values may cause damage to soil. The SAR may be determined by the formula (epm):

$$SAR = \frac{Na^{+}}{\sqrt{\frac{Ca^{++} + Mg^{++}}{2}}}$$

According to SAR classification, all the samples were found to be excellent type <10 epm (Table 3) which can be used for irrigation on almost all soils. A more detailed analysis for the suitability of water for irrigation was made by plotting the data on US Salinity Laboratory Diagram (USSL 1954).

3.2.4 USSL diagram

According to a method formulated by the US Salinity Laboratory (1954), water used for irrigation can be rated based on salinity hazards and sodium or alkali hazard. When the sodium hazard ratio and Electrical conductivity of water are known, classification of water for irrigation can be done by plotting these on the diagram. Low salinity water ($C1^{-}$< 250ì –250 mho cm^{-1}) can be used for irrigation of most crops on most soils with little likelihood that soil salinity will develop. All samples from the study were found to be C_1S_1 and C_2S_1 except one sample fall in C_3S_1 (Fig 2). It indicates that all samples are found to be low Salinity and low sodium Hazard with medium Sodium also except three sample (C_3S_1) which are in high salinity and low sodium hazard. High salinity water (C_3) cannot be used in soil with restricted drainage.

3.2.5 Gibb's diagram

Gibb's (1970) discussed the relationship between water composition and aquifers lithology. Viswanathaiah et al (1978) have studied the mechanisms controlling the groundwater composition of Karnataka. Gibb's plots distinguish the interaction of groundwater due to precipitation or rock evaporation by demarcating these fields. According to Gibbs plots (Fig 3) all samples (points) fall in Rock weathering Dominance Zone. It may be due to the negligible contribution of calcic minerals (e.g. calcite in limestones) within this luminosilicate terrain and/ or an additional input of sodium from a possible anthropogenic source.

3.2.6 Permeability Index

The soil permeability is influenced by long term use of irrigation water containing Sodium and Bicarbonates. Permeability Index is calculated using the formula (Donean, 1966)

$$PI = \frac{Na^{+} + \sqrt{HCO_3^{-}}}{Ca^{2+} + Mg^{2+} + Na^{+} + K} \times 100$$

Table 3 Suitability of Groundwater for Irrigation Base on Several Classifications

Classification	*Water Class*	*Number Of Samples*
Based on Residual Sodium Carbonate (RSC) (Richards, 1954		
< 1.25	Good	13
1.25 - 2.50	Doubtful	1
>2.5	Unsuitable	1
Base on Sodium Adsorption Ratio (SAR) (Richards, 1954		
< 10	Excellent	15
10-18	Good	0
18-26	Fair	0
> 26	Poor	0
Base on Na% - (Wilcox, 1955)		
Excellent	<20	3
Good	20-40	5
Permissible	40-60	3
Doubtful	60-80	4
Unsuitable	>80	0
Base on Gibbs diagram		
Evaporation	Longer residence time	0
Rock Interaction	Dissolution mixing of mineral matter	15
Precipitation	Recent recharge water	0
Based on EC(μS/cm) (Wilcox, 1955)		
Excellent	<250	1
Good	250-750	11
Permissible	750-2250	3
Doubtful	2250-5000	0
Unsuitable	>5000	0
% Na (Eaton 1950)		
Safe	<60	11
Unsafe	>60	4

The permeability Index of the study areas varied from 56.0776 to 126.128 epm. According to permeability index values, majority of water samples (53%) falls under class III. 26% of the water samples fall in class I and 21% are in class II (Fig 4)

3.2.7 Sodium percent (%Na)

Sodium content expressed in terms of sodium percentage or soluble sodium percentage defined as

$$\%\text{Na} = \frac{(\text{Na}^+ + \text{K}^+) \times 100}{(\text{Ca}^{2+} + \text{Mg}^{2+} + \text{Na}^+ + \text{K}^+)}$$

Where, all ionic concentrations are expressed in meq L^{-1}.

According to Wilcox (1955), the Na% indicates that the water samples are in excellent, good and permissible except four samples which are doubtful (Table 3). Most of the water falls in safe and four samples are in unsafe water samples given by Eaton (1950). When the concentration of

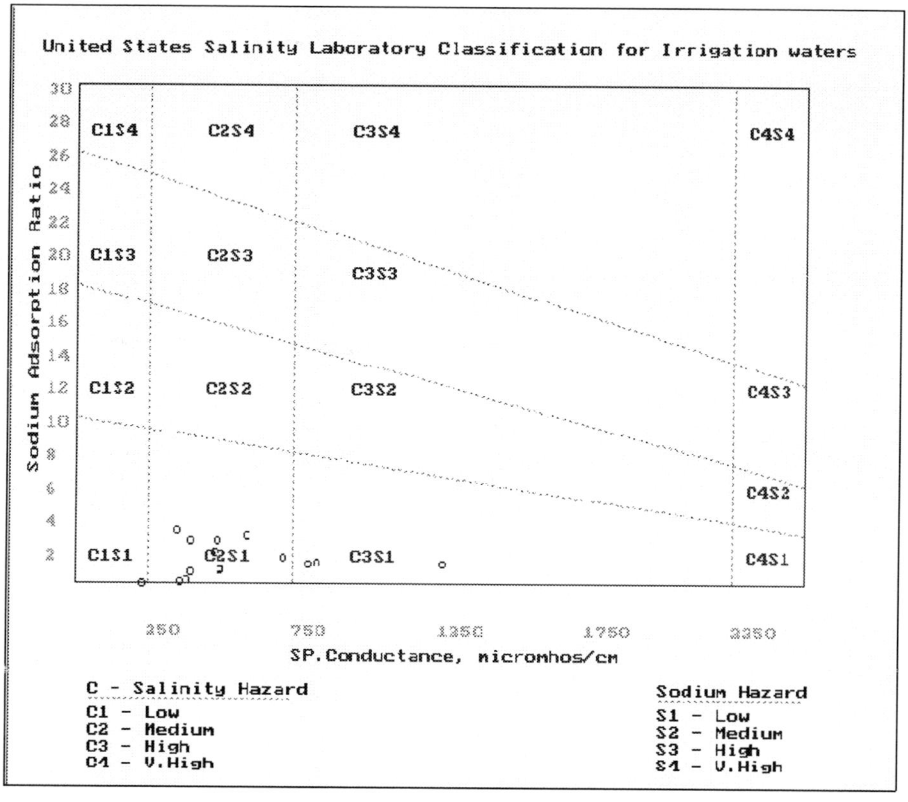

Fig. 2 USSL Classification of groundwater samples collected from Dakshina Kannada

sodium is high in irrigation water, sodium ions tend to be absorbed by clay particles, displacing Mg^{2+} and Ca^{2+} ions. This exchanged process of Na^+ in water for calcium and Magnesium in soil with poor internal drainage (Manish et al. 2007).

3.2.8 Residual Sodium Carbonate

In addition to the SAR, the excess sum of carbonate and bicarbonate in groundwater over the sum of calcium and magnesium also influences the suitability of groundwater for irrigation An excess quantity of sodium bicarbonate and carbonate is considered to be detrimental to the physical properties of soils as it causes dissolution of organic matter in soil, which in turn leaves a black stain on the soil surface on drying. This excess is denoted by Residual Sodium Carbonate (RSC) and is calculated as follows (Ragunath 1987):

$$RSC = (CO^{2-}_3 + HCO_3^-) - (Ca^{2+} + Mg^{2+})$$

Where, all ionic concentrations are expressed in meq L^{-1}.

According to Residual Sodium Carbonate (RSC), most of the water samples are good except two samples which fall in doubtful class (Table 3).

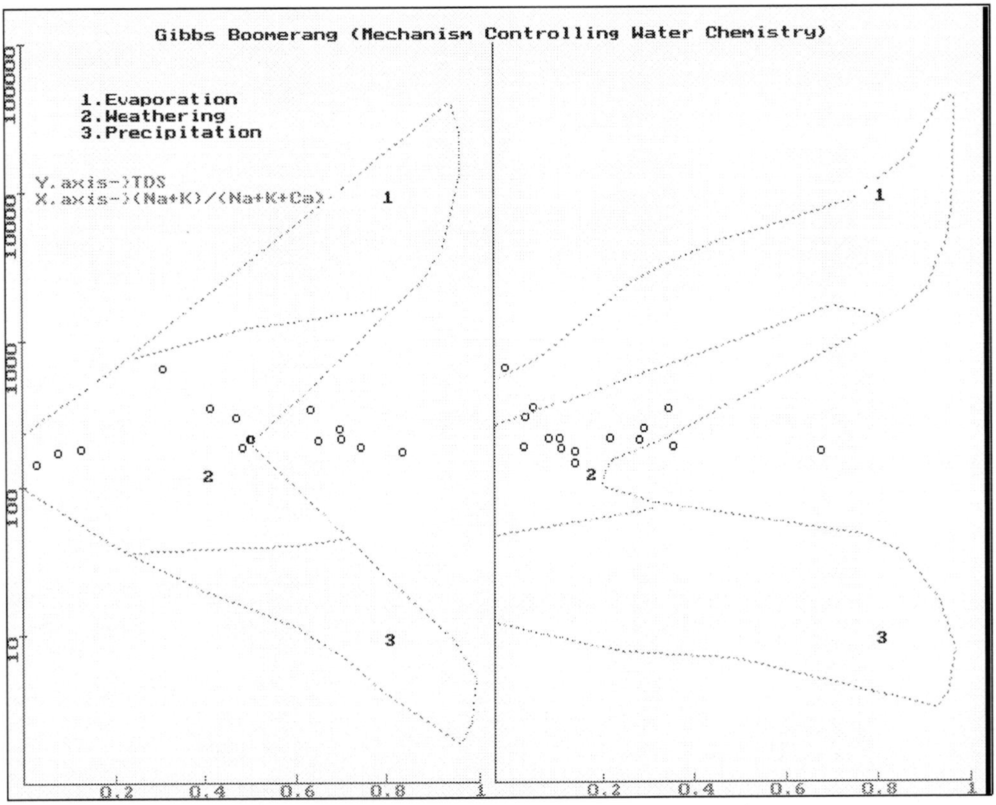

Fig. 3 Gibbs Diagram of the groundwater samples

3.3 Graphical Representation of Hydrochemical Data of Groundwater

In the present study, water quality data is analyzed through statistical distribution diagrams such as Piper diagram, Ternary diagram and Durov plot (Durov 1948) to gain better insight into the hydrochemical processes operating in the groundwater flow system that resulted in the observed spatial and temporal variation in the groundwater quality. The geochemical evolution of groundwater can be understood by plotting the concentrations of major cations and anions in the Piper trilinear diagram (Fig 5). The plot shows that most of the samples fall in mixed type, Ca-Mg-HCO_3^-. Some of the samples also represented Na+K–HCO_3^- Most of the water samples are coming in unpolluted to slightly polluted category, however there has been an inclination of sodium enrichment in groundwater. It may be because of waste water discharge or return flow from irrigation. Some samples are also representing mixed type.

The ternary plot for the study area shows higher amount of bicarbonate indicates that the groundwater falls in the region of carbonate weathering. Durov plot (Fig 7) indicates that most of the samples are anion dominant in the area. For further, it may be supported by the dominance of alkali and alkaline earth metal.

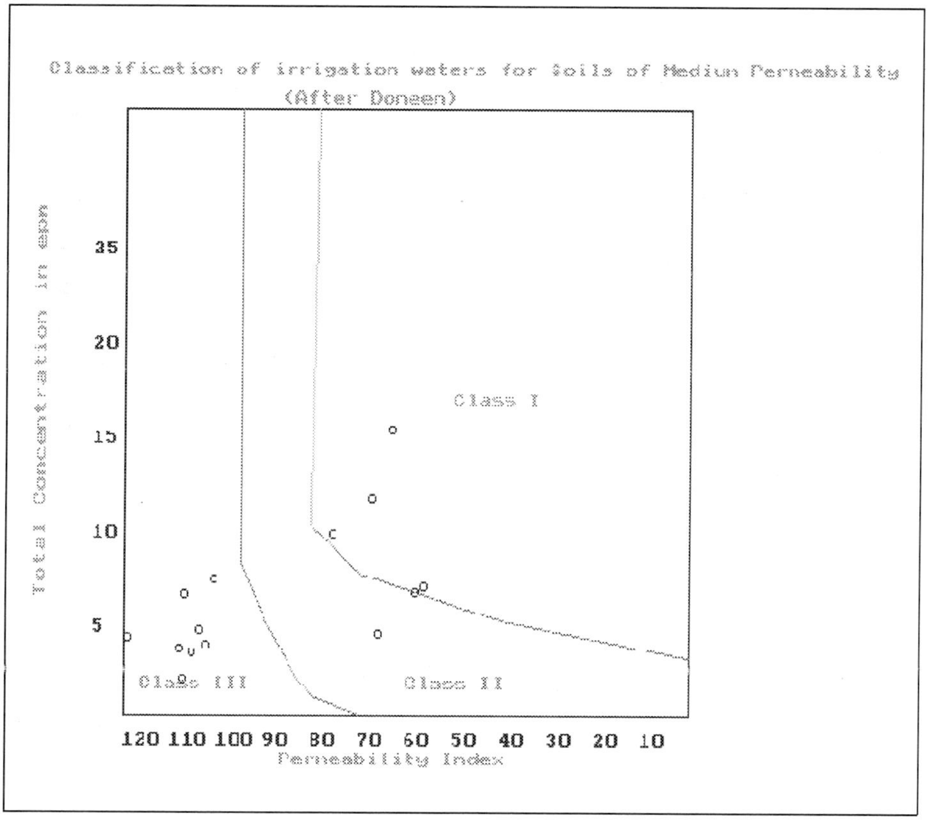

Fig. 4 Permeability index of the groundwater

Ionic strength is measure of total concentration of ions which emphasizes increased contribution of species with charges greater than one to solution non-ideality.

$$I = 0.5\ S\ m_1 z_1$$

Where m_1 is the atomic/ molecular weight and z_1 is the valance of the respective ion. The ionic strength of the groundwater samples range from 0.001 to 0.01(Fig 8). The lesser ionic strength of the samples from groundwater reflects that the dissolved ions is lesser may be due to the removal by precipitation or by dilution from the nearby source due to the induced extensive extraction of ground water.

3.4 Statistical Analysis

3.4.1 Factor Analysis

Statistical package for Social Sciences (SPSS) version 15.0 was used to performed factor analysis. Factor analysis has been widely used for the Hydrochemical interpretation of commonly collected groundwater and it is applied for the explanation of groundwater chemistry origin (Krzysztof Dragon, 2006; Gupta and Subramanian, 1998; Subharao et al.,

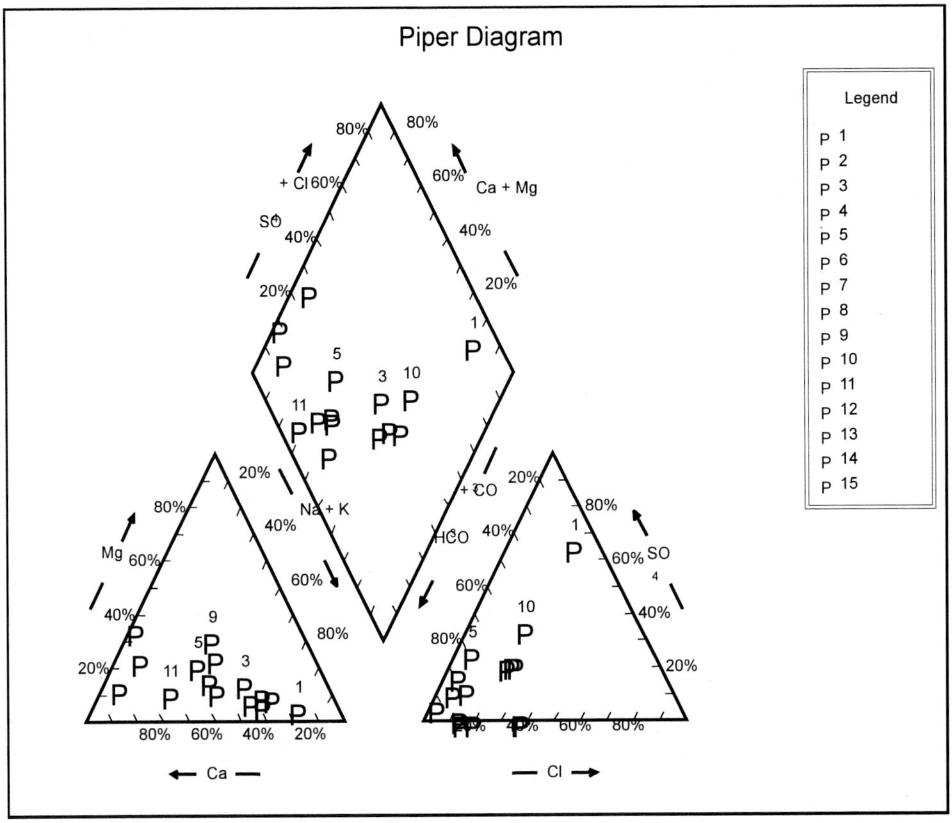

Fig. 5 Piper Diagram of the groundwater

1996). Factor analysis is also useful for Hydrochemical facies investigation and interpretation of their origin.

For groundwater samples, five factors with eigenvalue >1 have been extracted from verimax rotation (Table 4). Factor 1 represented with 38.08% of Variance by Total Hardness, EC and TDS indicating secondary leaching and saline water intrusion. In correlation data matrix DO does not show association with any other variable studied. Factor 2 represent with 21.63% of Variance by Magnesium, Fluoride and Nitrite indicating secondary leaching and Mg derived from the source rock. Although Nitrate is not included in this factor. Factor 3 accounts for 12.58% variance in the data matrix. This factor consists of variables Sulphate, Phosphate and Sodium. They are grouped together because of their origin from similar sources of anthropogenic impact. Factor 4 represented with 8.94% of total variance by Carbonate indicating weathering. This factor also explains pH controlled dissolution of Sodium minerals. Factor 5 which explain with 6.304% of total variance consists of Chloride and COD. In this factor contains negatively correlated pH with Chloride. The factor indicates that lower pH increases dissolution oh Chloride minerals and vice versa. This factor may be termed Chloride factor.

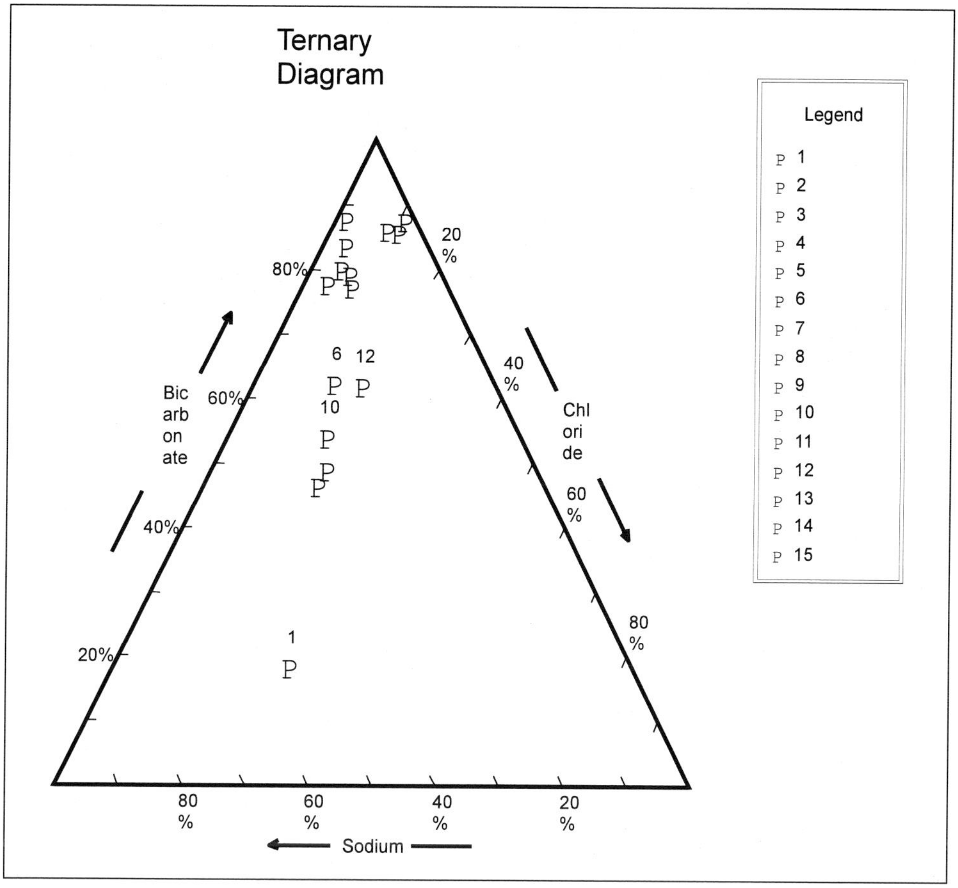

Fig. 6 Ternary Diagram of the groundwater

3.4.2 Correlation

A correlation analysis is a bivariate method applied to describe the degree of relation between two hydrochemical parameters. Using a correlation matrix resulted in improved definition of the principal components at the expense of a decreased proportion of explained variance and the number of eigenvalues to be used was selected on the basis of the mean value (Vidal et al. 2000).The result of the correlation analysis is considered in the subsequent interpretation. The highest at positive correlation (r=1.000, p<0.01) was found between EC and TDS (Table 5). It indicates the direct proportionality between TDS and EC and in turn appraises the dependence of EC and TDS too. Calcium is showing positive correlation with HCO_3^- (r= 0.961; p<0.01), Total hardness (r= 0.784; p<0.01), TDS and EC (r= 0.727; p<0.05). These results indicate that there is an increase in Ca^{2+} concentration with increase in the above parameters. HCO_3^- is found positively correlated with Total Hardness (r= 0.884; p<0.01), Mg^{2+} (r= 0.727; p<0.01), TDS (r= 0.717; p<0.01), EC(r= 0.717; p<0.01), and negatively correlated with DO (r= -0.673; p<0.01).

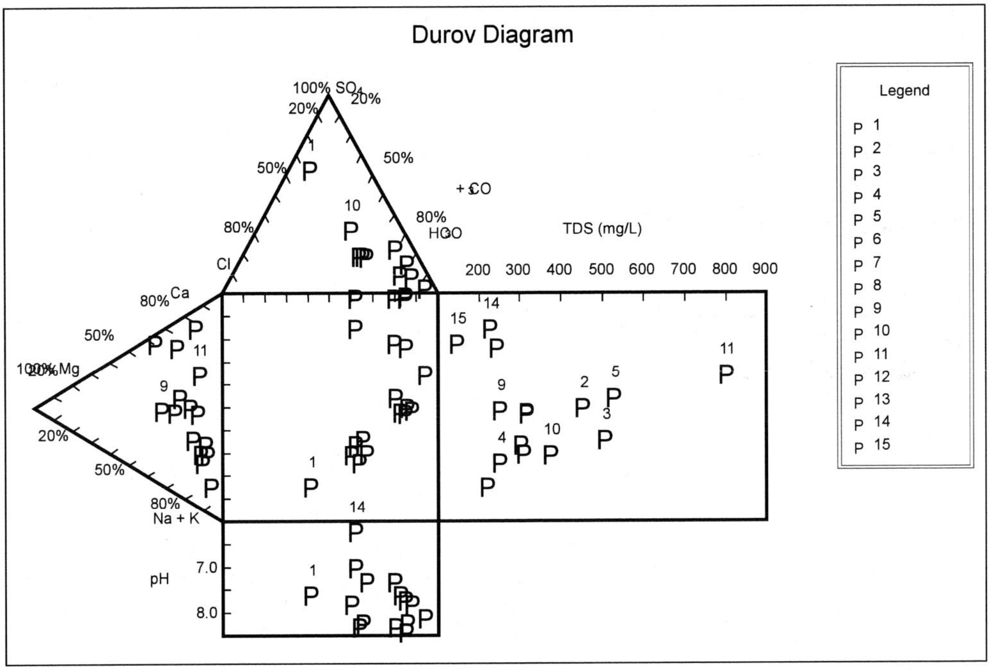

Fig. 7 Durov Diagram of the groundwater

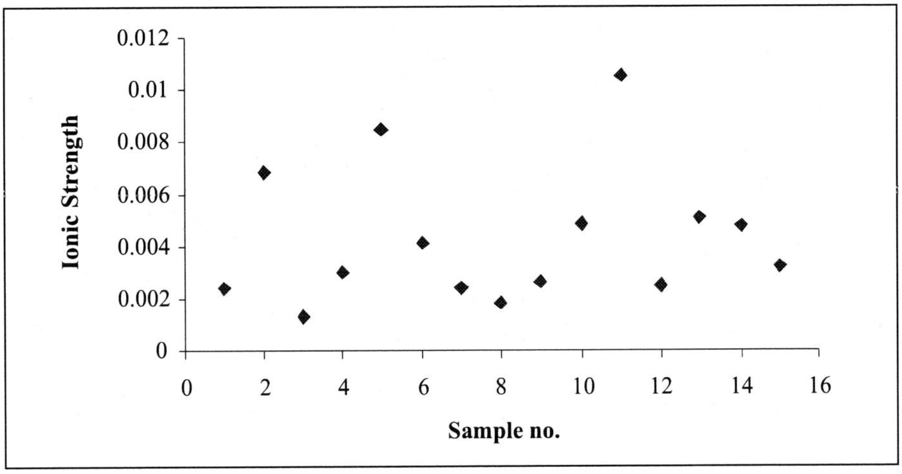

Fig. 8 Ionic Strength of groundwater samples

DO is found negatively correlated with Total Hardness($r = -0.733$; $p<0.01$), TDS ($r = -0.713$; $p<0.01$) and EC ($r = -0.713$; $p<0.01$). The higher negative correlation is shown among $Na^{+}+K^{+}$ and NO_3^{-} ($r=-0.645$; $p<0.01$). Therefore, it can be postulated that the simultaneous increase/

Table 4 Factor Analysis

Variable	*Varimax rotated factor Matrix*				
	Factor 1	*Factor 2*	*Factor 3*	*Factor 4*	*Factor 5*
Calcium	0.755	0.522	0.04	−0.072	0.117
Magnesium	0.461	0.769	0.235	0.092	−0.12
Sodium	0.458	−0.348	0.721	0.242	0.137
Bicarbonate	0.85	0.407	0.096	−0.096	0.056
Carbonate	0.002	0.014	−0.082	0.94	0.101
Chloride	-0.368	−0.097	0.456	−0.081	0.735
Nitrate	0.022	−0.843	0.223	−0.182	0.103
Sulphate	0.113	−0.103	0.899	0.11	−0.14
TDS	0.918	−0.075	0.197	0.052	−0.201
EC	0.918	−0.074	0.198	0.052	−0.2
pH	0.237	0.119	0.291	0.699	−0.536
Fluoride	0.379	0.668	0.088	−0.092	0.296
Phosphate	0.194	0.103	0.889	−0.206	0.105
Nitrite	-0.293	0.716	−0.37	−0.236	0.341
DO	-0.847	−0.095	0.036	−0.272	−0.084
COD	0.197	0.487	−0.192	0.146	0.667
TH	0.932	0.115	0.282	0.012	−0.047
Eigenvalue	**6.474**	**3.677**	**2.14**	**1.521**	**1.072**
Variance %	**38.083**	**21.63**	**12.588**	**8.949**	**6.304**
Cumulative %	**38.083**	**59.713**	**72.301**	**81.25**	**87.554**

decrease in the cations is the result mainly of dissolution/precipitation reaction and concentration effects.

4 CONCLUSION

The hydrogeochemical characteristic of ground water in Dakshina Kannada district of Karnataka, based on Handa, Schoeller, USSL, Gibbs, Piper and Stuyfzand methods revealed that the present status is better for irrigation and drinking purposed except for few location. The physico- chemical analysis results show that, the ground water sample in the study area was well within the permissible limits except one sample contains high fluoride but it may deteriorate in future. All samples in study areas, corrosivity ratio was less than 1 epm except two samples. For one of the samples (sample no. 1) from Bantwal Taluk, the corrosivity ratio was 7.176 epm. It has been found that the study area, maximum number of ground water is good for agricultural and domestic purposes except three samples was to be high salinity (C_3). Such areas need special salt attention as far as irrigation is concerned. In the present area of study the salt water intrusion into the groundwater system, may be due to over extraction. Such water can be used either after soil treatment or flushing the soil with excessive of irrigation water.

Table 5 Correlation coefficient (r) among various water quality parameters (Dakshina Kannada)

Ca	Mg	Na+K	Fe	HCO_3^-	CO_3^{-2}	Cl	NO_3	$S0_4$	TDS	EC	pH	F	PO_4	NO_2	DO	COD	TH	
Ca	1.00																	
Mg	.727**	1.00																
Na+K	.225	.102	1.00															
Fe	.741**	.437	.411	1.00														
HCO_3	.961**	.727**	.354	.803**	1.00													
CO_3	.007	.081	.162	−.232	-.067	1.00												
Cl	−.162	-.313	.296	-.294	−.226	−.037	1.00											
NO_3	−.391	-.508	.367	-.200	−.297	−.108	.198	1.00										
$S0_4$	.076	.220	.755**	.261	.118	.019	.227	.260	1.00									
TDS	.625*	.368	.547*	.793**	.717**	.025	−.347	.068	.300	1.00								
EC	.625*	.369	.548*	.793**	.717**	.025	−.346	.068	.301	1.00**	1.00							
PH	.119	.361	.357	.260	.170	.544*	−.368	−.243	.367	.441	.441	1.00						
F	.591*	.616*	-.050	.274	.523*	.016	.069	−.406	−.060	.345	.345	-.002	1.00					
PO_4	.179	.381	.608*	.210	.267	−.264	.371	.202	.706**	.318	.319	.131	.327	1.00				
NO_2	.246	.237	-.645**	−.197	.059	−.094	.161	−.595*	−.454	−.462	−.462	-.470	.489	−.286	1.00			
DO	−.577*	-.523*	−.404	-.568*	−.673**	−.228	.396	.092	−.129	−.713**	-.713**	−.300	−.371	−.144	.196	1.00		
COD	.376	.403	-.122	.006	.328	.115	.202	−.438	−.275	−.052	−.052	-.179	.466	.020	.444	−.373	1.00	
TH	.784**	.598*	.597*	.877**	.884**	−.038	−.245	−.036	.354	.891**	.891**	.366	.388	.412	−.329	−.773**	.179	1.00

References

American Public Health Association (APHA), 1995. Standard methods for the examination of water and wastewater,: 20th edn. American Public Health Association, Washington DC.

Andres Navarroa, Montserrat Carbonell (2007).Evaluation of groundwater contamination beneath an urban environment: The Beso' s river basin (Barcelona, Spain). Journal of Environmental Management. 85, 259–269

Ayed, R. K., 2002. Ecological Studies on two Freshwater bodies near Mysore, Karnataka, Ph. D.Thesis. Univ. of Mysore, Mysore.

Badrinath, S. D., Raman, A., Gadkari, S. K., Mhaisalkar, V. A., Deshpande, V. P., 1984. Evaluation of carbonate stability indices for Sabarmati river water. Ind. Water Assn. 16, 163-168.

Balasubramanian, A., 1986. Hydrogeological Investigation of Tambraparani River Basin, Tamil Nadu. Unpublished Ph.D. Thesis, Mysore University 349p.

Balasubramanian, A., Subramanian, S., Sastri, J.C.V., 1991. Hydrochemical Facies Classes Software, Basic Computer Program for Hydrogeolocal Studies.

Bhattacharjee, S., Chakravarty, Maity, S., Dureja, V., Gupta,K. K., 2005. Metal contents in the groundwater of Sahebgunj district, Jharkhand, India, with special reference to arsenic. Chemosphere. 58(9), 1203-1217.

Bilgehan Nas, Berktay Ali, 2006. Groundwater contamination by nitrates in the city of Konya,(Turkey): A GIS perspective. Journal of Environmental Management. Journal of Environmental Management 79, 30–37

Bureau of Indian Standards (BIS)., 1991. Specification. Indian Standard for Drinking Water, New Delhi.

Donean, L.D., 1966. The Salinity of irrigation waters, Trans. Am. Geophys. Un, 35, 943-950.

Durov SA., 1948., Natural waters and graphical representation of their composition. Dokl Akad Nauk USSR 59, 87–90.

Eaton FM., 1950. Significance of carbonate in irrigation water, Soil Science, 69.

Gero Vinzelberg , Jan Schwarzbauer, Ralf Littke., 2005. Groundwater contamination by chlorinated naphthalenes and related substances caused by activities of a former military base. Chemosphere. 61(6), 770-782

Gibbs, R.J., 1970. Mechanisms Controlling World's Water Chemistry, Sciences 170, 1088-1090.

Gupta, L.P., Subramanian, V., 1998. Geochemical factors controlling the chemical nature of water and sediments in the Gomti River, India. Environmental Geology 31 (1/2), 102–108.

Handa, B.K., 1964. Modified Classification Procedure for Rating Irrigation Water. Soil Sci. 98, 246-269.

Indian Council of Medical Research (ICMR), 1975. Manual of Standards of Quality of Drinking Water Supplies. ICMR, New Delhi.

Jayanthi, S.K., 1993. Studies on Industrial water pollution of Cauvery river near Belugol*a,* Mysore.Ph.D. Thesis, Univ. of Mysore, Mysore.

Krzysztof Dragon., 2006. Application of factor analysis to study contamination of a semi-confined aquifer (Wielkopolska Buried Valley aquifer, Pol*and*). Journal of Hydrology. 31, 272– 279.

Mahadev, J., Nagarathna, Hosmani, S.P., 2004. Studies on Water Quality with Significance of calcium Carbonate Saturation index in two water bodies of Mysore. Indian J. Env. Prot., 24(9), 678-683.

Manish Kumar, Kalpana Kumari, Ramanathan, AL., Rajinder Saxena, 2007. A comparative evaluation of groundwater suitability for irrigation and drinking purposes in two intensively cultivated districts of Punjab, India. Environmental Geology (In Press).

Meenakumari, H.R., Hosmani S. P., 2003. Physico-Chemical and Biological Quality of ground water in Mysore City, Karnataka, India. Indian J. Environ. & Ecoplan. 7(1), 79-82.

Piper, A.M., 1944. A Graphic Procedure in the Geochemical Interpretation Of Water Analysis. AM. Geology. Union. Trans., 25, 914-923.

Ragunath, H. M., 1987. Groundwater, Weley Eastern, New Delhi, pp 563.

Richards, L. A., 1954. Diagnosis and improvement of saline alkali Soil US department of Agriculture, Hand book. 60,160.

Schoeller, H., 1965. Hydrodynamique dans lekarst (Ecoulement et emmagasinement), Actes Colloques Doubronik. I, pp 3-20. AIHS et UNESCO.

Schoeller, H., 1967. Qualitative Evaluation of Ground Water Resource. (Method and Techniques of Ground Water Investigation and Development) Wat.Res. Series- 33 UNESCO, pp 44-52.

Subbarao, C., Subbarao, N.V., Chandu, S.N., 1996. Characterisation of groundwater contamination using factor analysis. Environmental Geology 28 (4), 175–180.

Thirumala S., Kiran B.R., Puttaiah E.T., 2004. Variation of water Quality in Ayyankere Tank. IJEP. 24(12), 917-920.

Trivedi, R.K., Goel .P.K., 1986. Chemical, Biological Methods of Water Pollution Studies. Environmental Media Publications, Karad..

USSL., 1954. Classification of Irrigation Waters. U.S. Department of Agriculture, Circu 969. Washington.

Venkateswaran, S., M. S. Jagathusan, A. Balasubraminian, 1993. Hydrogeochemical studies of Cumbam Vallay Madurai District. Tamil Nadu. Journal of Applied Hydrology. VI (1-4), 57- 68.

Vidal, M., Melgar, J., Opez A. L., Santoalla. M. C., 2000. Spatial and temporal hydrochemical changes in groundwater under the contaminating effects of fertilizers and wastewater. Journal of Environmental Management. 60, 215–225.

Viswanathaiah, M.N., Sastri, J .C.V., 1978. Capacity of wells in some Hard rocks of Karnataka, Jour, Geol. Sec. India. 19, 426-430.

WHO., 2004. Guideline for Drinking-Water Quality, Third Edition (Recommendations), World Health Organization, Geneva.

Wilcox, LV., 1955. Classification and use of irrigation waters, USDA. Circular 969, Washington, DC, USA.

Remedy for Arsenic Poisoning? A Quest in an Unconventional Area of Research

33

A.R. Khuda Bukhsh
Department of Zoology, University of Kalyani, Kalyani-741235, India

ABSTRACT

We conducted a human trial on the efficacy of a potentized homeopathic drug, Arsenicum Album, in amelioration of arsenic (As) toxicity and characteristic symptoms of arsenicosis in random populations of two high risk arsenic villages in West Bengal. Both "controlled" and "uncontrolled" experiments were performed. In the short-term controlled group, the As contents of blood and urine and various toxicity biomarkers like acid and alkaline phosphatases, alanine amino transferase, asparatate amino transferase, lipid peroxidation activity and reduced glutathione level and hematological parameters like total count of RBC and WBC, hemoglobin content, ESR, blood sugar level, ANA, etc. were examined before and periodically after administration of the homeopathic remedy. The data were compared with the "placebo fed" group and another group of people from a village not contaminated with groundwater arsenic. In the uncontrolled group, the subjects were fed only "verum" and long-term effects were studied. Initial results are highly encouraging.

1 INTRODUCTION

Primarily as a sequel to drawing up huge quantities of groundwater by shallow tube wells for agricultural purposes, and also for some other factors, many aquifers in some countries have now been contaminated with various toxic metal compounds including Arsenicals. Prolonged exposure to arsenic (As), a toxic metalloid, has caused various illnesses to millions of people globally distributed over some 20 countries, of which in Bangladesh and its adjoining part of West Bengal (India) alone, about 100 million people are at risk (Chowdhury *et al.*, 2001) of As poisoning. Drinking groundwater has been reported to be contaminated with As (concentration ranging from 60 to 560.23 μg/l) much exceeding the maximum permissible limits laid down by both WHO guidelines (1996) and US Environmental Protection Agency (2001). The attempts made so far by both Governmental and non-governmental organizations to provide As-free drinking water to the highly affected areas remain grossly inadequate. Further, arsenic enters into the body of unsuspecting villagers not only through drinking water alone, but a substantial amount also enters through food and vegetables grown in contaminated land around the village (Chowdhury *et al.*, 2001). Therefore, providing arsenic free drinking water cannot totally address the problem of As entering through other sources like food, cereals, vegetables, meat, etc. in the contaminated areas. We, therefore, feel that serious effort should also be directed to remove As after it enters the body. However, such efforts through orthodox medicines (e.g. chelating agents like DMSA, DTPA, etc. and some anti-oxidants) have not yet been reported to be successful by and large (Chowdhury *et al.*, 2001; Guha Mazumder *et al.*, 1997; Ratnaike, 2003). On the other hand, if there is no sustained effort to find suitable drugs that are capable of outward mobilization of As from living bodies, the situation will lead to large-scale human casualties within a few years, because chronic exposure generally leads to various ailments and dysfunctions of several vital organs like liver, kidney, lung, etc. (Patrick, 2003). The situation is further aggravated when there is an accompanying nutritional/dietary deficiency (Spallholz, 2004). Most of the affected people in general complain of muscle and joint pains and are highly depressed with various gastric problems and general weakness. Many people look pale and anemic and seem resigned to fate. Treatment with orthodox medicines is not only quite costly, but is also reported to have toxic side-effects. With all these perspectives in view, our efforts were directed to find out a remedy that is cheap, easy to administer, effective in low doses and has no toxic effect of its own. The encouraging results of early attempts made (Boiron *et al.*, 1983; Cazin *et al.*, 1987) to mobilize As in rats through micro doses of Arsenicum Album, and of our extensive studies in mice (Mitra *et al.*, 1998, 1999; Datta *et al.*, 1998 a, b; Kundu *et al.*, 2000 a, b; Mallick *et al.*, 2003) in this regard suggested that Arsenicum Album-30 has potentials of being used for alleviating sufferings of human victims as well. The favorable responses received for conducting an early trial (BBC News 27, Oct., 2003; New Scientist 22, Oct., 2003) prompted us to undertake a human trial, first of its kind, to assess the suitability of this potentized homeopathic drug, as a potential future medicine in combating arsenic poisoning or of its prophylactic use in areas contaminated with groundwater arsenic.

In the first phase of the trial, we monitored the changes in As concentration in urinary excretion and blood and also the activities of several relevant enzymes and toxicity biomarkers viz., aspartate amino transferase (AST), alanine amino transferase (ALT), acid phosphatase (AcP), alkaline phosphatase (AlkP), lipid peroxidation (LPO) and reduced glutathione (GSH) after administration of the ultra-low doses of the Arsenicum Album–30 and published our preliminary results (Khuda-Bukhsh *et al.*, 2005). Any increase in the concentration of As in the urinary excretion accompanied/followed by a decrease in the blood was viewed as a consequence removal of As from part of the body in response to the drug administration. Similarly, any increase in the activities of the favorable biomarkers and the decrease of that of the toxicity indicating enzymes were taken as supporting evidence of alleviation of As toxicity.

Further, as many people complained of joint pains in bones and muscles, a study was also conducted in the second phase to examine the status of ANA titer in the arsenic exposed subjects vis-à-vis those not known to be exposed, and a correlation to their As status in blood and urine (Belon *et al.*, 2005a) was attempted. Subsequently, experiments were also designed to know if the potentized homeopathic remedy, Arsenicum Album, could revert the alarmingly high incidence of elevated ANA titers observed in subjects of high–risk arsenic contaminated villages back to negative titer within a reasonable period of time, with the examination of concomitant changes, if any, in certain blood parameters like the total count of RBC and WBC, packed cell volume, hemoglobin content, level of blood sugar, etc. (Belon *et al.*, 2005b) in these subjects after administration of the homeopathic remedy, because anemia, diabetes, etc. are known to be associated with various systemic autoimmune disorders (Chakravarty, 2003).

2 MATERIALS AND METHODS

2.1 Work Site

The human trial on the efficacy of the potentized homeopathic remedy, Arsenicum Album, has been carried out in several phases.

In the first phase, the village Ghetugachhi (under Chakdaha Block, District Nadia, West Bengal) one of the worst affected and earliest known (since 1995) villages in West Bengal contaminated with groundwater arsenic was selected. In the next phase, another village, Dakshin Panchpota situated some 8 km away under the same block and district, where arsenic has been detected in tube well waters only in 2002, was included. Arsenic-free water plants have been installed only in November 2002, and December 2004, respectively, in these two villages, from where the villagers mainly drink water now, but generally use other water sources for cooking and other purposes. Therefore, the people of Ghetugachhi were known to be subjected to arsenic exposure for a greater period of time since detection than those of the Dakshin Panchpota village. Mainly because of its proximity to the laboratory (within 30 to 40 km), and the availability of the arsenic-free water to the villagers, these two villages were initially selected as site of our work.

Another village, Padumbasan, under Tamluk Subdivision of Purba Medinipur district, about 160 km away from these two villages, reported to be free of groundwater arsenic (also confirmed by us by testing water samples collected from several tube wells in that village), was selected for scoring data which could serve as negative controls. Most of the volunteers had generally a good health, although a few of them complained of some minor gastric problem, while two of them complained of occasional muscle and joint pains, including one diagnosed with arthritis.

2.2 The Subjects

2.2.1 Controlled experiments

(i) Positive controls: In the phase I of the trial, a group of 55 people comprising 32 males and 23 females (Control Group-I) living in one of the worst arsenic affected villages, Ghetugachhi, under Chakdaha block, Nadia, West Bengal were persuaded to become subjects of a double-blind placebo controlled trial only for study of urine As content for a few days (11 days only). The persons of this group took a dose of either "verum" (Arsenicum Album-30 soaked globules) or "placebo" (Alcohol-30 soaked globules) twice daily for 10 consecutive days, each dose containing 8 small globules (No. 20) soaked with the medicine/placebo. The information whether they were taking the placebo or the drug was kept withheld from them. At the end of the experiment when the codes were deciphered it was found that 22 and 33 persons received "placebo" and "verum", respectively, on a random basis. The urine samples were taken daily from day 0 up to 11 days since the first administration of drug/placebo and analyzed for As content. These data were critically analyzed and were utilized as controls for another group of people, who were persuaded to undertake the trial for a longer period and were requested to give not only urine samples periodically but blood as well, so that not only the relationship of As concentration in urine and blood could be fairly extrapolated, but also some other toxicity biomarkers could also be studied from their blood sera (Khuda-Bukhsh *et al.*, 2005).

In the next phase, however, another group of 43 volunteers were selected from Dakshin Panchpota (Control Group II), who agreed to undertake double blind placebo trial for one month and all the parameters of study as listed below were done in this group.

(ii) Negative controls: A group of 26 volunteers (of the same socio-economic group as that of the village Dakshin Panchpota in district Nadia) were chosen on a random basis from the village Padumbasan in Purba Medinipur district, to provide data which could serve as negative control (Control Group III).

The male volunteers as well as unmarried women volunteers were permanent residents of the three villages since birth, but the womenfolk married to the villagers became residents only after their marriage.

2.2.2 Uncontrolled "verum" fed experiments

Since greater number of volunteers did not agree to subject themselves to "placebo" trial, but were only eager to take the "verum" on their own, initially 68 persons of the Ghetugachhi village, and later 83 volunteers from Dakshin Panchpota village received only Arsenicum Album 30 and their blood was drawn at a monthly interval for two months after drug administration for studying the various blood parameters except for ESR and PCV.

As the inhabitants of Ghetugachhi had been exposed to groundwater arsenic for a longer period than those of Dakshin Panchpota, 47 volunteers from this village, who were initially treated with Arsenic Alb-30, were subsequently (after 3 months without medicine) administered Arsenicum Album 200, a higher potency, which is claimed in homeopathic doctrine to have better efficacy in case of chronic diseases (or long-standing disorders) and also for the purpose of examining relative efficacy after two months of administration of Arsenicum Album-200 vis-à-vis Arsenicum Album-30. All parameters were studied in them except for PCV and ESR.

2.2.3 Source and administration of "verum" and "placebo"

The potentized homeopathic drugs, Arsenicum Album-30 and Arsenicum Album-200 (verum) and potentized ethyl alcohol ("placebo") used in this study were prepared by Boiron Laboratory, Lyon, France following the homeopathic procedure of dilution and succussion. Sugar globules (No. 20) soaked with either verum or placebo were given to subjects, 8 globules forming a dose, twice daily for ten days for Arsenicum Album-30, and then withdrawn for ten days, and then given for 10 days again, and so on. However, Arsenicum Album-200 was given a single dose everyday for 5 consecutive days and withdrawn for 20 days to repeat again for 5 days and this cycle was continued for two months.

2.3 Methods

2.3.1 Blood collection

Volunteers of different age groups and sex were told not to take any food before draw of blood in the morning. Blood was drawn from their superficial vein around the forearm region by the routine procedure using sterile disposable syringe and needle. Blood was collected in two vials: one containing anti-coagulant (EDTA) and the other without EDTA and brought to the laboratory in flask containing ice. In the laboratory, blood was centrifuged at 8000 g for 10 minutes and serum was obtained from blood without EDTA, which was used for ANA test. Blood with EDTA was also used for determination of ESR, PCV, Hb, and TC of RBC and WBC. Plasma was isolated from blood with EDTA by centrifugation at 8000 g for 10 minutes for determination of blood sugar level.

Blood samples were kept for 5–6 hrs at 37°C in the laboratory for isolation of serum. Serum was centrifuged at 10,000 g for 30 minutes and 0.2 ml of supernatant (serum) was taken and diluted with 19.8 ml of phosphate buffer saline.

All standard protocols for quantitative estimation of the enzymes were followed, the details of which can be had from Khuda-Bukhsh *et al.* (2005) for As determination and toxicity biomarkers and Belon *et al.* (2005, a, b) for different blood parameters including anti-nuclear antibody (ANA) test.

2.3.2 Statistical analysis

Levels of significance between data of two fixation intervals in either placebo fed or drug fed series have been analyzed by performing student's "t" test. The differences between data of subjects before feeding of drug were compared with that of data after feeding of drug for different time intervals. Further, the data of subjects living in arsenic contaminated villages were also compared with that of arsenic-free village (negative control) before administration of drug, in order to show if the differences were statistically significant or not at the time of starting the experiments.

3 RESULTS AND DISCUSSION

3.1 Changes in As Content

3.1.1 As content in water collected from different tube wells

The As content of water from three tube wells of Padumbasan village tested was found to be below detectable limit. However, although the As content of water collected from the arsenic-free plants of both Ghetugachhi and Dakshin Panchpota tested As content below 10 μg/L (within permissible range up to 50 μg/L for third world countries) on an average of three tests done on three different days, some other tube wells in both the villages had water samples measuring an As content between 65.9 μg/L and 330.37 μg/L in Ghetugachhi (10 tube wells tested) and between 94.35 μg/L and 339.38 μg/L in Dakshin Panchpota village (5 tube wells tested), well above the recommended safe level. However, the tube wells in Padambasan villages tested negative for As content.

Results of Phase-I trial showed that there was a fluctuation of As concentration in the urine of both "placebo" and "verum" fed subjects on a day-to-day basis but the concentration in the latter group was always higher compared to that in the former one. The differences in the concentration between the two groups (drugged vs. placebo) were generally a little higher during the first week, but subsequently the differences were not so palpable, particularly at the 11th day (Khuda-Bukhsh *et al.*, 2005). Similarly, in the later phase, a verum vs placebo study was maintained for 30 days, the result of which showed some similarity in the urinary excretion of As for the first few days, but later on, the excretion of As in urine was rather variable in both verum fed and placebo fed series and the differences in urinary excretion rates were not very remarkable, showing a kind of wavy nature in the histogram. From these studies, it was suggestive that the efficacy of the "verum" in the mobilization of As was somewhat greater than that of the "placebo", particularly during the first few days after administration of the homeopathic remedy, but afterwards the differences diminished and the As content in the drug fed subjects appeared to be less than that of the placebo fed subjects.

The mean As level in urine and blood of the patients (Phase-I) before administration of the drug was higher (being 43.75 μg/L in urine and 24.13 μg/L in blood) than the normal permissible range (3.33-25.55 μg/L and 0.3-2.0 μg/L, respectively) (Chowdhury *et al.*, 2000; Briggs et al., 1997). The content in urine at 7d after administration of the drug was found to shoot up (Histogram 1) to a strikingly high level (62.30 μg/L) but thereafter it steadily declined to reach almost the lower limit of normal range by 45d (10.78 μg/L) (Khuda-Bukhsh *et al.*, 2005). As regards the As content in blood, it recorded a significant decline at 30d and came down to the normal range at 60d after drug administration (Histogram 2). Since the first estimation was taken only after 30 days, it was not known if it also shot up like that in the urine at any point of time during the intervening period.

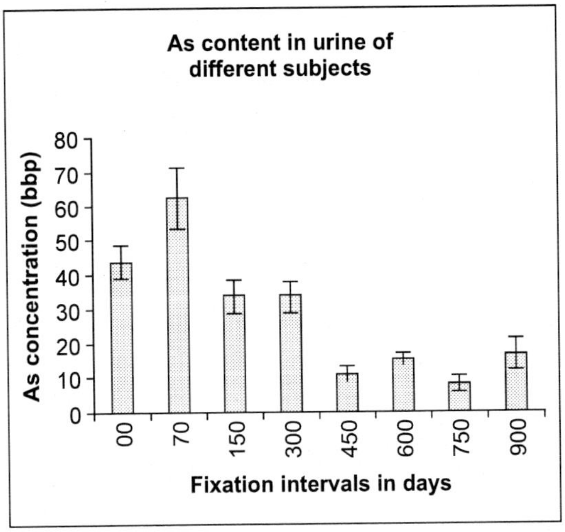

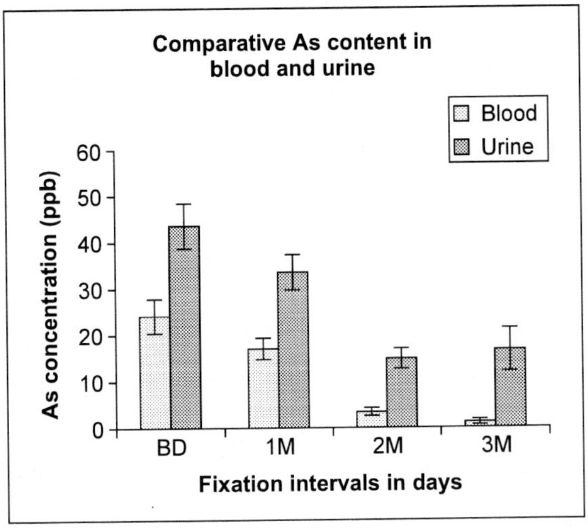

3.2 Changes in Toxicity Biomarkers

Interestingly enough, the activities of all the toxicity denoting enzymes in blood viz., AST, ALT, AcP and AlkP showed a steady decline and attained almost normal levels by 2 months time after drug administration (Khuda-Bukhsh *et al.*, 2005). Concomitantly, the LPO activity also decreased with an increase in the GSH levels in the subjects. Extensive toxicological studies have now established that an increase in lipid peroxidation, alkaline and acid phoshatase activities along with a decrease in level of glutathione denotes cytotoxicity and hepatocellular dysfunctions associated with As toxicity (Plaa *et al.*, 1986; Comporti, 1985; Banerjee et al., 1993). Therefore the favourable changes noted in the activities of these biomarkers (an increase in GSH and decreases in ALT, AST, AcP, AlkP and LPO) on drug administration would suggest its potential in alleviating the As toxicity in human.

3.3 Any Preference of the Remedy for Subjects with Different As load or for different sexes?

A critical analysis of the data on changes in As content in urine and blood and favorable changes in the activities of the enzymes revealed that the drug was apparently more effective in those with high and medium than with low level of body As load, but sex-wise no such difference was observed. The changes both in urine and blood As and the enzyme activities brought about by the drug were statistically significant in most cases.

3.4 Improvements in Other Health Aspects

Indeed, along with the changes in these parameters, the drug administration also improved considerably the overall physical strength, appetite, and food digesting ability and also mental cheerfulness in these subjects with a general feeling of better health. Even within two months, some volunteers with severe skin symptoms also started showing up some improvement in their skin symptoms (Fig. 1a, b).

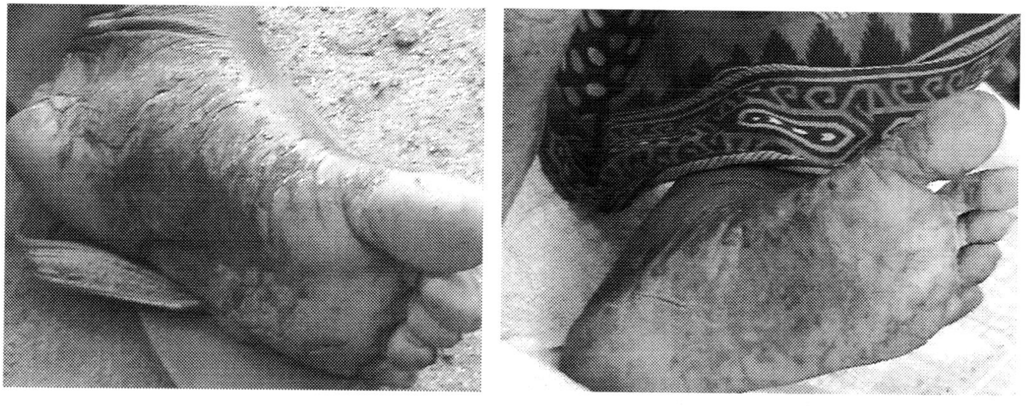

Fig. 1 Foot of a patient showing typical skin lesion symptom: (a) before administration; (b) after two months of administration of the remedy.

3.5 ANA Positivity and Correlation with As Content in Urine and Blood

There was a high incidence of ANA positivity in both the arsenic contaminated villages, but it was markedly more pronounced in Ghetugachhi subjects. On the other hand, the As contents of urine and blood in all volunteers from Padumbasan were below detectable limit and all subjects except one tested negative for ANA.

For a critical analysis of any possible relationship to exist between ANA titer to either age or sex, the data have been shown in a classified manner (Table 1). However, the data did not reveal any specific relationship per se (Belon *et al.*, 2005 a).

Table 1 Classified data on expression of ANA in relation to age and sex. Total population studied 156, Padumbasan 26 (Male 23, Female 3), Ghetugachhi 47 (Male 33, Female 14), Dakshin Panchpota 83 (Male 50, Female 33). B = borderline; + = positive; – = negative.

Name of villages	*Padumbasan*						*Ghetugachhi*						*Dakshin Panchpota*					
Age	Male			Female			Male			Female			Male			Female		
Group	+	B	–	+	B	–	+	B	–	+	B	–	+	B	–	+	B	–
<20 yrs	0	0	0	0	0	0	9	1	1	3	1	0	2	0	9	0	0	3
20-40 yrs	0	0	18	1	0	0	13	1	2	7	0	0	4	4	14	2	1	12
>40 yrs	0	0	5	0	0	2	5	1	0	3	0	0	4	1	12	4	1	10

In a bid to determine if As content of urine and/or blood has any relationship to the ANA positive titer, the data have been analyzed, but again no specific relationship could be revealed, because while some ANA positive subjects had a higher urinary or blood As content, others had their As content in the normal range. Some subjects, even those with normal As range in both blood and urine showed their sera to be ANA positive. Alternatively, some subjects showing both urinary/blood As range exceeding the higher limit, still had ANA negative titers. Therefore, the incidence of ANA positive titer might be attributable to the chronic exposure to As for a longer period of time. Further, some subjects apparently did not show any external symptoms of arsenicosis, yet they were ANA positive (Belon *et al.*, 2005 a). Therefore, ANA test could help in the determination of the extent of possible internal damage of some organs of people living in high-risk zone.

3.6 Conversion of ANA Titer

Out of 26 controls (negative control Group III) living in Padumbasan village (arsenic free), only one woman having symptoms of arthritis tested ANA positive while all others tested negative (see Table 1).

The controlled study for one month (Table 2) did not reveal any clear cut results. There was some indication of positive alteration of titer towards negativity in the drug fed subgroup. Interestingly, in the placebo fed ones also, there was a little decrease in the level although the change was not so remarkable (Belon *et al*, 2005b).

However, in the uncontrolled group (Phase-II) in Paschim Panchpota village, all ANA positive subjects tested negative after receiving Arsenicum Album-30 for two months. In this group, out of a total of 83 subjects (50 males and 33 females), 10 males and 6 females tested

Table 2 Reversal of expression of ANA in Group II populations from village Dakshinpanchpota. Total populations studied 43. Placebo fed 17 (Male 16, Female 1), Drug Ars Alb 30 fed 26 (Male 9, Female 17); NA = blood not available, BD = before administration of drug; AD = After administration of drug; + = positive; – = negative.

Phase II	*Placebo*												*Drug-AA30*											
	BP						AP 1M						BD						AD 1M					
Sex	M			F			M			F			M			F			M			F		
Age Gr.	+	B	–	+	B	–	+	B	–	+	B	–	+	B	–	+	B	–	+	B	–	+	B	–
<20 yrs	1	0	12	0	0	1	1	0	12	0	0	0	0	1	3	2	0	4	0	0	4	1	0	4
																						NA-1		
20-40 yrs	2	0	1	0	0	0	2	0	1	0	0	0	1	0	2	4	3	3	0	0	3	0	0	1
>40 yrs	0	0	0	0	0	0	0	0	0	0	0	0	1	1	0	0	1	0	0	1	1	0	0	0

ANA positive, while 6 males and two females were on the borderline of positivity, while the rest tested negative, before administration of the remedy.

Similarly, in the other uncontrolled subjects living at Ghetugachhi village, all of the ANA positive cases tested negative for ANA titer after two months of administration of the remedy, Arsenicum Album-200. In this group, out of 47 volunteers (33 male and 14 female), as many as 27 males and 13 females had tested positive while ANA titer of 3 males and 1 female were in the borderline of positivity before administration of Arsenicum Album-200. Therefore, both potencies of the remedy showed tremendous potential of bringing the ANA positive titer back to normal levels, i.e. to ANA negative titer.

3.7 Scl-70 Titer

Out of a total of 56 ANA-positive cases examined, only one subject was found to be Scl-70 positive. Therefore, it was clear that the ANA positivity was not actually contributed much by the Scl titer.

3.8 Blood Parameters

In general, the TC of RBC and WBC, PCV and Hb content of blood were elevated while the ESR level and blood sugar level fell considerably in the drug fed series (both Ars Alb-30 and 200 fed series), as compared to various controls. For further details of data, the reader may consult Belon *et al.* (2005b).

3.9 Possible Mechanism of Action

How the ultra-low doses of the homeopathic remedy could bring about such hematological and immunological changes remain unclear. But one hypothesis to explain such changes could be that signals carried by potentized homeopathic drugs might be able to trigger gene action of a cascade of relevant affected genes (Khuda-Bukhsh, 1997; 2003; Pathak *et al.*, 2005) to bring them back to their normal level of functioning and thereby bring about such multitude of changes involving different systems and organs.

4 CONCLUSION

The results of the various studies together suggest quite convincingly that the potentized remedy has a tremendous potential of alleviating sufferings of millions of poor people affected by and at risk of arsenic poisoning through groundwater contamination. The remedy not only helped in outward mobilization of arsenic, but also acted deep into the hematological system and had a multifaceted action at the physiological and molecular levels on other systems as well. But more trials (preferably double-blind placebo controlled trials) will be necessary before recommending large-scale use of this potentized drug to the arsenic affected people. Further, it is also pertinent to learn about how long a particular potency will remain effective and when and whether another remedy, or a constitutional one, will give even better results. At least, poor arsenic victims of the third world countries livingh in remote areas where provision for safe As-free drinking water has not yet been made or no suitable medical help is available can find some amelioration of their acute sufferings by taking this remedy, at least till better conditions of living and medical facilities can be provided to them. However, the efficacy of the remedy has not yet been tested in people residing in areas where no facility for drinking arsenic free water is available. The next phase of our program aims to carry out that study in such an area. Works are in progress at our lab. But more importantly, works by other independent workers are highly encouraged for verification and confirmation (or refutation) of our results.

Acknowledgements

The greater part of this work was supported by a grant from Boiron Lab, Lyon, France.

References

Banerjee, K.K. Bishayee, A. Chatterjee, M. 1993. Elevated lipid peroxidation, decreased glutathione levels and changes in glutathione related enzymes in rats treated with human placental extract. Acta Med. Okayama 47: 223–227.

B.B.C. News. 27, Oct, 2003. Remedy cures arsenic poisoning. http://www.bbc.co.uk/radio

Belon, P. Banerjee, P. Chaki Choudhury, S. Banerjee, A. Biswas, S.J. Roy Karmakar, S. Pathak, S. Guha, B. Chatterjee, S. Bhattacharjee, N. Das, J.K. & Khuda-Bukhsh A.R. 2005. Can administration of potentized homeopathic remedy, Arsenicum Album, alter anti-nuclear antibody (ANA) titer in people living in high risk arsenic contaminated areas?: I. A correlation with certain hematological parameters. eCAM (in revision).

Belon, P. Biswas, S.J. Roy-Karmakar, S. Banerjee, P. Banerjee, A. Guha, B. Das, J.K. Pathak, S Bhattacharjee, N. Choudhury, S.C. Bhadra, S. Chatterjee, S. & Khuda-Bukhsh A.R. 2005. Is an alarming prevalence of ANA positive titer in inhabitants of two arsenic contaminated villages in West Bengal (India) a time-dependent effect of arsenic poisoning?: Results of a preliminary survey. Chemosphere (in review).

Boiron, J. Abecassis, J. Belon, P. 1983. A pharmacological study of the retention and mobilization of arsenic, as caused by Hahnemanian potencies of Arsenicum Album. In: Aspects of research in homeopathy vol. I: 19–37.

Cazin, J.C. Cazin, M. Gaborit, J.L. Chaoui, A. Boiron, J. Belon, P. Cherruault, Y. Papapanayotou, C. 1987. A study of the effect of decimal and centesimal dilutions of arsenic on the retention and mobilization of arsenic in the rat. Human Toxicol 6: 315–320.

Chakravarty, A. K. 2003. Immunology. In: 2nd Edition: 678-724. Publishers Siliguri: India.

Chowdhury, U.K. Biswas, B.K. Chowdhury, T.R. Samanta, G. Mandal, B.K. Basu, G.K. Chandra, C.R. Lodh, D. Saha, K.C. Mukherjee, S.C. Roy, S. Kabir, S. Quammruzzaman, Q. Chakraborti, D. 2000. Groundwater arsenic contamination in Bangladesh and West Bengal- India. Env. Health Pers 108: 393–397.

Chowdhury, U.K. Rahman, M.M. Mondal, B.K. Paul, K. Lodh, D. Biswas, B.K. Basu, G.K. Chanda, C.R. Saha, K.C. Mukherjee, S.C. Roy, S. Das, R. Kaies, I. Barua, A.K. Palit, S.K. Quamruzzaman, Q. & Chakraborti, D. 2001. Groundwater arsenic contamination and human suffering in West Bengal, India and Bangladesh. Environ Sci; 8: 393–415.

Comporti, M. 1985 Lipid peroxidation and cellular damage in toxic liver injury. Lab. Invest 53: 599-623.

Datta, S. Mallick, P. & Khuda-Bukhsh, A.R. 1998. Efficacy of a potentized homoeopathic drug (Arsenicum Album-30) in reducing genotoxic effects produced by of arsenic trioxide in mice. I. Comparative studies of pre-, post- and combined pre- and post- oral administration of and comparative efficacy of two microdoses. Comp Ther Med 7: 62–75.

Datta, S. Mallick, P. & Khuda-Bukhsh, A.R. 1998. Efficacy of a potentized homoeopathic drug (Arsenicum Album-30) in reducing genotoxic effects produced by of arsenic trioxide in mice. II. Comparative efficacy of an antibiotic, Actinomycin D alone and in combination with either of two microdoses of the homeopathic drug, Arsenicum Album. Comp Ther Med 7: 156–163.

Guha Mazumder, D.N. Das Gupta, J. Santra, A. Pal, A. Ghosh, A. & Sarkar, S. 1997 Chattopadhyaya, N. Chakraborti, D. Non-cancer effects of chronic arsenicosis with special reference to liver damage. In CO Abernathy, RL Calderon, WR Chappell (eds), Arsenic: Exposure and Health Effects: 112–123. Chapman & Hall, London.

Khuda-Bukhsh, A.R. 1997. Potentized homeopathic drugs act through regulation of gene expression: a hypothesis to explain their mechanism and pathways of action *in vivo*. Com Ther Med 5: 43–46.

Khuda-Bukhsh, A.R. 2003. Towards understanding molecular mechanisms of action of homeopathic drugs: An overview. Mol Cell Biochem 253: 339–345.

Khuda-Bukhsh, A.R. Pathak, S. Guha, B. Roy-Karmakar, S. Das J.K. Banerjee, P., Biswas, S.J. Mukherjee, P. Bhattacharjee, N. Choudhury, S.C. Banerjee, A. Bhadra, S, Mallick, P. Chakraborti Mallick, J. & Mondal, B. 2005. Can homeopathic arsenic remedy combat arsenic poisoning in human exposed to groundwater arsenic contamination? : a preliminary report on first human trial. eCAM doi: 10.1093/ecam/neh124 http://www3:oup.co.uk/jnls/list/ecam

Kundu, S.N. Mitra, K. & Khuda-Bukhsh, A.R. 2000.Efficacy of a potentized homoeopathic drug (Arsenicum Album-30) in reducing cytoxic effects produced by of arsenic trioxide in mice. III. Tissue damage recovery, and enzymatic changes in liver. Comp Ther Med 8: 76-81.

Kundu, S.N. Mitra, K. & Khuda-Bukhsh, A.R. 2000. Efficacy of a potentized homoeopathic drug (Arsenicum Album-30) in reducing cytotoxic effects produced by of arsenic trioxide in mice. IV. On certain pathological conditions, gel electrophoretic protein profiles, DNA and RNA. Comp Ther Med 8: 157–165.

Mallick, P. Chakrabarti Mallick, J. Guha, B. & Khuda-Bukhsh, A.R. 2003. Ameliorating effect of microdoses of a potentized homeopathic drug, Arsenicum Album, on arsenic induced toxicity in mice. BMC Comp Alternative Med Vol. 3/7 p. 1–19.

Mitra, K. Kundu, S.N. & Khuda-Bukhsh, A.R. 1998. Efficacy of a potentized homoeopathic drug (Arsenicum Album-30) in reducing toxic effects produced by of arsenic trioxide in mice. I. On rate accumulation of arsenic in certain vital organs. Comp Ther Med 6: 178–184.

Mitra, K. Kundu, S.N. & Khuda-Bukhsh, A.R. 1999. Efficacy of a potentized homoeopathic drug (Arsenicum Album-30) in reducing toxic effects produced by of arsenic trioxide in mice. II. On alterations of body weight, tissue weight and total protein. Comp Ther Med 7: 24–34.

New Scientist.com news service. 22, Oct, 2003. Homeopathy reduces arsenic poisoning in mice. http://www.newscientist.com/news/

Pathak, S. Das, J.K. Biswas, S.J. & Khuda-Bukhsh, A.R. 2005. Protective potentials of a potentized homeopathic drug Lycopodium 30, in ameliorating azo dye induced hepatocarcinogenesis in mice. Mol Cell Biochem (in press)

Patric, L. 2003.Toxic metals and antioxidants: Part II. The role of antioxidants in arsenic and cadmium toxicity. Alt. Med. Rev 8: 106–127.

Plaa, G.L. 1986.Toxic responses of the liver. In: Casarett and Doulls' Toxicology, the basic science of poisons. In C.D. Klaassen, M.O. Amdur, J. Doull (eds.), MacMillan Press: New York.

Ratnaike, R.N. 2003. Acute and chronic arsenic toxicity. Postgrad Med J; 79: 391–396.

Spallholz, J.E. Boylan, L.M. & Rhaman, M.M. 2004 Environmental hypothesis: is poor dietary selenium intake an underlying factor for arsenicosis and cancer in Bangladesh and West Bengal, India? Science of the Total Env 323: 21–32.

U.S E.P.A. 2001. http://www.epa.gov/safewater/ars/quickguide.pdf.

W.H.O. 1996. In WHO guidelines for drinking water quality, WHO, (eds. 2), 940-949. Geneva.

Adsorptive Removal of Arsenate and Arsenite Ions from Groundwater Using Iron (III)-Loaded Resin

34

T.S. Anirudhan[1] and I.G. Shibi[2]

[1]Department of Chemistry, University of Kerala, Kariavattom, Trivandrum–695 581, Kerala, India
[2]Department of Chemistry, Sree Narayana College, Chempazhanthy, Trivandrum–695 587, Kerala, India

ABSTRACT

The aim of the study was to examine the usability of a newly developed adsorbent, hydrous ferric oxide-loaded cation-exchange resin (HFOLR) in a granulated form to remove arsenate [As(V)] and arsenite [As(III)] anions present in water by batch technique. Maximum removal of As(V) and As(III) ions onto HFOLR occurred at a solution pH range 4.0–6.0 and 6.0–8.0 respectively. Intraparticle diffusion model was used to understand the sorption kinetics. Isotherm studies were performed using four different isotherm equations. The maximal adsorption capacities of As(V) and As(III) were 36.24 and 25.79 mg/g respectively at 30°C. The groundwater samples were treated by HFOLR to demonstrate its efficiency in removing As(V) and As (III) ions from aqueous solutions. Arsenic adsorbed to the HFOLR was desorbed by 0.1 M NaOH in less than 3 h suggested the recyclability of the adsorbent.

1 INTRODUCTION

The United States Environmental Protection Agency has classified arsenic as a carcinogen and recommends that the maximum contaminant level for arsenic in drinking water be reduced to 0.01 ppm or less by 2006 (Kim *et al.*, 2004). In natural waters, arsenic is primarily found in its inorganic oxyanionic forms of arsenate ($HAsO_4^{2-}$) and arsenite (H_3AsO_3). As(V) predominates in oxygenated waters (e.g. surface water) whereas under anoxic conditions (e.g. anaerobic groundwaters) As(III) comprises the main form. The toxicity of arsenic varies greatly according to its oxidation state. As(III) is more toxic than As(V).

Adsorption/co-precipitation with iron oxyhydroxides is the most commonly adopted method for arsenic removal from contaminated water because the method is both effective and cheap compared with other methods such as ion exchange, membrane separation (reverse osmosis), bio-reduction and electrolysis (Zhang *et al.*, 2004). Hydrous iron oxide and iron oxyhydroxide have been known to have a remarkable selectivity to arsenic. The adsorbent such as iron-coated sand (Gupta and Jain, 2005) and Fe(III)-loaded chelating resin (Matsunaga *et al.*, 1996) have been tested for removing arsenic from wastewaters. DeMarco *et al.* (2003) incorporated hydrated Fe oxide to a macroporous cation exchange bead and developed a polymeric/inorganic hybrid sorbent for the removal As(III) and As(V) present in water.

The objective of this study is to investigate a possible strategy for the removal of arsenic by employing hydrous ferric oxide-loaded macroporous resin as adsorbent by utilizing both the ion exchange nature of the cation exchanger, Duolite C26 and the high affinity of Fe towards inorganic As. The effects of various experimental parameters, i.e. pH, contact time, adsorbent dose and the concentration of adsorbate are also investigated to identify their influences on As adsorption.

2 EXPERIMENTAL

2.1 Materials

A stock solution of As(V) and As(III) was prepared separately by dissolving dibasic sodium arsenate ($Na_2HAsO_4.7H_2O$) and sodium arsenite ($NaAsO_2$) respectively in distilled water. Ferric chloride was used for preparing Fe(III) stock solution. Working solutions were prepared by dilution of required volumes of stock solutions with deionised water. The concentrations of arsenic species were always given as elemental arsenic concentration in this study. The initial pH of the solutions was adjusted with either 0.1 M HCl or 0.1 M NaOH solution. The sulphonic acid type cation exchange resin Duolite C26 (bead size 520 μm) supplied by Sigma Aldrich, USA, was used to prepare the adsorbent.

2.2 Preparation of the Adsorbent

The Duolite C26 was first converted into Na^+ saturated form by washing several times with 0.1M NaOH and HCl. Loading of Duolite C26 with Fe was performed in batch conditions by shaking 100 mg of the sodium form resin with 50 mL of iron(III) solutions of varying concentrations ranging from 50 to 600 mg/L in tightly stoppered flasks for 3h at room temperature using a constant temperature water bath shaker at pH 2.0. The maximum loading capacity for ferric ions at pH 2.0 was found to be 56.10 mg/g from the adsorption isotherm. For the bulk preparation of iron-loaded Duolite, 10 g of the sodium form resin was agitated with 1 L ferric chloride solution (3g/L) at pH 2.0. The resulting material was estimated to contain 49.24 mg/g of ferric ions based on the calculation from the residual concentration of ferric ion. The iron-loaded resin was washed with double-distilled water until the pH of the residual water was close to 5.0 and when no iron could be detected in the filtrate. The residue was then oven dried at 65°C for one day.

The adsorbent, hydrous ferric oxide-loaded resin (HFOLR) was prepared by the column operation technique. A glass column (40 cm height, 1 cm inner diameter) was filled with the iron loaded Duolite. It was then wetted with distilled water. For converting the material into

HFOLR beads, a mixture of 75 mL of 1 M NaOH solution and 75 mL of 0.75 M NaCl solution was passed through the column in downflow method at a rate of 10 drops/minute. The bed was subsequently washed with deionized water to remove the excess alkali and chloride ions. The off-white colour of Duolite changes to reddish brown due to the formation of hydrous ferric oxide (HFO). The mesoporous character of the resin facilitates encapsulation of HFO.

2.3 Adsorbent Characterization

The FTIR spectra of adsorbents were recorded between 4000 and 400 cm^{-1} using a KBr method with a Perkin Elmer IR - 180 Spectrophotometer at 2 cm^{-1} resolution. Specific surface areas were determined using a Quantachrome NOVA Automated Gas Sorption System. N_2 adsorption/desorption experiments were carried out and pore size distributions were calculated using the Barret-Joyner-Halenda (BJH) model on the desorption branch of the isotherm. Potentiometric titration method was carried out to determine the surface charge density as a function of pH of the adsorbent. This allows, the pH of point of zero charge (pH_{pzc}) to be determined.

2.4 Batch Adsorption Experiments

In the adsorption studies, 50 mL arsenic solutions were added to the weighed HFOLR (0.1 g) placed in a conical flask. A suitable volume of HCl or NaOH solution was added to adjust the pH, which was measured with a Systronics microprocessor digital pH meter (model µ362). Optimum contact time for the dispersion of sorbent and metal to reach equilibrium conditions was allowed as found during preliminary experiments. The flasks were shaken in a temperature controlled water bath shaker at 200 rpm. At the end of the predetermined contact periods, the mixtures were centrifuged and arsenic concentrations of the supernatants were determined using atomic absorption spectrophotometry (GBC Avanta A 4100). In order to ascertain the reproducibility of results, the experiments were performed in duplicate and the mean values were considered.

2.5 Desorption Experiments

To investigate the possibility of recycling the adsorbent, consecutive adsorption-desorption cycles were repeated three times using the same adsorbent. Desorption tests were carried out using sodium hydroxide solution in batch operation. HFOLR pre-adsorbed with a known amount of As(V) or As(III) ions, was added to NaOH solutions of varying concentrations. The mixture was shaken for 3 h followed by a centrifugation for separation. The desorption efficiency was evaluated by determination of As(V) and As(III) concentrations in the supernatant. The HFOLR thus regenerated was reused for adsorption process. The loading and regeneration cycles were repeated three times.

3 RESULTS AND DISCUSSION

3.1 Nitrogen Adsorption Experiments

HFOLR showed a pore size distribution with a pore diameter of 120.62 Å, a surface area of 1.51 m^2/g), a total pore volume of 2.27×10^{-3} mL/g for pores smaller than diameter of 1622.5 Å, average pore radius of 60.31 Å and the micropore volume as 4.441×10^{-4} mL/g. The observed micropore area, having a value of 1.106 m^2/g, was much higher than the external surface area (4.042×10^{-1} m^2/g). The BET surface area based on nitrogen adsorption isotherms was thus not equal to the external surface area suggesting that sorption, in this case, occurred mainly in the internal surface.

3.2 Effect of pH on As(V) and As(III) Removal

The effect of pH on the removal of As(V) and As(III) is shown in Fig. 1. The maximum percentage removal of As(V) and As(III) was obtained in the pH range 4.0–6.0 and 6.0–8.0 respectively. The plateau values at an initial concentration of 10 and 25 mg/L were found to be 99.6 and 86.6% for As(V) and 90.8 and 75.5% for As(III) respectively. As(V) ions were found to be adsorbed onto HFOLR more efficiently than As(III) ions. In the pH range 4.0–6.0, $H_2AsO_4^-$ is the predominant As(V) species (Ghimire *et al.*, 2003) and apparently the major species adsorbed on the positively charged sites [$Fe(OH)_2^+$] available for the adsorption. The value of pH_{pzc} for HFOLR was found to be 7.8 and this value is higher than the naïve Duolite which is advantageous for the removal of As(V) through electrostatic attraction. When the pH is above the pH_{pzc} the surface of the HFOLR is negatively charged and anionic As(V) species must compete with the Coulombic repulsion. Oxyanionic species of As(V), adsorbs at iron oxyhydroxides surface by forming complex with surface sites. The beneficial role of iron oxyhydroxide on the formation of complex has been observed by earlier workers (Hiemstra and Riemsdijk, 1999; Zouboulis and Ioannis, 2002). In the pH range 6.0–8.0, H_3AsO_3 is the predominant As(III) species adsorbed on the HFOLR. The H_3AsO_3 adsorption onto HFOLR occurs by an inner-sphere complex mechanism where neutral H_3AsO_3 species forms complexes directly with the coordination environment of octahedral Fe(III) at the surface. The lowering of the pH_{pzc} value of HFOLR from 7.8 to 7.6 and 7.2 after the adsorption of As(III) from an initial concentration of 10 and 25 mg/L respectively; supports our hypothesis of adsorption of As(III) through inner-sphere complex mechanism.

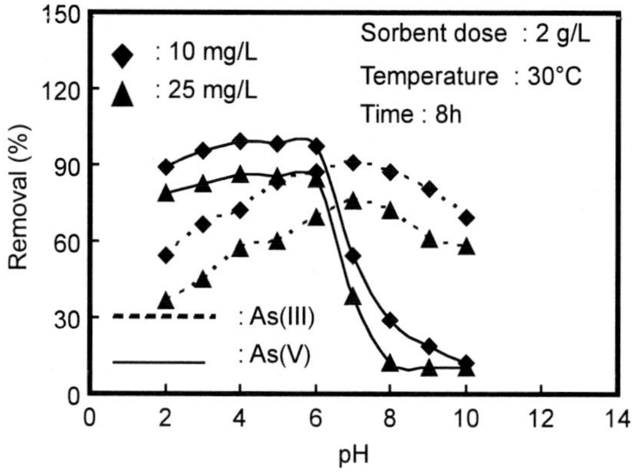

Fig. 1 Effect of pH on the removal As(V) and As(III) by HFLOR.

3.3 Effect of Ionic Strength

The effect of ionic strength (*I*) on arsenate and arsenite adsorption by HFOLR was checked by preparing adsorption curves at ionic strengths ranging from 0.001 to 0.1 M NaCl. It is noted that the arsenate adsorption on HFOLR showed no dependence on *I*. But it was also observed at optimum pH, that a slight increase in adsorption of As(III) occurs with increase in *I*. At an initial concentration of 10 mg/L, the percentage removal of As(III) at an ionic strength of 0.001, 0.005, 0.01, 0.05 and 0.1 M was found to be 86.7, 87.7, 88.9, 89.6, and 90.8 respectively.

McBride (1997) indicates that ions that form outer-sphere surface complexes show decreasing adsorption with increasing ionic strength. Ions that form inner-sphere surface complexes show little ionic strength dependence or show increasing adsorption with increasing ionic strength. Greater ion adsorption with increasing ionic strength is also explained due to the higher activity of the counter ions in solution available to compensate the surface charge generated by specific ion adsorption.

3.4 IR Spectral Studies

Positive identification of adsorbed arsenate/arsenite species was provided using the FTIR technique. The IR spectra of HFOLR containing adsorbed arsenate/arsenite and free HFOLR are given in Fig. 2. The characteristic absorption bands for $Na_2HAsO_4.7H_2O$ were reported at 1636 and 855 cm^{-1} (Deliyanni *et al.*, 2003). A shift of absorption bands should occur if the symmetry of the ion was lowered by interaction with HFOLR. Arsenate interacts with HFOLR surface by chemical bonding, as indicated by the band at 835 cm^{-1} which was rather weak in the original spectrum of HFOLR and the band at 1630 cm^{-1} which was enhanced in the spectrum of arsenate-adsorbed HFOLR. These bands shifted slightly from the 1636 and 855 cm^{-1} peaks

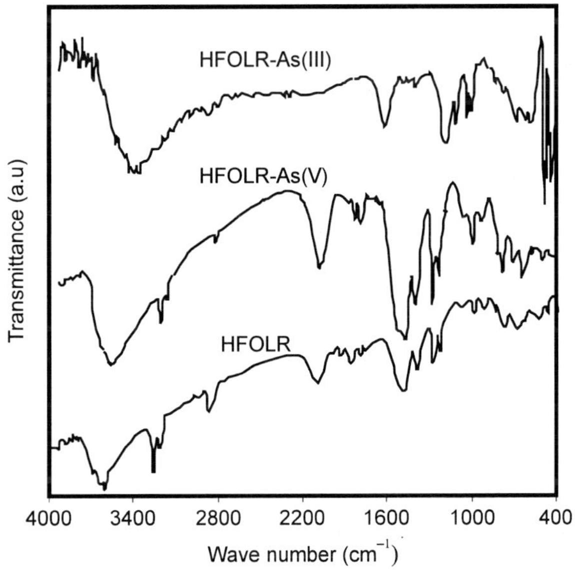

Fig. 2 IR spectra of HFOLR and HFOLR containing adsorbed arsenate/arsenite.

observed in the spectrum of sodium arsenate due to the fact that the symmetry of arsenate ion was lowered after the adsorption reaction.

The spectrum of arsenate-adsorbed HFOLR shows two bands around 3417 and 1640 cm^{-1} corresponding to O-H stretching vibration and H-O-H bending vibration of water respectively. The hydroxyl (≡Fe-OH) bending vibration band at 1120 cm^{-1} observed for the spectrum of HFOLR slightly shifted to 1131 cm^{-1} with reduced intensity for arsenate-adsorbed HFOLR. These changes in band level and band intensity indicate the involvement of hydroxyl group from ≡Fe–OH for arsenate adsorption, probably through complexation. The As–O groups possibly were involved in surface complexation, and band at 835 cm^{-1} is assigned to the Fe–O As groups (Zhang *et al.*, 2003). The bands around 624 cm^{-1} and the doublet with maxima around 451 and 410 cm^{-1} are caused by the Fe–O stretching and Fe–OH symmetric and asymmetric stretching vibrations, respectively. After arsenate adsorption the band at 624 cm^{-1} almost vanishes (Zhang *et al.*, 2003). A band near 835 cm^{-1} observed was reported in an earlier IR study of arsenate sorbed to goethite as assigned to the ν(As–OH) of As–O–Fe groups. This assignment is supported by another IR study of several metal-containing arsenate salts by Myneni *et al.*, (1998).

3.5 Effect of Initial Concentration and Contact Time

The adsorption was measured as a function of time of contact in order to check whether this adsorption system could be used for the practical operation. Figure 3 shows that it takes at least 8 h to attain the adsorption equilibrium for As(V) under the experimental conditions. On the other hand, the adsorption of As(III) can be equilibrated within 6 h.

Adsorption curves of As(V) on HFOLR for four concentrations show that for each of these initial concentrations, a relatively rapid adsorption takes place during the first 1 h, followed by a slow sorption which lasts from 4 to 8 h. Adsorption solely due to electrostatic processes is usually very rapid, in the order of seconds (Pierce and Moore, 1982). The adsorption of arsenic in the present study was in the order of hours, which may indicate a specific adsorption. A sharp increase in removal at initial stage may be due to the availability of sufficient number of

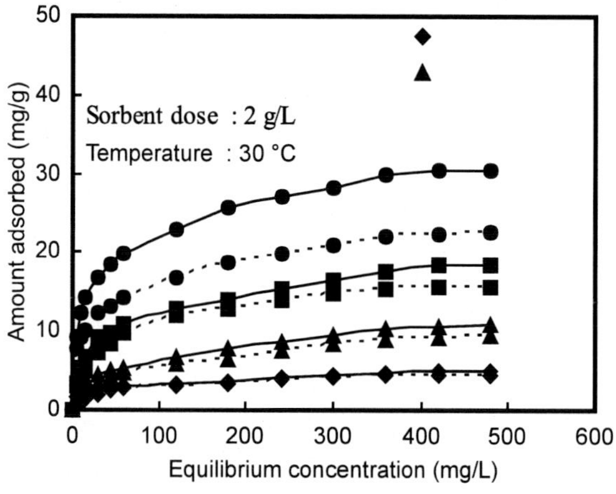

Fig. 3 Effect of contact time and initial concentration on the adsorption of As(V) and As(III) on HFOLR.

exchange sites. Based on the results in Fig. 3, the equilibrium period in forthcoming experiments was chosen to be 8 h. The adsorption density increased from 4.9 mg/g (99.9%) to 30.4 mg/g (60.8%) for As(V) and from 4.5 mg/g (90.8%) to 22.5 mg/g (45.0%) for As(III) by increasing the initial concentration from 10 to 100 mg/L suggesting the removal process to be concentration dependent.

3.6 Kinetic Modeling

The increase in sorption capacity with increasing residence time indicates that the sorption process is controlled by intraparticle mass transfer, rather than by external mass transfer, thus the critical stage in the arsenic removal kinetics is the diffusion of liquid into the pores (Zouboulis and Ioannis, 2002). Intraparticle diffusion model was used by many researchers to understand the sorption kinetics (Wan Ngah *et al.*, 2005; Ayyappan *et al.*, 2005; Wu *et al.*, 2001). The rate parameters for intraparticle diffusion (k_p) at different initial concentrations are determined using the following equation.

$$q_t = k_p t^{1/2} \quad (1)$$

where, q_t = amount adsorbed.

As shown in Fig. 4, the external surface adsorption (stage 1) occurs initially and is completed before 15 min, and then the stage of intraparticle diffusion control (stage 2) is attained and continues from 60 min to 6 h. The final equilibrium adsorption (stage 3) starts after 6 h. The first, a small sharper portion is the external surface adsorption or instantaneous adsorption stage. The second portion is the gradual adsorption stage, where intraparticle diffusion is rate-controlled. The third portion is the final equilibrium stage where intraparticle diffusion starts to slow down due to extremely low adsorbate concentrations in the solution. The As(V) is slowly transported via intraparticle diffusion into the particles and is finally retained in the micropores. In general, the slope of the line in stage 2 is called as intraparticle diffusion rate constant, k_p which increases from 0.019 to 0.122 mg/g $s^{1/2}$ for As(V) and from 0.019 to 0.094 mg/g $s^{1/2}$ for As(III) with the increase in concentration from 10 to 100 mg/L.

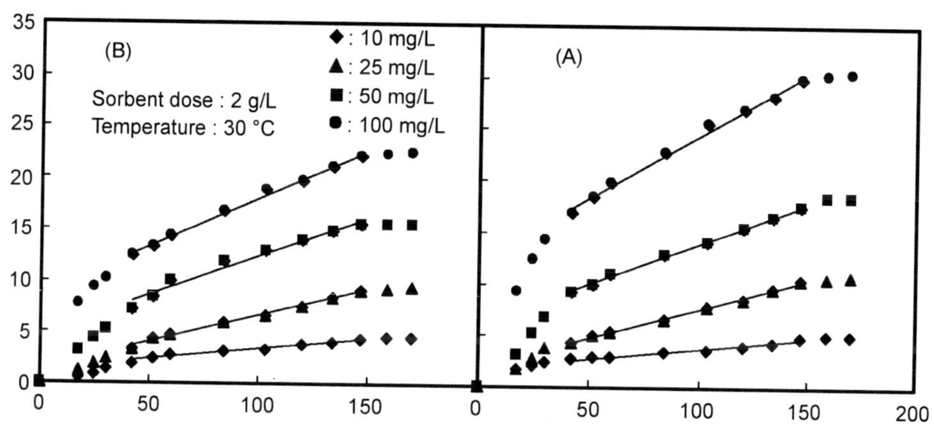

Fig. 4 Intraparticle diffusion plots for the adsorption of (A) As(V) and (B) As(III) at different concentrations.

Arsenic ions are generally present in a hydrated form in aqueous solution (Kim *et al.*, 2004). Arsenate ion has a tetrahedral shape, while arsenite ion has a plane-triangular shape. The radius of an arsenate ion is smaller than that of an arsenite ion: $HAsO_4^{2-}$ – 3.97 Å, $H_2AsO_4^-$ – 4.16 Å, $H_3AsO_4^-$ – 4.16Å, $H_2AsO_3^-$ – 4.80Å, and H_3AsO_3 – 4.80Å. Therefore, arsenate ions are able to diffuse onto the pore channel of HFOLR more easily than arsenite ions thus take longer period for attaining equilibrium.

3.7 Equilibrium Isotherms

The theoretical analysis of isotherm data describes how adsorbates interact with adsorbents and so are significant in optimizing the use of adsorbents and necessary for the convenient design and operation of adsorption systems. The Langmuir, Freundlich, Temkin, and Dubinin-Radushkevich isotherm models were used to describe the equilibrium isotherm data. The parameters corresponding to each of the isotherm models obtained from the linearized graphs are shown in Table 1.

Table 1 Langmuir, Freundlich, Dubinin-Radushkevich and Temkin isotherm constants for the adsorption of As(V) and As(III) on HFOLR at 30°C.

Metal	Langmuir			Freundlich			Dubinin Radushkevich			Temkin		
	Q°	b	Δq	K_F	1/n	Δq	q	B X	Δq	A_T	b_T	Δq
	(mg/g)	(L/mg)	(%)			(%)		10^6	(%)			(%)
As(V)	36.2	0.104	3.5	5.8	0.47	6.9	25.8	3.8	11.2	0.96	308.9	4.4
As(III)	25.8	0.079	3.6	4.3	0.42	4.1	22.4	3.6	8.8	0.78	446.8	3.7

Langmuir isotherm is based on monolayer coverage and is represented by:

$$qe = \frac{Q^0 bC_e}{1 + bC_e} \tag{2}$$

where q_e = the amount sorbed, Q^0 = the monolayer coverage, C_e = the equilibrium concentration, and b = binding constant. For verifying the validity of the Langmuir isotherm plots of specific sorption (C_e/q_e) against the equilibrium concentration (C_e) are constructed. The maximum uptake capacities (Q^0) for As(V) and As(III) are 36.24 and 25.78 mg/g, and binding constants (b) are 0.0791 and 0.104 L/mg respectively at 30°C.

Freundlich isotherm is an empirical expression of the following nature:

$$qe = K_F C_e^{1/n} \tag{3}$$

where K_F and $1/n$ are Freundlich isotherm constants. Using the linear form, Eq. (3), the logarithmic plots of the Freundlich expression for the adsorption of As on HFOLR was followed. The Freundlich constant, n indicates the degree of favorability of adsorption and should have values lying in the range of 1 to 10 for classification as favourable adsorption (Raji and Anirudhan, 1997). A smaller value of $(1/n)$ indicates a stronger bond between adsorbate and adsorbent, while a higher value for K_F indicates rate of adsorbate removal is high. The values are listed in Table 1.

The Dubinin-Radushkevich (DR) isotherm is generally expressed as follows (Dubinin, 1960):

$$qe = q_D \exp\left(-B_D\left[RT\ \ln\left(1+\frac{1}{C_e}\right)\right]^2\right) \tag{4}$$

where B_D and q_D are Dubinin-Radushkevich isotherm constants. The constant, B_D, is related to the mean free energy of sorption per mole of the sorbate as it is transferred to the surface of the solid from infinite distance in the solution and this energy, E can be computed using the following relationship:

$$E = \frac{1}{\sqrt{2B_D}} \tag{5}$$

The linear Dubinin-Radushkevich isotherm plots for the sorption of the As(V) and As(III) onto HFOLR on examination showed that the Dubinin-Radushkevich isotherm did not provide an accurate description of the data for both the ions over the concentration range studied. The q_D values are not consistent with the Q^0 values previously determined for the Langmuir isotherm.

The Temkin isotherm has been used in the form as follows:

$$q_e = \frac{RT}{b_T}\ \ln\ (A_T C_E) \tag{6}$$

where A_T and b_T are Temkin isotherm constants. The sorption data can be analyzed according to Eq. (6). Therefore, a plot of q_e versus ln C_e enables to determine the constants A_T and b_T and the values are listed in Table 1. Temkin isotherm considers the effects of indirect adsorbate-adsorbate interactions and assumes that the heat of adsorption of all the molecules in the layer would decrease linearly with coverage due to adsorbate-adsorbate interactions. Unlike the Langmuir isotherm, the Temkin isotherm contains a factor that explicitly takes into account adsorbing species-adsorbate interactions in the Temkin isotherm, the heat of adsorption of all the molecules in the layer decreases linearly with coverage due to adsorbate-adsorbate interactions. This model assumes that adsorption is characterized by a uniform distribution of binding energies, up to some maximum binding energy.

The experimental data of the amount of sorbate adsorbed on the sorbent are substituted into each equilibrium isotherm model to determine the best-fit model for the sorption system. For each of these parameter sets the q_e^{cal} was found out and the summation of the squares of the deviation of the q_e^{cal} from the q_e^{exp} values were found out. The normalized deviation $\Delta q(\%)$ for the set of values were obtained with respect to each of the models and was used to quantitatively compare the applicability of each model.

$$\Delta q(\%) = 100\sqrt{\frac{\left[\left(q_t^{exp} - q_t^{cal}\right)/q_t^{exp}\right]^2}{n-1}} \tag{7}$$

Based on correlation coefficients (r^2) and $\Delta q(\%)$ values (Table 2) it is noted that Langmuir and Temkin equations give best fit over the entire range of concentrations. $\Delta q(\%)$ values obtained for the other two isotherm equations are higher than that for the Langmuir and

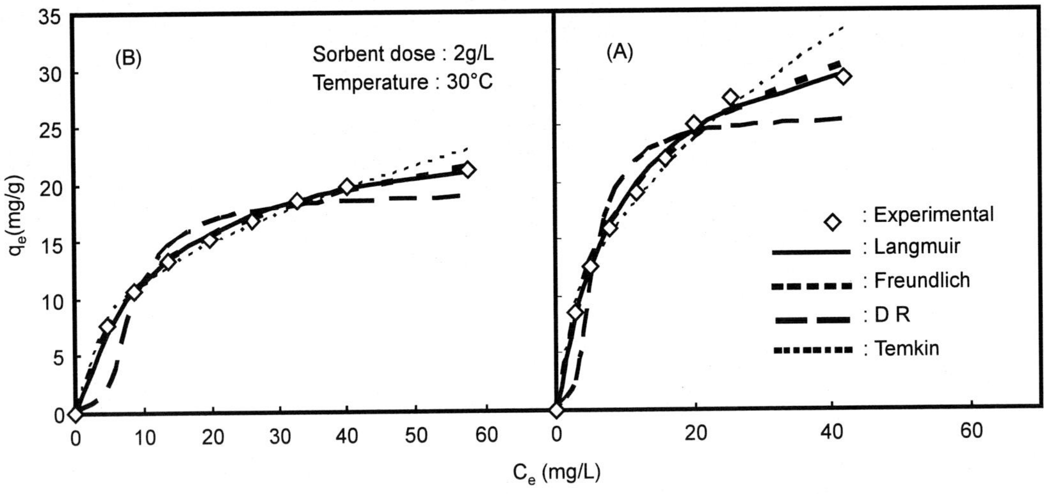

Fig. 5 Comparison of the model fit of various isotherms to the observed isotherm data for the adsorption of (A) As(V) and (B) As(III) on HFOLR.

Temkin equations. In order to assess the different isotherms and their validity to correlate experimental results, the theoretical plots for each isotherm have also been shown with the experimental data for the adsorption As(V) and As(III) in Fig. 5. Langmuir equation gives best fit over the entire range of concentrations for both As(V) and As(III) species.

3.8 Testing with Simulated Groundwater Samples

The present study examined the suitability of HFOLR for treating simulated groundwater sample for the purpose of remediation of arsenic. The sample is based on groundwater of New Jersey, USA, and contained apart from arsenic other metal ions based on Al, Mg, Fe, Ca and Mn as well as anions like fluoride, chloride, sulphate and carbonate (Bang *et al.*, 2005). Two different samples with 1000 µg/L (1.0 mg/L) each of As(V) and As(III) doped were prepared retaining the other reported features. The effect of the mass of HFOLR needed for the removal of As(V) and As(III) from water samples reveal that the removal of As(V) and As(III) was found to increase with increase in mass of adsorbent. Arsenic content could be brought down well below the maximum permissible levels as per WHO norms from 1.0 to 0.01 mg/L, with a HFOLR dose of 2.5 and 3.0 g/L for As(V) and As(III) respectively. The increase in adsorption with increase in mass of HFOLR is due to the increase in adsorbent surface area with increase of mass of adsorbent. Further work is now under way to establish the effectiveness of HFOLR at removing other toxic anions from wastewaters.

3.9 Desorption and Regeneration of the Adsorbent

To be economically attractive, the treatment of arsenic wastes using HFOLR, requires the regeneration of the exhausted sorbent. This regeneration was studied by determining the ability of the arsenic to be stripped selectively from the functionalized sorbent, using several eluents such as hydrochloric acid, citric acid and sodium hydroxide. These eluents are able to desorb As(III) and As(V) efficiently, however, in many cases iron is also released. A simultaneous release of iron and arsenate/arsenite ions does not actually allow the ions to be selectively separated, moreover, it also restricts the possibility to reuse the sorbent. With a neutral solution the metal-impregnated sorbent is stable, however, the desorption efficiency is very low. The best results are obtained with alkaline solutions. Quantitative elution of As(V) and As(III) was successfully attained with either by 0.1 M HCl or 0.1 M NaOH solution. Ferric ions are also liberated from the resin with HCl solution, but the amount of ferric ion released from the resin with 0.1M NaOH solution as regenerant is less than 0.5% of the total amount of ferric ion retained in the resin. Higher concentrations of NaOH than 0.1 M, however, gave a turbid solution, which is probably due to the generation of ferric hydroxide from the resin.

It is found during the three cycles the percentage of adsorption decreased from 99.7 to 92.7 for As(V) and from 90.8 to 79.2 for As(III) when an initial concentration of 10 mg/L was taken. During the three cycles the desorption decreased from 93.5 to 73.6% for As(V) and 87.1 to 75.4% for As(III). The observation that arsenic desorption was completed within 3 h reveals that sorption sites of HFO microparticles are accessible through the network of pores, i.e. obstruction in the pores and subsequent increase in the tortuosity of dissolved solutes did not result from the dispersion of submicron HFO particles within the porous beads. At high alkaline pH, HFO sorption sites are all deprotonated and negatively charged; so are all arsenite and arsenate species. Donnan co-ion exclusion effect is thus very predominant under this condition resulting in efficient desorption (DeMarco *et al.*, 2003). The results indicate that the regenerated resin can be used for the next adsorption-regeneration cycle for an acidic As(V) solution without additional treatment with ferric ions. Preliminary results have shown that arsenate/arsenite can be removed again on the regenerated sorbent.

4 CONCLUSIONS

The new adsorbent HFOLR is successfully prepared from a commercial cation exchanger, Duolite C26, for the removal of arsenate and arsenite ions from aqueous media. The most effective pH range for the removal of As(V) and As(III) was found to be 4.0–6.0 and 6.0–8.0 respectively. The influence of ionic strength was examined and commented. The adsorptive capacity of HFOLR was found to be 36.24 mg As(V) and 25.79 mg As(III) per g of HFOLR, which is higher in comparison with many other sorbents. The feasibility of the adsorbent in removing the arsenic content in groundwater samples is also tested. Alkali regeneration was tried for several cycles. The major advantages of the present adsorption process include (i) the reduction of arsenic levels to less than 0.01 mg/L (ii) the process needs no pre-oxidation step since the sorbent is able to remove As(III) as well as As(V) from aqueous medium effectively (iii) high adsorption capacity contribute to the removal of As from concentrated solutions (iv) big particle size makes it easy to be separated from aqueous system after adsorption activity (v) the adsorption system is expected to provide low capital and operating cost, and require minimal monitoring.

References

Ayyappan, R., Sophia, A.C., Swaminathan, K. & Sandhya, S. 2005. Removal of Pb(II) from aqueous solutions using carbon derived from agricultural wastes. *Process Biochemistry* 40: 1293–1299.

Bang, S., Patel, M., Lippincott, L. & Meng, X. 2005. Removal of arsenic from groundwater by granular titanium dioxide adsorbent. *Chemosphere* 60: 389–397.

Deliyanni, E.A., Bakoyannakis D.N., Zouboulis A.I. & Matis K.A. 2003. Sorption of As(V) ions by akaganeite-type nanocrystals. *Chemosphere* 50: 155–163.

DeMarco, M J., SenGupta, A. K. & Greenleaf, J. E. 2003. Arsenic removal using a polymeric/inorganic hybrid sorbent. *Water Research* 37: 164–176.

Dubinin, M. M. 1960. The potential theory of adsorption of gases and vapors for adsorbents with energetically non-uniform surface. *Chemical Reviews* 60, 235–266.

Ghimire, K.N., Inoue, K., Yamaguchi, H., Makino, K. & Miyajima, T. 2003. Adsorptive separation of arsenate and arsenite anions from aqueous medium by using orange waste. *Water Research* 37: 4945–4953.

Gupta, V.K. & Jain, S.N. 2005. Adsorption of As(III) from aqueous solutions by iron oxide-coated sand. *Journal of Colloid and Interface Science* 288: 55–60.

Hiemstra, T. & Van Riemsdijk, W.H. 1999. Surface structural ion adsorption modeling of competitive binding of oxyanions by metal hydroxides. *Journal of Colloid and Interface Science* 210: 182–193.

Kim, Y., Kim, C., Choi, I., Rengaraj, S. & Yi, J. 2004. Arsenic removal using mesoporous alumina prepared via a templating method. *Environmental Science & Technology* 38: 924–931.

Matsunaga, H., Yokoyama, T., Eldridge, R.J. &. Bolto, B.A. 1996. Adsorption characteristics of arsenic(III) and arsenic(V) on iron(III)-loaded chelating resin having lysine-Nα,Nα-diacetic acid moiety. *Reactive and Functional Polymers* 29: 167–174.

McBride, M. B. 1997. A critique of diffuse double layer models applied to colloid and surface chemistry. *Clays & Clay Minerals* 45: 598–608.

Myneni, S.C.B., Traina, S.J., Waychunas, G.A. & Logan, T.J. 1998. Experimental and theoretical vibrational spectroscopic evaluation of arsenate coordination in aqueous solutions and solids. *Geochimica et Cosmochimica Acta* 62: 3285–3300.

Pierce, M.L. & Moore, C.B. 1982. Adsorption of arsenite and arsenate on amorphous iron hydroxide. *Water Research* 16: 1247–1253.

Raji, C. & Anirudhan, T.S. 1997. Chromium(VI) adsorption by sawdust: Kinetics and equilibrium. *Indian Journal of Chemical Technology* 4: 228–236.

Wan Ngah, W.S., Ghani, S.A. & Kamari, A. 2005. Adsorption behaviour Fe(II) and Fe(III) ions in aqueous solution on chitosan and cross-linked chitosan beads. *Bioresource Technology* 96: 443–450.

Wu, F.C., Tseng, R.L. & Juang, R.S. 2001. Kinetic modelling of liquid-phase adsorption of reactive dyes and metal ions on chitosan. *Water Research* 35: 613–618.

Zhang, W., Singh, P., Paling, E. & Delides, S. 2004. Arsenic removal from contaminated water by natural iron ores. *Minerals Engineering* 17: 517–524.

Zhang, Y., Yang, M. & Huang, X. 2003. Arsenic(V) removal with a Ce(IV)-doped iron oxide adsorbent. *Chemosphere* 51: 945–952.

Zouboulis, A. I. &. Katsoyiannis, I.A. 2002. Arsenic removal using iron oxide loaded alginate beads. *Industrial & Engineering Chemistry Research* 41: 6149–6155.

Removal of Arsenate Ions from Groundwater using an Anion Exchanger Derived from Tamarind Fruit Shell

35

T.S. Anirudhan and P.G. Radhakrishnan
Department of Chemistry, University of Kerala, Kariavattom, Trivandrum–695 581, India

ABSTRACT

The objective of this study was to increase the anion exchange properties of the agricultural by-product, tamarind fruit shell (TFS) by chemical modification. The TFS was treated with epichlorohydrin and dimethylamine in the presence of pyridine and N, N'-dimethylformamide followed by treatment of HCl solution. The anion exchanger (TFS-AE) has been tested for its effectiveness in removing arsenate [As(V)] ions from aqueous solutions through batch experiments. The adsorbent exhibited the highest uptake capacity (>99.0%) for an initial concentration of 15.0 μmol/L at pH 8.0. The surface mass transfer coefficients were determined for the process under the influence of varying concentration. Arsenate adsorption was found to decrease with increase in ionic strength due to the expansion of the diffuse double layer. The L-type adsorption isotherm obtained in the present study indicated a favourable process and fitted the Langmuir equation model well. The adsorption efficiency towards As(V) removal was also tested using simulated groundwater samples. Adsorption isotherm experiments were also conducted for comparison using a commercial chloride form, Duolite A-7, a weak base anion exchanger. Acid regeneration was also tried for several cycles with a view to recover the adsorbed As(V) ions and also to restore the adsorbent to its original state. The results obtained point towards viable adsorbent which is both effective as well as economically attractive for As(V) removal from aqueous solution and groundwater.

1 INTRODUCTION

Arsenic is a toxic and carcinogenic element that can be found in natural waters as well as in various industrial waters, solid or liquid. It is usually found as an anion with acid characteristics in the trivalent and pentavalent forms. Arsenic(III) is more likely to be found in anaerobic groundwaters and arsenic(V) in aerobic surface waters (Rajakovic and Mitrovic, 1992). The removal of arsenic from groundwater and aqueous water streams is one of the most important environmental problems. Although the problem has been addressed for many years, an effective treatment option is still limited. The US Environmental Protection Agency has recommended that the maximum contaminant level (MCL) in drinking water be reduced to 0.01 mg/L or less by 2006. Therefore, effective treatment of arsenic is required to meet the MCL value.

Various techniques are employed for the removal of arsenic from water and wastewater. Precipitation, membrane process, ion exchange and adsorption are the most common. Of all these, adsorption and ion exchange are the best methods for the purification of water or they are often combined. A wide range of adsorbents have been used for removing arsenic from water and wastewater and have been reviewed by Maity *et al.* (2005). Lignocellulosic biomass wastes represent cheep and environmentally safe sources for the preparation of adsorbent materials that may be useful for the removal of heavy metals from solutions. There have been numerous studies on the adsorption of heavy metals from water by lignocellulosic materials, as reviewed by Shibi and Anirudhan (2005). Modifications such as co-polymerisation, cross-linking and quaternisation have been used for improving the adsorption characteristics of the original biomass (Noelin *et al.*, 2005). The potential of using chemically modified lignocellulosics for anion binding has not been explored. In the present study, a new adsorbent material prepared from tamarind fruit shell, a lignocellulosic residue has been employed for the removal of arsenate ions from water and groundwater.

2 EXPERIMENTAL

2.1 Preparation of the Adsorbent

Tamarind fruit shell (TFS) was used for the preparation of the adsorbent. For the synthesis of adsorbent, TFS was first cross-linked with epichlorohydrin into epoxy ether from which it was utilized for derivatisation. The sensitive component for cross linking might be the hydroxyl group of the cellulose unit (TFS-CH_2OH) present in TFS. Scheme 1 represents the general procedure adopted for the preparation of adsorbent. Ten grams of dried TFS was reacted with 100.0 mL epichlorohydrin in 120.0 mL of DMF at 100 °C for 1 h. Pyridine (40.0 ml) was added and the stirrng continued for 1 h at 100 °C. Amine groups were then introduced by adding 100.0 ml of 50.0% dimethylamine solution. The mixture was stirred for 3 h at 100 °C. The reaction product was washed with 2 L of diluted ethanol at 40 °C. The product was converted into anion exchanger by washing with 0.1 M HCl. The contents were repeatedly washed further with distilled water to free the excess acid. The product, hereafter designated as TFS-AE, was dried at 70 °C and sieved to 170-230-mesh size and used for all experiments.

2.2 Instrumental

The FTIR spectra of the adsorbent were obtained using the pressed disc technique in a Perkin Elmer IR – 180 spectrophotometer. The concentration of As(V) ions in the solution was determined using a GBC Avanta 5450 (Australia) atomic absorption spectrophotometer (AAS).

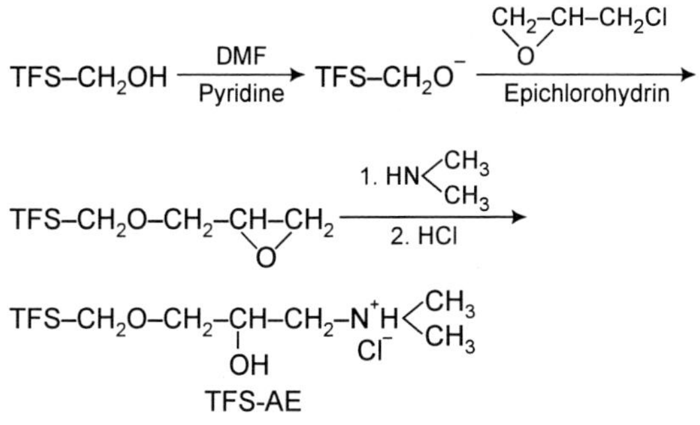

Scheme 1 Preparation of TFS-AE.

A Systronic microprocessor pH meter (model μ 362; India) was used to measure the potential and pH of the suspension.

2.3 Adsorption Experiments

The stock solution of As(V) was prepared by dissolving a weighed quantity of Na_2HAsO_4 7 H_2O in distilled water. All the required working solutions were prepared by diluting in distilled water. Batch experiments were carried out using a series of Erlenmeyer flasks of 100.0 mL capacity. Batch experiments were conducted to determine the effects of the medium pH, contact time, initial As(V) concentration, adsorbent dose and temperature on adsorption of As((V)). The effect of pH on adsorption was investigated in the pH range 2.0 – 8.0 at 30 °C. The initial pH of the solution was adjusted to the desired value with either 0.2 M HCl or 0.2 M NaOH solutions. For kinetic studies, 100.0 mg of the adsorbent was shaken with 50.0 mL of As(V) solution of varying concentration in different 100.0 mL Erlenmeyer flasks. The contents were continuously agitated in a temperature controlled flask shaker at 200 rpm. At the end of the pre-determined time intervals, the sorbent was centrifuged and As(V) concentration in the supernatant was determined using AAS. Adsorption isotherm experiments were also performed by agitating 100.0 mg of the adsorbent with 50.0 mL of varying concentration of As((V) form 15.0 to 275.0 μmol/L at 30 °C.

2.4 Desorption Experiments

To investigate the possibility of repeated use of the adsorbent, desorption and regeneration process was also studied. The adsorbent loaded with As((V) ions was obtained by applying the adsorption process for 3 h at pH 8.0. The adsorbent was separated from solution by filtration and washed gently with water to remove any unadsorbed As((V) . The spent adsorbent was then agitated with 50.0 mL of 0.1 M HCl, followed by filtration for separation. The desorbed As((V) was determined. The spent adsorbent thus regenerated was reused for adsorption purpose. The loading and regeneration cycles were repeated four times.

All the batch experiments were performed in duplicate and the results were reproducible with ± 3% and the mean values were taken for data analysis.

3 RESULTS AND DISCUSSION

3.1 Adsorbent Characterization

The FTIR spectra of TDF and TDF-AE are shown in Fig. 1. After comparing the spectrum of TFS and TFS-AE the broad absorption band observed around 3400 cm^{-1} which are attributable to the sum of the contribution from O-H stretching vibration from cellulose and hydroxyl groups of polyphenols. We found that the peak is relatively sharper with reduced intensity in case of the latter. The decrease in intensity may be due to the reaction of epichlorohydrin with -OH group of cellulose during epoxide formation. This reaction destroys a large number of -OH groups of cellulose structure. Additional peaks at 1433 and 1462 cm^{-1} in TFS-AE indicate the presence of aliphatic CN vibration and $-CH_2-N^+HR_2$ type nitrogen. These observations clearly indicate the formation of chain (backbone) and the presence of $-CH_2-N^+HR_2$ group in TFS-AE.

The physical and surface properties of the TFS and TFS-AE were carefully determined by standard methods described elsewhere (Shubha *et al.*, 2001). The surface area obtained from the N_2 adsorption isotherm was found to be 21.2 and 23.5 m^2/g for TFS and TFS-AE respectively. Although TFS is cross-linked with epichlorohydrin, the surface area of TFS remains considerably higher. This indicates that most of the pores with larger diameter are not clogged with dimethylaminohydroxypropyl group, presumably because the inner surfaces of the pores are uniformly covered with functional group. The values of pH_{zpc} for TFS and TFS-AE were found to be 7.2 and 7.8 respectively. The increase in pH_{zpc} after chemical modification indicates that the surface became more positive and this helps to adsorb negatively charged adsorbate ions through electrostatic attraction.

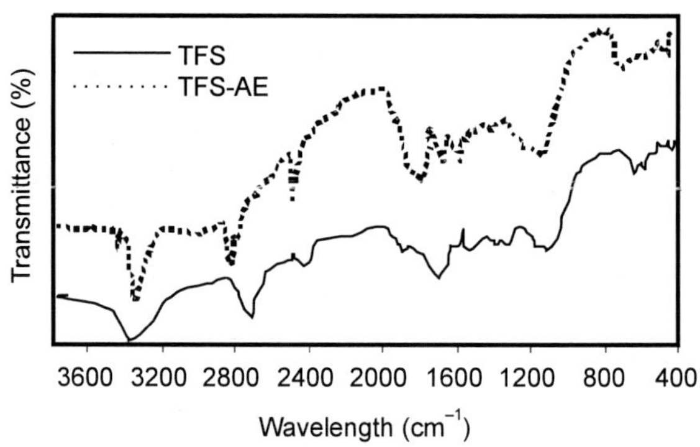

Fig. 2 FTIR Spectra of TFS and TFS-AE.

3.2 Effect of Adsorbent Concentration

Figure 2 shows the effect of adsorbent dose on the adsorption of As(V) by TFS and TFS-AE. From the figure it was observed that, the percentage adsorption increases with increasing adsorbent doses. This is to be expected, for a fixed initial solute concentration, increasing adsorbent dose provides greater surface area and adsorption sites. The results show that for the complete removal of As(V) from an initial concentration of 60.0 μmol/L a maximum adsorbent

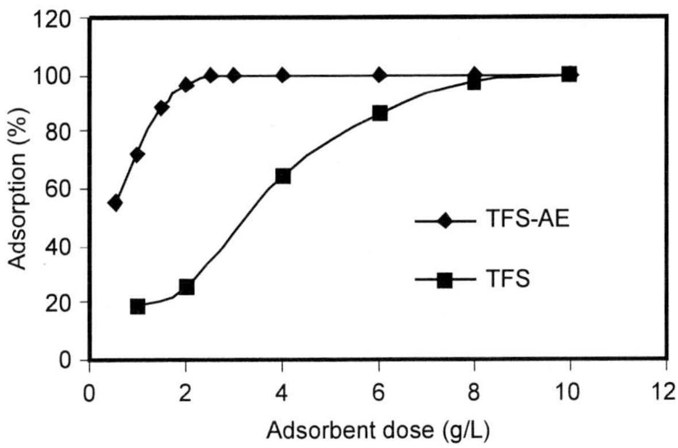

Fig. 3 Effect of absorbent dose on the adsorption of As(V) onto TFS and TFS-AE.

dosage of 3.0 g/L of TFS-AE and 10.0 g/L TFS is required. This indicates that TFS-AE is 3.3 times more effective than TFS for the removal of As(V) from aqueous solutions. This may be due to the moderate ion exchange capacity of TFS-AE. Hence, subsequent adsorption experiments with As(V) were performed only on TFS-AE.

3.3 Effect of pH

As the speciation of As(V) and the surface characteristics of the adsorbent are influenced by the pH value, the adsorption of AS(V) by the ion exchanger strongly depends on the pH value, as shown in Fig. 3. At pH less than 2.0, the adsorption onto the sorbent is negligible, and then increases rapidly with increasing pH value. In the final pH range 6.0–8.0, the adsorption percentage of As(V) on the sorbent reaches a maximum and then decreases with an increase of pH value from 8.0 to 11.0.

This result can be explained by the anion exchange mechanism. TFS-AE contains-CH_2-$N^+H\ (CH_3)_2Cl^-$ as ion exchange group. At low pH value, AS(V) predominantly exists, as H_3ASO_4 and this neutral species are difficult to attach to exchange sites of the sorbent for anion exchange. It has been reported that at pH ranges between 3.0 and 6.0, As(V) ion occurs mainly in the form of $H_2ASO_4^-$, while a divalent anion $HASO_4^{2-}$ dominates at higher pH values (between 8.0 and 10.5); in the intermediate region, i.e. at pH range between 6.0 and 8.0 both species co-exist (Matis,1999). It is evident that both species $H_2ASO_4^-$ and $HASO_4^{2-}$, can be effectively adsorbed by TFS-AE in the optimum pH range 6.0–8.0. When the pH value is more than 8.0, the concentration of OH^- in solution becomes higher, which may compete with the oxyanionic species of As(V) for ion exchange sites on the adsorbent, leading to a decrease in adsorption.

3.4 Effect of Contact Time and Initial Concentration

The effect of contact time on the adsorption of As(V) by TDF-AE is shown in Fig. 4. The results showed that increasing the contact time increased the adsorption of metal ions, and it remained constant after the equilibrium time of 3 h. Consequently the contact time sufficient to attain equilibrium was set at 3 h in each equilibrium experiments. The amount of As(V) adsorbed

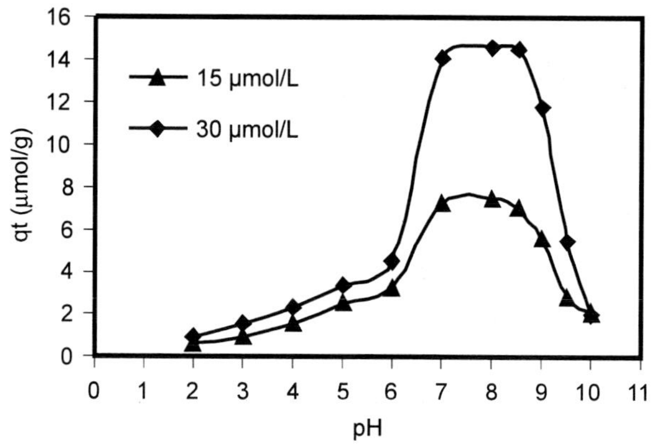

Fig. 4 Effect of pH on the adsorption of As(V) onto TFS-AE.

increased from 14.63 to 46.22 μmol/g with an increase in initial concentration from 30.0 to 125.0 μmol/L. The amount adsorbed at low As(V) concentrations is smaller than the amount adsorbed at high As(V) concentration. This is due to the decrease in the driving force of the concentration gradient, as on decreased with initial concentration. At low initial concentrations, the ratio of the initial number of moles of As(V) to the available sites of adsorption is low and hence the fractional adsorption becomes independent of initial concentration. But as the initial concentration increases, the ratio becomes higher and at high initial concentrations fractional adsorption depends on the initial concentration.

3.5 Adsorption Dynamics

The removal of As(V) from the liquid to solid phase can be considered a reversible reaction with equilibrium established between the two phases. A simple kinetic analysis of adsorption is the Lagergren equation. The adsorption follows pseudo-first-order reaction for different initial concentrations and temperatures. The adsorption rate constant for As(V) for different initial concentrations and temperatures was calculated using the following equation

$$\log (q_e - q_t) = \log q_e - \frac{k_{ad} t}{2.303} \quad \text{...(1)}$$

where q_e and q_t are the amounts of As(V) adsorbed (μmol/g) at equilibrium and time 't' and k_{ad} is the Lagergren rate constant of adsorption (min^{-1}). The plot of $\log(q_e - q_t)$ vs. t gives straight lines (Fig. 5) and agree with the linearity of the Lagergren equation. (The value of k_{ad} for different initial concentrations and temperatures were obtained from the slope of the plots for different initial concentrations and temperatures (Table 1). For different initial concentrations the value of k_{ad} remains almost constant (1.25×10^{-2} min^{-1}) clearly indicate that this parameter is totally independent of initial concentration. The values of k_{ad} increases from 1.08×10^{-2} to 1.61×10^{-2} min^{-1} when the temperature decreases from 40 to 10 °C. Mention fixed temperature for concentration variation expriment and fixed conc. for T variation experiment). This clearly indicates that adsorption process is exothermic. The activation energy E_a for the adsorption of As(V) on TDF-AE can be determined using Arrhenius equation

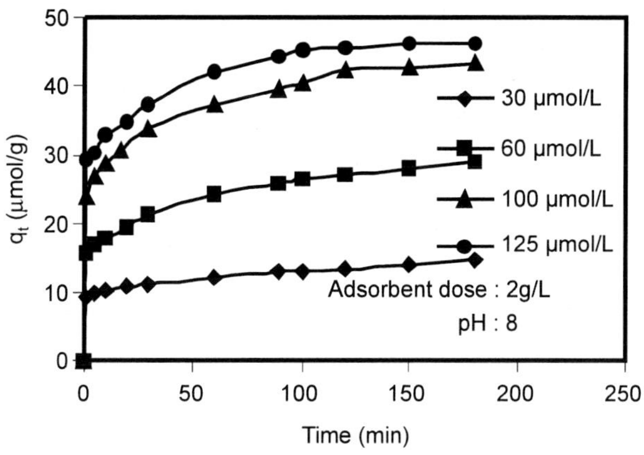

Fig. 5 Effect of contact time and initial concentration on the adsorption of As(V) onto TFS-AE.

Table 1 Kinetic parameters for the adsorption of As(V) on TFS-AE at different initial concentrations and temperatures.

Concentration (μmol /L)	*k_{ad} (min^{-1})*	*r^2*	*Temperature (°C)*	*k_{ad} (min^{-1})*	*r^2*
30	1.26×10^{-2}	0.99	10	1.61×10^{-2}	0.99
60	1.26×10^{-2}	0.99	20	1.42×10^{-2}	0.99
100	1.24×10^{-2}	0.98	30	1.26×10^{-2}	0.99
125	1.24×10^{-2}	0.99	40	1.08×10^{-2}	0.99

$$\ln k_{ad} = \ln k_0 - \frac{E_a}{RT} \qquad ...(2)$$

A plot of $\ln k_{ad}$ vs. I/T was found to be linear. E_a value calculated from the slope of the plot was found to be −9.66 kJ/mol. The relatively low E_a value suggests favourable As(V) adsorption onto TFS-AE.

The mass transfer analysis of As(V) during the adsorption process is studied by the kinetic model developed by McKay *et al.* (1985)

$$\ln\left(\frac{C_t}{C_o} - \frac{1}{1+mK_L}\right) = \ln\left(\frac{mK_L}{1+mK_L}\right) - \left(\frac{1+MK_L}{mK_L}\right)B_L S_s t \qquad ...(3)$$

where C_o is the initial concentration of the solute and C_t is the concentration of solute at time t. B_L is the mass transfer coefficient, m is the mass of the adsorbent per unit volume of particle free adsorbate solution ($m = w/v$), K_L is the Langmuir constant (obtained by multiplying $Q°$ and b), S_s is the specific surface area per unit volume of particle free slurry and is calculated as

$$S_S = \frac{6m}{d_P \rho_P (1 - \varepsilon_P)} \qquad ...(4)$$

where dp is the particle diameter (cm), ρ_p is the density of the adsorbent particle (g/mL) and ε_p (mL/g) is the porosity of the adsorbent particle. The values of B_L were determined from the

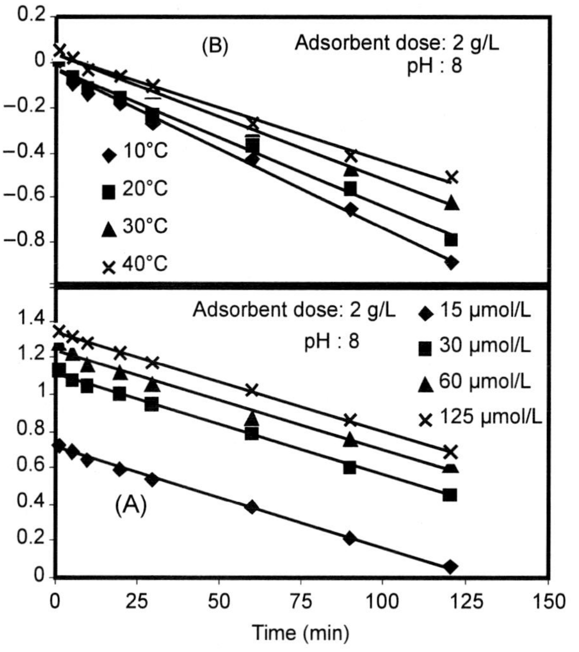

Fig. 6 Lagergren plots for the adsorption of As(V) onto TDF-AE at different (A) initial concentrations and (B) temperatures.

slope and intercept of the plots of ln $[(C_t/C_o - 1/(1 + mK_L)]$ vs t (Fig. 6) using regression analysis as 1.3×10^{-4}, 1.2×10^{-4}, 7.5×10^{-5} and 6.4×10^{-5} cm/s, for an initial concentration of 30, 60 and 125 μ mol/L respectively. As expected the B_L values suggest that the velocity of As(V) transport is rapid enough to use TFS-AE for the treatment of water and wastewater containing As(V).

3.6 Adsorption Isotherm

The adsorption isotherm of As(V) at 30 °C was generated in batch experiments using different initial concentrations ranging from 15 to 275 μ mol/L (Fig. 7). It can be seen that the adsorption isotherm exhibits an *L*- type which corresponds to the classification of Giles (1960). In this classification, type *L* assumes a monolayer formation in the active sites of the surface, and all the adsorption sites are supposed to be equivalent. The equilibrium isotherm data can be tested using Langmuir isotherm model.

$$\frac{C_e}{q_e} = \frac{1}{Q^\circ b} + \frac{C_e}{Q^\circ} \qquad ...(5)$$

where C_e is the equilibrium concentration of As(V) ions in solution, q_e is the adsorption capacity at equilibrium, Q° and b are Langmuir constants related to adsorption capacity and binding energy of adsorption respectively. The Langmuir isotherm parameters were determined by least-squares fit of the isotherm data in Fig. 7. The value of Q° and b determined from the slope and intercept were found to be 74.62 μ mol/g and 0.18 L/μ mol. The Langmuir isotherm

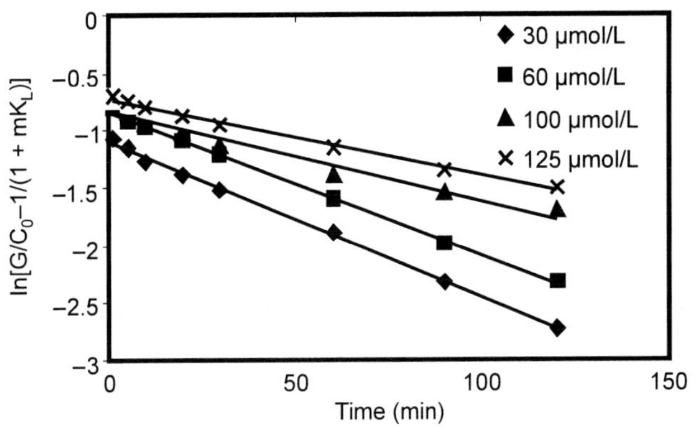

Fig. 7 McKay plots for the adsorption of As(V) onto TFS-AE at different concentrations.

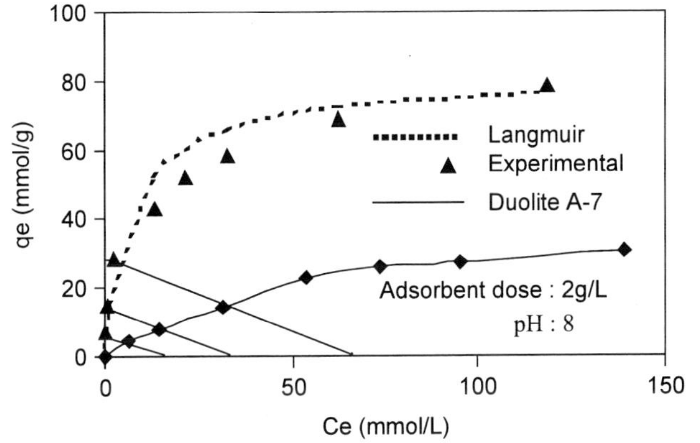

Fig. 8 Comparison of the experimental and Langmuir isotherm plots for the adsorption of As(V) onto TDF-AE.

equation provided an excellent fit to the equilibrium sorption data, giving correlation coefficient of 0.99. The reported $Q°$ values for the adsorption of As(V) on activated bauxasol coated sand (Fuhrman *et al.*, 2005) Chelating resin (Balaji *et al.*, 2005), modified granular activated carbon (Yang and Chen, 2005), iron hydroxide granulates (Daus et al. 2004) and Portland cement (Kundu *et al.*, 2004) were found to be 28.60, 64.72, 16.43, 30.60 and 53.08 μmol/g respectively. The comparison of $Q°$ value of TFS-AE used in the present study with those obtained with literature shows that TFS-AE is more effective for this purpose.

Adsorption experiments were also performed using a commercial chloride form weak base ion exchanger, Duolite A-7. Adsorption isotherm data for As(V) are shown in Fig. 7. The Langmuir constants were also calculated using regression analysis. The values of $Q°$ and b were found to be 38.61 μmol/g and 0.02 L/μmol at 30 °C, which are considerably lower than those calculated for TFS-AE. The amount of adsorbed As(V) is high enough for TFS-AE to be able to effectively remove As(V) from water and wastewater. Figure 7 also shows the operating lines which are generated with the slope of –V/m. The operating lines connecting (C_o, q_o) to

(C_e, q_e) at equilibrium. The final equilibrium solution concentration is predicted from the point of intersection of the operating lines and Langmuir isotherm for the adsorption of As(V). The operating line having a slope V/m = 0.5 was drawn through the initial concentrations Co: 15, 30 and 60 μmol/L. The values of q_e obtained from the operating lines were found to be 4.11, 12.32 and 25.17 μmol/g respectively. The corresponding q_e values obtained from the Langmuir isotherm equation and experiments were 4.40, 10.91 and 24.79 μmol/g and 4.7, 13.13 and 26.17 μmol/g from an initial concentration of 15, 30 and 60 μmol/L respectively. The operating lines also help to determine the theoretical number of stages for the removal of As(V) from aqueous solutions. For example, it was found that As(V) was reduced from 15 to 0.1 μmol/L (below MCL value) in the third stage.

3.7 Effect of Ionic Strength

The effect of ionic strength on the adsorption of As(V) onto TFS-AE was studied over the range between 0.001 and 0.1 M. At an initial concentration of 125.0 μmol/L, the percentage of As(V) adsorption was found to be 94.34, 92.84, 65.97, 59.67 and 51.64% at an ionic strength of 0.001, 0.005, 0.01, 0.015 and 0.1 M respectively. This effect may be imparted to the changes in As(V) activity or in the properties of the electrical double layer. Adsorption is sensitive to the changes in ionic strength if electrostatic attraction plays an important role in the removal of anions. According to the surface chemistry theory developed by Gouy Chapmann, when solid adsorbent is in contact with sorbate species in solution, they are bound to be surrounded by an electrical diffused double layer, the thickness of which is significantly expanded by the presence of electrolyte. Such expansion inhibits the adsorbent particles and As(V) species from approaching each other more closely and, through the decreased electrostatic attraction, leads to the decreased uptake of As(V) ions.

Table 2 Chemical composition and characteristics of the groundwater used in the batch experiments.

Alkalinity: 130 mg/L $CaCO_3$, Al: < 2 μg/L, As: 300 μg/L, Ca: 40 mg/L, Cl: 8.3 mg/L, F: < 0.1 mg/L, Total hardness: 140 mg/L, Fe: < 0.02 mg/L, Mg: 12 mg/L, Mn: < 2 μg/L, pH: 7.9, Silica total: 20 mg/L, SO_4^{2-}: 62 mg/L, Total Dissolved Solid: 0.3 mg/L, Turbidity: 0.16 NTU.

3.8 Test with Groundwater

The utility of TDF-AE was demonstrated by using it with a groundwater sample. The composition of the groundwater sample is given in Table 2 The effect of adsorbent dose on As(V) removal was studied. Almost complete removal of As(V) was possible with 150 mg of TDF-AE. This confirms that TDF-AE can effectively remove As(V) ions even in presence of other ions.

3.9 Desorption

Desorption of the adsorbed As(V) from the spent adsorbent was also studied. Acid regeneration was tried for three cycles with a view to recover the adsorbed As(V) ions and also to restore the adsorbent to its original state. An efficiency of 99.1% desorption of As(V) was obtained using 0.1 M HCl and is therefore suitable for regeneration of As(V) from spent TFS-

AE. After three cycles, the sorption capacity of the TFS-AE decreased from 99.1% to 93.2%. The recovery of As(V) by 0.1 M HCl decreases from 96.2% in the first cycle to 89.3% in the third cycle. The small fraction of adsorbed As(V) not recoverable by regeneration presumably represents the anions, which are bound through strong interaction and, as a result, sorption capacity is reduced in successive cycles.

4 CONCLUSIONS

The following conclusions may be drawn on the basis of the present studies. TFS-AE demonstrated a better performance for adsorption of As(V) ions than TFS. The sorption of As(V) is depended onto the pH of the solution and the maximum sorption occurs between 6.0 and 8.0. The equilibrium time for As(V) adsorption was reached within 3 h. Adsorption process was found to be concentration dependent. Mass transfer coefficient as a function of initial As(V) concentration has been determined. The equilibrium data fitted well with Langmuir isotherm. Comparison of TFS-AE with commercial ion exchanger and other materials reported in the literature shows that the former has the highest sorption capacity. The percentage of adsorption decreases with increase in ionic strength. The applicability of the adsorbent has been tested using groundwater. The adsorbed As(V) desorbed quantitatively by 0.1 M HCl solution and the adsorbent can be reused for three cycles consecutively. The results of this study showed that the TFS-AE can be used as an efficient adsorbent for the removal and recovery of As(V) from wastewaters. Studies are been developed to confirm the applicability of the TFS-AE under real conditions.

References

Balaji, T., Yokoyama, T. & Matsurage , H, 2005. Adsorption and removal of As(V) and As(III) using Zr-loaded lysine diacetic acid chelating resin. *Chemosphere*. 59: 1169–1174.

Daus, B., Wennrich, K. & Weiss, H. 2004. Sorption material for arsenic removal from water: a comparative study. *Water Research*. 38: 2948–2954.

Fuhrman, H.G., Bregnhoj , H.& McChonchie, D. 2005 Arsenic removal from water using sand- red mud columns. *Water Research*. In Press.

Giles, C.H., McEwan, T.H.,Nakhwa, S.N. & Smith D. 1960. *Studies in adsorption*. Part.III. A system of classification of solution adsorption isotherm. Journal of Chemical Society.4: 3973–3993.

Kundu, S., Kavalaktt , S.S., Pal , A., Ghosh , S.K, Mandal , M. & Pal , T. 2004. Removal of arsenic using hardened paste of Portland cement: Batch adsorption and column studies. *Water Research*. 38: 3780–3790.

Maity, S., Chakravarty , S., Bhattacharjee, S. & Roy , B. C. 2005. A study on arsenic adsorption on polymerized sea nodule in aqueous medium. *Water Research* 39: 2579–2590.

Matis, K.A., Lehmann, M. & Zouboulis. A.I., 1999. Modeling of sorption of metals from aqueous solutions onto mineral particles: the case of arsenic ions and goethite ore. In: Misaelides, P. et al (Eds)., *Natural Microporous Material in Environmental Ttechnology*. Kluwer Academic Publishers, The Netherlands, P.P. 463–472.

McKay, G., Otterburn, M.S. & Aga J. A. 1985. Fuller's earth and fired clay as adsorbent for dyestuffs, external mass transport process using adsorption. *Water Air and Soil Pollution*. 26: 149–161.

Noeline, B.F., Manohar, D.M. & Anirudhan, T.S. 2005 Kinetic and equilibrium modeling of lead(II) sorption from water and wastewater by polymerized banana stem in a batch reactor. *Separation and Purification Technology*. 45: 131–140.

Rajakovic, L.J.V. & Mitrovic , M.M. 1992. Arsenic removal from water by chemisorption filters. *Environmental Pollution*. 75: 279–287.

Shibi, I. G. & Anirudhan , T.S. 2005. Adsorption of Co(II) by a carboxylate -functionalised polyacrylamide grafted lignocellulosics. *Chemosphere*. 58: 1117–1126.

Shubha, K.P., Raji, C. & Anirudhan, T.S. 2001. Immobilization of heavy metals from aqueous solutions using polyacrylamide grafted hydrous tin(IV) oxide gel having carboxylate functional group. *Water Research*. 35: 300–310.

Yang, I & Chen, P.J. 2005. Chemical modification of granular activated carbon for enhancement of arsenic adsorption. *Advances in Environmental Technology*. Annual Meeting (Cincinnati, OH) Etryery.

Arsenic and Fluoride in Groundwater: An Emerging Problem in North Eastern India

36

A.K. Singh
North Eastern Regional Institute of Water and Land Management, Tezpur-787027, Assam, India

ABSTRACT

In North East India, a groundwater quality survey was conducted in the year 2003-05. Arsenic concentrations at elevated levels (above 50 μg/L) are encountered in groundwater of alluvial aquifers in Assam, Tripura, Manipur, Nagaland and Arunachal Pradesh. Arsenic distribution was mainly confined to the area adjacent to foothills bounded by Himalayan Mountains viz. Jorhat, Lakhimpur, Nalbari, Nagaon, Barpeta, Dhemaji, Dhubari, Darrang and Golaghat district of Assam; West Tripura, Dhalai and North Tripura district; Thoubal district of Manipur; Mokokchung and Mon districts of Nagaland, and Papumpare, East Kameng, West Kameng, Lower Subansiri, Dibang Valley districts of Arunachal Pradesh. The study indicates that the affected aquifers are generally shallow. In the area under investigation, fluoride concentration in groundwater is well above the permissible limit of 1.5 mg/L in the districts of Karbi Anglong, Nagoan, Sivasagar,Tinsukia, Goalpara, Kamrup, Cachar, Golaghat of Assam, and Lower Subansiri, Dibang district of Arunachal Pradesh causing threat in the groundwater occurring at deeper depth.

1 INTRODUCTION

Adequate water resources for future generations are not only regional issues but also a global concern. It needs no emphasis that our country's freshwater wealth is under serious threat due to variety of natural and human influences. Arsenic and fluoride occurs as minor constituents of groundwater in all categories of hydrogeological framework in India. The high concentration of fluoride and arsenic is of concern as large amount of groundwater is abstracted by drilling water-wells both in rural and urban areas for drinking and irrigation purposes. Thirteen States in India–Rajasthan, Andhra Pradesh, Karnataka, Tamil Nadu, Madhya Pradesh, Maharashtra, Gujarat, Punjab, Haryana, Uttar Pradeh, Bihar, and Delhi have already been identified endemic to fluorosis. Arsenic contamination of groundwater in eight districts of West Bengal is well documented and more cases of groundwater arsenic are reported from eastern part of Bihar, Gorakhpur, Balia, western part of Utter Pradesh and Chhattisgarh.

Arsenic has acquired an unparalleled reputation as a poison, with arsenic trioxide, a tasteless and odorless inorganic arsenic compound constituting a convenient agent for homicide (Peterson 1983). Arsenic could cause liver, lung, and kidney/bladder cancer in addition to skin cancer (Smith *et al.*, 1992). The main health risks of arsenic (WHO-1993, EEC-1975, U.K-1989 and BIS-1991, Canada-1976) are considered to be severe poisoning, and carcinogenic, specially, cancer of respiratory system and gastrointestinal tract. The main potential health risks from fluoride (EEC-1980, U.K-1989, US EPA -1998, BIS-1991, WHO-1993) are considered to be fluorosis or bone disease.

2 THE PRESENT STUDY

North Eastern States of India accounts for the largest storehouse of replenishing freshwater, but the people of the hill region often lack access to safe and adequate drinking water and sanitation for agriculture and livestock. In the rural areas of North Eastern (NE) India, drinking water was primarily obtained from surface water reservoirs like ponds, tanks, rivers, streams, springs and sometimes from lakes prior to sixties. Thereafter, groundwater started being tapped by sinking shallow and sometimes deep tube wells mainly to obviate the diseases caused by pathogenic organisms brought into these surface water bodies due to organic contamination. This led to the installation of millions of handpumps and tube wells all over the region as a means to provide bacteriologically safe water to rural population. This new arrangement was initially thought to be working well till some of the problems caused by this mining of groundwater started becoming understood. There were, amongst others, large ingestion of iron, lowering of water table, sinking of ground level causing marshy conditions at places during the monsoons, neglect of surface water bodies causing a decline in pisciculture, etc. The latest one is the appearance of fluoride (F) and arsenic (As), a very mobile and carcinogenic heavy metal, in tube well water in some places. The presence of excess fluoride and endemic *Fluorosis* was reported in May 1999 in Karbi Anglong district of Assam, though the disease was prevalent for last twenty years. Subsequently, because of intensified water quality testing and health survey conducted, problem of excess fluoride is getting detected in more and more areas of the State like Nagaon, Karbi Anglong and Kamrup district (Akoijan 1997; Sushella 2001). Similarly, problem of arsenic have been detected recently in the year 2003 (Singh 2004).

Recognizing the enormity and severity of the problem, the author had conducted groundwater quality survey for the entire North Eastern States of India in the year 2003–05 to identify the suitability or otherwise of groundwater quality for drinking and irrigation purposes. It was observed from the analysis of the groundwater samples that iron, fluoride and arsenic contamination are predominant in the states. This paper illustrates the arsenic and fluoride concentration in drinking water and its possible sources in groundwater.

2.1 Location and Climate

North Eastern region comprising eight states viz. Arunachal Pradesh, Assam, Manipur, Meghalaya, Mizoram, Nagaland, Sikkim and Tripura lies between 22° and 29° 30′ north of equator and between the longitudinal lines of 88° and 97° 30′ E. This region is bounded on its sides by international borders with China, Myanmar, Bangladesh, Nepal and Bhutan. This region covers an area of 262179 km^2 having a total population of approximately 42.695 million. The land surface is generally mountainous, uneven and the whole terrain abounds in rivulets, rivers, hills, valleys, plains and lakes. The altitude of this region is of diversified magnitude and the elevation extends upto 7000 m asl (in Arunachal Pradesh). The hills are generally dissected by a number of seasonal and perennial streams with valleys in between. The Brahmputra, the Barak and the Tista (in Sikkim) are the three major rivers of this region with broad valleys. The region is situated in the Eastern Himalayas and is considered the richest bio-geographical province of the Himalayan zone.

The climate of this region is influenced and characterized by its hilly terrain and climatic condition varies from hot humid to temperate. This region receives high precipitation during monsoon period even up to 400 cm. Due to heavy rainfall, groundwater reserves are generally adequate in some of the areas, while acute scarcity of water is observed in most of the hilly areas. Elevations of 7000 meters and above remain snowbound throughout the year whereas places as low as 3000 meters come within the snowline during the winter. The temperature varies between 4° to 35° C.

2.2 Sample Collection

The groundwater samples were drawn from pre-installed (5-8 year old) bore wells during post-monsoon period (Jan-Feb) in the year 2003 and 2004 from 1520 locations in Assam, 117 in Tripura, 181 in Meghalaya, 133 in Mizoram, 132 in Nagaland, 128 in Sikkim and 60 in Manipur covering entire NE States. Tube wells were operated at least 10 minutes before collection to flush out the stagnant water inside the tube and to get fresh groundwater. The water samples were collected in plastic container (PVC 250 ml) and sealed. For arsenic analysis in the laboratory, preservative (1:1 HNO_3 solution to pH < 2, about 3 ml/L sample) was added to each water samples at the time of sampling. The samples without preservatives were also collected for analysis of fluoride. These samples were tested for arsenic, fluoride and other water quality parameter like iron, pH, etc.

2.3 Methods of Analysis

For analysis of arsenic, As (V) was pre-reduced to As (III) prior to determination to prevent the interference. Pre-reduction was performed with KI solution (KI^+Ascorbic acid) in semi-concentrated (5 mol/L) HCl solution. Time for pre-reduction was 30 minutes. As was estimated in 10 ml of pre-reduced water using atomic absorption spectrophotometer (Perkin Elmer Analyst 200) with MHS-15 (Mercury Hydride Generation System) at 193.7 analytical wavelengths and 0.7 nm slit width. Radiation source was Electrodeless Discharge Lamp (EDL) for Arsenic with 50 seconds pre-reaction purge time and 30 seconds post-reaction purge time. The Argon gas and Sodium tetraborohydrate were used for hydride generation.

Iron in groundwater samples was tested using atomic absorption spectrophotometer. Ion Selective Electrode method (Cyberscan Ion meter) was used for fluoride analysis. The pH was tested in the field at the time of sample collection using potable pH meter (Eutech Instrument).

2.4 General Characteristics of Groundwater

Groundwater from the alluvial aquifers is characterized by slightly acidic to alkaline pH, high iron (>0.3 mg/L) concentrations, low sulphate (<55 mg/L) and nitrate (<1 mg/L). Dissolved oxygen is high and maximum amount of iron was observed in Assam, Meghalaya, Mizoram, Nagaland and Tripura (Table 1).

Table 1 Some important characteristics of groundwater in NE States of India.

NE State	*Population (Person)*	*pH*	*Nitrate (mg/L)*	*Sulphate (mg/L)*	*Iron (mg/L)*
Arunachal Pradesh	1091117	6.0-8.3	0.10-65.6	0.28-41.68	0.071-11.94
Assam	26638407	5.5-7.4	0.01-44.46	0.01-0.94	1.01-16.26
Manipur	2388634	6.4-8.7	0.10-98.0	1.66-12.50	0.40-3.20
Meghalaya	2306069	4.0-8.0	0.08-72.5	0.83-19.73	0.04-13.68
Mizoram	891058	4.8-7.4	0.10-95.0	0.06-18.06	0.11-17.29
Nagaland	1988636	5.8-8.0	0.05-10.9	0.28-21.95	0.09-20.60
Sikkim	420000	5.1-7.2	0.10-16.90	2.2-11.67	0.23-1.89
Tripura	3191168	4.3-8.3	0.00-81.5	1.11-3.61	0.12-19.28

3 DISTRIBUTION OF ARSENIC AND FLUORIDE IN GROUNDWATER

3.1 Arsenic Distribution

Arsenic concentration at elevated levels (above 50 μg/L) are encountered in groundwater of alluvial aquifers in Assam, Tripura, Manipur, Nagaland and Arunachal Pradesh. Arsenic concentration was found higher in the area adjacent to foothills bounded by Himalayan mountains in parts of Jorhat, Lakhimpur, Nalbari, Nagaon, Barpeta, Dhemaji, Dhubari, Darrang and Golaghat district of Assam; West Tripura, Dhalai and North Tripura district; Thoubal district of Manipur; Mokokchung and Mon districts of Nagaland, and Papumpare, East Kameng, West Kameng, Lower Subansiri, Dibang Valley districts of Arunachal Pradesh. The study indicates that the affected aquifers are generally shallow (15-40 m deep).

The name of the location where arsenic was found at higher level (>200 mg/L) is given below:

1. *Assam*: Jorhat (Titabor, Dhakgorah, Selenghat and Moriani Block), Dhemaji (Sissiborgoan and Dhemaji Block), Golaghat district (Podumani Block) and Lakhimpur (Boginodi, Lakhimpur Block) (Tables 1 and 2).
2. *Tripura*: West Tripura (Jriania Block), Dhalai (Salema Block) and North Tripura (Dharmanagar Block) districts.
3. *Manipur*: Thuobal (Kakching Block).

Table 2 Arsenic and fluoride in groundwater in different districts of NE States of India.

NE State	*Percentage of Fluoride contaminated area (Above BIS Permissible level of 1.50 ml L^{-1})*	*Percentage of Arsenic contaminated area (BIS Permissible level of 50 µL^{-1})*
Arunachal Pradesh	Lower Subansiri (5%) Dibang Valley (25%)	Papum Pare, East Kameng, Lower Subansiri Dibang Valley (3-5 % in all cases) Jorhat (21%),
Assam	KarbiAnglong (10%), Nagoan (8%), Tinsukia (6%), Goalpara (6%), Golaghat (5%), Kamrup (4%), Cachar (4%) and Sivasagar (2%),	Nalbari (19 %), Lakhimpur (21.2%), Nagoan (13 %), Golaghat (13 %), Dhubari (19%), Barpeta (18%), Dhemaji (22%), Darrang (7.6%) Sonitpur (12.9%), Cachar (8.35%), Hailakandi (2.9%), Karimganj (15.8%), Goalpara (11.8%), Kamrup (11.1%), Sibasagar (9.3%), Dibrugarh (8.9%), Bongaigoan (4.2%), Kokrajhar (3.4%) and Tinsukia (2.5%)
Manipur	Trace	Thuobal district (25%)
Meghalaya	Trace	West Garo Hills and West Khasi Hills district (3-5 % in all cases)
Mizoram	Trace	Trace
Nagaland	Trace	Mokokchong (31.8%), Mon (30%) and Zunheboto (6 %) district
Tripura	Trace	West Tripura (27.19%), Dhalai (36.12%) and North Tripura (24.47%) districts.
Sikkim	Trace	Trace

In floodplain area of Assam viz. Barpeta, Dhemaji, Dhubari, Darrang and Golaghat, the arsenic content was found between 100 and 200 µg/L. In remaining 11 districts of Assam where arsenic was detected contained arsenic in the range of 50–100 µg/L. Only three districts, namely, Karbi Anglong, NC Hills and Morigaon (flood plain) were found free from arsenic contamination.

Very high concentration of arsenic was found in four district viz. Jorhat, Lakhimpur, Nalbari and Nagoan. In Jorhat district, the contamination of arsenic was highest in the range of 228-657 µg/L. Altogether 80 samples were collected and 21 percent of the samples were found contaminated with arsenic. In Lakhimpur district, total 76 samples analysed and 21.2 percent of the samples detected with arsenic. The concentration of arsenic in Lakhimpur district was in between 210 and 550 µg/L. In Nalbari district, 19 percent samples (72 samples) contained arsenic with the value in between 234 µg/L and 422 µg/L. In Nagaon district, 76 samples were collected and 13 percent of the samples were found contaminated with arsenic. The range of arsenic concentration in Nagaon was in between 217–601 µg/L.

In Tripura state, the distribution of arsenic content in groundwater varies from 1.38–191.91 µg/L in West Tripura (27.19 per cent), 4.49–444.48 µg/L in Dhalai (36.12 percent), and 3.23–215.01 µg/L in North Tripura (24.47 percent). The maximum arsenic in drinking water was found in Jirania and Bishalgarh Block of West Tripura, Salema and Kamalpur Block in Dhalai and Dharmanagar Kailasahar, Kanchanpur and Jampui Block in North Tripura. Only South Tripura was found to be free from arsenic contamination.

In Thoubal district of Manipur, about 25% of the collected samples found with arsenic in the range of 798–986 µg/L.

The concentration of Arsenic was relatively high in shallow tube well as compared to deep tube well and ring wells. Dilution effect was observed in distribution of arsenic in the monsoon season in most of the locations where arsenic study was done. No report of *Arsenocosis* from any area of the region has been known till date. Hydrogeochemical information of Assam suggests that elevated arsenic concentrations in groundwater is governed by the presence of organic matter in the sediments and reduced alluvial aquifers with high iron, low sulphate and nitrate concentrations. Special care need to be taken to avoid oxidation of the sediment in reduced aquifers.

3.2 Fluoride Distribution

In the area under investigation (Table 2), fluoride concentration in groundwater is well above the permissible limit of 1.5 mg/L in the districts of Karbi Anglong, Nagaon, Sivasagar,Tinsukia, Goalpara, Kamrup, Cachar, Golaghat of Assam, and Lower Subansiri, Dibang Valley, Midland district of Arunachal Pradesh causing threat in the groundwater occurring at deeper depth. In other States of north east, the fluoride content was found within the permissible limit.

The name of the location where fluoride was found above 1.5 mg/L is given below:

1. *Arunachal Pradesh:* Lower Subansiri (Hapoli), Dibang Valley (Chatin Village, Midland Block)
2. *Assam:* Golaghat (Podumoni), Karbi Anglong (Lumbajong, Hawraghat), Sibasagar (Nazira),Tinsukia (Guijan, Margherita), Goalpara (Rangjuli), Kamrup (Dimaria), Nagoan (Kandura, Nilbagan, Dabaka), Cachar (Udarbond) districts.

Assam falls in moderately endemic zone with very high fluoride-groundwater occurring in pockets in certain areas. Granites, Granitic gneisses and phyllitic quartzite are the main hard rock aquifers that contain fluorine bearing minerals such as fluorite, biotite and muscovite. The highest fluoride concentration in groundwater is found near intrusive granites and near a fault/lineament trending along NW-SE direction.

4 POSSIBLE SOURCES OF ARSENIC AND FLUORIDE IN GROUNDWATER

4.1 Arsenic

From the works of the different organizations in arsenic field it is now known that the arsenic originates in the Himalayan head waters of the Ganga and Brahmaputra rivers and has lain undisturbed beneath the surface of the region's deltas for thousands of years in thick layers of fine alluvial mud smeared across the area by the rivers.

So far, the study indicates that the affected aquifers are generally shallow (15–40 m deep). Deep wells, tapping depths greater than 100 m, almost invariably have low arsenic concentrations, usually less than 50 $\mu g\ L^{-1}$. It also appears from analytical results that groundwater adjacent to foothills is more contaminated. This area lies within an alluvial basin bounded by lower Himalayan mountains in north and north east. High intensity of rainfall in the catchment and plain areas has contributed to high sediment loads which developed the valley into a long stretch of recent and old alluvium. In Assam valley, the depth of alluvium decreases from the rivers towards the hills on the north and south. The alluvium near the river is more sandy and periodic fluviatile action keeps the alluvium stratified. But its influence has been gradually obliterated by climate as one moves from recent floodplains to old floodplains and then to upland. As a result there is deposition of coarse sand and more coarser river-borne materials along with plant cells and other organic materials cells which may contain considerable amount of arsenic and other toxic elements. During course of time, these toxic elements get released in the reducing environment and get shelter within silty and clayey sediments. Bhattacharya *et al.* (1997) and Nickson *et al.* (1998) also put forward the hypothesis that the burial of the sediments, rich in organic matter, led to strongly reducing conditions in groundwater aquifer, which is facilitated by high water table, fine-grained surface layers and widely practised wetland paddy cultivation, as well as microbial oxidation of sedimentary organic matter, depleting thereby the dissolved oxygen in groundwater. Arsenic is released when arsenic-rich iron oxyhydroxides, which are efficient arsenic-scavengers, are reduced in anoxic groundwater. Such reduction is driven by concentrations of sedimentary organic matter.

In most arsenic contaminated areas, the mobilisation has probably occurred by a complex combination of redox changes brought on by rapid burial of the alluvial sediments, including reduction of the solid-phase As to As(III), desorption of As from iron oxides, reductive dissolution of the oxides themselves and likely changes in iron-oxide structure and surface properties following the onset of reducing conditions. This sequence begins with the consumption of oxygen and an increase in dissolved CO_2 from the decomposition of organic matter. Next, hydrous ferric oxides are reduced to Fe^{2+} followed by SO_4^{2-} reduction to S^{2-}. During sulphate reduction, the consequent sulphide reacts with any available iron to produce FeS and ultimately pyrite, FeS_2. Iron is often found more abundant than sulphur so that there is 'excess iron' beyond that which can be converted to pyrite. Arsenic (V) reduction would normally be expected to occur after Fe(III) reduction but before SO_4^{2-} reduction. There is considerable evidence from several studies that arsenic is released from soils following flooding and the development of anaerobic conditions (Deuel and Swoboda 1972; McGeehan 1996). This release is associated with iron oxide dissolution. Deuel and Swoboda (1972) found that reducing an untreated clay soil led to the release of As and that the amount released was related to the total arsenic content of the soil and the redox potential. They proposed that the release was primarily due to reduction (and dissolution) of 'ferric arsenates' rather than to changes in the As speciation. It is also reported that (Islam *et al.*, 2004) severe arsenic pollution of groundwater in most alluvial aquifers worldwide is driven by the microbially-mediated metabolism of organic matter, with FeOOH acting as the source of oxygen: the oxide is reduced during the process and its sorbed arsenic is released to groundwater. In Assam valley, sediment sequence is capped by a layer of clay or silt (of variable thickness) which effectively restricts entry of air to the aquifers and abundance of recent solid organic matter deposited with the sediments, has resulted in the development of highly reducing aquifer conditions which might be favouring the mobilisation of As.

Several other studies have shown that the groundwater in the region is generally in a reducing state (presence of relatively high concentration of sedimentary organic matter) and suggest that arsenic is being released when arsenic-iron bearing minerals in the sediments are reduced by oxygen deficient groundwater.

4.2 Fluoride

Fluoride bearing igneous rocks may contribute fluoride to the later formed sedimentary and metamorphic rocks which ultimately passes to the resultant soil and to groundwater.

Generally, porphyritic granites contain more fluoride bearing minerals followed by limestone, dolomite and quartzite. The basaltic rocks also contain more fluoride bearing minerals in it.

The hard rocks which contain minerals like quartz, feldspar, micas, etc are subjected to weathering and as a result of which a weathered mantle is formed. This weathered mantle forms water table aquifer and also the principal media for infiltration of rainwater into the groundwater. During the process of weathering the minerals rich in fluoride such as micas, flurapatite, cryolite, etc. come into solution and remain in groundwater without much precipitation.

Fluoride forms a dissociated weak ionic bond in aqueous solution. Its solubility in water is less than chloride. The mobilisations of fluoride through different media like soil and water results from certain physico-chemical processes. The solubility is governed by certain factors, e.g. the extension of fluoride bearing weathered rocks over an area, the accessibility of circulating water through the source rocks, the period of contact of circulating water in the source area and also the presence of other ions in soil, rock and percolating water. Water with high alkalinity and low calcium and magnesium usually dissolve excessive amount of fluoride.

Major part of the Karbi Anglong and Nagaon districts of Assam where maximum fluoride was found are underlain by Precambarian hard rocks, viz. granites, granite gneisses, quartizites, phyllites and schists and some parts underlain by territiary soft rocks, namely, sandstones, shales, siltstones, limestones, etc., followed by thin veneer of alluvium.

Since data regarding the total depth of the tube-wells is lacking, it is difficult to ascertain whether these wells are tapping the hard-rock aquifer or the top alluvial aquifer. However, since the dug wells have lower fluoride content compared to the bore-wells, it is apparent that the source of fluoride are the hard rocks present in the affected area.

5 ACTION NEEDED FOR MITIGATION

1. To identify and map the high-risk arsenic and fluoride zones in the NE States by water testing and GIS technologies including hydro-geological studies on water quality aspects.
2. Monitoring and surveillance of arsenic and fluoride on a regular basis.
3. Provision of safe drinking water in arsenic and fluoride endemic areas by selecting a viable option out of multiple options.
4. To conduct and support operational research on new mitigation options for water composition.
5. Water quality control options should include:
   (a) Studies on variety of treatment technologies for removing arsenic and excessive fluoride from contaminated groundwater.
   (b) Identifying the upper arsenic bearing aquifers and mapping of safe aquifers below 100 m bgl and to ensure that groundwater from those aquifers is safe from arsenic and fluoride.
   (c) Utilization of subsurface water of shallow or dry river bed for safe drinking water.
6. Systematic epidemiological survey of prevalence of arsenicosis disease and fluorosis in high risk arsenic fluorides endemic zones through cluster sampling or systemic surveys.

7. Sensitise Health officers of primary health care centres on the health effects of arsenic and fluoride.
8. Awareness generation needs to be carried on the arsenic and fluorosis disease prone population regarding the health effects of the disease, the need for taking safe water for prevention of the disease and the need for a balanced diet.
9. Organizing a policy workshop on arsenic and fluoride contamination and their related health effects for NE states for needs assessment and strategic planning with inputs from WHO, UNICEF, GOI and other partners for the creation of joint action plan and nodal agencies.

Acknowledgements

The author gratefully acknowledges North Eastern Council, Shillong for sponsoring the project "Characterisation of Drinking Water for its Quality in North Eastern States of India". Prof. P.V. Seethapathi, Former Director, NERIWALM is acknowledged for his kind cooperation, encouragement and providing all necessary facilities to carry out this work. Thanks are due to Laboratory Assistant, Project Assistants and Data Entry Assistant for their help during the course of this study.

References

Akoijam, 1997. Distribution of fluoride in groundwater and pollution of shallow aquifers in parts of Imphal valley, Manipur, India. *Chemical Environment Research* 6: 301.

Bhattacharya, P., Chatterjee, D. and Jacks, G. 1997. Occurrence of arsenic-contaminated groundwater in alluvial aquifers from delta plains, Eastern India: Options for safe drinking water supply. Water Resources Development 13: 79–92.

Deuel, L.E. and Swoboda, A.R.1972a. Arsenic solubility in a reduced environment. *Soil Science Society of America Journal* 36: 276–278.

Deuel, L.E. and Swoboda, A.R.1972b. Arsenic toxicity to cotton and soybeans. *Journal of Environmental Qual.* 1: 317–320.

Islam, F.S., Gault, A.G., Boothman, C., Polya, D.A., Charnock, J.M., Chatterjee, D. and Lioyd, J.R. 2004. Role of metal-reducing bacteria in arsenic release from Bengal delta sediments, Nature, 430: 68–71.

McGeehan, S.L. 1996. Arsenic sorption and redox reactions: Relevance to transport and remediation. Journal of Environmental Science and Health, A31: 2319–2336.

Nickson, R., McArthur, J., Burges, W., Ahmed, K. M., Ravenscroft, P. and Rahman, M. 1998. Arsenic poisoning of Bangladesh Ground water. *Nature* 395: 338.

Peterson, M.L. and Carpenter, R. 1983. Biogeochemical processes affecting total arsenic and arsenic species distributions in an intermittently anoxic fjord. *Mar Chem* 12: 295 –321.

Singh, A.K. 2004. Arsenic Contamination in Ground water of North Eastern India. Pages 255-262, In: Jain, C.K., Trivedi, R.C. and Sharma, K.D. (Editors). Proceedings of 11th National Symposium on Hydrology with Focal Theme on Water Quality, Roorkee, India.

Smith AH; C.Hopenhayn-Rich; MN Bates, HM Goedien, I. Hertz Picciotto, HM Duggan, R. Wood, MJ Kosnett, and MT Smith, 1992. Cancer risk from arsenic in drinking water: *Environmental health perspectives* 97: 259–267.

Sushella, A.K., 2001. A treatise on Fluorosis, Fluorosis Research and Rural Development Foundation, New Delhi. 15 pages.

A Simple Electrocoagulation Setup for Arsenite Removal from Water

37

P.D. Nemade and Sanjeev Chaudhari

Centre for Environmental Science and Engineering, IIT, Bombay, Powai, Mumbai – 400076, India

ABSTRACT

Electrocoagulation is an emerging water treatment technology that has been applied successfully to treat various wastewaters. The available arsenic removal methods in general, have quite low efficiency for As(III) removal and thereby chemical oxidation of As(III) followed by adsorption is suggested. Electrocoagulation offers possibility of anodic oxidation and *in situ* generation of adsorbents (such as hydrous ferric oxides). In the EC process with iron electrodes, As(III) gets oxidized to As(V) and gets adsorbed on to the metal hydroxides (HFO) generated in the process. Therefore, electrocoagulation seems to be a better choice for arsenic removal from water. A simple setup of electrocoagulation along with cloth filter has been used for arsenite removal. The arsenic laden water is passed through a container (volume 2 L) in which electrocoagulation (current: 0.2 amp) is done and then it passes through double layer cotton cloth at flow rate 44ml/min. Experiments were carried out with initial arsenic concentration of 1 mg/L and 2 mg/L with varying current flow. Results show that arsenic levels below 50 μg/L could be achieved, which is the drinking water standard in India and Bangladesh.

1 INTRODUCTION

Arsenic (As), a common trace inorganic contaminant in drinking water and well-known toxic element around the world is extremely detrimental to human beings. Recent epidemiological evidences on arsenic carcinogenicity suggest that the standard of 50 ppb may not be sufficiently protective to human health (Pontius *et al.*, 1994). Therefore, the U.S. EPA has reviewed the benefits and costs associated with implementing a MCL of 3–20 ppb and recently they have finalized a new U.S. standard of 10 ppb (U.S. EPA 2001). Leaching from geological formations and anthropogenic activities such as uncontrolled industrial discharge from metallurgical, smelting and mining industries, and organo-arsenical pesticides is a source of arsenic in the groundwater (Krishna *et al.*, 2001). In groundwater and surface water environments, arsenic is present as As(V) (arsenate) in oxidizing environments, while As(III) (arsenite) is the predominant arsenic in reducing environments (Ferguson and Gavis, 1972). Chronic health effects of arsenic include development of various skin lesions such as hyper-pigmentation (dark spots), hypo-pigmentation (white spots), and keratosis of the hands and feet (Jain and Ali, 2000; Karim, 2000). Skin cancers and internal cancers like, kidney, liver, and bladder can appear due to high arsenic exposure (Farrel *et al.*, 2001). Adsorption on low-cost media is an attractive means for trace metal removal from water: Adsorption-based processes are reliable and efficient for removal of complex inorganic and organic metals than many conventional treatment methods (Benjamin *et al.*, 1996). Amorphous iron oxide or ferrihydrite is a common coating of subsoil particles and has a high capacity for different anions including arsenic (Pierce and Moore, 1982). Iron containing salts have also been used for coating of quartz sand for the removal of arsenic from groundwater (Joshi and Chaudhuri, 1996). Electrocoagulation has been reported to offer various advantages over conventional coagulation in conjunction with other processes (Mills, 2000), such as dissolved air floatation (Pouet and Casmick, 1995), electrofloatation (Chen *et al.*, 2000). Various authors reported the numerous treatment types for wastewater such as domestic grey water reuse (Lin *et al.*, 2005), laundry wastewater (Ge *et al.*, 2004), domestic wastewater (Vlyssides *et al.*, 2002).

In the present study, arsenic is removed by electrocoagulation—an innovative process using iron as an electrode which is cheaply available and is non-toxic to human beings. As(III) was chosen for this study due to its predominant occurrence in the Bengal Delta Basin.

2 MATERIALS AND METHODS

2.1 Preparation of Samples

The experiments were performed at Centre for Environmental Science and Engineering Laboratory, IIT Bombay, at ambient temperatures ranging from 26 to 28° C. The chemicals were of analytical reagent grade and were used without any further purification. All glassware were cleaned with water and 1N H_2SO_4 and then rinsed with distilled water. Stock solutions of arsenite were prepared by dissolving appropriate quantity of arsenic trioxide, As_2O_3, (S.D. Fine Chem. Ltd. India) in distilled water containing 1% (w/w) NaOH and the solution was then diluted up to 1 L with distilled water before use. The working solutions containing arsenic were prepared by dissolving appropriate amount of arsenic from stock solutions in tap water. Tap water was tested for the pH, alkalinity, and the presence of arsenic, iron and phosphate. It was found that the pH of the water varied from 7.2 to 7.5, bicarbonate alkalinity was approximately 45-50 mg/l as $CaCO_3$, the dissolved iron, phosphate and arsenic concentration were not detectable in tap water.

3 PROCEDURE

Preliminary experiments were conducted with iron electrode. Two electrodes of same material were placed 0.5 cm apart in a 2 l beaker and arsenic containing water was added in this beaker.

Table 1 Experimental conditions employed in continuous mode electrocoagulation.

Expt. Run No.	Arsenic type	As conc. (mg/l)	Current (mA)	Current Density (mA/cm^2)	pH ml/min	Flow rate (ml)	Reactor Volume
1	As(III)	1	200	1.66	7±0.2	44	2000
2	As(III)	1	240	2.00	7±0.2	44	2000
3	As(III)	2	200	1.66	7±0.2	44	2000

The total submerged surface area of each electrode was 36 cm^2. Before each experiment, the electrodes were abraded with sand paper to remove scale and then cleaned with successive rinses of water and 1 N H_2SO_4.

A direct current by stabilized power supply (0-15 V, 2 A) was applied to the terminal electrodes in which electrical current was controlled by a variable transformer. Experiments were conducted at pH 7.0. The pH of the solution was adjusted by adding either dilute HCl or NaOH.

3.1 Analytical Methods

The residual arsenic in water sample was determined using molybdenum blue complex method (Johnson and Pilson, 1972). The method was used to estimate As (III) and As (V) concentrations in treated water samples to assess the efficiency of the oxidation step and the subsequent removal of arsenic. Spectrophotometric measurements were made at a wavelength of 865 nm using absorbance cells of 5 cm path length for arsenic determination. Calibration curve for total arsenic was prepared using solutions containing As (III) dissolved in distilled water.

4 RESULTS AND DISCUSSION

In any electrochemical process, electrode material has significant effect on the treatment efficiency. The electrode material for drinking water treatment should also be non-toxic to human health. Hence iron was chosen as electrode material as these are non-toxic and readily available instead of aluminum (Kumar *et al.*, 2004).

A study (Fig. 1) was conducted to evaluate the arsenic removal by the electrocoagulation using 2 litre glass beaker and double cotton cloth to assess the feasibility of this process in domestic filters in arsenic affected areas. Arsenic concentration up to 2 mg/L was treated by EC cloth filter with 44 ml/min flow rate for 3 hours. This result shows that treated water contain arsenic concentration below 50 μg/L. This finding is significant for arsenic removal in Bengal Delta Plain (BDP) areas.

Hydraulic retention time (HRT) is calculated by = Volume of Reactor/flow rate

= 2000 ml/44 ml = 45 minutes.

In electrocoagulation, iron dissolves from the anode and hydrogen gas is developed at cathode. Faraday's law can be used to describe the current density (A/ cm^2) and the amount of iron which goes into the solution (g Fe/ cm^2) (Vik *et al.*, 1984),

$$w = \frac{itM}{ZF}$$

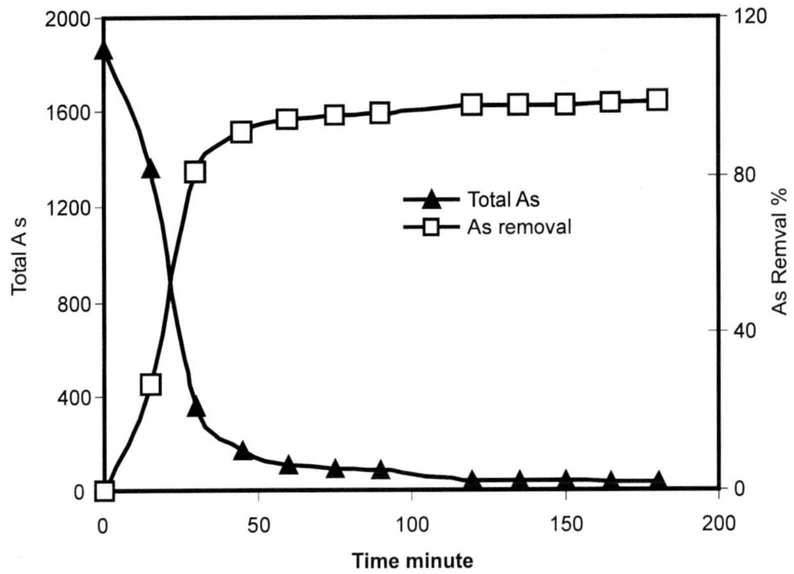

Fig. 1 Effect of time on arsenic removal (experimental run no. 3) As = 2 mg/L, double layer cloth; Current = 0.2 Amp, Volume = 2 lit. in beaker, pH = 7 flow rate 44 mL /min, (Run No.3).

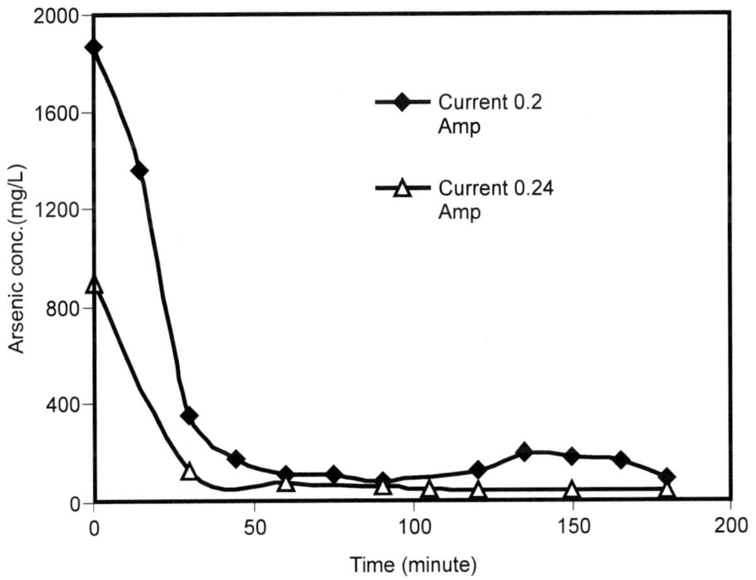

Fig. 2 Effect of current density on arsenic removal (experimental run no. 2, 3) symbols indicate (D) As = 2 mg/L and Current = 0.24 Amp, and (•) As = 2 mg/L and current 0.2 Amp, double layer cloth; Volume 2 lit, pH =7, flow rate 44 mL /min.

where,

w = metal dissolving (g Fe/Fe cm^{-2}); i = current density (A cm^{-2}); t = time (sec);
M = molecular weight of Fe (55.84); Z = number of electrons involved in the oxidation/reduction reaction (Z = 3); F = Faraday's constant, 96500.

Initially, at hydraulic retention time (HRT) of 45 minutes arsenic removed was found 90%, further Hydrous Ferric Oxide was generated and residual arsenic concentration was below 50 µg/L after 2 hours as shown in Fig. 1. Theoretically iron dissolved in to the solution at HRT of 45 minute is 104 mg removing 90% total arsenic treating 1.98 lit of water with adsorbing capacity of 32.23 µg As/mg iron at a flow rate of 44 mL/min. At steady state condition after 2 hrs, total arsenic removed was found below 50 µg/L achieving adsorbing capacity of 96 µg As/mg iron at equilibrium arsenic concentration of 45 µg/L treating 5.28 litre of water, as shown in Fig. 1. Further, this curve remains horizontal till the end of experiment. Pinisakul *et al.* (2002) found that 1.0–1.1 gm of Fe is required in the electrochemical process to remove 1000 mg of Arsenic. Arsenic removal capacity of zero-valent iron was found to be approximately 7.5 mg As/g Fe (Lien and Wilkin, 2004) which is much lower than the arsenic sorption capacity obtained in this study.

Figure 2 shows that up to 85–90% of the initial concentration decreased within 30 min of the process and the residual arsenic concentration in water was found 358 µg/L and 122 µg/L at pH 7, for current of 0.2 Amp and 0.24 Amp respectively. After 2 hrs residual concentrations was found 48 µg/L and 44 µg/L respectively, which is below Indian standard of 50 µg/L.

At the beginning of process the arsenic removal is rapid and later it gradually decreases over the entire process examined due to arsenic ions which are more abundant at the beginning of the EC process, and the generated iron hydroxides due to corrosion of the anode at that time will form complexes with arsenic and therefore rapid removal of arsenic was observed (Kumar *et al.*, 2004). As the experiment proceeds the aqueous phase arsenic concentration goes on reducing and simultaneously hydrous ferric oxides concentration increases, so the curves nearly remains horizontal at the end of experiment.

The pH increase in electrocoagulation is attributed to the formation of hydrogen gas at the cathode (Vik *et al.*, 1984; Chen *et al.*, 2000). Also a slight increase in pH may be expected because of sorption reactions of As(V) and As(III), which release OH^- groups from sorbents as a result of ligand exchange (Arienzo *et al.*, 2002). Kumar *et al.* (2004), reported that there is no effect of pH on both As(III) and As(V) removal up to 97-99%. Therefore, the filter runs at pH 7.0.

5 CONCLUSIONS

Since As (III) is more mobile and more toxic than As (V), it is advantageous to convert As (III) to As(V). Also due to poor As(III) removal from water by many conventional processes, oxidation of As(III) to As(V) as the first step is recommended and then subsequent removal of As(V).

Electrocoagulation process was able to remove 99% of arsenic from arsenic contaminated water and met the drinking water standard of 50 µg/L with iron electrode. As(III) was more efficiently removed in electrocoagulation than conventional methods.

As electrocoagulation process is very cheap and simple, that can be used in any household in developing countries like India, Bangladesh and other parts of world, due to the robustness and convenience in preparation EC filter which may become a viable and cost-effective alternative as a filter medium for arsenic removal from contaminated drinking water.

References

Arienzo, M., Paola, A., Chiarenzelli, J., Maria, R.B., Martino, A.D. 2002. Retention of arsenic on hydrous ferric

Ge Jiantuan, Qu Jiuhui, Lei Pengju, liu Huijuan. 2004. New bipolar electrocoagulation- electroflotation process for the treatment oxides generated by electrochemical peroxidation. *Chemosphere* 48: 1009-1018.

Benjamin, M. M., Slatten, R. S., Bailey, R. P., Bennet, T. 1996. Sorption and filtration of metals using iron-oxide-coated sand. *J. Environ. Eng.* 125: 782–784.

Chen, X., Chen, G., Po, L.Y. 2000. Separation of pollutants from restaurant wastewater by electrocoagulation. *Sep. Purif. Technol.* 19: 65–76.

Farrell, W.J., Peggy, O.D., Colklin, M. 2001. Electrochemical and spectroscopic study of arsenate removal from water using zero-valent ion media. *Environ. Sci. Technol.* 35: 2026–2032.

Ferguson, J.F., Gavis, J. 1972. Review of the arsenic cycle in natural waters. *Water Res.* 6: 1259–1274.of laundry wastewater, *Separation and Purification Technology* 36: 33–39.

Jain, C. K., Ali, I. 2000. Arsenic: occurrence, toxicity and speciation techniques, *Water Res.* 2000, 34: 4304-4312.

Johnson, D.L., Pilson, M. 1972. Spectrophotometric determination of arsenite, arsenate and phosphate in natural waters. *Anal. Chim. Acta* 58: 289–299.

Joshi, A., Chaudhuri, M. 1996. Removal of arsenic from ground water by iron-oxide coated sand. *J. Environ. Eng.* 122(8): 769–776.

Karim, M. 2000. Arsenic in groundwater and health problems in Bangladesh *Water Res.* 2000, *34*: 304–310.

Krishna, B., Chadrasekaran, K., Karunasagar, D., Arunachalam, J. 2001. A combined treatment approach using Fenton's reagent and zero-valent iron for the removal of arsenic from drinking water. *J. Hazard. Mater.* B 84: 229- 240.

Kumar P. R., Chaudhari S., Khilar K. C., Mahajan S. P. 2004. Removal of arsenic from water by electrocoagulation, *Chemosphere,* 55: 1245–1252.

Lien Hsing-Lung, Wilkin Richard T. 2005. High-level arsenite removal from groundwater by zero-valent iron, *Chemosphere,* 59 (3): 377–386.

Lin Chin-Jung, Lo Shang-Lien, Kuo Chao-Yin, and Wu Chung-Hsin. 2005. Pilot-Scale electrocoagulation with bipolar aluminum electrodes for on-site domestic greywater reuse. *J. Environ. Eng.,* 131(3): 491–495.

Mills, D. 2000. A new process for electrocoagulation. *J. Am. Water Works Assoc.* 92, 35–43.

Pierce, M. L., and Moore, C. B. 1982. Adsorption of arsenite and arsenate on amorphous iron hydroxide. *Water Res.,* 16(7): 1247-1253.

Pinisakul A, Polprasert C, Parkplan P, Satayavivad J. 2002., Arsenic removal efficiency and mechanisms by electro-chemical precipitation process, *Water Science and Technology,* 46 (9): 247–254.

Pontius, F. W., Brown, K. G., and Chen, C. J. 1994. Health implications of arsenic in drinking water. *J. Am. Water Works Assoc.,* 86(9): 52–63.

Pouet, M.T., Grasmick, A. 1995. Urban wastewater treatment by electrocoagulation and flotation. *Water Sci. Technol.* 31: 275–283.

Vik, E.A., Carlson, D.A., Eikum, A.S., Gjessing, E.T. 1984. Electrocoagulation of potable water. *Water Res.* 18: 1355- 1360.

Vlyssides A. G., Karlis P. K., Rori N., Zorpas A. A. 2002). Electrochemical treatment in relation to pH of domestic wastewater using TI/Pt electrodes. *Jr. Hazardous Materials* B95: 215–226.

Variation in Concentration of Fluoride in Groundwater of South West District of NCT Delhi: A Case Study

38

Shashank Shekhar, S.K. Mohiddin and P.N. Singh
Central Ground Water Board, 18/11 Jamnagar House, Mansingh Road, New Delhi –110 011, India

ABSTRACT

The SW district of Delhi has considerable dependency on groundwater for drinking purposes. It has been notified by Central Ground Water Authority for regulated development of groundwater resources on account of rapidly declining water levels and predominance of poor groundwater quality areas. Groundwater of the district has considerable fluoride contamination to the extent that more than 50% of groundwater samples collected randomly in the district showed fluoride concentration beyond permissible limit (concentration > 1.5 mg/liter). The predominance of fluoride contamination is found in samples collected from older alluvium areas underlain mostly by saline water and in north-eastern part of the hard rock areas of the district. The groundwater samples from rest of hard rock areas and mid-eastern part of the district has shown concentration of fluoride below permissible limit. The probable reasons for distribution of fluoride contamination have been discussed focusing on the residence time of groundwater.

1 INTRODUCTION

The South West district of the National Capital Territory (NCT) of Delhi (Fig. 1) has been notified by Central Ground Water Authority (CGWA) of India for regulated development of groundwater resources. The step was taken on account of rapid decline in the water levels of the district and because of the fact that majority of the groundwater resources available in the district is saline and has poor quality. The district has strong predominance of rural belt and considerable unauthorized clusters which are mostly dependent on groundwater for drinking

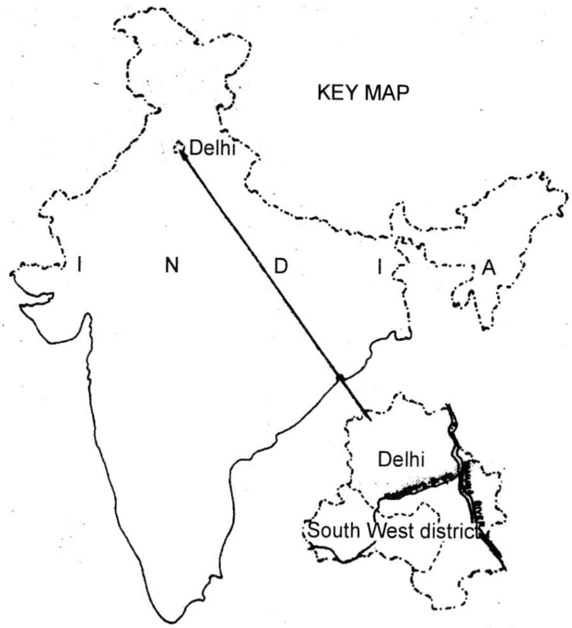

Fig. 1 Location of Delhi and South West District in India.

purposes, even the planned settlements and other localities have considerable dependency on groundwater for drinking purposes. Fluoride concentration in drinking water in the range of 0.8 to 1.0 mg/L is essential for teeth and bones but when the concentration of fluoride in drinking water is found above 1.5 mg/L, it can have adverse impact on human health like mottled teeth, pain and tenderness of bones, still births, deformed children, etc. The present paper tries to study fluoride concentration in groundwater of the district and have an insight in the areas having anomalous concentration of fluoride (> 1.5 mg/L) and think of the reasons thereof. This will provide an useful input in managing the groundwater resources of the district in general and the drinking water sources in particular. In the perspective of hydrogeology, the district has simple geology comprising Alwar quartzite in the east and fluvioaeolian deposits in the rest of the district (Fig. 2). The borehole data of Central Ground Water Board (CGWB) shows alternation of fluvial bands with aeolian bands in sporadic manner. The sediments are mostly silt with medium to fine sand, clay and kankar. The depth to bedrock data indicates towards steep subsurface escarpment on western side of the Delhi ridge (Alwar Quartzite) exposed in the district, thereby hinting at presence of a fault on the western side of the hard rock exposure.

The hydrogeological conditions of the area have to be incorporated, the WL, its variation with season, the aquifer, transmissivity and their spatial variation in the study area helps to get the idea of residence time, relationship of GW potential with respect to geomorphology and to that of the structure has to be added.

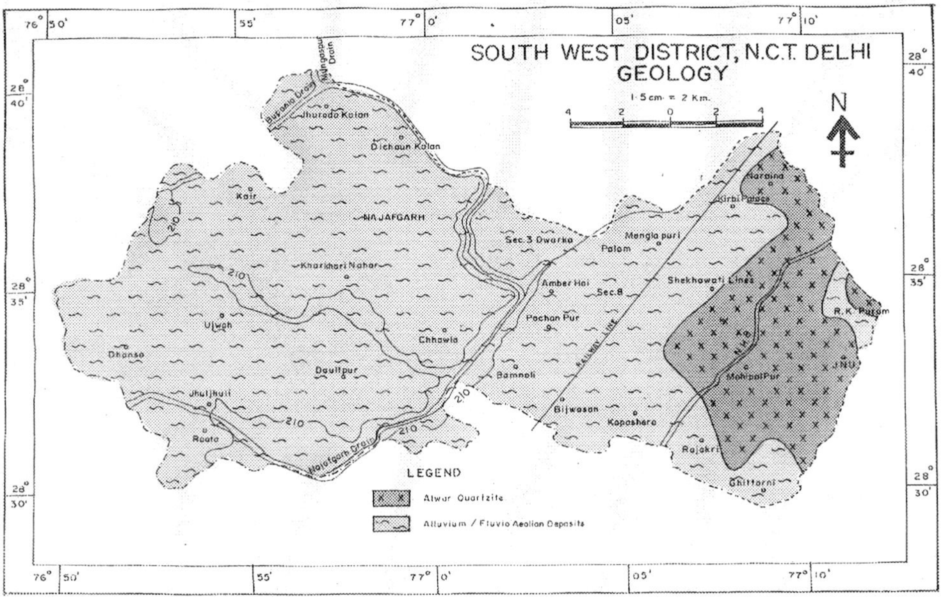

Fig. 2 Geology of South West District, Delhi.

2 FLUORIDE CONCENTRATION IN THE GROUNDWATER OF THE DISTRICT

In order to assess the concentration of fluoride in the groundwater of the district random sampling throughout the district was taken up (sampling location shown on Fig. 3). More than 50% of the groundwater samples collected from the district showed concentration of fluoride above 1.5 mg/L. It has been generally observed that the older alluvium areas of the district underlain by saline waters have groundwater having concentration of fluoride beyond permissible limit. The concentration of fluoride in the groundwater of the district can be studied under the following subheadings.

2.1 The Hardrock Areas of the District

The hardrock areas of the district (Fig. 2) on eastern side shows fluoride concentration in groundwater (Fig. 3) mostly within permissible limit except three sampling locations of Naraina, Basant Gaon and University of Delhi, South Campus. The above mentioned localities are situated (Fig. 3) in the north-eastern part of the hardrock area. Amongst these three localities, except Basant Gaon which has electrical conductivity of 1360 μS/cm the other two localities have electrical conductivity greater 2000 μS/cm.

2.2 The Mid-eastern Part of District Closely Adjacent to the Hardrocks

The mid-eastern part of the South West district just adjacent to the hardrock areas consisting of localities like Kirbi Palace, Shekhawati Lines, Bijwasan and Samalkha (Fig. 3) have shown concentration of fluoride within permissible limits. The regions are also supposed to have

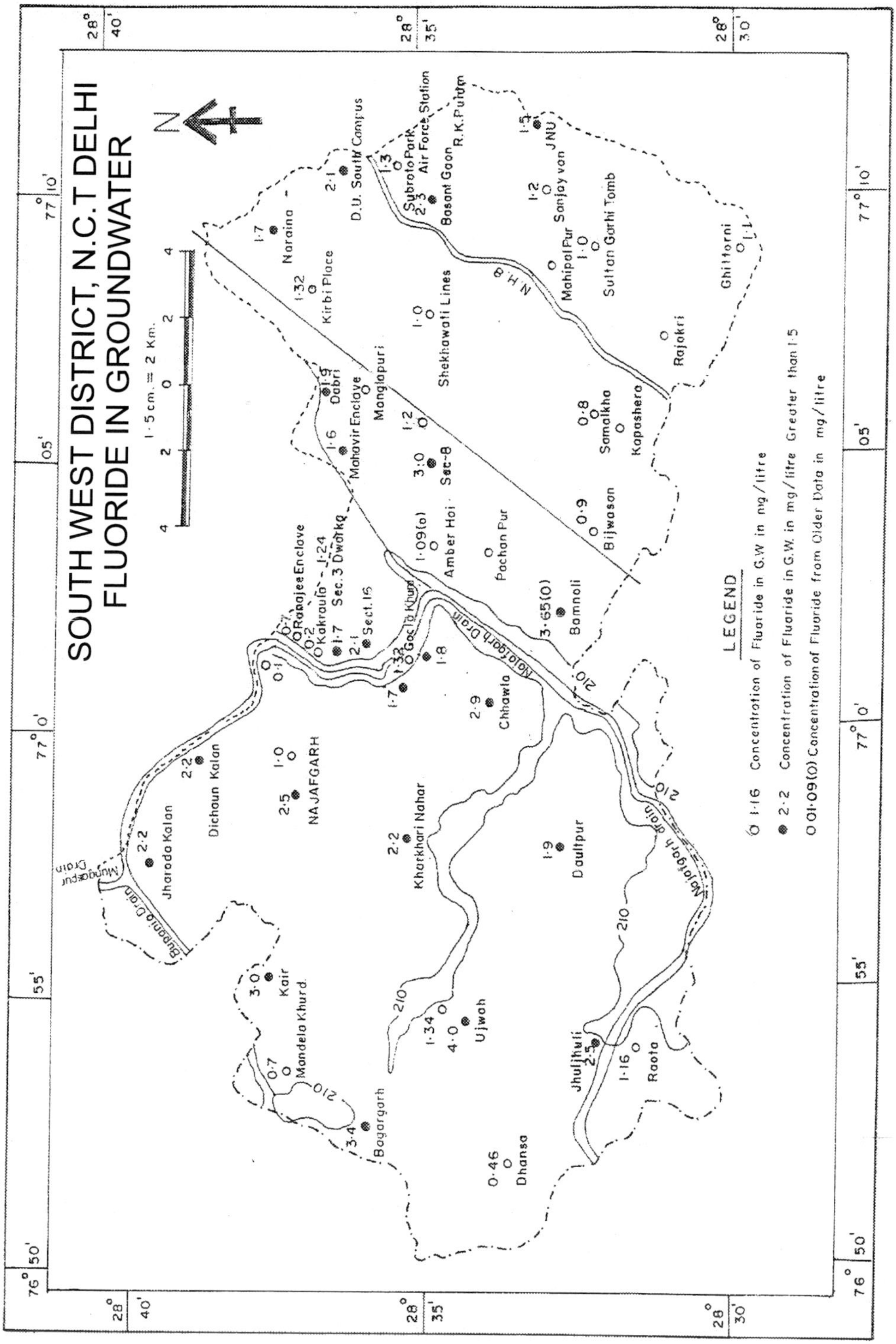

Fig. 3 Map showing fluoride concentrations in groundwater of South West District, Delhi.

considerable thickness of freshwater. The electrical conductivity of groundwater from the comparatively deeper aquifers in the locality ranges between 800 and 1900 micro siemens/cm.

2.3 Dwarka and Surrounding Localities to East of the Najafgarh Drain

Just adjacent to the earlier described mid-eastern part of the district and to the east of the Najafgarh drain, lies the sub-city of Dwarka and the surrounding localities of Manglapuri, Amberhai, Bamnoli, etc. (Fig. 3). The area, in general, has isolated patches showing concentration of fluoride beyond permissible limit. The samples from Mahavir Enclave and Dabri locality have shown concentration of fluoride beyond permissible limit. The hand pump from Sector 8 of Dwarka sub-city has shown concentration of fluoride beyond permissible limit (Fig. 3). In the earlier sampling done by CGWB, the Bamnoli village samples have shown fluoride concentration beyond permissible limit. The two hand pump samples from Sector 16 of Dwarka near Kakrola adjacent to the Najafgarh drain have shown fluoride concentration beyond permissible limit (Fig. 3). The rest of the areas in the above mentioned region of the district have fluoride concentration within the permissible limit.

2.4 Rural Part of South West District to West of the Najafgarh Drain

In the area to the west of Najafgarh drain consisting predominantly of saline waters, the sampling points of groundwater samples have mostly shown fluoride concentration beyond permissible limit except a few localities. The samples taken from near Ojwah piezometer on the fringes of Ojwah depression have shown fluoride below permissible limit and the sample from Goela Khurd area near to the drain have shown fluoride below permissible limit. The sampling points here showing lower concentration of fluoride also have the electrical conductivity of the groundwater on the lower side.

3 REASONS FOR THE OBSERVED PATTERN OF FLUORIDE CONCENTRATION IN GROUNDWATER OF THE DISTRICT

The pattern of fluoride concentration in the groundwater of the district has been reasoned under the subheadings below.

3.1 The Hardrock Areas of District

The occurrence of anomalous concentration of fluoride in random samples of groundwater of the hardrock area could be due to the presence of pegmatite veins in vicinity or en route to the paths traversed by groundwater. The pegmatite veins have fluoride rich minerals prone to leaching, thereby contributing excess fluoride in to groundwater. The same pegmatite vein when weathered to clay pockets could account for higher values of electrical conductivity in groundwater samples from these localities. The fractures occupied by the pegmatite veins should have high permeability when the clay minerals are washed out. When connected to surface water sources, they are supposed to provide conduit with greater permeability. Thus, the aspect of chance introduction of fluoride from anthropogenic sources is not ruled out in conjunction with the nitrates. This logic follows from the account that these hardrock sampling points also has higher concentration of the nitrates.

3.2 The Mid-eastern Part of District Closely Adjacent to the Hardrocks

The mid-eastern part of South West district just adjacent to the hardrock areas consisting of localities like Kirbi Palace, Shekhawati Lines, Bijwasan and Samalkha (Fig. 3) have shown concentration of fluoride within permissible limits. As mentioned in Shekhar *et al.* (2005) the regions covered here receives considerable surface and groundwater runoff from the adjacent hardrock areas. Besides they have predominance of comparatively coarser sediments and are traversed by a fault zone. Thus, this region has high permeability and steep hydraulic gradient transverse to the hardrock exposures, this leads to continuous flushing in the zone by fresh groundwater. Therefore, it has considerable thickness of the freshwater with a much diluted concentration of fluoride below permissible limit.

3.3 The Remaining Areas of District With Predominance of Older Alluvium Having Finer Sediments

The concentration of fluoride in groundwater of older alluvium in rest of the district has been mostly found to be above permissible limit, especially in the areas having saline waters in shallower aquifers. In perspective of Shekhar *et al.* (2005) views the higher salinity in groundwater of the older alluvium can be accounted for by the fact that topography is such that maximum surface runoff is away from these regions, the localities are away from major recharge sources like depressions, drains and the subsurface formation is predominantly finer sediments. On account of very low permeability there is greater residence time for groundwater. The greater residence time of groundwater leads to dissolution of salts from soil formation and enrichment of fluoride in the dissolved salts takes place by geochemical changes. In general, it has been observed that fluoride contamination is predominant in groundwater with higher salinity.

4 SUMMARY AND CONCLUSIONS

The groundwater in South West district of NCT Delhi has considerable fluoride contamination. More than nearly 50% of the groundwater samples collected randomly in the district showed fluoride contamination beyond permissible limit. The hard rock areas of the district in general have fluoride concentration below permissible limit except a few localities. The anomalous concentration of fluoride in hard rock areas could be due to presence of pegmatite veins or from anthropogenic sources through the paths formed by weathered pegmatite veins. In perspective of the views put forward in Shekhar *et al.* (2005) for the factors controlling depth to fresh/saline water interface in groundwater of the district, the comparatively lower concentration of fluoride in mid-eastern part of the district has been accounted for on account of considerable surface runoff from the adjacent hard rock areas, predominance of coarser sediments, higher permeability along the fault zone and steep hydraulic gradient transverse to the hard rock areas in the sediments fringing the hardrock area. The high permeability and steep transverse hydraulic gradient in the area leads to continuous flushing. This produces considerable thickness of freshwater with diluted concentration of fluoride. The higher salinity in older alluvial areas has been thought of on account of greater retention of groundwater in the formations because of low permeability. The greater residence time of groundwater leads to dissolution of salts from formation by chemical reaction, and enrichment of fluoride in the dissolved salts take place. There is further scope of work for ascertaining the nature and dynamics of prevailing geochemistry in saline water areas. Moreover, the nature of chemical reaction in area specific perspective leading to enrichment of fluoride in groundwater is yet to be understood in totality, in this particular type area.

References

Basic Data Reports (BDR) of the boreholes drilled by Central Ground Water Board (CGWB) in South West District of NCT Delhi., India.

Shekhar, S. (2004) Integrated groundwater management study of South West district with special reference to fresh/saline water interface, Report, Central Ground Water Board, Delhi, Ministry of Water Resources.

Shekhar, S., Singh, S.B. & Romani, S. (2005) The controls to the variation in depth to fresh/saline interface in the groundwater of South West District, NCT, Delhi – a case study. Jour. Geol. Soc. India, 66: 17–20.

39 Newly Developed Absorbent for Arsenic Removal from Groundwater

S.A. Deowan, J. Hoinkis[1] and S. Evers[2]

[1]University of Applied Sciences Karlsruhe Moltkestr. 30, 76133, Karlsruhe, Germany

[2]EVERS e.k. , Water Technology – Filter Materials, Rheiner Str. 14a , 48496 Hopsten, Germany

ABSTRACT

This topic represents the appropriateness of a reliable, simple and cost-effective technology for removal of arsenic from groundwater. Dangerously high levels of naturally occurring arsenic, a colourless, odourless, tasteless poison, have been found in underground water wells, threatening thousands of people to death. A promising technology was looked for to face this naturally occurring maniac. The newly developed adsorbent focused on elimination of arsenic from drinking water. This technology has been justified in lab scale unit under different operating conditions. The results found from different experimental setups have been compared with the results of commercially available adsorbents. One part of this work also discusses about the disposal criteria of waste arsenic residue. For this purpose all the activities were supported by company EVERS (Water Technology - Filter Materials). Based on the lab oriented experimental results the appropriateness of this technology was verified at the end depending on the quality of arsenic spiked water.

1 INTRODUCTION

1.1 Overview

For community water systems, the US Environmental Protection Agency (EPA) has promulgated a more stringent drinking water standard for arsenic from 50 µg/L to 10 µg/L. Some of the most promising technologies for selective arsenic (As) removal from groundwater supplies are adsorption processes using metal oxide media such as activated alumina and ferric hydroxide. EVERZIT®As offered by company EVERS is one kind of adsorption media which can be applied for arsenic rejection from drinking water. This study will focus on the performance of this adsorption media in different experiments and will compare these values with the other available adsorption media for arsenic rejection.

1.2 Adsorbents Developed by Company EVERS

The activities of this paper are accelerated testing methods those can be used to determine adsorptive characteristics by measuring the performance of small diameter media at higher hydraulic loading rates. The main motivation for this work was driven by company EVERS. We have good cooperation with this company. This company provided all the financial and mental support for these works. This company provides a new adsorbent called EVERZIT® As to find out the quality and performance of this adsorbent for arsenic elimination from drinking water.

So, EVERZIT® As was the main adsorptive media for this work and Bayoxide E33 (Kennedy B, 2005) developed by company BAYER had been kept as a reference.

1.2.1 Material Properties and Technical Data

Some of the important properties and technical data for Bayoxide E33 and EVERZIT® As are mentioned in Table 1 below.

Table 1 Properties and technical data of the used adsorbents (EVERS, 2005).

Properties	*Bayoxide E33*	*EVERZIT® As*
Chemical designation	Synthetic iron oxide hydroxide	Iron hydroxide
Formula	α-FeOOH	$Fe(OH)_3$
Granule size	App. 0.5-2 mm	App. 0.5-2 mm and 2-4 mm
State/ shape	Granular	Granular
Colour	Red brown	Red brown
Density	App. 0.5 g/ml	App. 0.625 ± 0.5 g/ml (0.5-2 mm) App. 0.600 ± 0.5 g/ml (2-4 mm)
Specific surface area	App. 300 m^2/g	300 m^2/g
Porosity of grains	App. 70-80 %	Min. 70%
Back flush	NA	Once in a week.
Expected adsorbing capacity	NA	7-9 g As/kg
Standard packaging unit	500 kg bulk bag	NA

In this work EVERZIT® As has different vesions like

- EVERZIT® As 2-4 mm
- EVERZIT® As 0.5-2 mm
- EVERZIT® As 0-5-2 mm (grinded).

EVERZIT® As 0.5-2 mm (grinded) to the size of 1 mm was derived from EVERZIT® As 0.5-2 mm in lab in this study by a sieving process since this size brings better result.

1.2.2 Selection of Proper Adsorbent from Different EVERZIT®As

We did some screening column experiments at pH 7, temperature 20 ± 2°C, flow rate 1 L/h (0.6 m/h), spiked Karlsruhe tap water[1], 500 μg/L as feed with different types of EVERZIT® As and we realised that EVERZIT® As 0.5-2 mm and EVERZIT® As 0.5-2 mm (grinded) were suitable for the application of arsenic rejection for this study and Bayoxide E33 was taken as reference in all respects.

2 ANALYTICAL METHOD FOR ARSENIC DETECTION

Detection of exact level of arsenic in drinking water is very important since the recovery is dependent on this value. For this case study we followed an analytical method called DIN 38405 D 18 (German Standard). This is a low operating and maintenance cost detection method compared to USEPA approved methods. With this method dissolved As(V) and As(III) is transformed into arsine which is stripped into a silverdiethyldithiocarbamate solution forming a coloured complex of arsine which can be analysed by using spectrophotometry.The standard deviation of this method had been observed in our experiments as ± 1 μg/L.

3 LAB SCALE EXPERIMENT

3.1 Experimental Procedure

In order to determine the performance of the adsorbents lot of experiments were done. The main part of the experiment was batch experiment and column experiment.

3.1.1 Batch Experiment

From different literature studies we know that batch experiment is performed to reach the thermodynamic equilibrium because the maximum adsorption is attained under this condition. After this adsorption process the obtained data are fitted with some isotherm models like Langmuir and Freundlich to find out the fitness of the experiment and adsorbing capacity.

In the batch experiment, As(III) removal was studied at the normal pH 7 of Karlsruhe tap water. In this case 100 ml arsenic spiked feed water was transferred to a 500 mL Erlenmeyer flask containing 1 g of adsorbent and the flask was filmed with parafilm. The sample was placed on a mechanical shaker and shaken at 32 RPM for 24 hours. All the experiments were done at the room temperature of 20 ± 2°C. At a time 3 Erlenmeyer flasks were used containing the adsorbents. The initial feed concentrations of these experiments were: 20 mg/L, 50 mg/L, 100 mg/L, 150 mg/L, 200 mg/L, 250 mg/L. After 24 hours of running time the samples were collected. The fine particles of the samples were removed by a filtration process with 0.45 μm filter and analysed for equilibrium concentration of arsenic.

3.1.1.1 Data Analysis

The data obtained from the batch experiment were used to analyse the adsorption isotherms in order to estimate the constants and adsorption maxima. The experimental results were fitted to both the Langmuir and Freundlich isotherms.

3.1.2 *Column Experiment*

Four types of column filtration systems as mentioned below using different adsorbents were developed in our lab:

- Non-Automized Column Filtration (NACF)
- Automized Column Filtration (ACF)
- Unique Column Filtration (UCF)
- Column Filtration With Integrated Flow meter (CFWIF)

These experimental setups were developed for different types of adsorbents to determine their adsorbing capacity under different operation conditions. Among these column filtration systems only automized column filtration (ACF) system is described here and the result is presented in the following section. Details of other filtration systems can be found from the publication in our department, University of Applied Sciences Karlsruhe (UASK), Germany (Deowan S.A, J. Hoinkis, 2005).

3.1.2.1 Automized Column Filtration (ACF)

We designed an automized column filtration system at the department of Mechatronics and Natural Science in our University of Applied Sciences, Karlsruhe. There are three columns containing different types of adsorbents as shown in the following Table 2 below.

Table 2 Different adsorbents in ACF.

No.	Content	Packed volume (L)	Amount (g)
1.	Bayoxide E33	0.087	40
2.	EVERZIT® As 0.5- 2 mm	0.087	50
3.	EVERZIT® As 0.5-2 mm (grinded)	0.087	50

Bayoxide E33 and other adsorbents were taken in different amount to equalize the bed volume. An automized column filtration system was developed with the adsorbents shown in the table above. For this purpose the feed tank containing 1 m^3 of arsenic spiked water with feed concentration of 500 μg/L was placed at the outside of the building and was connected with a small tube to the columns. From this main tube three channels were made for the three columns. Three small pumps were installed to supply water to each column and each pump was controlled by inductive sensor integrated with the system. Water level in each column was kept at the upper surface of the adsorbent to attain uniform water distribution. The column is shown in the Fig. 1 below.

Feed water flow rate through the adsorber was more or less constant (1 ± 0.L/h) and water could get uniform chance to be in touch with the adsorbent. The effluent coming out from each column was collected in three individual effluent tanks. The effluent level was measured everyday to determine the amount of water passing through each column. pH of feed water was around 7. To check arsenic level in the effluent, samples were collected and chemical analysis were done in every second day. This column filtration system was running continuously for whole day and night and even throughout the weekend. The operation conditions for all the columns were the same.

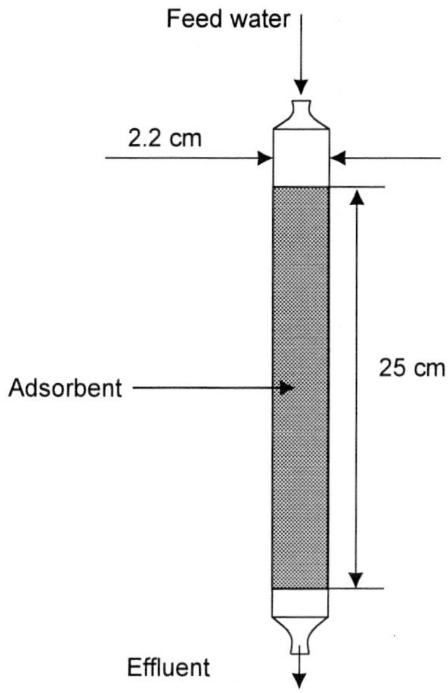

Fig. 1 Sample column of experimental setup of automized column filtration (ACF).

3.2 Experimental Result

Since a lot of experiments were done through column filtrations and batch experiments it is not possible to present all the results in this section. So the important results are shown in the following part. The rest of the results can be obtained from the department of our university.

3.2.1 Batch Experiment

In batch experiment different adsorbents were tested at the same time to find out their adsorbing capacity. To observe if the arsenic adsorption data obtained experimentally, using a wide range of adsorbate concentrations fit the Langmuir and Freundlich isotherms properly or not, we performed some experiments. To perform fitting we transferred the Langmuir and Freundlich isotherms into linear equations (Deowan S.A, J. Hoinkis, 2005). Some graphs were developed for estimation of Langmuir and Freundlich isotherms parameters and these parameters are given in Table 3 below to compare with the results of other available adsorbents.

A larger value of K and Q_0 indicate a larger adsorption capacity. A smaller value of b and 1/n indicate a stronger adsorption bond. Based on the results given in Table 3 regarding to Langmuir isotherm parameters we can observe that Bayoxide E33 has the largest capacity, than those of other adsorbents. For Freundlich isotherm parameters Bayoxide E33 has also the higher adsorbing capacity than the rest of the adsorbents and the following is EVERZIT® As 0.5-2 mm (grinded) and the rest.

3.2.2 Column Experiment

Most of the experiments were done in column filtration under different operation conditions. The results obtained from ACF is described in this section. Automized Column Filtration was

Table 3 Langmuir and Freundlich isotherm parameters in batch experiment.

Adsorbents	*Langmuir isotherm*		*Freundlich isotherm*	
	Q_o (mg/g)	*b (l/mg)*	*K (l/mg)*	*1/n*
Bayoxide E33	238	2.4	138	0.51
EVERZIT® As 0.5-2 mm (Grinded)	163	1.2	69	0.54
EVERZIT® As 0.5-2 mm	160	0.7	50	0.59

performed for As(III) and As(V) and the chemical analysis were done for determination of arsenic level in the effluent. Performance of different adsorbents for As(III) and As(V) are described below.

For As(III)

For As(III) ACF was running continuously until arsenic level in different adsorbents exceeded the maximum contaminant level (MCL) at different bed volume. The operation condition was influent 500 μg/L, temperature 20 ± 2° C and pH 7. The result of this filtration system can be shown in the following Fig. 2.

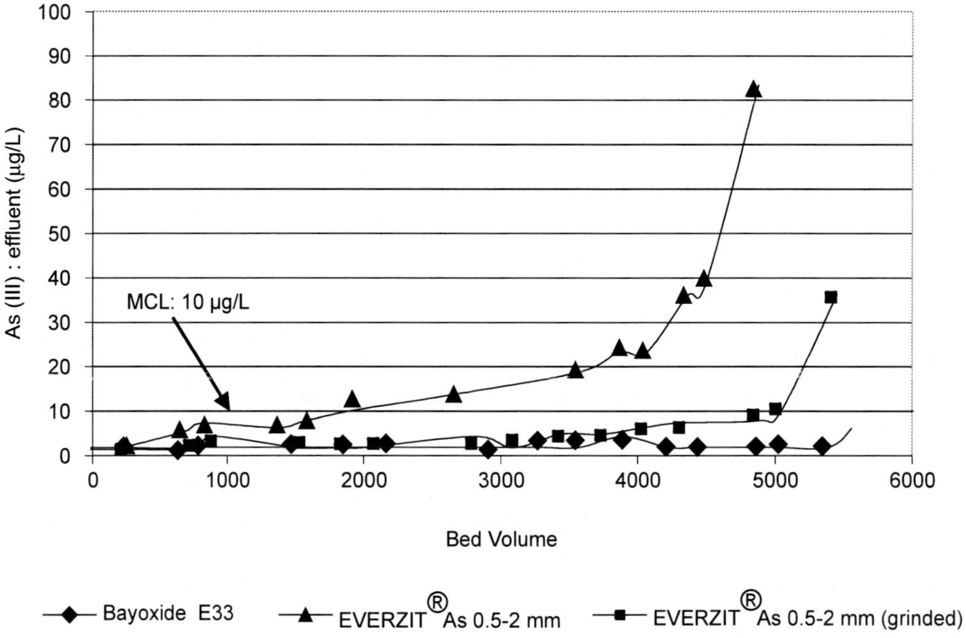

Fig. 2 Performance of different adsorbents for As(III) in ACF.

From the Fig. 2 above we can observe that EVERZIT® As 0.5-2 mm has bed volume of around 2700 with arsenic level below 10 μg/L where grinded version of this adsorbent brings this value upto 5000 μg/L but Bayoxide E33 can be continued further following the permissible limit.

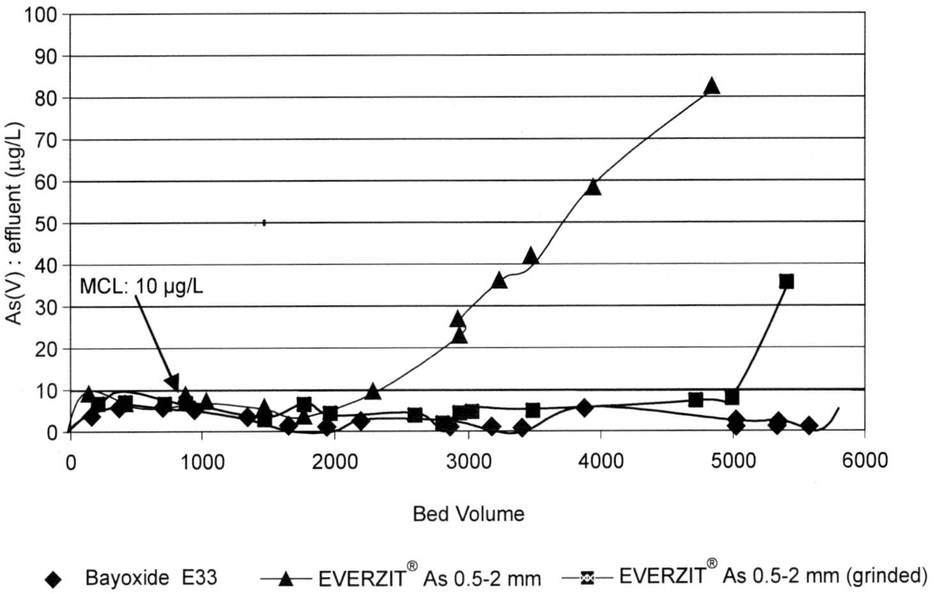

Fig. 3 Performance of different adsorbents for As(V) in ACF.

For As(V)

Under the same operating conditions the whole experiment was repeated for As(V) and the result can be shown in the Fig. 3 below.

The results obtained from Fig. 3 indicate more or less the similar results found from As(III). This means that the adsorbents we applied have the same adsorbing capability for As(III) and As(V).

3.3 Comparison of Results of Column Experiments

The results obtained from different gravity flow column filtration experiments can be compared with the results found from different publications (Fig. 4). There are many parameters to compare the results. For all the experimental results temperature 20 ± 2°C and pH 7 had been taken as operating condition. From the column of equivalent bed volume (BV) we can distinguish that Bayoxide E33 has the highest BV among all the adsorbents and different versions of EVERZIT® As has similar or even better BV than GFH and much more better than AA. The saturation bed volume was calculated based on the linearly normalized concentration of 21 µg/L, the lowest concentration among all experiments.

3.3.1 Based on Bed Volume

From the Fig. 4 it is clear that Bayoxide E33 has highest bed volume and the following is the EVERZIT® As (grinded). This information indicates that EVERZIT® As is comparable with the commercially available adsorbents.

3.3.2 Based on Adsorbing Capacity

From Fig. 4, we got the idea of the adsorbing capacity of different adsorbents based on saturated bed volume. Figure 5 brings the numerical values of their adsorbing capacity in mg

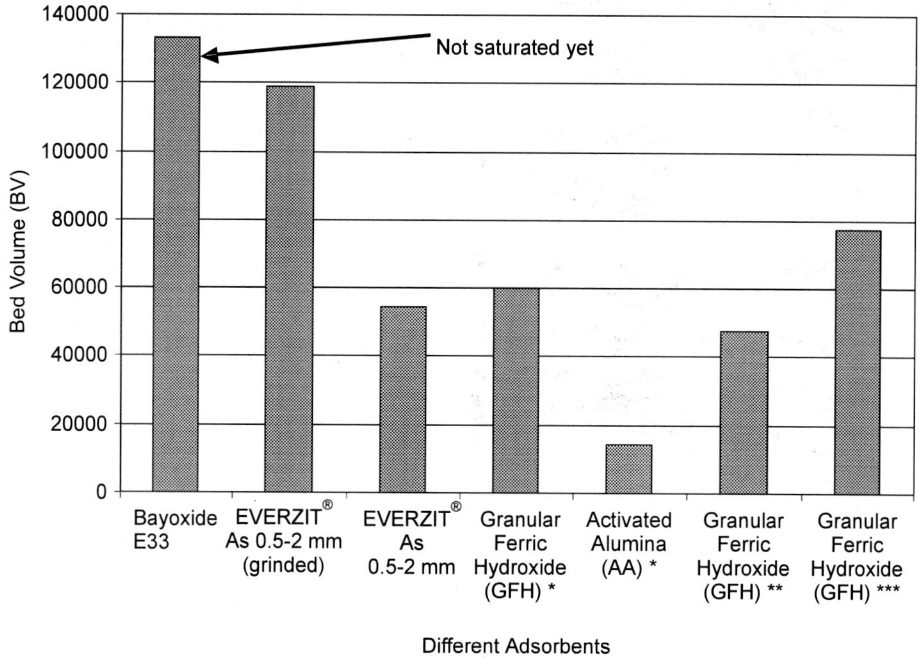

Fig. 4 Comparison of adsorbents based on Bed Volume (*: Thirunavukkarasu, OS, *et al.*, 2003, **: Pal, B.N., ***: IPEC, 2002).

arsenic adsorbed per g of adsorbent and these values have been compared with other results obtained from different publications following the similar column filtration systems.

From Fig. 5 it is clear that Bayoxide E33 has higher adsorbing capacity in every type of experiment followed by EVERZIT® As 0.5-2 mm (grinded) and after that EVERZIT® As 0.5-2 mm. These values specially obtained by Bayoxide E33 and EVERZIT® As 0.5-2 mm (grinded) are quite high compared to the values shown in the figure below. From this table it is also clear that the adsorbing capacity of EVERZIT® As 0.5-2 mm seems to be in the range of the available adsorbents like granular ferric hydroxide (GFH) and activated alumina (AA) and can be applied in the practical field.

4 TOXICITY CHARACTERISTIC LEACHING PROCEDURE (TCLP)

The handling and disposal of arsenic-laden residuals may be a problem because various handling and disposal methods may release arsenic back to the environment. Because arsenic removal is sensitive to both the pH of precipitation and the oxidation state, any process that changes pH or results in a reducing environment may release arsenic from the solid phase. These processes, including chemical conditioning during dewatering, storage and lagooning, and ultimate disposal options such as landfilling, land application, discharge to sewer, and coagulant recycle, may all contribute arsenic back to the environment (Michael J. MacPheeGail E. Charles David A. Cornwell, 2001).

Toxicity Characteristic Leaching Procedure (TCLP) is an internationally accepted standard procedure to evaluate the concentration of the contaminant in the waste and determine if it will

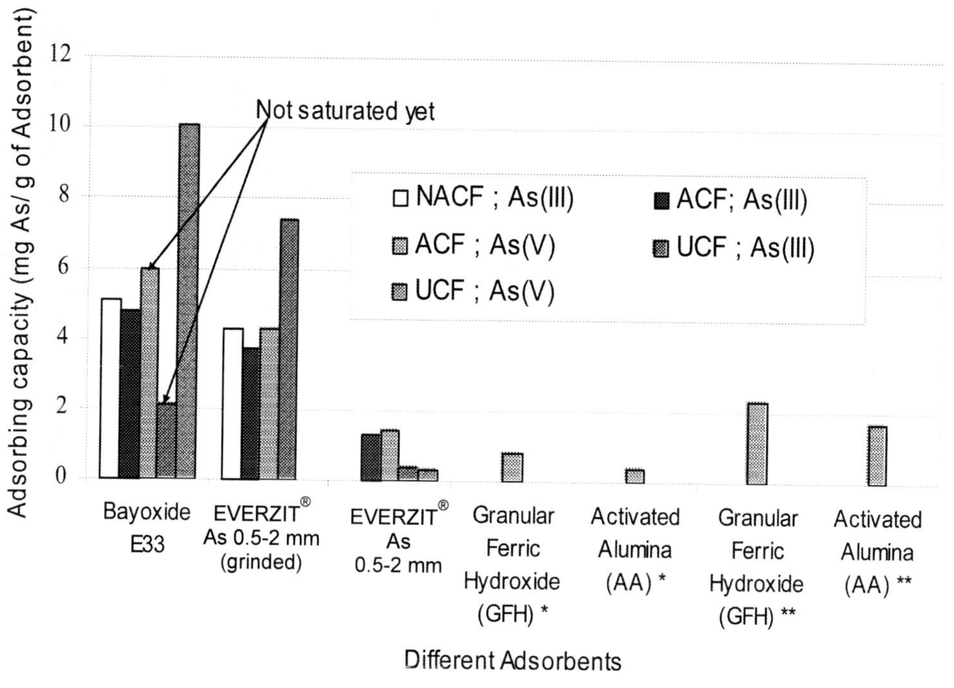

Fig. 5 Adsorbing capacity of different adsorbents (*: Thirunavukkarasu, OS. *et al.*, 2003, **: New England Water Treatment Center, 2004).

satisfy the leachate criteria. Except TCLP there are also some other standard procedures for different countries. The TCLP analysis simulates landfill conditions. Over time, water and other liquids percolate through landfills. The percolating liquid often reacts with the solid waste in the landfill, and may pose public and environmental health risks because of the contaminants it absorbs. The TCLP analysis determines which of the contaminants identified by the United States Environmental Protection Agency (USEPA) are present in the leachate and their concentrations (Nevada Small Business Development Center, 2003).

To check if the waste residue of EVERZIT® As leaches arsenic back to the environment or not we performed TCLP test following the operating conditions according to USEPA method 1311. 10 g of waste EVERZIT® As was mixed in 200 ml of distilled water in a 500 ml Erlenmeyer flask and pH value was adjusted to 2.88 adding acetic acid. After tightly closing the stopper of this flask, it was running for 18 h with 32 RPM in the same way as we did the batch experiment. The temperature of the flask was 20 ± 2 °C. After 18 h of operation time the mixture was filtrated with 0.45 μm micro filter to remove the turbidity. This filtrated water was diluted in some factors and the arsenic level was detected by chemical analysis. Based on the arsenic levels in the waste residue of different adsorbents according to the TCLP test mentioned above lot of data were found and one is shown in Fig. 6.

From Fig. 6 we find that leached arsenic concentration of Bayoxide E33, EVERZIT® As 0.5 – 2 mm (grinded), EVERZIT® As 0.5 – 2 mm were 130 μg/l, 450 μg/L and 150 μg/L respectively. We also noticed that there is no big difference between As(III) and As(V) in this case.

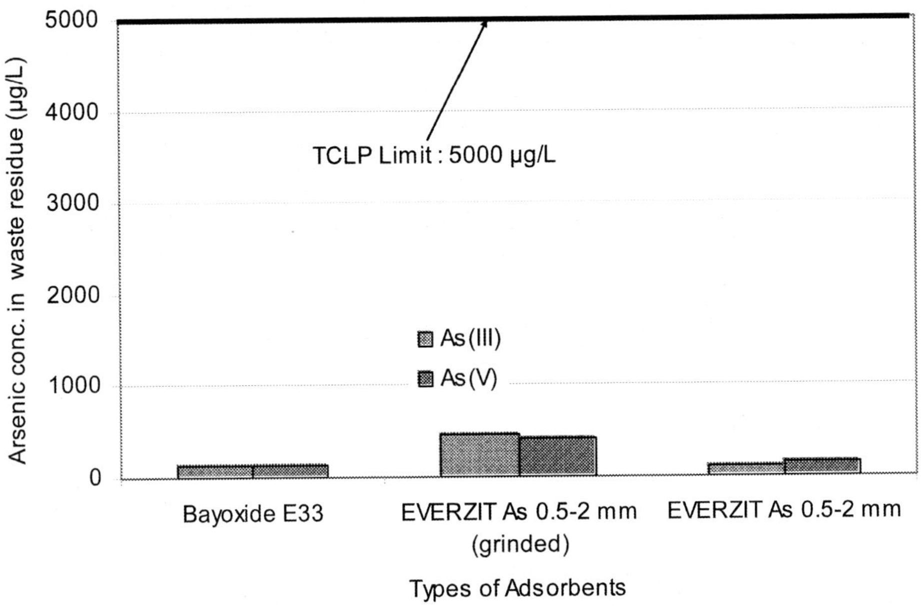

Fig. 6 TCLP test result of ACF.

According to the USEPA, if the extract from a representative waste sample contains any of the listed contaminants at a concentration equal to or greater than the respective value, the waste itself is considered to be toxic. For arsenic this concentration is 5000 µg/L (USEPA Method 1311, 1992). Since the arsenic concentration in the TCLP extract of the solid samples range between 50 – 450 µg/L, it is apparent that none of the solid sludge of EVERZIT® As and Bayoxide E33 collected from different column filtration experiments exhibits toxicity as per the definition of the USEPA. So, no hazardous leachate was identified from the sludge produced from these work. Therefore, the solid sludge of EVERZIT® As and Bayoxide E33 appear to be safe and not to contribute to recontamination of the environment.

5 SUMMARY AND OUTLOOK

After verification of the results obtained through this study based on adsorbing media, we can summarize the following points.

Bayoxide E33 has an extraordinary quality as an adsorbent for elimination of As(III) and As(V) under any operating conditions.

After analysing the results of different column filtrations and comparing these results with those of different publications (Deowan and Hoinkis 2005; Evers 2005) we have come to the point that different version of EVERZIT® As 0.5-2 mm has nearly the same adsorbing capacity like other available acceptable field-used adsorbents like Granular Ferric Hydroxide (GFH), Activated Alumina (AA), and so on.

Modified version of EVERZIT® As 0.5-2 mm i.e. EVERZIT® As 0.5-2 mm (grinded) has an elevated adsorbing capacity. May be the grinding process has increased its specific surface area and its adsorbing capacity has been lifted. This prediction should be further investigated.

After comparing the results of different types of EVERZIT® As 0.5-2 mm with the other results of different publications and with the fitted curves of Langmuir and Freundlich isotherm we found that these parameters are in the range of the available adsorbents like Granular Ferric Hydroxide (GFH), Activated Alumina (AA), etc. So we can point out that EVERZIT® As 0.5-2 mm has the potentiality to apply in the practical field for elimination of arsenic to comply with new MCL like other ordinary adsorbents.

Company EVERS has developed a simple filter system called EVERS WATER WONDER (EWW). By this filter filtration takes place by means of sintered active carbon and microfiltration membrane. This filter system shall be modified with adsorbent to apply for arsenic rejection from drinking water. There is another new adsorbent called ADSORBSIA GTO, a Titanium based adsorption media developed by company DOW (Greenberg A, 2005). The technologies are kept open for investigation in near future to find out their potentiality.

References

Deowan S.A., J. Hoinkis (2005), Elimination of Arsenic Applying Membrane Technology and Adsorption Media. Master Thesis : Karlsruhe University of Applied Sciences, Germany.

EVERS (2005) We repair Water -Water Technology from EVERS URL: http://www.evers.de/lieferprogramm/produkt.php?pid=19&PHPSESSID=502a21023958b9369803582b764a5b44 (Accessed on Feb. 15, 2005) //www.dow.com/liquidseps/pdf/ADSORBSIA_Tech_Presentatopm.pdf (Accessed on Nov. 8, 2005) IPEC (2002), URL: http//www.ipec.utulsa.edu/Ipec/Conf2002/aragon_thomson_chwirka_50.pdf (Accessed on Aug. 15, 2004).

Knnedy B. (2005) Severn Trent Services URL: http://www.severntrentservices.com/press_room/2005_pr/pdf/SacramentoArsenic.pdf (Accessed on Nov.,11, 2005).

Michael J. MacPheeGail E. CharlesDavid A. Cornwell (2001) Treatment of Arsenic Residuals from Drinking Water Removal Processes URL http://www.epa.gov/ORD/NRMRL/pubs/600r01033/600R01033chap1.pdf (Accessed on Nov. 10, 2005).

New England Water Treatment Technology Assistance Center (2004) Evaluating Arsenic Removal Adsorbents URL: http://www.unh.edu/erg/wttac/Project_Summaries/, B.N : Granular Ferric Hydroxide for Elimination of Arsenic from Drinking Water. URL: http://www.unu.edu/env/Arsenic/Pal.pdf (Accessed on Jan. 10, 2005).

Thirunavukkarasu, OS, Viraraghavan, T. and Subramanian, K.S. (2003) Arsenic Removal from drinking Water using Granular Ferric Hydroxide. URL: http://www.wrc.org.za/archives/watersa%20archive/2003/april/7.pdf (Accessed on Jan. 15, 2005).

USEPA Method 1311 (1992) Toxicity Characteristic Leaching Procedure URL: http://www.epa.gov/epaoswer/hazwaste/test/pdfs/1311.pdf (Accessed on Nov. 10, 2005).

40

Arsenic Linkage in the Irrigation Water-Soil-Rice Plant Systems

M.R. Islam*, M. Jahiruddin and S. Islam
Department of Soil Science, Bangladesh Agricultural University, Mymensingh-2202, Bangladesh

ABSTRACT

This investigation reports the levels of As in irrigation waters, soils and Boro (dry season) rice grain from 100 shallow tube well (STW) over the sadar upazila of Chapai Nawabganj, Bangladesh. The STW water As concentration ranged from 0.015 to 0.352 µg/ml with a mean of 0.075 µg/ml, The levels of total As in soils over the locations ranged from 5.8 to 17.7 µg/g with a mean of 11.2 µg/g. The rice grain–As concentration was in the range of 0.24 to 1.30 $\mu g\ g^{-1}$ having a mean of 0.76 µg/g. Thirteen per cent of the grain samples had As level <0.5 µg/g, 39% in the range of 0.50–0.75 µg/g, 36% 0.75–1.0 µg/g and the rest 12% more than 1.0 µg/g. Ninety nine per cent of the irrigated rice in Chapai Nawabganj lead to an intake of more than 100% maximum tolerable daily intake (MTDI) for an adult.of 60 Kg body weight.

1 INTRODUCTION

Bangladesh is currently facing the challenge of high arsenic concentration in shallow aquifers (Nickson *et al.*, 1998). A large number of hand pumps and shallow tube wells (STW) in some of the localities of 59 districts out of 64 districts have been identified to have As concentration above the Bangladesh standard of 0.05 As mg/l (Alam *et al.*, 2002). More than 80% of the population depends on agriculture for their livelihood. The agricultural sector employs about 90% of the rural males of the country (WDI, 1998). To become self-sufficient in food grain, the Bangladesh Government encourages production of high yielding varieties of rice which requires a large volume of irrigation water. The use of groundwater for irrigation has increased markedly over the last three decades. About 86% of the total groundwater withdrawn is utilized in the agricultural sector (WRI, 2000). Irrigating the paddy fields with As contaminated groundwater may lead to result in increasing soil As status in Bangladesh (Meharg and Rahman, 2003). Irrigation with arsenic contaminated groundwater to rice increases its concentration, and eventually arsenic enters into food chains through crop uptake and poses long term risk to human health (Marin *et al.*, 1992, Abedin *et al.*, 2002a, Meharg and Rahman, 2003, Islam *et al.*, 2004 a). Most of the hand and STWs in the Chapai Nawabganj sadar upazila are contaminated with arsenic (Chowdhury *et al.*, 1999; Islam *et al.*, 2004a). But reports on arsenic in irrigation water-soil-plant systems are limited in Bangladesh. Therefore, the aim of this study is to assess the linkage of As in irrigation water-soil-rice plant systems and also to determine the contribution of arsenic ingestion to an adult through the rice grown in Chapai Nawabganj sadar upazila.

2 MATERIALS AND METHODS

One hundred geo-referenced samples for each of irrigation water, soil and rice grain were collected from 100 STW command areas of Chapai Nawabganj sadar upazila, Bangladesh during dry season of 2003. Chapai Nawabganj is located on 24° 25′ N latitude and 88° 58′ and 88° 38′ E longitude. Water samples were collected from STW after 15 minutes of operation. Ninety millilitre of water sample was filtered through 0.45mm membrane filter and kept in acid washed plastic bottles containing 10 ml of 2M HCl (ANALAR grade) to arrest the microbial growth and to prevent precipitation of metals. The water samples were brought to the laboratory and kept in the refrigerator until analysis by HG-AAS (Unicam model 969).

Five soil samples (0–15 cm depth) were randomly collected from the adjacent field of STW after 15–20 days of transplanting of rice. During the sampling time, there was 2-3 cm water in the rice fields. Bamboo sticks were placed on the sampling site to identify the location for subsequent rice grain collection at harvesting stage. The soil samples were air dried, ground and passed through a 2-mm sieve, and finally stored in plastic pots.

Ten rice plants at maturity of the crop were collected from the rice field from the soil sampling sites. Rice grains were separated from the panicles by hand. The grain samples were stored in the polythene bags. Grains were dried for 72 h at 65°C in oven. Husks were removed from the rice grains by grinding using a ceramic mortar. The samples were finely ground and stored for chemical analysis. For concentrations of total arsenic in rice grains, milled sub-samples (0.5 g) were weighed into glass digestion tubes, steeped in 5.0 mL of nitric acid and allowed to pre-digest/stand overnight at room temperature. The samples were digested at 120°C for 2 hours and 2.5 mL H_2O_2 was added and digested until it became clear. In case of soil samples, a sub-sample of 1.0 g was taken and followed the same procedure to that of rice grain. Total As in the digest was determined followed by flow injection hydride generation atomic absorption spectrophotometer (HG-AAS) with Unicam model 969 and MHS-10 hydride generator assembly using matrix-matching standard (Welsch *et al.*, 1990).

3 RESULTS AND DISCUSSION

3.1 Arsenic in STW Water

The As levels in STW water showed a wide spatial variation in samples from different STW command areas (Table 1). The water As concentration ranged from 0.015–0.352 μg ml with a

Table 1 Arsenic concentration in water, soil, rice grain and straw samples (n = 100).

Parameters	*Range*	*Mean*	*S.D*
Water As (μg g^{-1})	0.015– 0.352	0.075	0.061
Soil As (μg g^{-1})	5.8– 17.7	11.2	2.6
Grain As (μg g^{-1})	0.24– 1.30	0.76	0.23
Straw As (μg g^{-1})	1.48– 17.60	5.88	3.4

mean of 0.075 μg/ml. Forty-four per cent of the samples had As level below 0.050 μg/ml, 33% of the samples had As level within 0.051– 0.100 μg/ml, 15% samples 0.101–0.200 μg/ml, 2% 0.201–0.300 μg/ml and the rest 6% had As level >0.3 μg/ml (Fig. 1). As in STW water was lower in shallow well depth, reaching to a maximum at 25 m depth and then decreased with increase in depth of the wells (Fig. 2). It appears that increasing the depth of STW may partially overcome the inflow of As through the irrigation water into the rice fields.

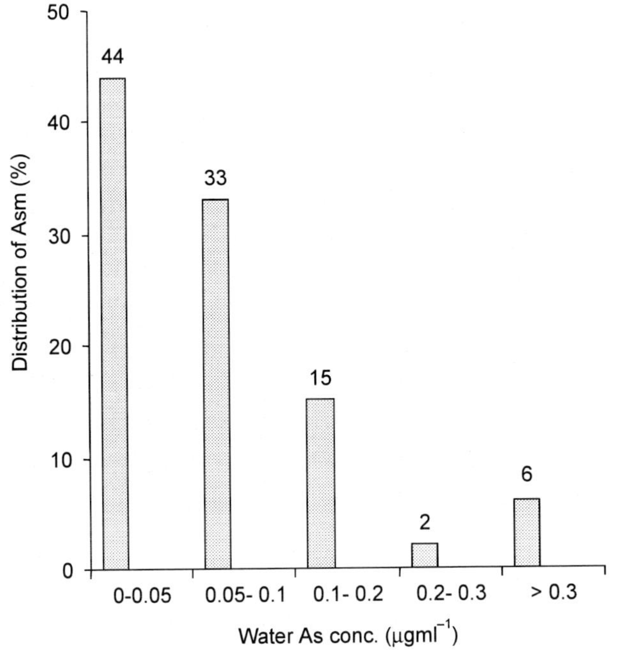

Fig. 1 Frequency distribution of water As (n = 100).

3.2 Arsenic Levels in Soils

There was a wide variation in total As concentrations in the soil samples from different STW command areas of Chapai Nawabganj upazila. The levels of total As in soils over the locations ranged from 5.8 –17.7 μg/g with a mean of 11.2 μg/g. Of the 100 samples, 32% of the soils had total As from 5.1 – 10.0 μg/g, 63% soils 10.1 – 15.0 μg/g, and the rest 5% soils 15.1–20.0 μg/g.

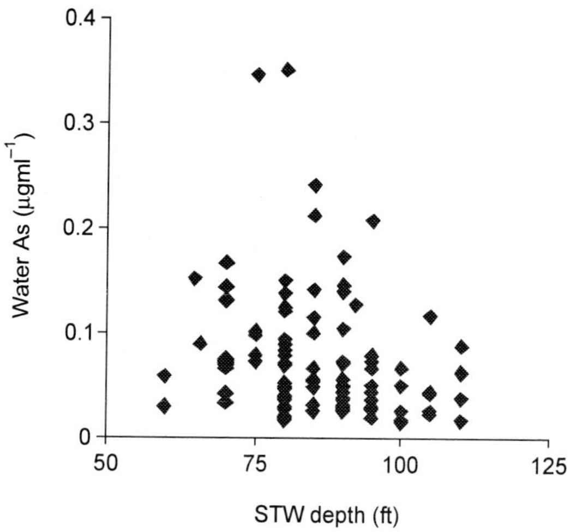

Fig. 2 Relationship between STW depth and water As.

Jahiruddin *et al.* (2000) studied the As levels of 20 calcareous and 20 non-calcareous soils of Bangladesh and reported that the mean As level of calcareous soils was 17 μg/g while for non-calcareous soils was 7.65 μg/g. Total As content in soils was positively correlated with irrigation water As (Fig. 3) indicating a possibility of build up of arsenic in soils due to irrigation with As contaminated irrigation water. Cultivation of dry season rice in Bangladesh requires about 1000 mm of irrigation water for 1 ha land (Meharg and Rahman, 2003). Irrigating the fields with water of 0.075 μg As ml^{-1} would add 0.75 kg As ha^{-1}, the potential level of As would increase by 0.375 μg/g per annum up to 15 cm soil depth. But there are a number of ways arsenic is lost from the soils. The average yields of Boro rice grain and straw in

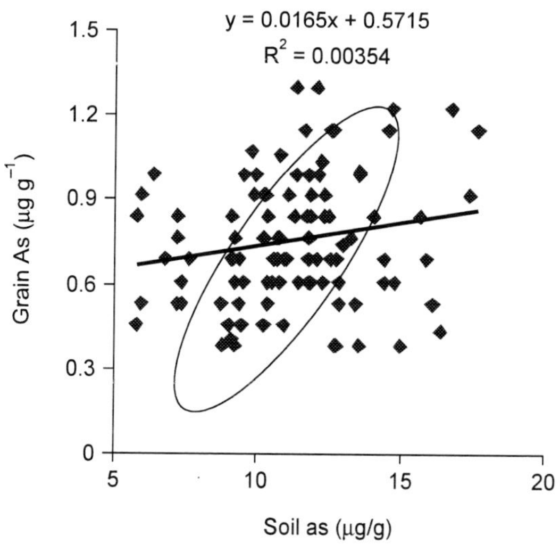

Fig. 3 Relationship between water As and soil As.

Bangladesh are 3.2 and 4.0 t ha^{-1}, respectively (BBS, 2003). The mean As concentrations of As in Boro rice grain and straw in this study are 0.759 and 5.88 µg/g, respectively. The amount of As removed by the boro rice crop would be 0.026 kg ha^{-1}. The rest 0.724 kg ha^{-1} will be in the rice fields. In addition, the use of phosphatic fertilizers, manures and pesticides, and sediment deposition during the flooding season can also increase the As content in surface soils. Data are not available about specific contribution of the above processes in the study area. When irrigation water comes in contact with air, the soluble ferrous iron is rapidly oxidised to ferric form precipitating arsenate present in water (unpublished data). Therefore, the level of arsenic in soils near the STW will be higher compared to those fields at far end of the command area (Jung *et al.*, 2002, Keegan *et al.*, 2002). During monsoon season, a good amount of arsenic may be lost from the soil with the horizontal flow of water and also through volatilisation as gaseous arsines under anaerobic condition by soil microbes (Cullen and Reimer, 1989; Turpeinen *et al.*, 1999). The percolation of irrigation water in paddy field is low, generally 1-2 cm a day (Brammer, 1996) due to the presence of a 3-4 cm hard plough plan common to rice fields of Bangladesh developed over the years due to continuous cultivation. Khan (2005) reported from a leaching study on four soils of Bangladesh that the amount of leaching loss through the undisturbed soil column ranged from 1-3% of the applied As through irrigation water. Besides, the upper 1-2 cm of a flooded paddy field is aerobic where As will be adsorbed by the soil particles. The As accumulated in the top layers may also be subject to losses through water erosion due to heavy monsoon rainfall. The cultivation of Aman (wet season) rice during the monsoon season will also uptake some of the As from the soil. The yield and concentration of As in T. aman rice is lower compared to that of Boro rice (Duxbury *et al.*, 2003; Islam *et al.*, 2004 b). Hence, the uptake of As by Aman rice will be lower compared to that of Boro rice. However, all these need to be taken into consideration before final quantification of building up of arsenic in paddy field soils. Build-up of As in paddy soils through irrigating with As-contaminated irrigation water was also reported by some of the authors (Roychowdhury *et al.*, 2002, Meharg and Rahman 2003, Ali *et al.*, 2003). The variation in soil As levels in different locations of Chapai Nawabganj may be due to variation of the amount of irrigation water having different levels of As used in different soils and also with the time period for irrigation with contaminated waters. Panaullah *et al.* (2003) reported that spatial variability for As contents was more prominent in case of soil compared to irrigation water and rice grain, however, it was not clear whether this variability of soil As depends on irrigation water or background As.

3.3 Arsenic Levels in Rice Grain

The grain As concentration varied widely ranging from 0.24 to 1.30 µg/g with a mean value of 0.76 µg/g (Table 1). One per cent of grain samples had As level below 0.25 µg/g, 12% samples within the range of 0.25-0.50 µg/g, 39% samples within the range 0.50-0.75 µg/g, 36% samples 0.75–1.0 µg/g, and the rest 12% had As levels beyond 1.0 µg/g (Fig. 4). Grain-As was poorly correlated with soil As and irrigation water As (Figs. 5 and 6) indicating that the amount of available As to plants is related to total soil and irrigation water As in a complex manner. The availability and uptake of As by plants may vary with the extent of As contamination of soil (Islam *et al.*, 2004a, Jahiruddin *et al.*, 2004, Duxbury *et al.*, 2003), the forms of As species (Abedin *et al.*, 2002a,b), phosphate level (Meharg and Macnair, 1994), season (Duxbury *et al.*, 2003, Islam *et al.*, 2004b), rice variety (Meharg and Rahman, 2003, Islam *et al.*, 2004b; Jahiruddin *et al.*, 2004). Meharg and Rahman (2003) tested grain As levels of nine Bangladesh rice genotypes of rice

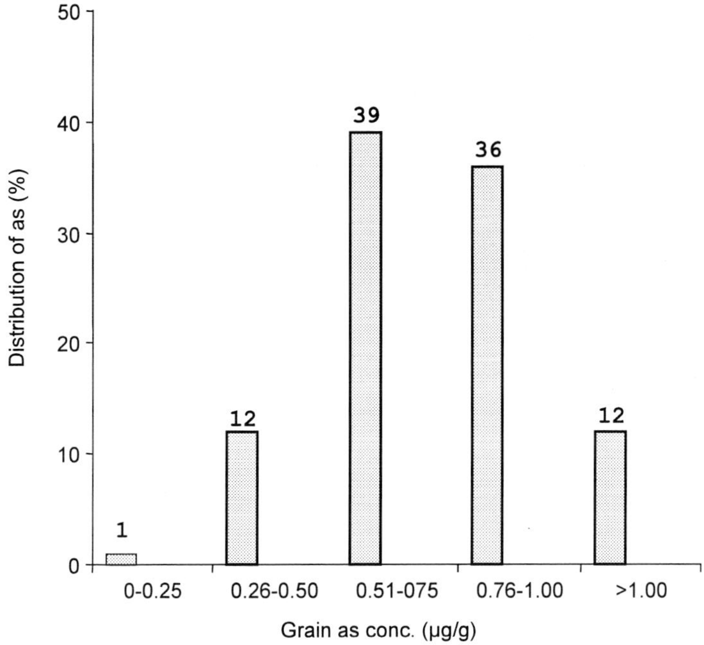

Fig. 4 Frequency distribution of grain As (n = 100).

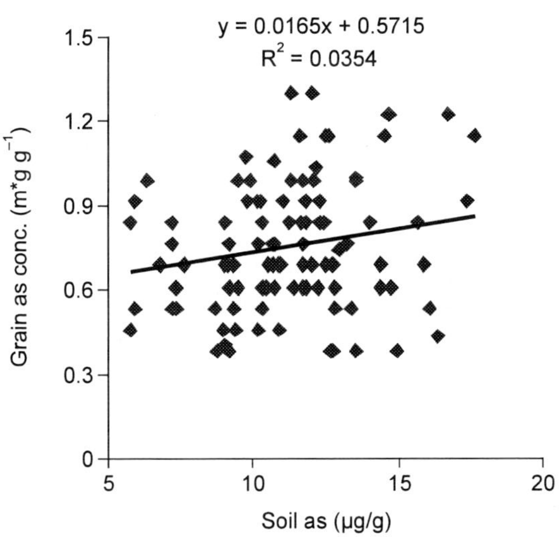

Fig. 5 Relationship between soil As and grain As.

grown in BRRI farm (As-non-contaminated soil) and found that the As levels ranged from 0.043 to 0.206 μg/g while rice grain samples collected from As contaminated areas of Chapai Nawabganj and Naogaon were in the range of 1.747 to 1.835 μg/g. Duxbury *et al.* (2003) analysed 150 paddy rice samples from Barishal, Comilla, Dinajpur, Kaunia, and Rajshahi

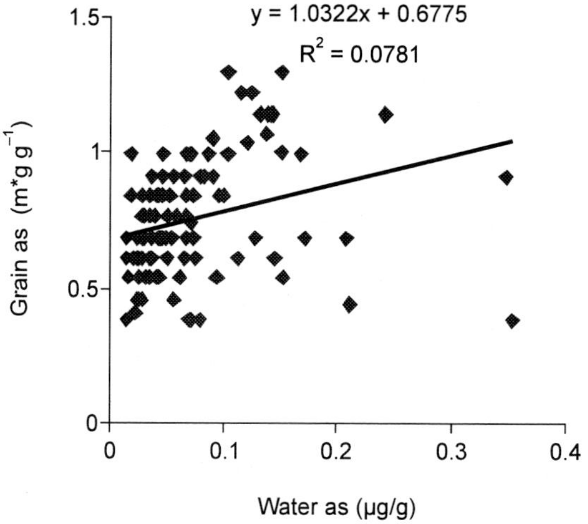

Fig. 6 Relationship between water As and grain As.

districts and also from the BRRI experimental station at Rajshahi in the boro (dry season) and Aman seasons (wet season) of 2000 and reported that mean concentration of As in unprocessed paddy rice ranged from 0.01 to 0.42 µg/g at 14% mositure content. Roychowdhury *et al.* (2002) reported that the mean As levels in rice grains from Jalangi and Domkal Blocks of West Bengal, India were 0.23 and 0.25 µg/g, respectively. Islam *et al.* (2004a) reported that grain As concentration of samples collected from five districts of Ganges River floodplain of Bangladesh varied from 0-2.05 µg/g. Lin *et al.* (2004) analysed 407 rice samples from Taiwan and the mean As content was 0.08 µg/g.

3.4 Human exposure to arsenic via rice

Rice is the staple food for Bangladesh people, who consume about 450 g of rice per day by an adult (CIMMYT, 2004). The total As content in Chapai Nawabganj sadar upazila ranges from 0.24–1.30 µg/g. Consuming 450 g rice from these samples will lead to an intake of total As ranging from 108–584 mg day^{-1}. Inorganic As in rice grains varies widely depending on varieties and countries of origin. Schoof *et al.* (1998) reported that the inorganic As in Taiwanese rice were 69–73% of the total arsenic whereas the inorganic arsenic in the United States rice were 24% (Schoof *et al.*, 1999) and 44% (Douglas *et al.*, 2001). The percentage of inorganic arsenic in Bangladesh rice varies from 70–90% depending on the varieties and growing conditions (Andy Meharg, personal communication). If we assume that 80% of the total arsenic in our studied samples is inorganic, an amount of 86–447 µg As day^{-1} will be ingested by an adult of 60 kg body weight. The maximum tolerable daily intake (MTDI) for an adult of 60 kg body weight is 120 µg kg^{-1} day^{-1}. Results in Table 2 show the distribution of daily arsenic ingestion from the 100 rice samples from Chapai Nawabganj sadar upazila. Ninety-nine per cent of rice grain samples would lead to an intake of more than 100% MTDI for an adult. Besides, the people are also exposed to arsenic ingestion through other routes like drinking water and vegetables consumption. As concentrations of the vegetables grown in these areas are quite high ranging from 0.312–0.812 µg dry weight^{-1} (Islam *et al.*, 2005). Most of the hand

Table 2 Comparison of total arsenic intake from rice samples (n = 100).

Rice samples (%)	Rice grain As concentration ($\mu g\ g^{-1}$)	Total As ingestion through rice ($\mu g\ day^{-1}$)	Inorganic AS ingestion ($\mu g\ day^{-1}$)	% of PMTDI
1	0–0.25	108	86	66
12	0.251–0.50	172–206	138–165	115–138
39	0.501–0.75	240–333	192–266	160–222
36	0.751–1.00	343–450	275–360	229–300
12	>1.00	464–584	372–467	310–389

and tube wells of these areas are also contaminated with As (Chowdhury *et al.*, 1999). Therefore, the dietary ingestion of As by the people in these areas will be very high.

4 CONCLUSION

Arsenic status in soils of Chapai Nawabganj are below the maximum permissible limit of 20 µg/g (Kabata-Pendias and Pendis, 2001) but irrigating the rice crops with elevated As concentration may lead to the accumulation of As in soils as well as uptake by the crops. It may not be cost-effective to remediate contaminated rice field soils. There may be a number of options for the reduced inflow of arsenic in the rice fields. Results of this study show that the concentration of arsenic in water was lower in deeper aquifers compared to shallow aquifers. Therefore, pumping irrigation water from the greater depths might lead to a decrease in the arsenic load to the rice fields. Another option is the economic use of irrigation water by (i) reducing the depth of pounded water, (ii) keeping the soil just saturated, or (iii) alternate wetting/drying (Guerra *et al.*, 1998). Water savings under saturated soil conditions were on average 23% with a yield reduction of only 6% (Bouman and Tuong, 2001). Rice varieties also differ in their ability to tolerate arsenic toxicity (Dasgupta *et al.*, 2004). Therefore, selecting or breeding rice varieties suitable for growing on arsenic contaminated soils might be also a good option to reduce the entry of arsenic in rice grain and ultimately food chain.

Acknowledgements

We are grateful to the IRRI-PETRRA, Bangladesh for providing financial support to carry out this research.

References

Abedin, M. J., Cotter, H.J. & Meharg, A. A. 2002a Arsenic uptake and accumulation in rice (*Oryza sativa* L.) irrigated with contaminated water. *Plant and Soil* 240(2): 311–319.

Abedin, M. J., Fieldman, J., Meharg, A. A., 2002b Uptake kinetics of arsenic species in rice plants. *Plant Physiol.* 128: 1120–1128.

Alam, M. G. M., Allison G, Stagnitti F, Takane A, Westbrooke M 2002 Arsenic contamination in Bangladesh groundwater: a great environmental and social disaster. *Int. J. Environ. Health Res.* 12(3): 235–253.

Ali, M. M., Ishiga, H., Wakatsuki, T., 2003 Influence of soil type and properties on distribution and changes in arsenic contents of different paddy soils in Bangladesh. *Soil Sci. Plant Nutr.* 49(1): 111–123.

BBS (Bangladesh Bureau of Statistics) 2003 *Monthly Statistical Bulletin* (December 2003). Government of the Peoples Republic of Bangladesh.

Brammer, H. 1996 The Geography of the Soils of Bangladesh. Univ. Press Ltd., Dhaka, Bangladesh p 287.

Bouman, B. A. M., Tuong, T. P., 2001 Field water management to save water and increase its productivity in irrigated lowland rice. *Agril Water Mgt* 49: 11–30.

CIMMYT 2004 Annual report 2002-2004. Arsenic and Agriculture. Coping with the "Largest Mass Poisoning in History .www.cimmyt.org/english/docs/ann- report/ recent/ health/ arsenic.htm

Chowdhury, T. R., Basu, G. K., Mandal, D. K., Samanta, G., Chowdhury, U. K., Chanda, C. R., Lodh, D, Lal Roy, S., Saha, K. C., Roy, S., Quamruzzaman, Q., Charaborti, D., 1999 Arsenic poisoning in the Ganges delta. *Nature* **401** 545–546.

Cullen,. W. R., Reimer, K. J., 1989 Arsenic speciation in the environment. *Che Rev* **89** 713–784.

Dasgupta, T., Hossain, S. A., Mehrag, A. A., Price, A. A., 2004 An arsenic tolerance gene on chromosome 6 of rice. *New Phytologist* **163:** 45–49.

Douglas, T. H., Nohora, P.V., Kirsten, R. S., Craig, S. W., 2001 Determination of total and speciated arsenic in rice by ion chromatography and inductively coupled plasma mass spectrometry. *J Analyt Atomic Spectrometry* **16** : 299–306.

Duxbury, J. M., Mayer, A. B., Lauren, J. G., Hassan, N., 2003 Food chain aspects of arsenic contamination in Bangladesh: Effects on quality and productivity of rice. *J Environ Sci Health* **38**(1): 61–69.

Guerra, L. C., Bhuiya, S. I., Tuong, T. P., Barker, R., 1998 Producing more rice with less water from irrigated systems. SWIM paper 5.IWMI/IRRI., Colombo, Sri Lanka p24.

Islam, M. R., Jahiruddin, M., Islam,. S., 2004a Assessment of arsenic in the water-soil-plant systems in Gangetic floodplains of Bangladesh. *Asian J Plant Sci* **3**(4) : 489–493.

Islam, M. R., Islam, S., Jahiruddin, M., Islam, M. A. 2004b Effects of irrigation water arsenic in the rice-rice cropping system. *J Biol Sci* **4** (4): 542–546.

Islam, M.R., M. E. Hoque, M. Jahiruddin, S. Islam. (2005) Heavy metal contamination of vegetables grown in Chapai- Nawabganj, Bangladesh and its implication to daily intake for human health. *Bangladesh J. Agric.Environ.* 1(1): 37–47.

Jahiruddin, M., Harada, H., Hatanaka, T., Islam, M. R. 2000 Trace element status in agricultural soils and factors affecting their concentration. *Soil Sci Plant Nutr* **46**(4): 963–968.

Jahiruddin, M., Islam, M. A., Islam, M. R. Islam, S. 2004 Effects of arsenic contamination on rice crop. *Environtropica* **1**(2) 204–210.

Jung, M. C., Thornton, I. and Chon, H. T. 2002. Arsenic, Sb and Bi contamination of soils, plants, waters, and sediments in the vicinity of the Dalsung Cu-W mine in Korea. Sci. Total Environ. 295: 81–89.

Kabata-Pendias, A., Pendis, H. 2001 Trace Elements in Soils and Plants. 3rd Ed. CRC Press, London. pp 225–239.

Keegan, T., Hong, B., Thornton, I., Farago, M., Jakubis, P., Jakubis, M. 2002 The EXPASCAN Study Group. Assessment of environment arsenic levels in Prievidza district. *J Expo Anal Environ Epidemiol* **12:** 179–85.

Khan, M. A. 2005 Movement of arsenic in undisturbed soil columns with irrigated rice culture. Proc. Symp. Behavior of arsenic in aquifers, soils and plants: Implications for management. Dhaka, Bangladesh, January 16–18, 2005.

Lin Haw-Tarn, Wong Sue-Sun, Li Gwo-Chen. 2004 Heavy metal content of rice and shellfish in Taiwan. 2004. *J Food Drug Anal* **12**(2) :167–174.

Marin, A. R., Masscheleyn, P. H., Patrick, W. H. 1992 The influence of chemical form and concentration of arsenic on rice growth and tissue arsenic concentration. *Plant and Soil* **139:** 175–183.

Meharg, A. A., Rahman, M. M. 2003 Arsenic contamination of Bangladesh paddy field soils: Implications for Rice Contribution to Arsenic consumption. *Environ Sci Technol* **37**(2): 229–234.

Meharg, A. A., Macnair, M. R. 1994 Relationship between plant phosphorus status and the kinetics of arsenate influx in clones of *Deschampsia cespitosn* (L) Beauv that differ in their telorance to arsenate. *Plant and Soil* **162:**99–106.

Nickson, R., McArthur, J. M., Burgess, W., Ahmed, K. M. Ravenscroft P, Rahman M 1998 Arsenic poisoning of Bangladesh groundwater. *Nature* 396 : 338.

Panaullah, G. M., Ahmed, Z. U., Rahman, G. K. M. M., Jahiruddin, M., Miah, M. A. M., Farid, A.T. M., Biswas, B. K., Lauren, J. G., Loeppert, R. H., Duxbury, J. M., Meisner, C. A. 2003The arsenic hazard in the irrigation water-soil-plant system in Bangladesh: A preliminary assessment. *Proc. 7th International Conference on the Biogeochemistry of Trace Elements*, Ed. Gobran, G.R. and Lepp N., vol. 2, pp 104–105.

Roychowdhury, T., Uchino, T., Tokunaga, H., Ando, M. 2002 Survey of arsenic in food composites from an arsenic affected area of West Benga, India. *Food Chem Toxicol* **40**: 1611–1621.

Schoof, R. A., Yost, L. J., Crecelius, E., Irgolic, K., Goessler, W., Guo, H. R. Greene, H. 1998 Dietary arsenic intake in Taiwanese districts with elevated arsenic in drinking water. *Human Ecol Risk Assess* **4** :117–135.

Schoof, R. A., Yost, L. G., Eickhoff, J., Crecelius, E. A., Cragin, D. W. Meacher, D. M., Menzel, D. B. 1999 A market basket survey of inorganic arsenic in food. *Food Chem Toxicol.* **37:** 839–846.

Turpeinen, R., Pantsar-Kallio, M., Haggblom, M., Kairesalo, T. 1999 Influence of microbes on the mobilization, toxicity and biomethylation of arsenic in soil. *Sci. Total Environ.* **236:** 173–180.

Welsch, F. P., Crock, J. G., Sanzolone, R. 1990 Trace-level determination of arsenic and selenium using continuous flow hydride generation atomic absorption spectrophotometry (HG-AAS), pp. 38-45. In: *Arbogast B.F.* (Ed.) Quality assurance manual for the branch of Geochemistry. U.S. geological Survey. Denver Co.

WDI (World Development Indicator) 1998 World Bank, Washigton D.C.

WRI (World Resources Indicator) 2000 Peoples and Ecosystem: *The Fraying Web of Life.* World Resources Institute, Washington D.C.

Surveillance, Hydrogeochemistry, and Methods for Removal of Fluoride Contamination in Groundwater: Case Study from Kerala, India

41

P.S. Harikumar[a], Estifanos Haile[b], K. Madhavan[a], Vikas Chaudhary[b], Gunnar Jacks[b] and Prosun Bhattacharya[b,*]

[a]Centre for Water Resources Development and Management, Kozhikode–673571, Kerala, India

[b]KTH-International Groundwater Arsenic Research Group, Department of Land and Water Resources Engineering, Royal Institute of Technology (KTH), SE-10044, Stockholm, Sweden

ABSTRACT

Groundwater with elevated concentration of fluoride is an important environmental hazard in the world. Fluoride contamination is widely reported in groundwater in different parts of India. Ingestion of fluoride from drinking water has imminent effects on dental and skeletal health. A survey of dental and skeletal fluorosis among population was used to delineate the areas potentially affected by high fluoride concentrations in drinking water. Such information supplemented with hydrogeochemistry of fluoride contaminated groundwaters were used to identify, assess, and delineate the mechanisms of the mobilization of fluoride in groundwaters and also to propose an appropriate method of mitigation in severely affected areas. The present study was carried out in Alappuzha, a coastal district of the state of Kerala, located on the south-western part of India where fluoride concentration in groundwater was greater than the permissible drinking water limit of 1 mg/L established by the Bureau of Indian Standards (BIS). Surveillance of the fluoride problems in Alappuzha town was done through a survey of dental fluorosis among school children as well as through monitoring of groundwater quality. Dental fluorosis survey indicates that 14% of the surveyed school children are affected. The high concentration of bicarbonate, high pH and low calcium is found to

*Corresponding author. Tel.: +46 8 7907399; Fax: +46 8 411 0775

E-mail address: prosun@kth.se (P. Bhattacharya)

correlate positively with high fluoride concentration. A comparison of defluoridation techniques using the locally available chemicals indicate that alum is the most efficient and cost effective method that shows a removal efficiency of 98% fluoride.

Keywords: Fluoride, groundwater chemistry, dental fluorosis, defluoridation, Kerala, India

1 INTRODUCTION

Elevated concentrations of naturally occurring fluoride in groundwater are causing widespread health crisis in many parts of the world. Fluorine constitutes about 0.032% of the earth's crust (WHO, 1996) and can enter groundwater by natural processes through the weathering and leaching of bedrock with high content of fluorine. Natural concentration of fluoride in groundwater depends on the hydrogeological and geochemical characteristics of the aquifers. Weathering of the crustal rocks contributes to the dissolved fluoride concentrations in groundwater, but there can be significant influx of fluoride from atmospheric deposition in the form of soil dust and the emissions of fluoride from the industries (Jacks *et al.*, 2005).

The two principal health impacts of excess fluoride intake in human body are manifested as dental and skeletal fluorosis. Symptoms affecting the soft tissues such as muscles and ligaments are also reported (Kharb and Susheela, 1994). Fluoride toxicity may also cause adverse effects on the reproductive system of males living in fluorosis endemic areas (Susheela and Jethanandani, 1996). The sources of fluoride and the severity of these symptoms are governed by the age of the person and various climatic conditions such as high ambient temperature, chemical characteristics viz. high alkalinity, low calcium and magnesium contents in the drinking water, nutritional deficiencies (Viswanadham and Murty, 1974) and in certain cases high concentration of nitrate ions (Sharma and Rao, 1997). Water quality, especially the concentrations of calcium and magnesium and alkalinity seem to influence the incidence of fluorosis (Sharma and Rao, 1997). These elements reduce the amount of fluoride absorbed from the gastrointestinal tract. Epidemiological observations reveal that the nutritional status might influence chronic fluoride toxicity (Jolly *et al.*, 1969).

A large number of published studies deal with the non-skeletal and non-dental health effects associated with the exposure of fluoride from a variety of sources (McLaren, 1976; Waldbott, 1976; Susheela *et al.*, 1992; Li and Cao, 1994; Mullenix *et al.*, 1995; Zhao et al., 1996). The prevalence of fluorosis is reported among about 32% of the global population (Mella *et al.*, 1994). In Mexico, 5 million people (about 6% of the population) are affected by fluoride in groundwater (Fernando *et al.*, 1997). Fluorosis is prevalent in some parts of central and western China (Wang *et al.*, 1999). A survey conducted in 1990 revealed that, among a population of 300 million people living in endemic fluorosis areas of China, 3 million (1%) suffer from severe skeletal fluorosis and 40 million (13.3%) from dental fluorosis. (Li and Cao, 1994). High concentration of fluoride was also reported in many parts of East Africa (Nayaro *et al.*, 1984; Gaciri and Davies, 1993; Gizaw, 1996).

Endemic fluorosis is a public health problem in India. According to Government estimates, 66 million people in India which account for more than 5% of the total population, are at risk of fluorosis (Rajiv Gandhi National Drinking Water Mission, 1993). India is one of 21 nations with serious health problems due to consumption of fluoride-contaminated drinking water from groundwater sources, where concentrations of fluoride vary from 1.5 to 39 mg/L (Susheela, 2000; Jacks *et al.*, 2005).

Teotia and Teotia (1984) have reported the prevalence of dental and skeletal fluorosis among the rural populations of India consuming water containing 0.6 mg/L fluoride. Earlier epidemiological studies indicated that high fluoride concentration in drinking water supplies resulted in the prevalence of skeletal fluorosis at F^- concentration levels of 2.4 mg/L, while concentration above 3 mg/L was found to have caused crippling fluorosis (Jolly *et al.*, 1969).

In the last few years several studies on the issue of fluorosis as related with the high concentration of fluoride in Indian groundwater sources were reported. Significantly high levels of fluoride in groundwater within the states of Andhra Pradesh, Assam, Bihar, Delhi, Haryana, Gujarat, Maharashtra, Orissa, Punjab, Rajasthan and Tamil Nadu pose risk of fluorosis among inhabitants of these areas (Choubisa *et al.*, 1996; Datta *et al.*, 1996; Susheela and Jethanandani, 1996; Agarwal *et al.*, 1997; Chakraborti *et al.*, 2000; Kundu *et al.*, 2001; Kundu *et al.*, 2002). About 25 million people in these areas are seriously affected with endemic fluorosis, and about 40 million are at risk (Industrial Toxicology Research Centre, 1995). Depending on the endemic status of fluorosis, the Indian states were grouped into three categories (Table 1; Susheela, 1991).

Table 1 Status of fluoride endemicity in different states of India categorized on the basis of the percentage of districts affected with fluorosis.

Category	*States affected*	*Districts affected*
I	Jammu & Kashmir, Kerala, and Orissa	< 30%
II	Assam, Bihar, Haryana, Karnataka, Maharashtra, Madhya Pradesh and Punjab	30-50%
III	Andhra Pradesh, Gujarat, Rajasthan, Tamil Nadu and Uttar Pradesh	> 50%

Several methods are reported on the removal of fluoride, which involve mainly the processes of precipitation and adsorption. Precipitation with calcium and aluminium salts is the common method (Boruff, 1934; Saha, 1993). Adsorption on activated alumina, alum, charcoal ash, ion exchange and membrane processes such as reverse osmosis, nanofiltration, electrodialysis are some other methods attempted for the removal of fluoride (Chaturvedi *et al.*, 1988; Killedar and Bhargava, 1993; Yang and Dluhy, 2002). The National Environmental Engineering Research Institute (NEERI), Nagpur, India, has developed a method known as the Nalgonda Technique (Bulusu *et al.*, 1979) which uses alum for the removal of fluoride.

Drinking water sources of Kerala are influenced by the special physiographic and hydrologic conditions of the state. Kerala has about 1720 operational water supply systems (Economic Review, 1993) and 1219 villages receive protected water supply (CWDRM, 1995). Even though the state of Kerala is grouped under category-I (Table 1), Alappuzha district is reported to have endemic fluoride problems (Susheela, 1991). Drinking water supply to Alappuzha municipality area is provided by the Kerala Water Authority (KWA) through abstraction of groundwater from tube wells containing fluoride concentration above the drinking water standard (1 mg/L) suggested by the Bureau of Indian Standards (BIS, 1991). The objective of this study is to examine the extent of fluoride problem of this coastal district of

India by surveying the prevalence of dental fluorosis among school children and monitoring of groundwater quality with special reference to the distribution of fluoride. Efficiency of locally available chemicals and plant materials were investigated as the possible methods of defluoridation.

2 LOCATION AND HYDROGEOLOGY OF THE STUDY AREA

2.1 Location

Alappuzha district, in Kerala state, with an area of 1414 km^2, is located between latitudes 90° 05′ and 90° 52′ N and longitudes 76° 17′ and 76° 48′ E (Fig. 1). The district is bounded by Ernakulam and Kottayam districts towards the north-east, by districts of Pathanamthitta and Kollam towards east and south-east respectively and by the Arabian Sea in the west. The climate is moist and hot in the coast, and it is slightly cooler and drier in the interior of the district. The average monthly temperature of the district is approximately 18°C. As in the case of other parts of the state of Kerala, this district also gets the benefit of two outstanding monsoons.

According to 2001 census, Alappuzha district has a population of 2.1 million. A vast network of rivers, lakes and lagoons in the state are the sources of fresh water in the district. However, tube wells are provided extensively in Alappuzha district to meet the extended need for drinking water. Warkali aquifer is the main source of drinking water to Alappuzha town. The Warkali aquifer comprises alternate layers of fine to medium grained sand with clay and thin band of lignite.

2.2 Hydrogeology of Alappuzha District

Soils of Alappuzha district are classified as sandy, clayey and lateritic. In southern parts of the district sandy loam covers Karthikappally and Mavelikkara areas. On the coastal belt, sandy soil covers Cherthala and Ambalappuzha. The area around Kuttanad is covered by clayey loam and supports extensive agricultural activity. The Vembanad Lake, south of Alappuzha town, is the major freshwater source in the region. Peaty type kari soil occur as a small belt in the eastern regions of Cherthala and Ambalappuzha areas and the western part of Kuttanad. The characteristic soil type in the Chengannur and Mavelikkara areas are laterite. These lateritic soils are rich in kaolinite and are formed due to intense weathering of gneissic and granitic rocks under alternate wet and dry tropical conditions (CWDRM, 1995).

Groundwater in the Alappuzha region of Kerala predominantly occurs in unconfined aquifers of alluvial deposits of Holocene age. Most of the tube wells in Alappuzha town tap groundwater from Warkali aquifers (Jacob *et al.*, 1985). These aquifers are normally a few metres in thickness. Open wells of diameter about 2 m are common groundwater abstraction structures. In some places, the sandy aquifer may extend to depths of a few tens of metres. In such places, it is suitable to construct filter point wells. In some areas, groundwater is confined, mainly in the Vaikom, Kollam and Warkkali sedimentary formations of tertiary age, capped by laterite (CWDRM, 1995). The piezometric surface of these confined aquifers varies from 0.5 to 14.3 m above MSL (mean sea level). The Vaikom aquifer is considered to be the largest reservoir of groundwater with thickness varying between 25 and 79 m. These sedimentary aquifers in the

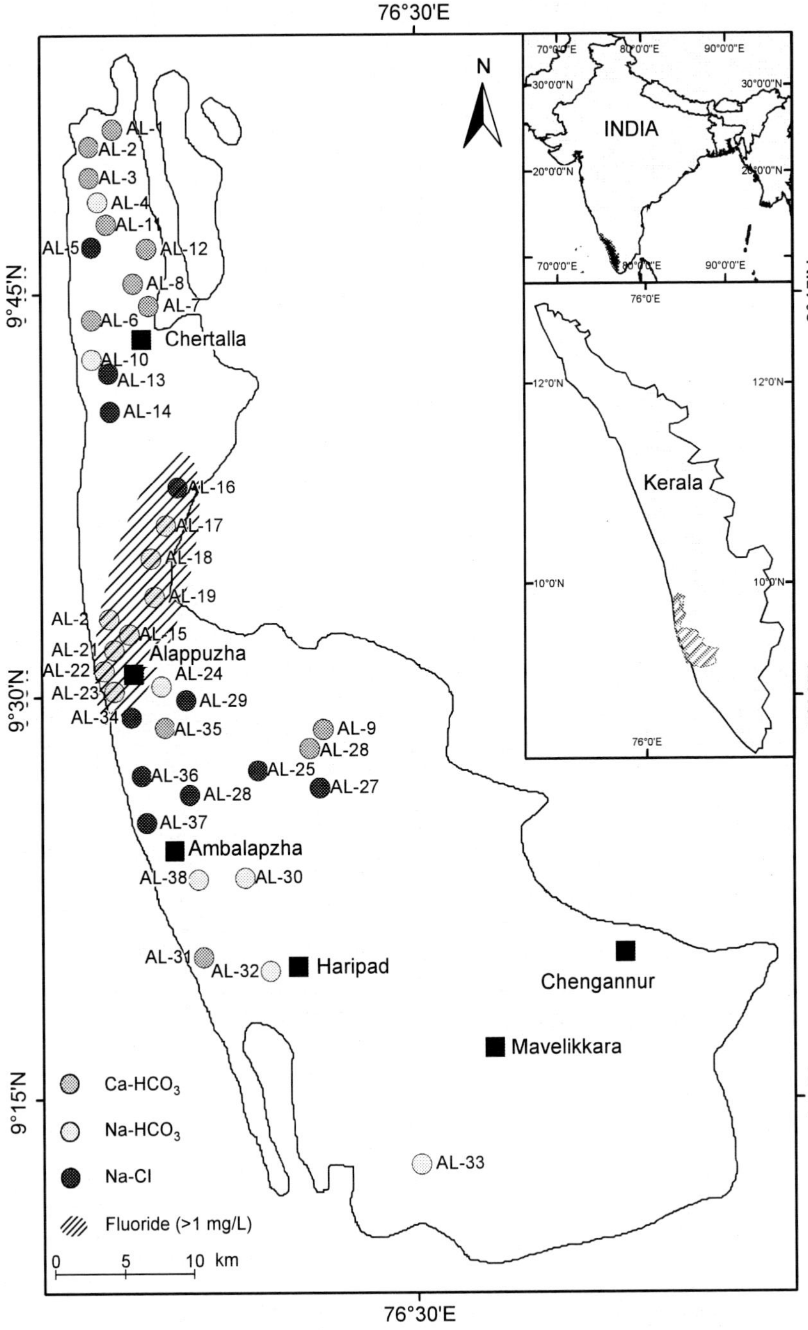

Fig. 1 Map of Alappuzha district in the south-western coastal region of the state of Kerala, India. For explanation of the names of the sampling locations, refer to Table 2. Water types are identified by green, yellow and red colors. The hash marked part shows, wells with significant amount (> 1 mg/L) fluoride levels.

region have a discharge of 300 to 1800 m^3/h, and is suitable for medium capacity tube wells (CWDRM, 1995).

3 METHODS

3.1 Dental Fluorosis Survey

The endemicity of fluorosis in Alappuzha was studied through a survey on the prevalence of dental fluorosis among school children. Detailed profile data sheets were prepared for about 750 children between age group 5-10 years studying in various schools. The surveillance of dental fluorosis among these children involved examination of 1) loss of lustre and shine of teeth enamel; 2) discoloration of teeth to yellow brown and black streaks or spots; 3) pitting and chipping of teeth; and 4) loss of teeth. Respective address of the children was collected in order to identify the sources of fluoride rich waters. The types of dental fluorosis were confirmed by standard procedure (Susheela, 1991).

3.2 Water Sampling and Analysis

Seasonal water sampling was done during pre-monsoon (March, 2002) and post-monsoon (July 2002) periods. Numerous groundwater samples were collected during monsoon and pre-monsoon periods respectively from different parts of Alappuzha district (Table 2). Sampling of groundwater, their preservation and analyses were carried out following the Standard Methods for Examination of Water and Wastewater (APHA, 1992). Water quality parameters such as pH and electrical conductivity were measured in the field. Analysis of bicarbonate, calcium, magnesium, total hardness, and sulphate were carried out at the chemical laboratories of the Centre for Water Resources Development and Management (CWRDM), Kozhikode. The concentration of fluoride was determined by SPANDS method using Hitachi UV-visible spectrophotometer at the laboratories of CWRDM, Kozhikode. Since no samples contained interfering ions above the prescribed limit, samples were directly used for the determination of fluoride.

3.3 Fluoride Removal Techniques

Laboratory scale study was conducted to determine the efficiency of various commercially and locally available materials for fluoride removal. The material used were different types of charcoal, activated alumina, alum and activated silica.

Standard solutions with different concentrations of fluoride were prepared by serial dilution of stock solution with 100 mg/L fluoride prepared from NaF. Adsorption studies on fluoride removal were conducted using charcoal (powder) supplied by various manufacturers viz. Merck, Nice and Himedia.

4 RESULTS AND DISCUSSIONS

4.1 Dental Fluorosis Survey

The outcome of the survey is out of 750 school children aged between 5 and 10 years, 152 of them were affected by dental fluorosis (Fig. 2). Chalky white and yellow brown discoloration

Table 2 Location of the wells and the well types sampled during the study in the high-fluoride affected regions of the Alappuzha district, Kerala, India.

Location/Sample ID	*Locality*	*Well type*
AL-1	Eramallur	Open well
AL-2	Kadamthruthi	Open well
AL-3	Thuravoor	Open well
AL-4	Pattanakad	Open well
AL-5	Chertalla	Open well
AL-6	Chertalla Govt.H.S	Open well
AL-7	Thaneermukkom	Open well
AL-8	Kokkathamangalam	Open well
AL-9	Velianad	Open well
AL-10	Chertalla Govt.U P School	Open well
AL-11	Kakkothamangalam	Open well
A L-12	Varanad	Open well
AL-13	S L Puram	Open well
AL-14	Kalavoor	Open well
AL-15	Pathirapally	Open well
AL-16	KWA Convent	Tube well
Al-17	KWA Vellkinar	Tube well
AL-18	KWA Aliseery	Tube well
AL-19	KWA Chudukkad	Tube well
AL-20	KWA Chudukkad	Tube well
AL-21	KWA Thookkalam	Tube well
AL-22	KWA Pazhavangadi	Tube well
AL-23	Vazhicherry	Tube well
AL-24	Kalarcode	Tube well
AL-25	Chambakulam	Open well
AL-26	Ramankary	Open well
AL-27	Mankombu	Open well
AL-28	Nedumudi	Open well
AL-29	Karumachi	Open well
AL-30	Thakazhi	Open well
AL-31	Thottapally	Open well
AL-32	Harippad	Open well
AL-33	Kayamkulam	Open well
AL-34	Kurvantheeli	Open well
AL-35	Punnapra	Open well
AL-36	Punnapra (Behind Milk Plant)	Tube well
AL-37	KWA Chandanakkavu	Tube well
AL-38	KWA Chathanad	Tube well

were the major types of dental fluorosis prevalent among the children living in this region (Fig. 3). About 10% of the children studying in the school investigated in Punnapra, exhibited different forms of dental fluorosis manifested as discoloration, pitting and chipping of the teeth. Students belonging to low-income households generally lack balanced nutrition that is another factor contributing to dental fluorosis.

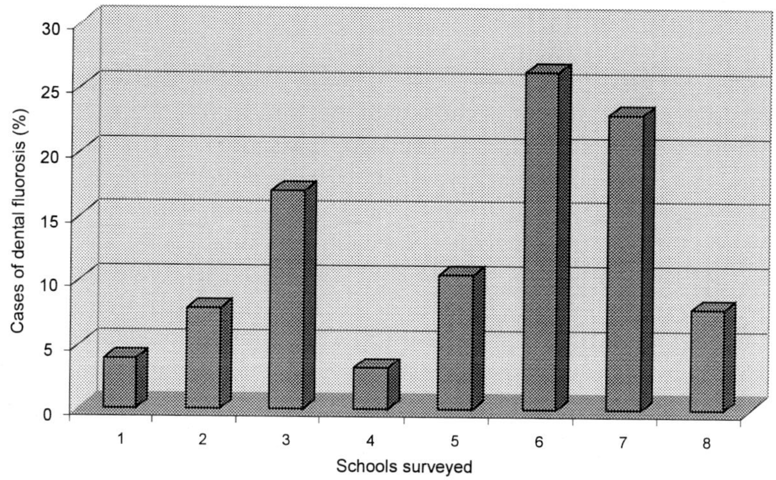

Fig. 2 Number of children affected with dental fluorosis in different schools of Alappuzha district, Kerala, India. Legend: 1. Government UP School, Veliyamkulam; 2. Government Harijan Welfare UP School, Varanad; 3. Government JB School, Punnapra; 4. Government WH School, Kadamthuruthi; 5. Government UP School, Chertalla; 6. Leo XIII LP School, Alappuzha; 7. Government MMAUP School, Vazhicherry; 8. Government School, Kokkathamangalam.

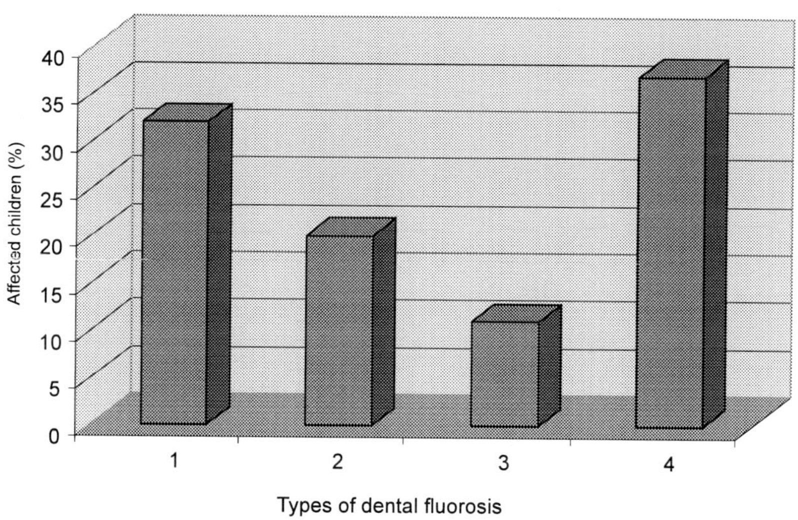

Fig. 3 Types of dental fluorosis detected among surveyed school children in different investigated schools of Alappuzha district, Kerala, India. Legend: 1. Discolouration; 2. Loss of lustre and enamel; 3. Loss of teeth; 4. Pitted and chipped teeth.

4.2 Groundwater Chemistry

4.2.1 General groundwater types

Chemical composition of the analysed groundwater samples and the major water types in the investigated aquifers of Alappuzha district are presented in Tables 3a and 3b. The groundwaters represent three dominant water types: Ca-HCO_3, Na-HCO_3 and Na-Cl. (Fig. 4). The fresh water of Ca-HCO_3 type occurs in areas south of Amblapuzha-Thakazhi zone. The chemical composition of water changes to $NaHCO_3$ Alappuzha area, northwards. Towards further north i.e. in the Alappuzha-Chertala zone, hard brackish water is found (see Fig. 1). These groundwaters are generally of Na-Cl type. The changes in the degree and type of mineralisation of the water in the Warkali aquifers from south to north direction were previously explained by Jacks and Rajagopalan (1996). Flushing of calcium rich groundwater through the saline aquifer has a natural softening effect due to the exchange of sodium from clay minerals through ion exchange processes.

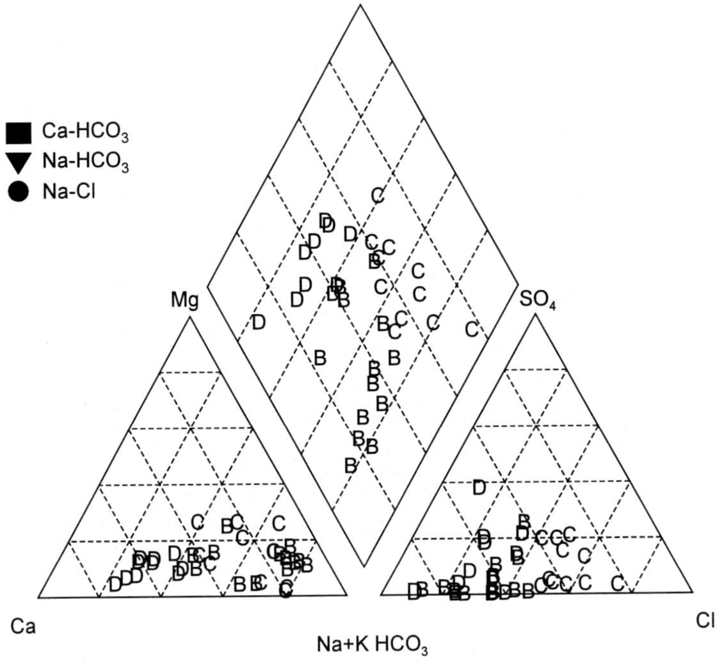

Fig. 4 Piper diagram showing the major ion composition of the groundwater samples and the different water types.

4.2.2 Groundwater chemistry and its relation to fluoride

Summary of data collected on fluoride levels in potable water with concentration grouped in three ranges: < 0.5 mg/L, 0.5–1.0 mg/L, and > 1.5 mg/L is given in Fig. 5. During pre-monsoon period, more than 44% of the samples indicate fluoride concentration exceeding 1 mg/L. After the monsoon period the number of samples exceeding 1 mg/L, decreased to around 30%. Such

Table 3(a) Basic chemical characteristics and concentrations of fluoride in groundwaters of Alappuzha district during pre-monsoon (March, 2002) sampling period.

Sample ID	*Locality*	*Well type*	*pH*	*EC*	*TH**	*Ca^{2+}*	*Mg^{2+}*	*F^-*
				µS/cm	mg/L	mg/L	mg/L	mg/L
AL-5a	Chertalla (T Prabhu)	Open well	7.6	570.0	42.0	14.4	1.46	0.1
AL-5b	Chertalla (Raju)	Open well	8.0	300.0	12.0	74.0	66.0	0.4
AL-5c	Chertalla (Joseph)	Open well	8.5	285.0	168.0	57.6	5.83	0.5
AL-6	Chertalla (MM UP School)	Open well	6.2	354.0	74.0	22.4	4.37	0.2
AL-12a	Nethaji School Varanad	Open well	8.3	225.0	28.0	110.0	106.0	0.2
AL-12b	HWLP School Varanad	Open well	6.9	405.0	10.0	14.9	8.0	n.d.
ALI-16	KWA Convent	Tube well	6.4	930.0	140.0	25.6	18.5	1.2
ALI-17	KWA Vellakinnar	Tube well	8.0	870.0	114.0	20.0	15.6	1.4
AL-18	KWA Aliserry	Tube well	8.7	900.0	72.0	16.0	7.78	1.5
AL-19	KWA Chudukkad a	Tube well	8.4	600.0	56.0	12.0	6.32	1.8
AL-20a	KWA Chudukkad b	Tube well	8.5	870.0	162.0	13.6	31.1	1.9
AL-20b	KWA Chudukkad c	Tube well	8.1	600.0	48.0	10.4	5.35	1.9
AL-21	KWA Thookalam	Tube well	8.1	231.0	98.0	19.2	12.2	1.9
AL-22	KWA Pazhavangadi	Tube well	8.5	954.0	114.0	34.4	14.1	1.4
AL-23	KWA Vazhicherry	Tube well	8.3	786.0	100.0	18.4	13.2	1.3
AL-35a	Punnapra	Open well	8.1	580.0	86.0	16.0	11.2	1.5
AL-35c	Punnapra (Balachandran)	Open well	7.8	390.0	122.0	8.8	24.3	0.4
AL-35b	Punnapra (Tilakan)	Open well	8.1	570.0	16.0	8.0	0.97	0.3
AL-37	KWA Chandanakkavu	Tube well	8.5	1554.0	124.0	24.8	15.1	2.6
AL-38	KWA Chathanad	Tube well	8.0	930.0	104.0	18.4	14.1	1.4

* TH: Total hardness

a decrease might be attributed due to the effect of dilution occurring through the infiltration of meteoric water during monsoon period.

Fluoride concentration in groundwater is governed by adsorption equilibria and by fluorite solubility. The concentration of fluoride in groundwater increases until the saturation level of fluorite is attained (Handa, 1975):

$$Ca^{2+} + 2F^- \leftarrow\rightarrow CaF_2 \qquad K_{fluorite} = 10^{-10.6}$$

Accordingly, the presence of elevated amount of Ca^{2+} lowers the amount of F^- in groundwater and vice versa. A plot between Ca^{2+} and F^- (Fig. 6) indicates that all the samples have these inverse characteristics although the concentration levels for pre-monsoon and post-monsoon periods differ slightly. Pre-monsoon period samples have higher amount of F^- and lower amount of Ca^{2+}. This can be due to the effect of high evapotranspiration during pre-monsoon period. Evapotranspiration causes precipitation of calcite and lowering of the activity of Ca^{2+}, leaving room for increase in fluoride levels. However, the groundwater remained to be undersaturated with respect to fluorite for all the samples. This is accompanied by a shift of pH from about 8 during pre-monsoon period to 7 during post-monsoon (Fig. 7). A decrease in the concentration of Ca^{2+} and concomitant increase of Na^+ in groundwaters are reported under alkaline conditions (Jacks *et al.*, 2005). Significant correlation was observed

Table 3(b) Detailed hydrochemical characteristics of groundwaters from Alappuzha district during post-monsoon (July, 2002) sampling period.

Sample ID	Locality	pH	EC $\mu S/cm$	Na^+ mg/L	Ca^{2+} mg/L	Mg^{2+} mg/L	Cl^- mg/L	SO_4^{2-} mg/L	HCO_3^- mg/L	F^- mg/L	Water Type
AL-1	Eramallur	7.1	153	19	20.0	1.94	14	14.5	58	n.d.	$Ca\text{-}Na\text{-}HCO_3\text{-}Cl$
AL-2	Kadamthruthi	7.5	275	950	39.2	14.6	620	35	328	0.38	Na-Cl
AL-3	Thuravoor	7.7	270	25	39.2	5	22	10	106	n.d.	$Ca\text{-}Na\text{-}HCO_3\text{-}Cl$
AL-4	Pattanakad	6.3	184	27	21.6	2.92	22	9.5	68	n.d.	$Na\text{-}Ca\text{-}HCO_3\text{-}Cl$
AL-5	Chertalla	6.4	306	56	17.6	1.94	54	4.0	48	n.d.	$Na\text{-}Ca\text{-}Cl\text{-} HCO_3$
AL-6	Chertalla Govt.H.S	6.4	122	11	21.6	1.5	8.0	32.5	54	n.d.	$Ca\text{-}Na\text{-}HCO_3\text{-}SO_4$
AL-7	Thaneermukkom	6.9	291	26	40.0	5.8	20	4.5	110	0.03	$Ca\text{-}Na\text{-}HCO_3$
AL-8	Kokkathamangalam	6.6	122	9	17.6	2.43	12	14.5	52	n.d.	$Ca\text{-}Na\text{-} HCO_3\text{-}Cl\text{-}SO_4$
AL-9	Velianad	7.4	418	40	40.0	4.86	46	0.5	122	0.10	$Ca\text{-}Na\text{-}HCO_3\text{-}Cl$
AL-10	Chertalla Govt.U P School	7.0	189	25	20.8	4.4	28	1.0	64	n.d.	$Na\text{-}Ca\text{-}HCO_3\text{-}Cl$
AL-11	Kakkothamangalam	5.0	221	n.d.	0.8	1.4	8.0	4.5	8.0	n.d.	$Ca\text{-}HCO_3$
AL-12	Varanad	6.6	250	26	24.0	n.d.	34	26	54	0.07	$Ca\text{-}Na\text{-}Cl\text{-}HCO_3\text{—}SO_4$
AL-13	S L Puram	6.1	500	40	30.4	14.1	99	30	82	n.d.	$Na\text{-}Ca\text{-}Mg\text{-}Cl\text{-}HCO_3$
AL-14	Kalavoor	6.2	291	46	9.6	0.49	48	27.5	60	0.09	$Na\text{-}Cl\text{-}HCO_3\text{-}SO_4$
AL-15	Pathirapally	6.3	122	16	32.0	3.9	16	8.5	38	0.07	$Ca\text{-}Na\text{-}HCO_3\text{-}Cl$
AL-16	KWA Convent	6.6	908	146	28.8	17.5	132	9.5	216	1.0	$Na\text{-}Cl\text{-}HCO_3$
AL-17	KWA Vellkinar	6.8	857	152	16.8	18.95	118	2.5	226	1.01	$Na\text{-}HCO_3\text{-}Cl$
AL-18	KWA Aliseery	7.5	581	122	20.0	7.76	32	5.0	218	1.09	$Na\text{-}HCO_3$
AL-19	KWA Chudukkad	7.2	571	134	10.4	9.2	38	2.5	208	1.13	$Na\text{-}HCO_3$
AL-20	KWA Chudukkad	7.3	510	95	9.6	7.8	18	3.0	202	1.14	$Na\text{-}HCO_3$
AL-21	KWA Thookkalam	7.0	602	98	16.8	10.7	62	1.0	188	1.02	$Na\text{-}HCO_3\text{-}Cl$
AL-22	KWA Pazhavangadi	7.1	826	164	24.8	15.5	98	0.5	452	1.08	$Na\text{-}HCO_3\text{-}Cl$
AL-23	Vazhicherry	7.3	730	166	20.0	13.1	76	0.5	236	1.09	$Na\text{-}HCO_3\text{-}Cl$
AL-24	Kalarcode	6.7	337	59	20.0	1.9	36	9.0	112	0.61	$Na\text{-}Ca\text{-}HCO_3\text{-}Cl$
AL-25	Chambakulam	6.2	352	44	16.0	11.7	50	20.5	58	0.15	$Na\text{-}Mg\text{-}Ca\text{-}Cl\text{-}HCO_3$
AL-26	Ramankary	6.4	337	29	32.8	6.3	36	30.3	80	0.09	$Ca\text{-}Na\text{-}HCO_3\text{-}Cl$
AL-27	Mankombu	6.5	714	77	50.4	9.2	132	9.5	154	0.27	$Na\text{-}Ca\text{-}Cl\text{-}HCO_3$
AL-28	Nedumudi	6.1	765	86	30.4	16.5	120	n.d.	176	0.41	$Na\text{-}Ca\text{-}Mg\text{-}Cl\text{-}HCO_3$
AL-29	Karumachi	7.4	908	81	61.6	13.6	102	65	162	0.38	$Na\text{-}Ca\text{-}Cl\text{-}HCO_3$
AL-30	Thakazhi	7.0	530	68	30.4	17.9	36	0.5	194	0.44	$Na\text{-}Ca\text{-}Mg\text{-}HCO_3$

Contd.

Contd.

AL-31	Thottapally	7.8	306	18	41.6	2.43	30	6.6	98	0.21	Ca-Na-HCO_3-Cl
AL-32	Harippad	7.1	1020	192	84.0	6.3	120	55	270	0.24	Na-Ca-HCO_3-Cl
AL-33	Kayamkulam	5.7	173	24	14.4	3.9	20	20.5	42	0.33	Na-Ca-HCO_3-Cl- SO_4
AL-34	Kurvantheeli	5.1	51	8.0	1.6	0.97	10	1.0	14	n.d.	Na-Cl-HCO_3
AL-35	Punnapra	7.1	360	5.0	13.6	0.486	14	1.0	210	1.3	Ca-HCO_3
AL-36	Punnapra (Behind Milk Plant)	6.6	194	30	16.0	5.35	24	n.d.	22	n.d.	Na-Ca-Mg-Cl-HCO_3
AL-37	KWA Chandanakkavu	7.2	1275	425	24.0	12.4	240	n.d.	258	1.15	Na-Cl-HCO_3
AL-38	KWA Chathanad	7.3	887	275	20.0	14.6	116	n.d.	228	3.95	Na-HCO_3-Cl

n.d.: not detected

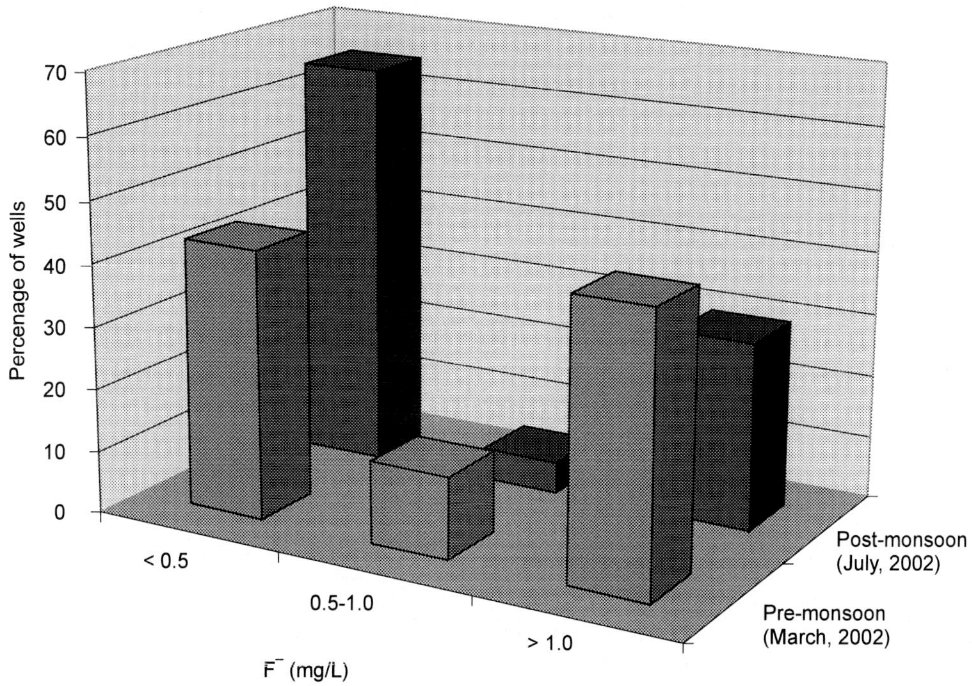

Fig. 5 Percentage of wells and the range of fluoride levels in groundwater from the wells in the investigated area around Alappuzha during pre-monsoon (March, 2002) and post-monsoon (July, 2002).

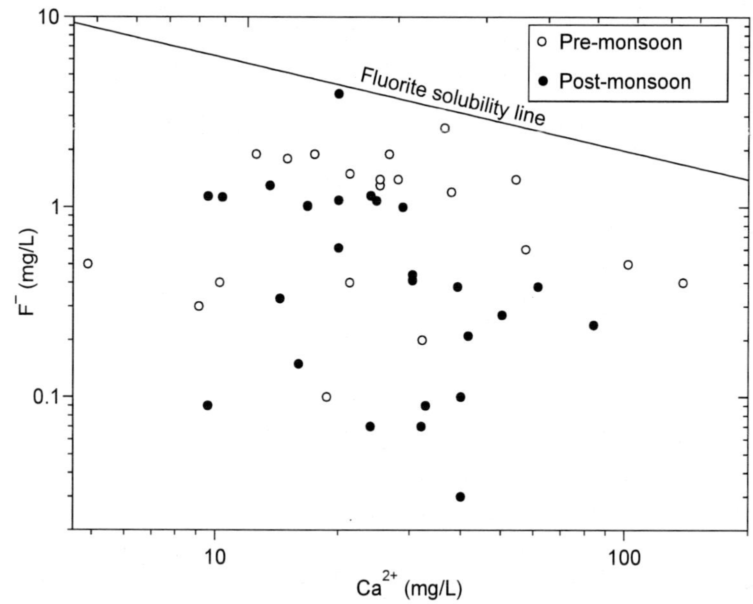

Fig. 6. Plot between Ca^{2+} and F^- in the groundwaters of Alappuzha district during the pre-monsoon (open circles) and post-monsoon (solid circles) periods showing inverse relationships. The samples during the pre-monsoon period have higher concentration of F^-.

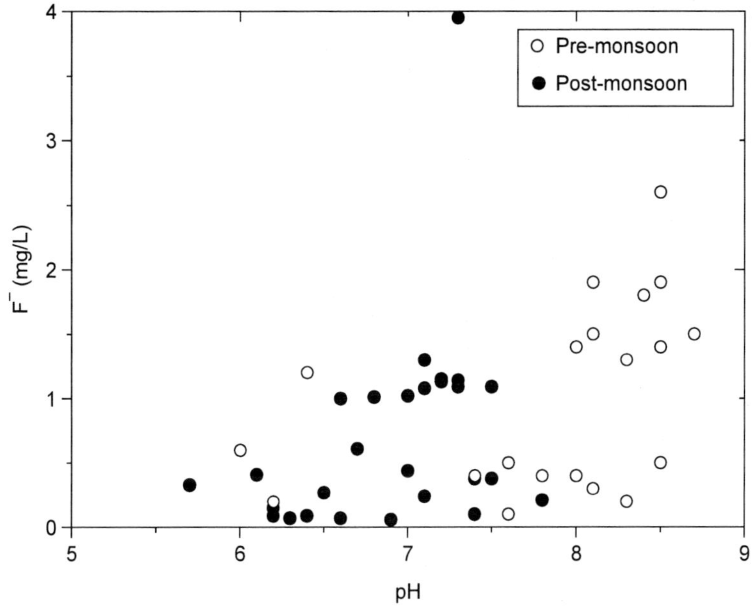

Fig. 7 Relationship between the distribution of F^- with pH in the groundwaters of Alappuzha district during the pre-monsoon (open circles) and post-monsoon (solid circles) periods of sampling. The pre-monsoon samples have a higher pH.

amongst F^- and HCO_3^- (Table 4). This reveals that high HCO_3^- and low Ca^{2+} concentration favours the dissolution of fluorite and precipitation of calcite in water.

Table 4 Correlation coefficients between F^- and HCO_3^- in groundwater from Alappuzha district, Kerala, India.

Sampling period	*Correlation coefficient* ($p < 0.01$)
March (pre-monsoon)	+ 0.84
July (post-monsoon)	+ 0.49

4.3 Comparison of Defluoridation Techniques

Comparison of the removal of F^- using different commercially available materials is presented in Table 5. When 2.8 g activated alumina was used, for an initial concentration of 5 mg/L of fluoride, 93% removal was observed. Acidic pH range was found to be more suitable and maximum adsorption of 96% was observed at pH 5 for 5 mg/L of F^- with addition of 1.4 g alumina.

The efficiency of F^- removal using alum (aluminium sulphate) was up to 99.6% when 0.1 g of alum was used to remove fluoride solution of 5 mg/L. Maximum percentage removal of F^- was possible at lowered/acidic pH levels. Removal efficiency 99.0% was attained at a pH of 4.1. This may be because of the fact that soluble Al-F complexes are formed at pH less than 7

Table 5 Percentage removal of fluoride by different types of commercially available adsorbents.

S.No.	*Materials used*	*Percentage removal*
1.	Charcoal (Himedia)	92.4
2.	Charcoal (Merck)	86.3
3.	Charcoal (Nice)	75.8
4.	Alumina (Qualigens)	92.8
5.	Alum (Nice)	99.6
6.	Silica Gel (IDPL)	94.8

resulting in the presence of aluminium ions in the treated water (Ghorai and Pant, 2005). The optimisation of pH on the removal of fluoride was done by studying the uptake of fluoride on activated alumina and alum as a function of pH (Fig. 8).

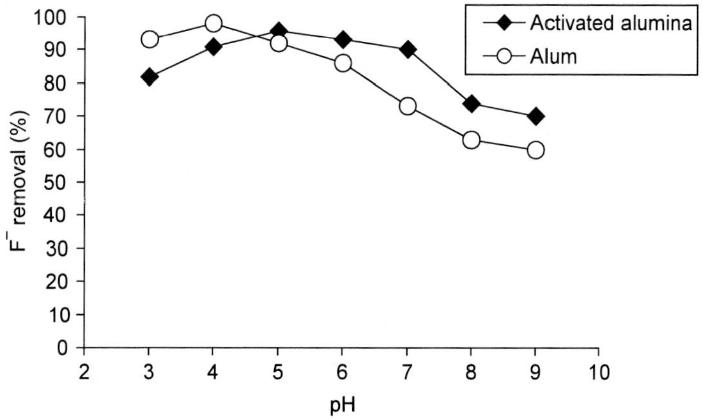

Fig. 8 Effect of pH on the adsorption of fluoride by activated alumina and alum.

An experiment with 24 g of silica gel removed 94.8% of F^- from a solution containing 5 mg/L of F^-. When the adsorbent quantity was reduced to 12 g at a pH value of 5.4, removal efficiency was lowered to 84%. Applying 2.8 g of activated alumina can remove 92.8% F^- from a solution with 5 mg/L F^-. Column studies indicate that 8.8 L of 5 mg/L fluoride can be treated by 100 g of alumina. The material can also be regenerated using acid and alkali washing.

The removal efficiency of various chemicals is indicated in Fig. 9. Factors like pH, temperature, initial concentration of fluoride and the amount of adsorbents influence the rate and the extent of fluoride removal by various adsorbents. Consequently, a brief comparison was made between charcoal, alumina, alum and silica gel, by performing adsorption studies under similar conditions. The results of the study indicates that, aluminium sulphate is the best material available in the market in terms of removal efficiency, quantity of material to be used and cost. The cost of alum is much cheaper than activated carbon, activated alumina and silica gel. A comparison of the cost of various adsorbents is listed in Table 6.

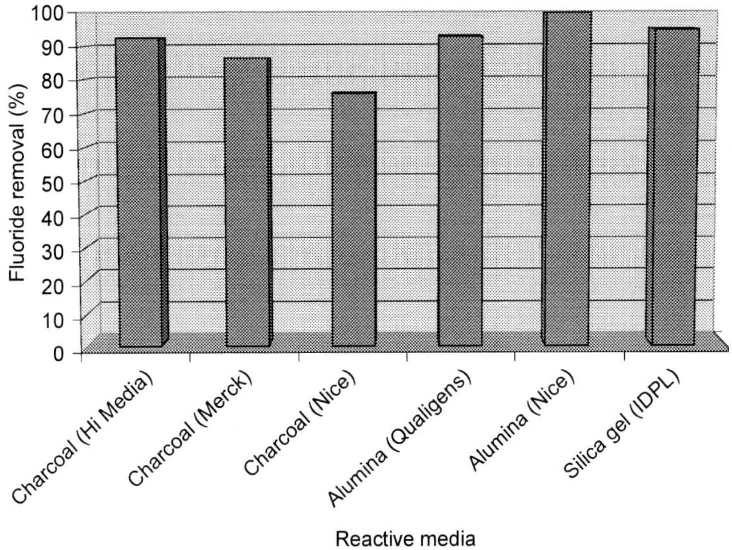

Fig. 9 A comparison of the efficiency of fluoride removal by different commercially available reactive media.

Table 6 Comparison of cost-effectiveness of defluoridation by different commercially available adsorbents.

Sl.No.	*Materials used*	*Amount of Material (gram)*	*Cost/100g of the material (INR)**	*Percentage removal*
1.	Charcoal (Himedia)	26.0	26	92.4
2.	Charcoal (Merck)	26.0	25	86.3
3.	Charcoal (Nice)	26.0	24	75.8
4.	Alumina (Qualigens)	2.8	21	92.8
5.	Alum (Nice)	0.1	17	99.6
6.	Silica Gel (IDPL)	24.0	34	94.8

*1 INR (Indian Rupee) ≈ 0.222862 US Dollars

5 DEVELOPMENT OF DOMESTIC FILTERS USING TESTED MATERIALS

Different filters were developed for defluoridation at the Centre for Water Resources Development and Management (CWDRM) for domestic use.

5.1 Filter-I

A low cost domestic filter of 2 L capacity was designed and checked for its fluoride removal efficiency. The system consists of two PVC tubes of diameter 11 cm and height 27 cm, which were, placed one over the other. A mesh is fitted at the bottom of the PVC tube, which is placed at the top. The fluoride solution is poured into this tube. The treated water is collected in the lower tube and is withdrawn through a tap fitted at the bottom tube.

5.1.1 *Use of activated alumina as filter material*

The filter was packed with 250 g of powdered activated alumina and defluoridation capacity of activated alumina was then calculated. The efficiency increased as the period of contact between the F^- solution and alumina is increased (Table 7).

Table 7 Defluoridation using Filter I containing 150 g powdered activated alumina (volume of treated water: 750 L).

S. No.	Initial concentration of F^- (mg/L)	Final concentration of F^- (mg/L)	Time of contact (hours)	Percentage removal
1.	5.0	3.9	2	23
2.	5.0	3.2	3	36
3.	5.0	2.3	5	43
4.	5.0	2.0	18	60

5.1.2 *Use of charcoal as filter material*

The filter was packed with powdered coconut shell charcoal (in 5 cm thickness). Here sand is used as the bearing medium. Fluoride solution of 5 mg/L was poured in to the upper chamber. The efficiency is found to be 64% when the treated water is analysed after 18 hours. The efficiency increased when the amount of charcoal is increased.

5.2 Filter-II

This system consists of stainless steel cylindrical column with the provision for input water reservoir at the top and the treated water collector at the bottom. A tap is connected at the bottom. In this filter the adsorbent is packed in a specially designed bottle. This bottle is provided with some grooves that can be punctured to allow the sample water to pass through. This bottle is fitted to the top cylindrical column. There is a fine mesh in the lowest portion of the bottle through which treated water is passed to the bottom cylinder. The capacity of this filter is 5 L. A study has been conducted to optimise the flow rate for the effective removal of fluoride. Effective removal of fluoride is possible up to a flow rate of 20 mL/minute.

5.2.1 *Use of alumina as filter material*

The bottle is filled with 250 g of powdered alumina and three grooves are punctured. The concentration of fluoride in treated water was analysed periodically and percentage removal of fluoride is calculated. The average flow rate is 20 mL/minute. The same experiment was carried out by increasing the flow rate and changing the initial concentration of fluoride and pH. The results are shown in Table 8. No further change in removal of fluoride was noticed with decreasing pH below 5.8. At lower pH and initial concentration of 5 mg/L F^-, alumina is a better adsorbent.

5.2.2 *Use of CATAL 2000™ as filter material*

CATAL 2000™ was another preparate used for defluoridation. It is a fine powder grade alumina having a bulk density of 900 kg/m^3 and manufactured by Indian Petrochemicals Corporation Limited. The filter is packed with this substance and the efficiency was studied by

Table 8 Defluoridation using Filter II containing 250 g of powdered alumina. (Average flow rate = 20 mL/minute, volume of treated water 1000 L).

S. No.	pH	Initial concentration of F^- (mg/L)	Final concentration of F^- (mg/L)	Percentage removal
1.	7.3	5.0	2.9	41
2.	6.1	3.0	0.5	83
3.	5.8	3.0	0.4	88

changing the initial concentration of fluoride. The results are shown in Table 9. The filter developed using CATAL 2000 is also found to be more effective at lower pH.

A filter was tested with a mixture of a plant material viz. clove and charcoal as filtering media. The unit was packed with 350 g-clove and 85 g charcoal and the treated water is analysed. The filter could remove 70-80% of fluoride for a fluoride contamination of up to 5 mg/L.

The filters used in above filters need regeneration depending upon the initial concentration of fluoride concentration.

Table 9 Defluoridation using Filter II containing CATAL 2000. (average flow rate = 20 mL/minute , volume of treated water 700 L).

S. No.	pH	Initial concentration of F^- (mg/L)	Final concentration of F^- (mg/L)	Percentage removal
1.	5.8	3.0	0.9	70
2.	5.8	5.0	1.7	65

6 CONCLUSIONS

The fluoride problem in the coastal district of Alappuzha, India was studied through a dental fluorosis survey and water quality assessment. The survey indicated that 14% of the surveyed school children had dental fluorosis. The problem is aggravated among children who do not get nutritional food and balanced diet. All the drinking water supply pumping wells in Alappuzha town has the fluoride concentration higher than the permissible limit. The fluoride concentration is significantly correlated with depth of the wells, pH, calcium and bicarbonate.

The defluoridation experiments using the locally available chemicals indicate that, treatment using alum is the most efficient and cost-effective method. Further the following conclusions were drawn from the experiments conducted on the comparison of various defluoridation techniques:

- Percentage removal of fluoride is found to increase with increase in amount of different adsorbents, but reaches a saturated value beyond which no further increase was observed.
- Among the three different types of charcoal used, the material supplied by Himedia is found to be the most efficient.
- The experiments indicate that neutral to acidic pH favours for maximum adsorption efficiency.

- From the adsorption studies, it is found that adsorption of fluoride obeys Freundlich adsorption isotherms. Studies using activated charcoal indicate that the adsorption obeys Langmuir adsorption isotherm.
- Aluminium sulphate is the best material available in the market in terms of removal efficiency, quantity of material to be used and cost. The materials can be used as a coagulant for which only simple technology is involved and can be carried out locally.

Acknowledgements

PSH is thankful to the Royal Institute of Technology (KTH), Sweden and Indian National Science Academy (INSA) for financial assistance under a Visiting Researchers Programme.

References

Agarwal V, Vaish A, Vaish P. Groundwater quality: focus on fluoride and fluorosis in Rajasthan. Curr Sci India 1997; 73: 743–746.

American Public Health Association. Standard methods for the examination of water and wastewater. APHA, Washington, 1992, 981 pp.

Boruff CS. Removal of fluorides from drinking waters. J Ind Eng Chem 1934; 26: 69–71.

Bulusu KR, Sundaresan BB, Pathak BN, Nawlakhe WG, Kulkarni DN, Thergaonkar VP. Fluorides in water,defluoridation methods and their limitations. J. Institute Engineers (India)-Environmental Engineering Division 1979; 60:1–25.

CWRDM. Water Atlas of Kerala. Centre for Water Resources Development and Management, Kozhikode, India.1995, 82 p.

Chakraborti D, Chanda CR, Samanta G, Chowdhury UK, Mukherjee SC, Pal AB, Sharma B, Mahanta KJ, Ahmed HA, Sing B. Fluorosis in Assam, India. Curr Sci India 2000; 78: 1421–1423.

Chaturvedi AK, Pathak KC, Singh VN. Fluoride removal from water by adsorption on china clay. Appl Clay Sci 1988; 3: 337–346.

Choubisa SL, Sompura K, Bhatt SK, Choubisa DK, Pandya H, Joshi SC, Choubisa L. Prevalence of fluorosis in some villages of Dungarpur district of Rajasthan. Indian J Environ Health 1996; 38: 119–126.

Datta PS, Deb DL, Tyagi SK. Stable isotope (O-18) investigations on the processes controlling fluoride contamination of groundwater. J Contam Hydrol 1996; 24: 85–96.

Economic Review. Kerala State Planning Board, Thiruvanathapuram. 1993.

Fernando D, Navarro-Quezada A, Grijalva M, Grimaldo M, Loyola-Rodriguez J, Deogracias O. Endemic fluorosis in Mexico. Fluoride 1997; 30: 233–239.

Gaciri SJ, Davies TC. The Occurrence and Geochemistry of Fluoride in Some Natural-Waters of Kenya. J Hydrol 1993; 143: 395–412.

Ghorai S, Pant KK. Equilibrium, kinetics and breakthrough studies for adsorption of fluoride on activated alumina. Sep Purif Technol 2005; 42:265–271.

Gizaw B. The origin of high bicarbonate and fluoride concentrations in waters of the Main Ethiopian Rift Valley, East African Rift System. J Afr Earth Sci 1996; 22: 391–402.

Handa BK. Geochemistry and genesis of fluoride containing groundwaters in India. 1975; 13: 275–281.

Industrial Toxicology Research Centre. Contributions of ITRC to Rajiv Gandhi National Drinking Water Mission-A consolidated Report 1986-1994. Lucknow, 1995.

Jacks G, Bhattacharya P, Chaudhary V, Singh KP. Controls on the genesis of some high-fluoride groundwaters in India. Appl Geochem 2005; 20: 221–228.

Jacks G, Rajagopalan K. Effects of past climate change on hydrochemistry and hydraulics in two large aquifers in S and SE Asia. In: 14th Salt water intrusion meeting, Malmö, 1996; 79–83.

Jacob V, Subramanian P, Balasubramaniyam, Thampi D, Satyanarayana-Rao SN. Preliminary Report on the Hydrogeology of the Project Area, Coastal Kerala Ground Water Project. CGWB, 1985;

Jolly SS, Sing BM, Mathur OC. Endemic fluorosis in Punjab (India). Am J Med 1969; 47: 553–563.

Kharb P, Susheela AK. Fluoride Ingestion in Excess and Its Effect on Organic and Certain Inorganic Constituents of Soft-Tissues. Med Sci Res 1994; 22: 43–44.

Killedar DJ, Bhargava DS. Effects of stirring rate and temperature on fluoride removal by fishbone charcoal. ndian J Environ Health 1993; 35: 81–87.

Kundu N, Panigrahi MK, Sharma SP, Tripathy S. Delineation of fluoride contaminated groundwater around a hot spring in Nayagarh, Orissa, India using geochemical and resistivity studies. Environ Geol 2002; 43: 228–235.

Kundu N, Panigrahi MK, Tripathy S, Munshi S, Powell MA, Hart BR. Geochemical appraisal of fluoride contamination of groundwater in the Nayagarh District of Orissa, India. Environ Geol 2001; 41: 451–460.

Li JX, Cao SR. Recent Studies on Endemic Fluorosis in China. Fluoride 1994; 27: 125–128.

McLaren JR. Possible Effects of Fluorides on Thyroid. Fluoride 1976; 9: 105–116.

Mella S, Molina X, Atalah E. Prevalence of endemic dental fluorosis and its relation with fluoride content of public drinking water. Rev Med Chile 1994; 122: 1263–1270.

Mullenix PJ, Denbesten PK, Schunior A, Kernan WJ. Neurotoxicity of Sodium-Fluoride in Rats. Neurotoxicol Teratol 1995; 17: 169–177.

Nayaro T, Aswathanarayana U, Mungere J, Lahermo P. A geochemical model for the abnormal fluoride concentrations in waters in parts of northern Tanzania. J Afr Earth Sci 1984; 2: 129–140.

Rajiv Gandhi National Drinking Water Mission. Prevention and control of Fluorosis in India 1986. 1993; 8–9.

Saha S. Treatment of Aqueous Effluent for Fluoride Removal. Water Res 1993; 27: 1347–1350.

Sharma DRR, Rao SLN. Fluoride concentration in ground waters of Visakhapatnam, India. B Environ Contam Tox 1997; 58: 241–247.

Susheela AK. Technical information for training and awareness camp of doctors, public health engineers and other officers on control of Fluorosis. New Delhi, 1991; 299–301.

Susheela AK. Fluorosis management in India: the impact due to networking between health and rural drinking water supply agencies. In: Proceedings of the Symposium: Interdisciplinary Perspectives on Drinking Water Risk Assessment and Management, Santiago de Chile, 2000;

Susheela AK, Jethanandani P. Circulating testosterone levels in skeletal fluorosis patients. J Toxicol-Clin Toxic 1996; 34: 183–189.

Susheela AK, Taposh K, K D. Fluoride ingestion and its correlation with gastrointestinal discomfort. Fluoride 1992; 25: 5–22.

Teotia SPS, Teotia M. Endemic fluorosis in India: a challenging national health problem. J Assoc Physicians India 1984;

Waldbott GL. Non-Skeletal Phase of Chronic Fluorosis - Introduction. Fluoride 1976; 9: 5–8.

Wang XC, Kawahara K, Guo XJ. Fluoride contamination of groundwater and its impacts on human health in Inner Mongolia area. J Water Serv Res Tec 1999; 48: 146–153.

Viswanadham C, Murty B. Fluoride in ground waters of Rajasthan and Gujarat. In: Proceedings of the Symposium: Fluorosis, Hyderabad, 1974.

Yang CL, Dluhy R. Electrochemical generation of aluminum sorbent for fluoride adsorption. J Hazard Mater 2002; 94: 239–252.

Zhao LB, Liang GH, Zhang DN, Wu XR. Effect of a high fluoride water supply on children's intelligence. Fluoride 1996; 29: 190–192.

Arsenic Contamination in Groundwater Resources in Kathmandu Valley

42

Makhan Maharjan, Bhushan R. Shrestha, Kabita B. Shrestha, Rosha Raut (Khadka) and Roshan R. Shrestha[1]

Suresh Das Shrestha[2]

Heihachiro Ishihara and Bikash Kafle[3]

[1]Environment and Public Health Organization (ENPHO), P.O Box: 4102, New Baneshwor, Kathmandu, Nepal.

[2]Central Department of Geology, Tribhuvan University, Kathmandu, Nepal.

[3]Japan International Cooperation Agency (JICA) Expert Office, Ministry of Physical Planning and Works (MPPW), Singh Durbar, Kathmandu, Nepal.

ABSTRACT

The present study was undertaken to assess arsenic vulnerability and seasonal variation in arsenic level in groundwater resources in Kathmandu Valley. Water samples from 149 shallow tube wells and 87 dugwells were tested for arsenic concentration in pre-monsoon, and 122 deep tube wells were tested in pre-monsoon and monsoon. Arsenic concentration in 11.4% of shallow tube wells and 11.5% of shallow dugwells exceeded 0.01 mg/L, while 71.3% and 57.4% of deep tube wells exceeded 0.01 mg/L of arsenic in pre-monsoon and monsoon, respectively. Of the tested deep tube wells, 13.1% and 8.2% of samples exceeded 0.05 mg/L of arsenic in pre-monsoon and monsoon, respectively. Arsenic concentrations in pre-monsoon and monsoon, strongly correlated for deep tube wells (r = 0.94, $p < 0.001$). Data show weak positive correlations between arsenic concentrations and tube well depths for both pre-monsoon and monsoon. Arsenic level in groundwater fluctuates with the time period though a definite trend of increase or decrease was not found.

1 INTRODUCTION

Over the last few decades, groundwater has become one of the major water sources of Kathmandu Valley in Nepal. In Kathmandu Valley, groundwater was first exploited for water supply in 1970 and the extraction increased rapidly with time. At present about 45% of the total municipal water supply is fulfilled by groundwater resources in Kathmandu Valley (Acres International *et al.*, 2002). As surface water sources are scarce, increasing demand of water for drinking, industry and irrigation will definitely increase extraction of groundwater in days to come. Therefore, quality of groundwater is a major issue to authorities who are responsible for providing access to safe drinking water and sanitation for population of about 1, 645, 091 in the Valley (CBS, 2002).

Arsenic contamination of natural waters has become an issue of growing concern around the world during the past decade. Several studies in past have shown that South Asian region is highly vulnerable to arsenic. Arsenic in groundwater in Nepal's Terai region has already become a challenging issue for His Majesty's Government of Nepal (HMG/N) and other agencies working in the field of water and sanitation in that region (Tandukar *et al.*, 2001; Bhattacharya *et al.*, 2003; Shrestha et al., 2003; Shrestha *et al.* 2004; Maharjan *et al.*, 2005) Of more than 400,000 water samples tested for arsenic so far, about 13% had arsenic concentration above WHO guideline value of 0.01 mg/L and nearly 3% samples above Nepal Standard of 0.05 mg/L. This is also a challenging issue for meeting the target of sustainable access to safe drinking water and sanitation of Millenium Development Goal.

Vulnerability of arsenic in deep tubewells in Kathmandu Valley is a major public health concern as groundwater is being extensively used. Therefore, national water and sanitation policy and strategy must clearly address the issues regarding arsenic in groundwater and appropriate actions need to be initiated immediately. Arsenic contamination studies, in Nepal, have so far been conducted only in Terai region and data are scarce in case of Kathmandu Valley. Policies developed on the basis of data and information available for Terai region may not effectively address issues related to Kathmandu due to differences such as geology, hydrogeology, settlement pattern and water supply in the Valley.

The present study has been conducted with the objectives of known arsenic levels in both shallow and deep aquifers and preparing an information base regarding arsenic contamination in deep aquifers of Kathmandu Valley. Information and recommendations made by this study will provide valuable input in planning, policy making, and implementing plans/programmes regarding extraction of groundwater and providing arsenic free water in the Valley.

2 MATERIALS AND METHODS

2.1 Study Area and Sampling Sources

The present study covered Kathmandu and Lalitpur municipality areas for dug wells (DW) and shallow tube wells (STW), and the whole Kathmandu Valley for deep tube wells (DTW). Sampled DTWs are distributed over about 450 sq. km (i.e. 18 km × 25 km) area of the central part of the Valley. In the present study, a well usually cased with bricks or concrete ring masonry with a diameter of 1-2 metres dug below the water table was considered as a DW.

While a boring or drilled well with a depth of < 50 m was considered as a STW and a boring or drilled well ≥ 50 m depth as a DTW.

2.2 Data Collection and Water Sampling

The study was conducted in two different periods, pre-monsoon and monsoon. Necessary information/data collection, water sampling and analysis for pre-monsoon were done in between March-May 2005, while it was done between August-September 2005 for monsoon. Raw water samples were randomly collected from 149 STWs and 87 DWs proportion to ward size in Kathmandu and Lalitpur municipal areas, and water samples from 137 DTWs were collected in Kathmandu Valley. Tube wells ≥ 50 m depth were listed and water sample was collected from all operating tube wells where permission was granted from the owners. Locations of the sampled DTWs were also identified during the first water sampling using GPS. Arsenic fluctuation monitoring of DTWs exceeding 0.05 mg/L of arsenic was done for two consecutive days in both pre-monsoon and monsoon to know and compare arsenic levels.

Water sample was collected in 160 mL pre-acid washed polypropylene bottle after running water for at least for 5 minutes, and immediately acidified with concentrated hydrochloric acid (HCl). The samples were carried in an ice-box and transported to the ENPHO Research Laboratory at New Baneshwor in Kathmandu for analysis.

2.3 Analysis of Water Samples

Measurement of total arsenic in water samples was conducted by an atomic absorption spectrophotometry (SOLAAR 969AA Spectrometer, Thermo Elemental, UK), equipped with a continuous hydride generator (HG-AAS). Samples with very high arsenic concentration were re-sampled and tested. Assay accuracy was ensured by inclusion of reference material, NIST SRM 1640 (National Institute of Standards and Technology, Gaithersburg, MD, USA) for trace elements in water (with 26.67 ± 0.41 μg/kg arsenic). The reference material produced the value within the certified range.

2.4 Creation of Data Sets and Map Preparation

A dataset of deep tube wells tested was created with all required attributes (well ID, owner's name, locality, VDC/municipality, use, depth, pre-monsoon arsenic concentration, monsoon arsenic concentration, geographic location (East) and geographic location (North) for spatial analysis, statistical analysis and mapping.

Basic features to prepare maps were identified and purchased from different sources based on their usefulness and availability. Topographic features covering Kathmandu Valley were acquired from Survey Department and geological features with well fields from Department of Mines and Geology. Spatial data layers acquired in several independent sheets were checked, cleaned and merged to form a single digital map of Kathmandu Valley. Geological map of the study area with different geological formation was also prepared. These maps later were used as base maps for preparing various output maps of the study.

Spatial data set for location of all sampled wells was created with reference to the location of wells collected through GPS, and used for vulnerability assessment of arsenic in DTWs in the

Valley. Arsenic test data set, results of geo-statistical analysis and spatial data layers created were integrated and different types of maps were produced with the help of Geographic Information System (GIS) for visualizing arsenic concentration issue in Kathmandu Valley.

3 RESULTS AND DISCUSSION

In pre-monsoon, out of 87 DWs, 11.5% of samples exceeded the WHO guideline value of 0.01 mg/L of arsenic, and out of 149 STWs, 11.4% of samples exceeded 0.01 mg/L of arsenic. Arsenic concentration in none of the STW or DW exceeded the Nepal Standard of 0.05 mg/L of arsenic. Compared to DWs and STWs, DTWs were found to be highly vulnerable to arsenic (Table 1). Based on the findings, in present study, detail results are presented only for DTWs.

Table 1 Distribution of water sources by arsenic (As) concentration.

Arsenic (mg/L)	*Number (%) of samples in pre-monsoon*		*Number (%) of samples in monsoon*	
	DTW	STW	DW	DTW
upto 0.01	38 (28.4)	132 (88.6)	77 (88.5)	54 (43.2)
0.011–0.05	80 (59.7)	17 (11.4)	10 (11.5)	61 (48.8)
> 0.05	16 (11.9)	0 (0.0)	0 (0.0)	10 (8.0)
Total	**134 (100)**	**149 (100)**	**87 (100)**	**125 (100)**

In pre-monsoon, water samples from 134 DTWs were tested and in monsoon water samples from 125 DTWs were tested. Some of the wells tested in pre-monsoon could not be tested in monsoon due to non-operation of the wells and some of the wells that were not tested in pre-monsoon were tested in monsoon. Out of total 137 tested DTWs, 122 DTWs were tested in both pre-monsoon and monsoon (Fig. 1).

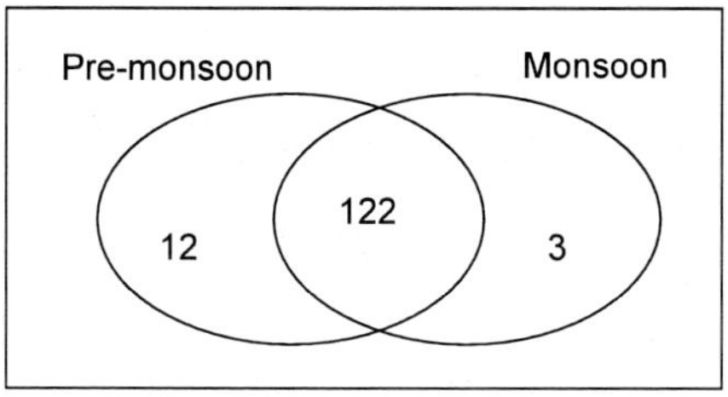

Fig. 1 Venn diagram showing number of deep tube wells tested in pre-monsoon and monsoon.

3.1 Arsenic Concentrations in Pre-monsoon

Among 134 tested DTWs in pre-monsoon, 71.6% wells had arsenic concentration above the WHO guideline value of 0.01 mg/L and 11.9% of wells exceeded Nepal Standard of 0.05 mg/L (Table 1). The highest arsenic concentration of 0.265 mg/L was detected in Kuleswore followed by 0.250 mg/L in Harihar Bhawan, 0.212 mg/L in Kalanki and so on. The lowest arsenic concentration of < 0.005 mg/L (i.e. not detected) was observed in 21 wells (15.7%) in different areas of the Valley.

3.2 Arsenic Concentrations in Monsoon

Out of 125 tested DTWs in monsoon, 56.8% of the wells had arsenic concentration above the WHO guideline value and 8.0% of the wells exceeded Nepal Standard (Table 1). Highest arsenic concentration observed in Kuleswore in pre-monsoon was reduced to 0.211 mg/L followed by 0.173 mg/L in Harihar Bhawan and 0.111 mg/L in Kalanki. The number of wells with lowest arsenic concentration of < 0.005 mg/L was increased to 28 wells (22.4%). Mean, median and maximum arsenic concentrations were 0.033 mg/L, 0.019 mg/L and 0.265 mg/L, respectively for pre-monsoon, while it was 0.022 mg/L, 0.013 mg/L and 0.211 mg/L, respectively for monsoon. The histogram of arsenic concentration (not shown), for both periods, shows that the distribution of arsenic is strongly positively skewed which indicates that only a few wells contain very high arsenic concentration. Arsenic concentration of almost all wells reduced from pre-monsoon to monsoon.

3.3 Arsenic Fluctuation Monitoring of Deep Tube wells Exceeding Nepal Standard

Arsenic fluctuation monitoring of 15 deep tube wells exceeding 0.05 mg/L of arsenic in pre-monsoon extraction hours. The monitoring of one well could not be done since the owner of the well did not allow for water sampling.

Three water samples were collected in between total water extraction hours of the sampling day for two consecutive days. Thus, six water samples were tested for arsenic from each well. Arsenic concentrations of monitored wells for pre-monsoon and monsoon are presented in Fig. 2. The minimum, maximum and mean arsenic concentrations for each well are plotted in Fig. 3 for pre-monsoon and in Fig. 4 for monsoon, which show that the fluctuation in arsenic is not much in the majority of tested wells, except for the Well ID 50. Statistically (e.g. Man-Whitney Test) there is no significant difference between the mean arsenic concentration of six samples and arsenic concentration of pre-monsoon or monsoon.

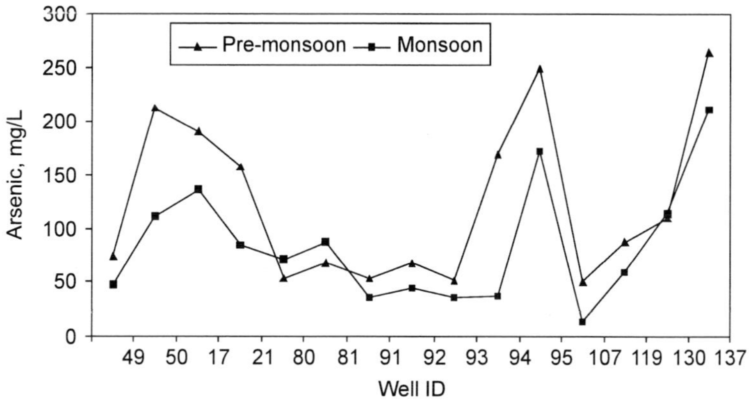

Fig. 2 Arsenic concentrations in DTWs exceeding 0.05 mg/L for pre-monsoon and monsoon.

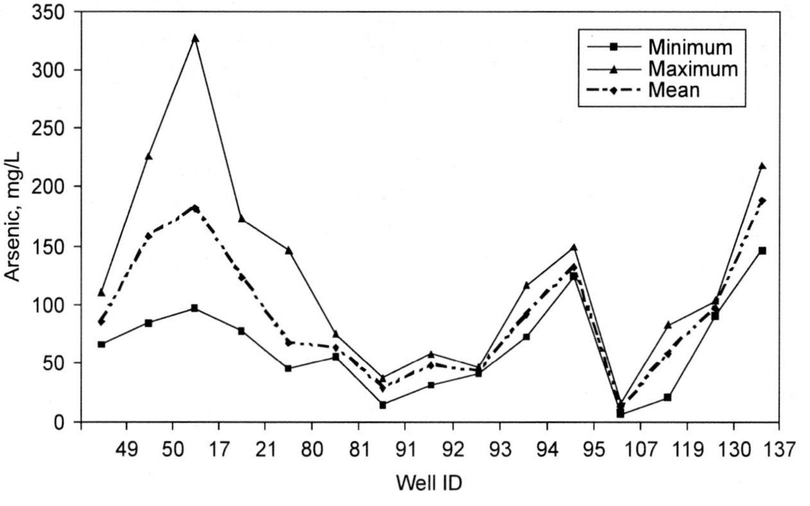

Fig. 3 Minimum, maximum and mean arsenic concentrations for pre-monsoon.

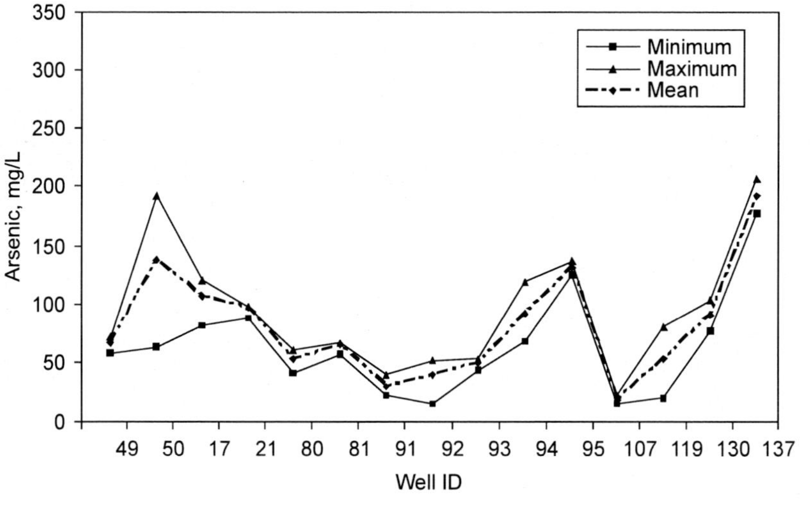

Fig. 4 Minimum, maximum and mean arsenic concentrations for monsoon.

3.4 Variation in Arsenic Concentration From Pre-monsoon to Monsoon

Of the total 122 DTWs tested in both pre-monsoon and monsoon, the arsenic concentration of almost all wells reduced from pre-monsoon to monsoon. Arsenic concentration in 71.3% of DTWs exceeded 0.01 mg/L in pre-monsoon, while it was 57.4% in monsoon. Similarly, 13.1% of DTWs with arsenic concentration above 0.05 mg/L in pre-monsoon was reduced to 8.2% in monsoon. In pre-monsoon; mean, median and maximum arsenic concentrations were 0.033 mg/L, 0.019 mg/L and 0.265 mg/L, respectively which reduced to 0.022 mg/L, 0.013 mg/L and 0.211 mg/L, respectively in monsoon. The decreasing pattern of arsenic from pre-monsoon

Table 2 Classification of arsenic concentration by pre-monsoon and monsoon.

Arsenic (mg/L)		*Monsoon*				
		upto 0.01	*0.011 – 0.05*	*> 0.05*	*Total*	*%*
Pre-monsoon	upto 0.01	35	–	–	35	28.7
	0.011 – 0.05	17	54	–	71	58.2
	> 0.05	–	6	10	16	13.1
	Total	52	60	10	122	100.0
	%	42.6	49.2	8.2	100.0	

to monsoon is clearly observed when the arsenic concentration is classified into three ranges (Table 2). Low arsenic levels in monsoon compared to that of pre-monsoon might be due to the effect of recharging of groundwater.

The scatter diagram plotted for arsenic concentrations in pre-monsoon and monsoon shows strong correlation ($r = 0.94$, $p < 0.001$), suggesting very similar distributions of arsenic for the two periods (Fig. 5). Similar close relation ($r = 0.84$, $p < 0.001$) was also found for 15 arsenic fluctuation monitored wells.

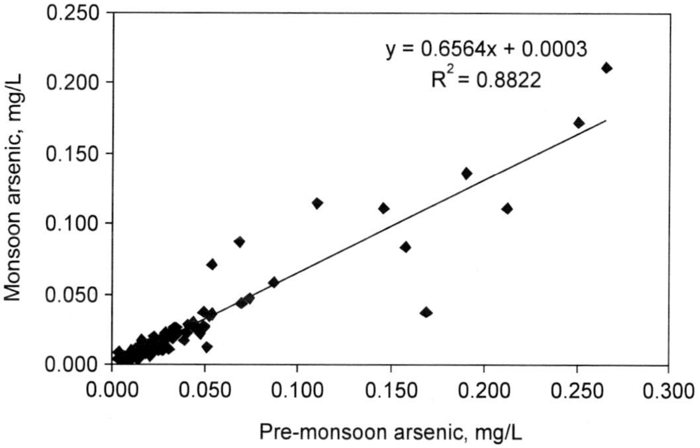

Fig. 5 Arsenic concentration in pre-monsoon and soon.

3.5 Relation of Arsenic Concentration by Depth

Regarding the depth of wells, 120 wells tested in pre-monsoon and 113 wells tested in monsoon have recorded depth. The depth ranges from 50 to 318 m. For pre-monsoon, the mean (SD) depth is 211 (75.7) m and for monsoon; it is 206 (76.4) m. Majority of the wells tested above 0.05 mg/L have depth more than 250 m in both periods. The depth of tested wells is classified into three equal groups such as 50–149 m, 150–249 m, and 250–349 m. Nearly half of the wells have depth more than 250 m, about 30% have depth between 150–249 m and about one-fourth of the wells have depth between 50 and 149 m in both periods.

The distribution of wells by three classes of arsenic concentration and depth groups of wells is given in Table 3. In pre-monsoon, the percent of wells containing more than 0.05 mg/L of arsenic is only 3.4% for the first depth group (i.e. 50–149 m) and this percent increases to 10.8% in the second depth group (i.e. 150–249 m) and further increases to 20.4% in the third depth group of 250–349 m. The percent of wells containing 0.011–0.05 mg/L of arsenic is also increasing from 48.3% in first depth group to more than 60.0% in the second and third depth groups. On the other hand, the per cent of wells below WHO guideline value is decreasing from 48.3% in first depth group to 24.3% in the second depth group and further decreases to 16.7% in the third group. This trend is also similar for monsoon. Weak positive correlations between arsenic concentration and depth of wells ($r = 0.15$ for pre-monsoon and $r = 0.13$ for monsoon) suggest that the deeper wells in Kathmandu Valley tend to have higher arsenic concentration compared to the lower depth wells. It contradicts with the situation generally found in the Terai region where deeper wells tend to have lower arsenic concentration. More study is needed to see the relation between depth and arsenic concentration.

Table 3 Arsenic concentration by depth for pre-monsoon and monsoon.

Seasons	*Arsenic (mg/L)*	*Depth group (m)*		
		50-149	150-249	250-349
		Number (%)	*Number (%)*	*Number (%)*
Pre-monsoon	upto 0.01	14 (48.3)	9 (24.3)	9 (16.7)
	0.011-0.05	14 (48.3)	24 (64.9)	34 (63.0)
	> 0.05	1 (3.4)	4 (10.8)	11 (20.4)
	Total	29 (100.0)	37 (100.0)	54 (100.0)
Monsoon	upto 0.01	18 (60.0)	15 (42.9)	14 (29.2)
	0.011-0.05	11 (36.7)	17 (48.6)	28 (58.3)
	> 0.05	1 (3.3)	3 (8.6)	6 (12.5)
	Total	**30 (100.0)**	**35 (100.0)**	**48 (100.0)**

3.6 Arsenic Concentration and the Screen Positions

The available lithologs of many deep tube wells show that the majority of wells have more than one screen at different horizons. The wells tap multiple aquifers to abstract maximum amount of water. In doing so, water of various qualities is mixed. Most wells with arsenic concentration above 0.05 mg/L have the screens placed on both the soft sediments and the basement rocks. It is not certain how this has affected the quality of water but two wells exceeding 0.05 mg/L of arsenic have their screens only in the basement rocks. A well in Sanepa (Well ID 130) has only a single screen and has the arsenic concentration of 0.110 mg/L in pre-monsoon and 0.115 mg/L in the monsoon. Another well of Pulchowk, near to Sanepa, (Well ID 081) has multiple screens placed in the basement rocks and have the arsenic concentration of 0.068 mg/L in the pre-monsoon and 0.087 mg/L in the monsoon.

The lithologs of 5 wells containing more than 0.05 mg/L of arsenic show that most of the wells are placed on the basement rocks and most of them extract water from both the rock as well as the sediment above consisting of mainly sand and gravel. In this region the basement rocks are at shallow depth. The basement rocks are basically limestone and the sandstone. Hence, high arsenic concentration shows direct relation with the placing of screens near or on the basement rock. The aquifer horizons tapped ranges from 998–1312 m above mean sea level. The number of screens used range from 1–5.

3.7 Mapping of Arsenic Contamination

3.7.1 *Spatial distribution of arsenic concentration*

Two types of maps, Topographic map and Geological map, are used to locate the wells. All tested wells are located in the topographic map with four features-boundary, road, river and land use and in the geological map with geological formation. The major purpose of mapping of arsenic concentration is to show the spatial distribution of arsenic in the Valley. The map provides the exact location of arsenic tested well in geographical coordinates with the arsenic concentration in one of the three ranges: upto 0.01 mg/L, 0.011 – 0.05 mg/L and > 0.05 mg/L.

Six types of maps are prepared to show the spatial distribution of arsenic concentration in Kathmandu Valley in pre-monsoon and monsoon separately. 134 DTWs tested for arsenic in pre-monsoon. It shows that most of the tested wells in the northern part of the Valley are arsenic safe regarding WHO guideline value. The majority of wells of the central part exceeded 0.01 mg/L. The wells located in the central settlement south of the Bagamati River in the Lalitpur area have high arsenic concentration exceeding 0.05 mg/L. In the far-western and far-eastern part of the Valley, the arsenic concentration was low. However, only a few wells in this part were tested. 125 DTWs tested for arsenic in monsoon. It also shows the same pattern of spatial distribution of arsenic concentration pre-monsoon. The map clearly shows that for both pre-monsoon and monsoon, there is a narrow belt of Balkumari-Pulchowk-Kuleswor-Rabibhawan in the central west of the Valley with high arsenic concentration exceeding 0.05 mg/L. A well in Kuleswor with the highest arsenic value of 0.265 mg/L in pre-monsoon also lies in this region. Some areas near to this region such as Thapathali, Tripureswor also have comparatively high arsenic concentration exceeding 0.01 mg/L. Variation in arsenic concentration from pre-monsoon to monsoon.

3.7.2 *Arsenic concentration and the geology*

Map 4 provides the locations of tested wells with respect to geological formation of the valley and arsenic concentration ranges. Of the 16 types of geological formation shown in the map of the study area, the arsenic tested wells are located only in 7 types of geological formations: Chandragiri, Chapagau, Gokarna, Kalimati, Lokundol, Recent floodplain, and Tokha. The distribution of tested wells (%) by geological formation are shown in Fig. 6 for pre-monsoon and in Fig. 7 for monsoon. The highest number of tested wells (63 in pre-monsoon and 57 in monsoon) lie in the Gokarna Formation, followed by Kalimati Formation (46 in pre-monsoon and 43 in monsoon). Very few DTWs are distributed over other formations.

Generally, arsenic concentration in any particular well, especially the DTW, is influenced by subsurface geology or the aquifer horizon exploited rather than the surface geology. Arsenic concentration in Terai region is much influenced by the type of deposits (Shrestha *et al.*, 2004). In general, coarse region with grained sediments have low arsenic, which is also observed in this study. DTWs in the northern most part, i.e. in Tokha Formation and northern half part of

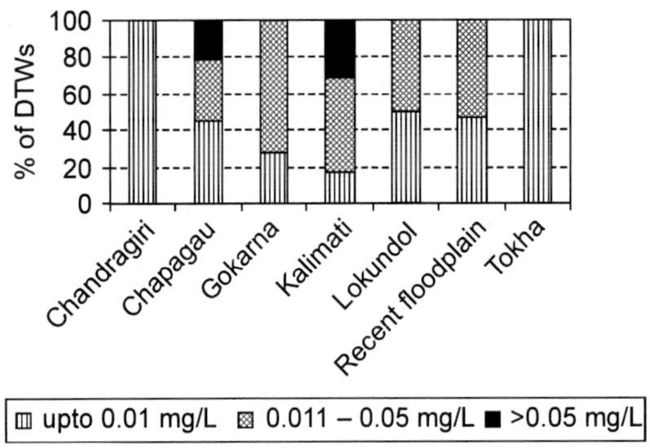

Fig. 6 Distribution of wells tested in pre-monsoon by arsenic concentration and geological formation.

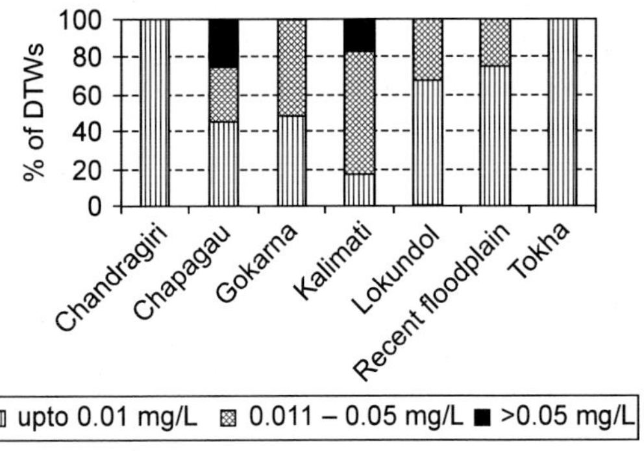

Fig. 7 Distribution of wells tested in monsoon by arsenic concentration and geological formation.

Gokarna Formation generally have arsenic concentration below 0.01 mg/L or only slightly higher. This is also true in the southern part where although there are only a few DTWs, all of them have arsenic concentration below 0.01 mg/L. In Terai, higher arsenic is associated with fine grained sediments. This is also true in this study. The thick Kalimati layer of the Kalimati formation does seem to influence the arsenic concentration in the DTWs to some extent.

3.7.3 *Arsenic concentration and the hydrogeological boundary*

As the distribution of hydrogeological boundary is based on the geology and the hydrogeological parameters, the influence is very much imminent in the arsenic concentration. Most wells in the northern part of the northern groundwater zone have arsenic concentration below 0.01 mg/L due to relatively coarser grained sediments which make up the aquifer material. Although the boundary between the northern groundwater zone and the central groundwater zone crosses right across the Gokarna formation in the west and the Kalimati formation in the east, the Kalimati layer is relatively thin and not continuous as compared to the south.

The majority of the arsenic tested wells exceeding 0.01 mg/L of arsenic concentration are restricted to the central groundwater zone. DTWs with very high arsenic are concentrated in a very small area of Hariharbhavan, Kalanki and Kuleswor area.

3.7.4 *Arsenic concentration and depth of deep tube wells*

120 DTWs tested for arsenic in pre-monsoon, whose depth were recorded. 113 DTWs tested for arsenic in monsoon. DTWs are classified into one of the three ranges of arsenic concentration (upto 0.01 mg/L, 0.011–0.05 mg/L and > 0.05 mg/L) and one of the three groups of depth (50–149 m, 150–249 m and 250–349 m).

3.7.5 *Zoning of kathmandu valley*

Zoning of the Kathmandu Valley with respect to arsenic concentration of the DTWs distributed over the valley by geostatistical analysis shows the pattern of spatial dependence quantitatively. Such pattern can be used to help in the setting of priority areas for emergency testing of arsenic and formulation of groundwater extraction policy. Due to the spatial distribution of arsenic, it is important to study behavior of arsenic concentration spatially taking into consideration the location of each well.

The spatial continuity of arsenic concentration in the Kathmandu Valley was measured by variogram for pre-monsoon and monsoon separately. The variogram for the $\log_{10}$-transformed arsenic concentration of pre-monsoon and monsoon were computed up to lag distance of 15 km at intervals of 500 m. As the distribution of arsenic data of pre-monsoon and monsoon were very similar, the variogram models also found very similar for pre-monsoon and monsoon. The spherical model with range of 9 km, sill of 0.99 $(mg/L)^2$, and nugget effect of 0.36 $(mg/L)^2$ was fitted to the arsenic data of pre-monsoon. Similarly, the spherical model with range of 9 km, sill of 0.68 $(mg/L)^2$, and nugget effect of 0.32 $(mg/L)^2$ was fitted to the arsenic data of monsoon. Then, the maps of arsenic concentration as an interpolated grid surface were created by Ordinary Kriging Method using geostatistical software GS+. The corresponding maps of standard errors of the estimates were also created for both pre-monsoon and monsoon. Thus, four types of maps were prepared for zoning of the Kathmandu valley with respect to arsenic concentration, which are: Maps 9-10 show zoning of the valley with respect to arsenic concentration into six zones of arsenic range: upto 0.01 mg/L, 0.011–0.02 mg/L, 0.021–0.03 mg/L, 0.031–0.04 mg/L, 0.041–0.05 mg/L and > 0.05 mg/L for pre-monsoon and monsoon.

3.7.6 Vulnerability assessment

Vulnerability of arsenic in groundwater resources were categorized into three classes - Highly vulnerable (> 0.05 mg/L), Vulnerable (0.011–0.05 mg/L) and Safe (upto 0.01 mg/L) according to arsenic concentration. Arsenic concentration in the vulnerable range (0.011–0.05 mg/L) in the map for zoning was further classified into four classes: 0.011–0.02 mg/L, 0.021–0.03 mg/L, 0.031–0.04 mg/L and 0.041–0.05 mg/L.

Vulnerable area in pre-monsoon is larger compared to that in monsoon with reference to the zoning maps for pre-monsoon and monsoon, therefore, the arsenic vulnerability map for Kathmandu Valley is produced on the basis of zoning map for pre-monsoon. Arsenic vulnerability map is a topographic map of Kathmandu Valley with arsenic concentration zones, NWSC's well fields, groundwater districts and location of DTWs using for different purposes. Geographic coordinates of the DTWs other than tested wells located in the map are taken from the report "Urban Water Supply Reforms in the Kathmandu Valley, Groundwater Monitoring Programme (UWSR)" (Metcalf & Eddy, Inc./CEMAT, 2000). It is important to note here that no verification has been done for the locations of these wells.

It shows that northernmost part of the north groundwater zone is arsenic safe whereas lower part of the zone is vulnerable. Most of the areas in the central groundwater zone are vulnerable, some part of this zone can also be classified as highly vulnerable, which is demarked in the map by red zoning line. Whereas southern part of the southern groundwater zone is found safe but northern part of this zone is vulnerable and small part of the northern part is highly vulnerable.

4 CONCLUSIONS AND RECOMMENDATIONS

The present study has confirmed the existence of arsenic in groundwater resources in Kathmandu Valley. Arsenic contamination is spatial and deeper tubewells in the Valley have higher arsenic concentration compared to the lower depth tubewells. Some of the areas in the valley particularly the areas within the central groundwater zone are vulnerable to arsenic contamination. Arsenic concentration is slightly lower in monsoon, indicating the possibility of dilution due to recharging or other reason(s) yet to be explored.

Recommendations

- Arsenic contamination in groundwater resources in Kathmandu Valley must be addressed in 'Urban Water and Sanitation Policy'.
- There is no authentic agency for regulating groundwater extraction. Licensing and control mechanism should be developed for permitting groundwater abstraction and regular monitoring of water quality of installed tube wells.
- A database system should be established to keep records and update lithologs of all deep tube wells installed and water quality data in the Valley. This will help in creating information base for identifying location and depth for good aquifer for arsenic safe groundwater for decision makers, planners and implementers and for future in-depth research.
- Information about arsenic vulnerable zone will help in proper management of groundwater resources in Kathmandu Valley. Safe zones identified by this study must be protected from haphazard extraction of groundwater and tubewells with high arsenic concentration should be regularly monitored.
- Drilling of deep tubewells should be practiced following standard procedures (e.g. use of professional drilling companies, not tapping water through multiple screens, use of bentonite or concrete, etc).
- Some areas identified as being vulnerable to arsenic contamination in present study, have several deep tube wells that supply drinking water to sensitive areas such as schools, hospitals and residential colonies. Awareness programs and appropriate mitigation measures are needed to safeguard public health.

Acknowledgements

We thank Ms. Sayako Tokuda of JICA Nepal Office for providing partial financial support to conduct this study. We are grateful to personnel in the hotels, industries/factories, offices, institutions who provided water samples and the information on water sources. We thank field workers and all those who helped in the study directly or indirectly.

References

Acres International, ARCADSI Euroconsult Land and Water Product Management Group, East Consult (P) Ltd., Water Asia (P) Ltd. 2002. Optimizing Water Use in Kathmandu Valley Project, Inception Report, pp. 11.

Bhattacharya P., Tandukar N., Neku A, Valero AA, Mukherjee AB, Jacks G (2003) Geogenic arsenic in groundwaters from Terai alluvial plain of Nepal. Jour. de Physique IV France 107: 173-176. CBS. 2002. Population of Nepal. Village Development Committees/Municipalities. Population Census 2001, Central Bureau of Statistics, HMG/N, UNFPA.

DMG/BGR. 1998. Engineering and Environmental Geological Map of the Kathmandu Valley, scale 1: 50,000.

DMG/DOI/BGR. 1998. Hydrogeological Condition and Potential Barrier Sediments in the Kathmandu Valley, Technical Cooperation Project, Environmental Geology, Final Report. Department of Mines and Geology Nepal, Department of Irrigation Nepal and Federal Institute for Geosciences and Natural Resources Hannover Germany. pp. 18.

HMG Nepal/CEMAT Consultant (P) Ltd Nepal. 1999. Urban Water Supply Reforms in the Kathmandu Valley, Asian Development Bank TA No-2998-Nep. Ground Water Licensing and Monitoring Workshop, pp. 3.

HMG Nepal/JICA. 1990. Groundwater Management Project in the Kathmandu Valley, Final Report, Executive Summary. pp. 3, 4.

HMG Nepal/JICA. 1990. Groundwater Management Project in the Kathmandu Valley, Final Report, Data Book. pp. B-1.

HMG Nepal/JICA. 1990. Groundwater Management Project in the Kathmandu Valley, Final Report, Supporting Report. pp B-1, D-3, D-9, D-11, D-15, E-1, E-31,

HMG Nepal/JICA. 1990. Groundwater Management Project in the Kathmandu Valley, Final Report, Main Report. pp. 9.

JICA. 1990. Groundwater Management Project in Kathmandu Valley. Journal, A.G. and Huijbregts, Ch. J. 1989. *Mining Geostatistics*, Academic Press Inc., San Diego, CA 92101, USA.

Metcalf & Eddy, Inc./CEMAT. 2000. Urban Water Supply Reforms in the Kathmandu Valley, Groundwater monitoring program.

Maharjan M, Watanabe C, Ahmad SA, and Ohtsuka R. Arsenic Contamination in Drinking Water and Skin Manifestations in Lowland Nepal: The First Community-based Survey. *Am. J. Trop. Med. Hyg.*, 2005, 73(2): 477–479.

Shrestha BR, Whitney JW and Shrestha KB (Eds.). 2004. The State of Arsenic in Nepal - 2003. Environment and Public Health Organization (ENPHO)/National Arsenic Steering (NASC), Nepal, USGS, Kathmandu.

Shrestha SD, Brikowski T, Smith L and Shei TC. 2004. Grain size constraints on arsenic concentration in Shrestha RR, Shrestha MP, Upadhyaya NP, Pradhan R, Khadka R, Maskey A, Maharjan M, Tuladhar S, Dahal BM, Shrestha K. 2003. Groundwater arsenic contamination, its health impact and mitigation program in Nepal. *J Environ Sci Health, Part A-Toxic/Hazardous Substances & Environmental Engineering,* A38(1):185–200.

Tandukar N, Bhattacharya P , Mukherjee AB (2001). Preliminary Assessment of Arsenic Contamination in Groundwater in Nepal. In: Arsenic in the Asia-Pacific Region — Managing Arsenic for our Future, Adelaide, South Australia (Book of Abstracts): 103–105.

WHO-Project Nepal 0025. 1973. Groundwater Investigation, Kathmandu Water Supply and Sewerage Scheme. Appendix 4.5. pp. 7–10, 15,16,21,35,48.

43

Groundwater Scenario of Kalahandi District, Orissa, India

R. Nayak, P.K. Mohapatra, G.C. Pati and N.C. Nayak
Central Ground Water Board, SER, Bhubaneshwar, Orissa, India

ABSTRACT

Kalahandi, one of the western districts of Orissa, has the dubious distinction of frequent occurrence of drought due to late onset of monsoon and erratic rainfall pattern. The district receives a normal annual rainfall of 1378 mm, most of which is received from the south-west monsoon. The total replenishable groundwater resource of the district is estimated to be 89520 ham but the stage of the groundwater development is only 12.83% leaving a huge amount for further utilization. Optimal utilization of groundwater holds the key to counter the age-old natural problem of frequent drought in the district. An attempt has been made in this paper to highlight the prevailing groundwater scenario and the need for scientific management of this vital resource. Judicious utilization of groundwater through properly designed abstraction structures constructed at suitable locale can mitigate the problems of water scarcity faced during summer. Rainwater harvesting, proper run off management measures and optimal groundwater conservation practices will be helpful for eradicating the water scarcity of the district.

1 INTRODUCTION

Kalahandi, one of the western districts of Orissa, has the dubious distinction of frequent occurrence of drought due to late onset of monsoon rainfall and its erratic and uneven distribution pattern. Located between the East longitudes of 82° 32′ 00″ and 83° 47′ 00″ and North latitudes of 19° 08′ 00″ and 20° 25′ 00″ the district is having a geographical area of 7920 km^2. The district headquarters, Bhabanipatna, is located about 418 km from Bhubaneswar. The share of the rural population is 92.5% of the total population and agriculture is the main occupation of the district. The net area irrigated is only 32% of the net area sown, out of which

68% is covered by canal irrigation. So the agriculture mainly depends upon monsoon rainfall. Although the district possesses significant groundwater development potential the present stage of groundwater development is very low. There is vast scope of development of groundwater through suitable groundwater structures. Optimal utilization of groundwater resources will be able to mitigate the effects of drought in Kalahandi District, Orissa (GWRD, 2001 and HFDP, 2001).

2 PHYSIOGRAPHY AND DRAINAGE

Physiographically the district can be divided into three distinct zones (DSH, 2001):

- Eastern Ghat Hills: The presence of structural and denudational hills forms the characteristic landforms with intermontane valleys (south-eastern part of the district covering Thuamul-Rampur and Lanjigarh blocks). The elevation of hill ranges from 953 m. amsl at Tottighar to 1229 m. amsl at Tongri Donger.
- Pediplain along with Residual hills/Inselberg: Major portions of undulating plains covering in parts of Dharamgarh subdivision, parts of Narla, Rampur-Madanpur and Bhawanipatna blocks are under this category. Pediplains with shallow to deep weathered zones are gently to moderately undulating. Residual hills/Inselbergs represent the erosional remnants rising over the pediments and pediplain areas.
- Lateritic Upland: Lateritic Upland (generally flat) are developed mostly on granitic basement and occur as thick cappings over the Khandalites near Karlapat. Laterites form duri-crust at places.

River Tel and Indravati form the major drainage system in the district. The general slope of the terrain is towards the north-east. The river Tel that originates from the Nawarangpur high lands flows NE in major portion of the district. The topographic slope mainly controls the flow of this river. Udanti, Indra, Sangada, Ret, Hatti and Uttei are important tributaries of river Tel. Another river Indravati rises from north of Thuamul-Rampur and drains a small area in the Southern part and then enters into Nabarangpur district. The river Vamsadhara drains a limited area of Kalahandi district in the South-eastern part and river Nagavalli originates from South-eastern hilly tract which drains very small part of the district. The drainage pattern is mainly trellis and angular and is controlled by initial slope and in parts by structure.

3 AGRICULTURE, IRRIGATION AND INDUSTRIES

Agriculture is the main occupation of the inhabitants. Paddy, wheat, pulses and oil seeds are sown in the district. The net sown area and net area irrigated are 290901 and 93053 ha respectively. The net area irrigated is 32% of the net area sown, out of which 68% is covered by canal irrigation. The main livestock of the district include bovine population-buffaloes, cattle and goats etc. The district is backward in terms of industrial development. There are small-scale cottage and handloom industries. There are number of working mines of graphite and quartzite. The Upper Indravati and the Uttei are the main irrigation projects in the district having irrigation command areas of 29881 and 9306 ha respectively.

4 CLIMATE

The district enjoys a subtropical humid climate characterized by three seasons: summer, winter, and rainy season. The district receives a normal annual rainfall of 1378 mm. More than 80% of the annual rainfall is received from the south-west monsoon during June to September. Due to late onset of monsoon rainfall and its erratic and uneven distribution pattern, occurrences of drought is frequent in the district. The frequency of drought conditions in the district in terms of percent of the years is between 45 and 80 that is matter of serious concern. It has been observed that the drought occurs in every two and half years and normal drought in every six years in the district. The highest temperature of 47°C in summer and lowest temperature of 4.5°C in winter are recorded in the district.

5 GEOLOGY

Major parts of the district are underlain by rocks belonging to Eastern Ghats group comprising of granite gneiss, khondalites, charnockites and quartzites. The Quaternary alluvium occurs as thin discontinuous pockets restricted to the close vicinity of major river channels. The generalized stratigraphic sequence of the district is given in Table 1 and geological map of the district is presented in Fig. 1 (Ravindra, 1985).

Table 1 Stratigraphic Sequence in Kalahandi district, Orissa.

Age	*Group/Formation*	*Lithology*
Recent to Sub-recent	Alluvium Laterite and Lateritic gravels	Sand, silt, clay and gravels of various grades
	Unconformity	
	Younger intrusives	Gabbro, diorite, norite, anorthosite
Eastern Ghats Groups (Archean)	Granite suite	Porphyritic granite gneiss garnetiferous granite, biotite granite gneiss, pegmatite and quartz veins.
	Charnockite suite	Acid and intermediate basic/ultra-basic charnockites
	Khondalite suite	Quartz-garnet-sillimanite-gneisses and schist, garnetiferous quartzites; calc-silicates and granulites

5.1 Archean Group

The Archean Group is represented by the Eastern Ghats series consisting of Khondalites, charnokites and granite suites of rocks in the district.

Khondalitie Suite: Khondalites are essentially grey and red foliated garnet-sillimanite-schists containing quartz, feldspar and graphite and are widely distributed in the district. Khondalities often contain graphite deposits of commercial importance. Feldspathic Khondalities on weathering give rise to laterite and bauxite.

Charnokite suite: Charnokites are hypersthene bearing granite composed of blue quartz, feldspar and hypersthene. Charnokite exhibits intrusive relationship with the Khondalite and forms hybrid gneisses both with the Khondalites and other ancient gneisses. Charnokite ranging in composition from basic to acidic varieties constitute a suite of rocks referred to as "Charnokite series". Basic types are essentially pyroxene granulites. Intermediate varieties range from homogeneous hypersthene diorites to pyroxene granulites and granulitic migmatites.

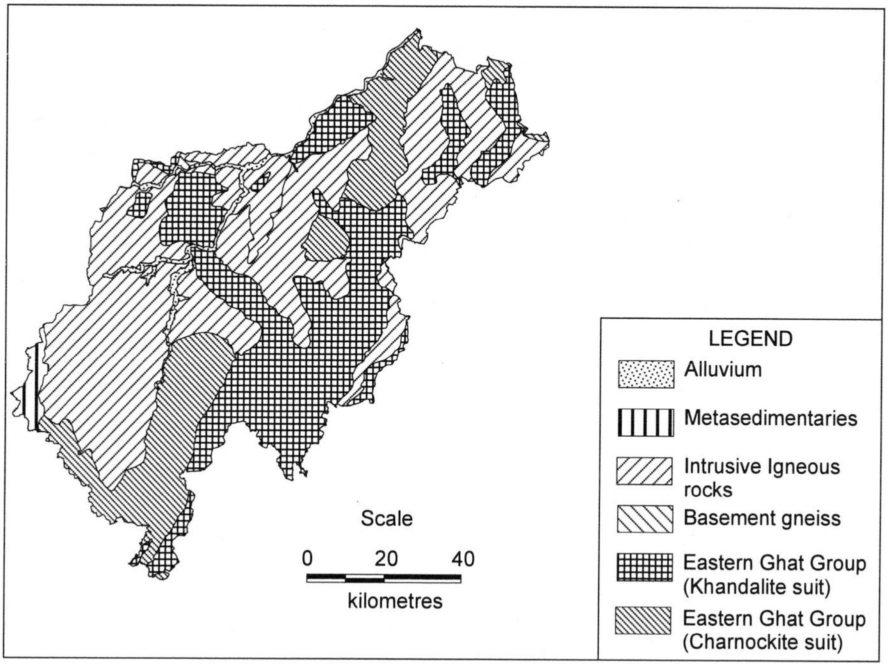

Fig. 1 Geological map of Kalahandi district, Orissa.

Granite suite: The most predominent rocks occurring in the district are porphyritic granite gneiss, garnetiferous granite gneiss, biotite granite gneiss, etc. These rocks are grey to greyish white in colour. Granite gneisses exibit wide variation in texture and composition due to different cycles of igneous activities and different grades of metamorphism. They occur as massive types with inclusions of metabasic rocks in Kalahandi district. In few places granite gneisses are intruded by dolerite dykes and intersected by veins of pegmatite and quartz.

5.2 Laterite

Laterite occurs as capping over the older formations and found extensively in the district. The Laterite covered plateau probably represents uplifted and dissected erosional surfaces possibly of different generations. The southern parts of the district occupied by Khondalite hills are capped with thick laterite ranging in thickness between 24 and 60 m.

5.3 Alluvium

Alluvium occurs as thin discontinuous pockets adjoining the rivers/stream courses throughout the district. The alluvial sediments are clays, sands and silts.

5.4 Structure

The rocks of the Eastern Ghats group are highly metamorphosed, structurally deformed to considerable extent and transformed to various degrees. The general trend of the formation is along NE-SW direction. There is a fault zone at the contact between the Eastern Ghats group and Chhatisgarh groups, fault plane running in N-S direction.

6 HYDROGEOLOGY

The Consolidated Formation (hardrocks) forms the main hydrogeological unit occurring in the district. It includes Granite, Charnokite, Khondalite suites of rocks of Eastern Ghats Group. These rocks lack primary porosity and are rendered porous and permeable by weathering and fracturing. The water yielding capability of these rocks depends entirely on the intensity of fracturing and weathering. The depth of dug wells ranges from 3.5 to 13 m. Dug wells and deep bore wells are the feasible groundwater structures in the consolidated formations. Groundwater occurs under phreatic condition in the weathered zone and under semi-confined to confined condition in the deeper fracture zones. Usually 3 to 5 water bearing fracture zones are encountered within a depth range of 200 meters. Granite suites of rocks are more promising aquifers having yield between 3 and 18.6 lps with moderate drawdown.

The unconsolidated formations in the form of Alluvium of Quaternary age occur as discontinuous patches in the floodplains of major rivers. Groundwater occurs under phreatic condition in the sand layers present in the alluvium. Dug wells and shallow tube wells are the most suitable groundwater abstraction structures in the unconsolidated formations. Yield of the shallow tube wells in the alluvium varies from 5 to 10 lps for a drawdown of 5 to 10 meters.

The weathered residue developed on the hardrock serves as good groundwater reservoir in most part of the district. In plateau areas lateritic capping is developed on the weathered residium. This is the zone commonly tapped by means of dug wells, which is the main groundwater abstraction structure in the district. The depth to water level in the dug wells during the premonsoon period varies from 0.68 to 8.19 meters below ground level. The post monsoon water level varies from 0.28 m above ground level to 4.69 m below ground level. The seasonal fluctuation of water level varies from 0.38 m to 5.82 m. The water table configuration generally follows the topography of the area.

7 GROUND WATER LEVEL BEHAVIOUR

The groundwater regime monitored in 54 hydrograph stations by CGWB (by our division) reveals the following facts. The depth to water level is shallow in most part of the district. The pre-monsoon depth to water level map of Kalahandi district is presented in Fig. 2. During pre-monsoon generally the northern part of the district is having deeper water level as compared to the southern part and remains between 5 and 8 meters below ground level.

In the southern part the same varies from 2 to 5 meter below ground level. In canal command areas and spring point the depth to water level is within 2 meter below ground level. The post-monsoon depth to water level map of Kalahandi district is presented in Fig. 3. In post-monsoon period the major part of the district is having depth to water level within 2 to 5 meter below ground level except in hilly areas where it is within 2 meter. The seasonal fluctuation of groundwater level is more than 4 meters in the north-eastern hilly areas of the district (parts of Rampur Madanpur and Narla block). Major part of the district have water level fluctuation within 2 to 4 meters. But valley and canal command areas has fluctuations within 2 meters. Comparison of pre-monsoon and post-monsoon water level data with decadal mean shows that

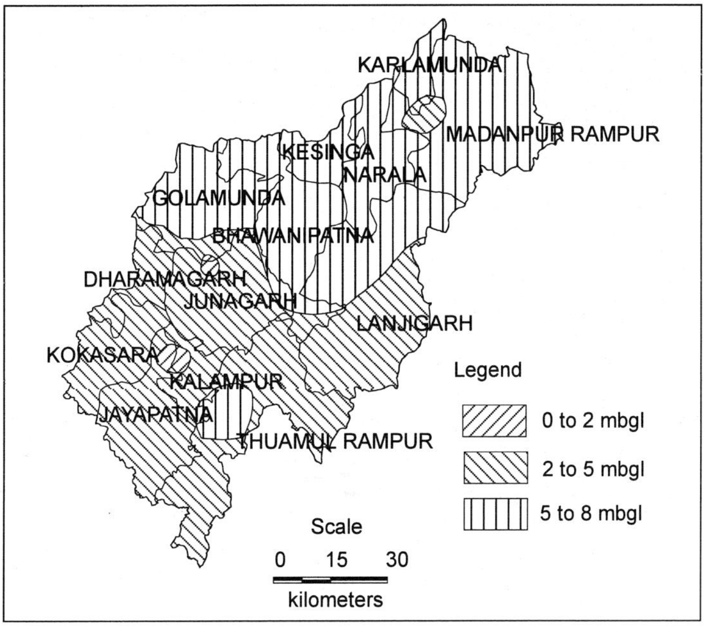

Fig. 2 Pre-monsoon depth to water level map of Kalahandi district, Orissa.

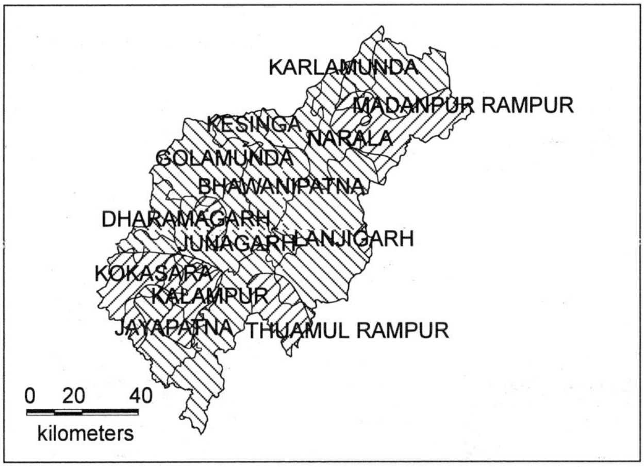

Fig. 3 Post-monsoon depth to water level map of Kalahandi district, Orissa.

there is no appreciable change in groundwater regime. Trend Analysis of the Hydrograph Stations also supports this phenomenon.

8 GROUNDWATER EXPLORATION

Ground Water Exploration Programme of CGWB (67 exploratory wells, 22 observation wells till date) in Kalahandi district reveals that depth of the wells generally varies from 61 to 220 m and the yield varies from negligible to a maximum of 18.6 lps (Biswanathpur). Generally, 40% of the wells are found to be successful having more than 3 lps discharge. Very high discharge more than 10 lps are recorded in five wells, i.e Biswanathpur, Nandool, Deypur, Shagartha and Ichapur. In general, Bhawanipatna, Koksara, Jaypatna, Kalampur blocks are successful from groundwater exploration point of view. Most part of Lanjigarh and Thuamul-Rampur blocks are hilly and are not suitable for deep drilling. Madanpur-Rampur block is also not feasible for deep bore wells due to unproductive formation (Khandalite, Graphite bearing Garnetiferous Granite Gneiss and Porphyritic Granite).

9 GROUNDWATER QUALITY

The quality of groundwater barring a few exceptions is generally good with Electrical Conductivity ranging from 500 to 1500 micro siemens/cm at 25°C in majority of hydrograph stations. However, six stations show more than 1500 micro siemen/cm EC value, i.e Badlimal, Koksara, Pastikudi, Dharamgarh, Badbasul, Daspur. 14 hydrograph stations out of 54 show chloride content in excess of 100 mg/l. From only three places high fluoride has been reported which exceeds the permissible limit of 1.5 mg/l (Ladugaon, Badbasul and Golmunda hydrograph stations). Ghungarbahal Village in Narla block is also having more than 1.5 mg/L Fluoride content in groundwater in shallow aquifer. More than 40% of the hydrograph stations show high values of nitrate above permissible limit (45 mg/L as prescribed by WHO Raghunath 1982) in groundwater. Kesinga, Sargigora, Dharamgarh, Baldiamal, Badbasul, Pastikudi, Daspur and Narla area are a few having higher nitrate concentration in groundwater.

10 GROUNDWATER RESOURCES

The Kalahandi district is having adequate groundwater resource potential out of which a small part is being utilized. The estimation of the groundwater resources indicate that the total replenishable groundwater resource is 89520 Ham out of which 82609 Ham is utilizable for irrigation and 6911 Ham is available for domestic and other purposes. The annual net draft for irrigation purpose is only 7832 Ham. The blockwise groundwater resources (balance groundwater resources, resources for irrigation, resources for domestic and industrial use and annual draft for irrigation) of Kalahandi district is presented in Fig. 4.The stage of groundwater development is 12.83% in the district with the highest at Bhawanipatna Block (23.63%) and the least at Thuamul-Rampur Block (4.94%). Blockwise stage of groundwater development in Kalahandi district is presented in Fig. 5.The Irrigation Potential of 28968 Ha created from groundwater sources exists in the district. Based on the balanced groundwater resources available an additional Irrigation Potential of 80377 Ha can be created in future. It is worth mentioning that the resource is calculated based on the April and November depth to water level data most of which is usually drained as base flow during post-monsoon periods.

11 CONCLUSIONS AND RECOMMENDATIONS

The district has the dubious distinction of frequent occurrence of drought due to late onset of monsoon and erratic rainfall pattern. Thus groundwater holds the key to counter this age-old natural problem. As groundwater development is quite low in the district, much stress is to be

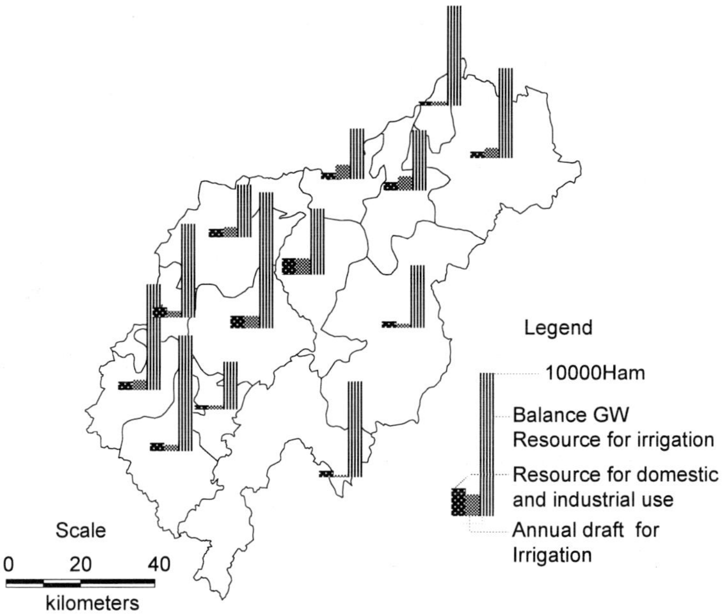

Fig. 4 Block-wise groundwater resources in Kalahandi district.

given on its optimal utilization. Groundwater has the added advantage over surface water irrigation as it is more cost effective and can be utilized at the place of requirement. However, the thin weathered zone found in some part of the district is responsible for scarcity of groundwater in summer, which is generally highlighted. As most of dug wells are shallow and not tapping the entire weathered zone, usually they get dried up during summer. Hence, deep dug well and if possible large diameter dug wells can be constructed in suitable locations like valley fills and pediplains to harness the groundwater resource. Good alluvial zone is available in some of the rivers and can be easily tapped by means of shallow tube wells and filter points. Especially the alluvium found in Tel river in Dharamgarh block can be easily utilized for the purpose. Rainwater harvesting and run off management through the construction of percolation and storage tanks, check dams, check and diversion weirs are to be taken up for the conservation of rainwater and for augmenting recharge to groundwater. As most of the groundwater resources are going waste as base flow during post-monsoon period, suitable groundwater conservation structures like Sub-Surface Dykes and cement plugs in specific locations are to be constructed. Few well designed check dams with proper arrangements for flood control can be constructed in some of the major rivers flowing in the district which will facilitate the use of base flow in lean season and reduce base flow from the aquifers.

Acknowledgements

The authors are very much thankful to Dr. Saleem Romani, Chairman, Central Ground Water board for his kind permission for preparation and presentation of this paper. Thanks to Shri S.

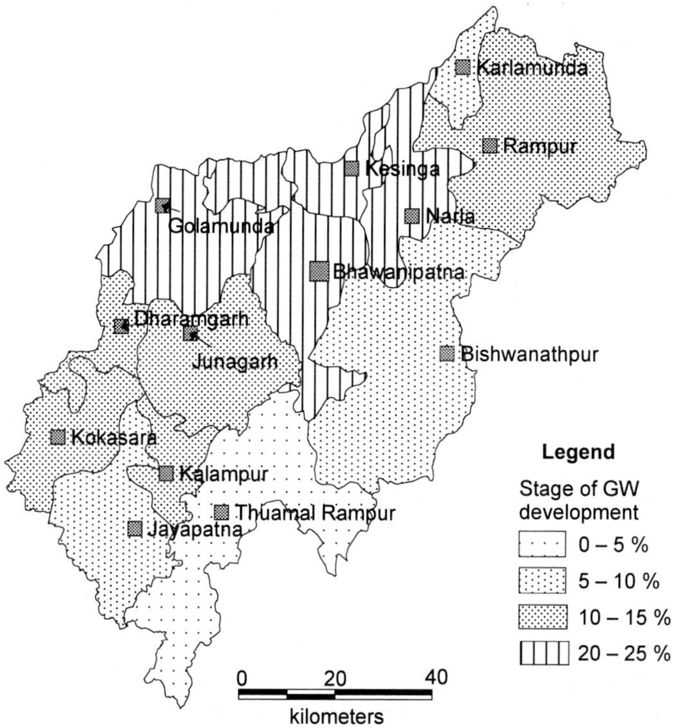

Fig. 5 Block wise stage of groundwater development in Kalahandi district.

Chakladar, Regional Director, C.G.W.B. (South Eastern Region), Bhubaneswar for his kind support during the paper preparation work.

References

District Statistical Handbook Kalahandi 2001 Directorate of Economics and Statistics, Orissa, Bhubaneswar.

Ground Water Resource and Development Prospects of Orissa 2001 GWS&I, Orissa.

Ground Water Resource Estimation Methodology. 1997. Central Ground Water Board. Ministry of Water Resources. New Delhi.

Hydrogeological Framework and Development Prospects of Kalahandi District, Orissa 1999 Central Ground Water Board, SER, Bhubaneswar.

Kumar Ravindra 1985 Fundamentals of Historical Geology and Stratigraphy of India.

Raghunath H M 1982 Ground Water New Delhi, New Age International Publisher.

A Decision Support Tool to Explore Conjunctive Use Options in Canal Command

44

A. Upadhyaya, A.K. Sikka, A.K. Singh and P. R. Bhatnagar
ICAR Research Complex for Eastern Region, WALMI Complex, Phulwari Sharif, Patna, Bihar, India

ABSTRACT

In order to convince the farmers about concept of conjunctive use and its economical benefits, a decision support tool is developed both in Hindi and English using Visual Basic platform. This interactive tool calculates: (i) annual fixed and operational costs of irrigation from tube well and canal, (ii) yield and total cost of produce, (iii) excess expenditure incurred in tube well irrigation over and above the canal water charges, and (iv) required yield increase needed to compensate for the additional cost of irrigation from tubewell. Along with canal irrigation the economic viability of tube well irrigation under three situations (i) own tube well, (ii) renting pumping sets to run tube well for irrigation, and (iii) purchasing water from tubewell owners, were analyzed. Analysis showed that compared to owning tube well, water purchasing is the most economical option followed by getting only pumping sets on rent basis to run tube wells for irrigation. The developed tool is quite effective and convincing in propagating the concept of conjunctive use among the farmers.

1 INTRODUCTION

Survival of the agrarian economy of the country largely depends upon the sustainable use of land and water resources. India is perceived to be endowed with large and reasonably good land and water resources, but with alarming increase in population since last 30 to 40 years per capita resource availability is progressively decreasing. On global basis, the irrigated agriculture covers only 16% of the total land under cultivation while it contributes about 36% of the total food production. This supports the fact that irrigation is the most potential and effective means of increasing agricultural productivity. It is however realized that water is a scarce resource, which goes into many competing and conflicting processes and requires a comprehensive planning for appropriate utilization. Development of these resources by individuals has not only created unfavourable conditions for crop production and environmental sustainability but also raised the energy uses in agriculture for irrigation. At the same time intensive irrigation in various canal commands has created the problems of water logging and salinity, making the soil unproductive while overexploitation of groundwater in certain parts of the country has resulted in lowering of water tables and causing decline in yield and productivity of wells, saltwater intrusion along the coast, drying of springs and shallow dug wells, increase in cost of lifting, etc. Hence, it is advantageous to look water as a single resource irrespective of its source of development, rain, surface or groundwater. Planning for conjunctive use requires basic understanding of the processes and principles involved in water resource management. A well-formulated conjunctive use programme, must consider fulfilling the irrigation requirement of crops during lean period by other sources of water when supply through canal is not enough to meet the irrigation requirement. In addition, it must also take into account the conservation of rainwater along with augmenting the groundwater storage through recharge.

Hence, conjunctive use management can be defined as the management of multiple water resources in a coordinated operation so that the total water use of the system over a period of time exceeds the sum of water uses of the individual components of the system resulting from uncoordinated operation. This reflects that conjunctive use planning has a major demand in areas, essentially facing a resource management problem and needs a proper planning at farm, project and regional level.

Analysis of conjunctive use programmes in India (Prasad, 1993) reveals that in most of the cases, what is being practised in the country at the moment does not really envisage the optimal use of rain, surface and groundwater resources. None of the states, so far have formulated clear-cut operational plan for conjunctive use in any irrigation projects though bulk of attempts through research and extension agencies have been undertaken and developed technologies have been disseminated in the country. Reviews indicate that no irrigation project, however, has either been planned, or designed and operated, on conjunctive use principles so far, which can incorporate the institutional framework, socio-economic factors of wide range of stakeholders and policy guidelines for development of water resources along with the technical aspects of canal management due to unresolved technological issues, absence of a methodological framework, and certain inhibiting policy and socioeconomic factors. Upadhyaya *et al.* (2004) developed a conceptual framework for a decision support system based on prevailing constraints faced by farmers and opportunities available to them.

In this paper, attempts have been made to present the constraints in conjunctive use identified through survey and interaction with farmers and to explore conjunctive use options in the canal command with the help of an interactive decision support tool.

2 STUDY AREA

In order to understand the constraints related to conjunctive use and explore possibilities for its promotion RP Channel V distributary of Patna Main Canal under Sone Canal system in Bihar was selected. In this area conjunctive use of rain, surface and groundwater is poorly developed, despite the chronic unreliability and shortage of canal water. Tubewell irrigation is practised mostly at the tail end of RP Channel-V, where canal water supply is poor. Although many wells and tube wells have been created, as shown in Table 1, yet farmers use them rarely.

Table 1 Wells and pumpsets in RP channel V.

Reach	*Number of wells*			*Pump sets*	
	Open	*Tube wells*	*Open + Tube wells*	*Motor*	*Diesel*
Head	57	26	2	–	28
Middle	69	46	4	2	45
Tail	131	101	14	17	98

The cost of tubewell water varies from Rs. 40/- to Rs. 70/- per hour depending on location, time and amount of water supply, whereas cost of canal water supply in the command varies from Rs. 217 to Rs 370.50 per ha depending on the crop. The cost of canal water supply is given in Table 2 below.

Table 2 Cost of canal water (Rs/ ha per crop season).

Sl. No.	*Crop*	*1985-95*	*1995-2001*	*2001-02 till date*
1.	Rice	89.41	172.9	217.36
2.	Wheat	51.13	148.2	185.25
3.	Sugar cane and vegetables	157.59	296.4	370.50

Low cost of canal water misleads cultivators into avoiding use of water sources that are more expensive. Hence, they avoid the cost of paying for water from wells or tube wells in the belief that this would be unnecessary in the event that canal water becomes available, or rain occurs. Because canal water is inadequate, irregular, untimely and uncertain, many farmers wait to buy water until it is too late and their crops fail.

3 CONSTRAINTS IN CONJUNCTIVE USE BASED ON FARMERS' SURVEY

Before propagating the concept of utilizing rain, surface and groundwater conjunctively in the canal command farmers were interviewed through a questionnaire and their responses towards conjunctive use were analyzed. Major constraints in adoption of conjunctive use as reported by farmers are:

3.1 No Land Consolidation

55% farmers in the head reach, 35% farmers in the middle reach and 32 % farmers in the tail reach of RP Channel V distributary reported that small and fragmented land holding and non-existence of land consolidation were the major problems in owning a tubewell and utilizing groundwater along with canal and rain water conjunctively.

3.2 Uneconomical Because it Needs High Initial Investment

All the farmers reported that utilization of groundwater has become uneconomical because initial investment ranges from Rs. 25,000/- to Rs. 40,000/- depending on the selection of pump, engine/motor and other accessories, whereas canal water though irregular and unreliable, but available at much cheaper rate.

3.3 High Recurring Expenditure Due to Increase in Price of Diesel

All the farmers were of the opinion that due to hike in price of diesel recurring expenditure increases and it discourages the farmers to withdrawal groundwater and utilize it for crop production.

3.4 Frequent Power Failure

Power supply in the villages is totally unreliable and there is a frequent failure so electric motors are very limited in use and farmers are compelled to use diesel engine in place of electric motors.

3.5 Lack of Awareness About Selection of Pump, Motor/engine as Per Farmers' Requirement

60% farmers reported that they don't have knowledge about selection of pump, motor/engine and other accessories and they purchase either on the basis of experience of other farmers or on the advice of supplier. Due to this they don't get the expected output of pump and face problem due to early failure.

3.6 Lack of Trained Manpower to Operate Efficiently and Maintain Timely the Pump, Engine/motor and Other Attachment

All the farmers reported that trained and efficient mechanic is not available and accessible in affordable price. Due to this repair and maintenance of pumping unit is not done at proper time, which results in frequent loss of time and money.

3.7 Transportation from one Place to Another is Difficult

Some people have boring in their fields but don't have pumping unit, whereas some people have pumping unit but no boring. These people face problem in transportation of pumping unit due to the weight of whole assembly.

3.8 Advantages of Conjunctive Use Quoted by Farmers

When farmers were asked about the advantages of conjunctive use, they reported that groundwater is an assured source of water supply and we can use it efficiently and economically as per requirement of the crop. 63% farmers in the head reach, 60% farmers in the middle reach and 95% farmers in the tail reach, who are also using groundwater, reported that use of groundwater only during nursery period, helped them in attaining higher yield of rice. The farmers requested to demonstrate the concept of conjunctive use and conditions under which it is economically beneficial to them.

This response of farmers inspired for development of a decision support tool in order to analyse the possibility of groundwater utilization and promotion of conjunctive use in the canal command.

4 DECISION SUPPORT TOOL FOR CONJUNCTIVE USE

An interactive decision support tool in Visual Basic was developed in English and Hindi versions. A pictorial view of English version with example is given below.

This tool calculates:

- Annual fixed and operational costs of irrigation from tube wells and canal.
- Yield and total cost of produce.
- Excess expenditure incurred in tube well irrigation over and above the canal water charges.
- The required yield increase needed to compensate for the additional cost of irrigation from tube wells.

The tool was demonstrated to water users and data was collected through the specified format from 150 farmers (50 farmers each at head, middle and tail reaches). The tool was found capable of making the required analyses and of convincing them to use it to make decisions about groundwater use under the prevailing constraints they face. Along with canal irrigation the economic viability of tube well irrigation under three situations were also analyzed to understand why conjunctive use is not so popular in this region :

- Own tube well.
- Renting pumping sets to run tube wells for irrigation.
- Purchasing water from tube well owners.

Analysis showed that farmers applied 2 to 3 irrigation during rice nursery and 2 to 3 irrigations during Rabi season to irrigate wheat, the required yield increase to compensate the additional cost of irrigation from tube well in case of the water users owning tube well varied in the range of 0.069 to 1.28 t/ha in head, 0.067 to 1.51 t/ha in middle and 0.13 to 1.32 t/ha in tail reach of the canal, whereas for the water users renting pumping sets to run tube wells for irrigation the required yield increase varied in the range of 0.078 to 0.8 t/ha in head reach, 0.052 to 0.81 t/ha in middle reach and 0.16 to 0.78 t/ha in tail reach, and in case of water users purchasing water from tube well owners it varied in the range from 0.13 to 0.26 t/ha in head reach, 0.16 to 0.35 t/ha in the middle reach and 0.18 to 0.52 t/ha in tail reach of canal. The results reveal that compared to owning tube well, water purchasing is the most economical option followed by getting only pumping sets on rent basis to run tube wells for irrigation.

5 CONCLUSIONS

Conjunctive use is defined as operation of surface and groundwater in such a way, which enhances their combined output. Since long, researchers and planners are trying to recommend the concept of conjunctive use in canal command as well as salt affected areas and many theoretical studies have been conducted in the past, but so far there is hardly any evidence

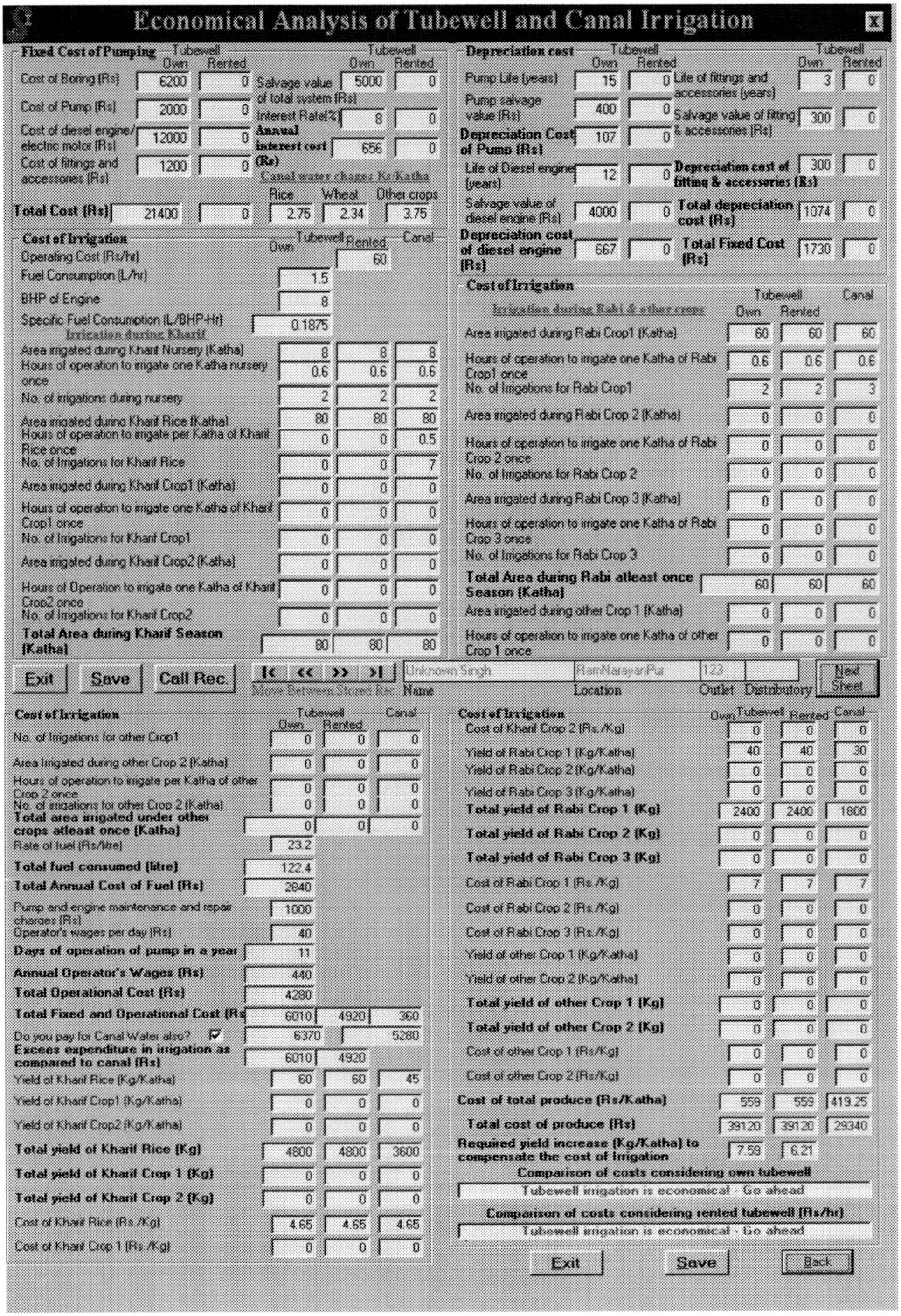
Economical Analysis of Tubewell and Canal Irrigation
Fixed Cost of Pumping — Tubewell Own Rented
Cost of Boring (Rs) 6200 0
Cost of Pump (Rs) 2000 0
Cost of diesel engine/electric motor (Rs) 12000 0
Cost of fittings and accessories (Rs) 1200 0
Total Cost (Rs) 21400 0
Salvage value of total system (Rs) 5000 0
Interest Rate(%) 8 0
Annual interest cost (Rs) 656 0
Canal water charges Rs/Katha
Rice 2.75 Wheat 2.34 Other crops 3.75
Depreciation cost — Tubewell Own Rented
Pump Life (years) 15 0
Pump salvage value (Rs) 400 0
Depreciation Cost of Pump (Rs) 107 0
Life of Diesel engine (years) 12 0
Salvage value of diesel engine (Rs) 4000 0
Depreciation cost of diesel engine (Rs) 667 0
Life of fittings and accessories (years) 3 0
Salvage value of fitting & accessories (Rs) 300 0
Depreciation cost of fitting & accessories (Rs) 300 0
Total depreciation cost (Rs) 1074 0
Total Fixed Cost (Rs) 1730 0
Cost of Irrigation — Tubewell Own Rented Canal
Operating Cost (Rs/hr) 60
Fuel Consumption (L/hr) 1.5
BHP of Engine 8
Specific Fuel Consumption (L/BHP-Hr) 0.1875
Irrigation during Kharif
Area irrigated during Kharif Nursery (Katha) 8 8 8
Hours of operation to irrigate one Katha nursery once 0.6 0.6 0.6
No. of irrigations during nursery 2 2 2
Area irrigated during Kharif Rice (Katha) 80 80 80
Hours of operation to irrigate per Katha of Kharif Rice once 0 0 0.5
No. of Irrigations for Kharif Rice 0 0 7
Area irrigated during Kharif Crop1 (Katha) 0 0 0
Hours of operation to irrigate one Katha of Kharif Crop1 once 0 0 0
No. of Irrigations for Kharif Crop1 0 0 0
Area irrigated during Kharif Crop2 (Katha) 0 0 0
Hours of Operation to irrigate one Katha of Kharif Crop2 once 0 0 0
No. of Irrigations for Kharif Crop2 0 0 0
Total Area during Kharif Season (Katha) 80 80 80
Cost of Irrigation — Tubewell Own Rented Canal
Irrigation during Rabi & other crops
Area irrigated during Rabi Crop1 (Katha) 60 60 60
Hours of operation to irrigate one Katha of Rabi Crop1 once 0.6 0.6 0.6
No. of Irrigations for Rabi Crop1 2 2 3
Area irrigated during Rabi Crop 2 (Katha) 0 0 0
Hours of operation to irrigate one Katha of Rabi Crop 2 once 0 0 0
No. of Irrigations for Rabi Crop 2 0 0 0
Area irrigated during Rabi Crop 3 (Katha) 0 0 0
Hours of operation to irrigate one Katha of Rabi Crop 3 once 0 0 0
No. of Irrigations for Rabi Crop 3 0 0 0
Total Area during Rabi atleast once Season (Katha) 60 60 60
Area irrigated during other Crop 1 (Katha) 0 0 0
Hours of operation to irrigate one Katha of other Crop 1 once 0 0 0
Exit Save Call Rec. |< << >> >| Move Between Stored Rec.
Unknown Singh RamNarayanPur 123
Name Location Outlet Distributory
Next Sheet
Cost of Irrigation — Tubewell Own Rented Canal
No. of Irrigations for other Crop1 0 0 0
Area Irrigated during other Crop 2 (Katha) 0 0 0
Hours of operation to irrigate per Katha of other Crop 2 once 0 0 0
No. of irrigations for other Crop 2 (Katha) 0 0 0
Total area irrigated under other crops atleast once (Katha) 0 0 0
Rate of fuel (Rs/litre) 23.2
Total fuel consumed (litre) 122.4
Total Annual Cost of Fuel (Rs) 2840
Pump and engine maintenance and repair charges (Rs) 1000
Operator's wages per day (Rs) 40
Days of operation of pump in a year 11
Annual Operator's Wages (Rs) 440
Total Operational Cost (Rs) 4280
Total Fixed and Operational Cost (Rs) 6010 4920 360
Do you pay for Canal Water also? 6370 5280
Excess expenditure in irrigation as compared to canal (Rs) 6010 4920
Yield of Kharif Rice (Kg/Katha) 60 60 45
Yield of Kharif Crop1 (Kg/Katha) 0 0 0
Yield of Kharif Crop2 (Kg/Katha) 0 0 0
Total yield of Kharif Rice (Kg) 4800 4800 3600
Total yield of Kharif Crop 1 (Kg) 0 0 0
Total yield of Kharif Crop 2 (Kg) 0 0 0
Cost of Kharif Rice (Rs./Kg) 4.65 4.65 4.65
Cost of Kharif Crop 1 (Rs./Kg) 0 0 0
Cost of Irrigation — Tubewell Own Rented Canal
Cost of Kharif Crop 2 (Rs./Kg) 0 0 0
Yield of Rabi Crop 1 (Kg/Katha) 40 40 30
Yield of Rabi Crop 2 (Kg/Katha) 0 0 0
Yield of Rabi Crop 3 (Kg/Katha) 0 0 0
Total yield of Rabi Crop 1 (Kg) 2400 2400 1800
Total yield of Rabi Crop 2 (Kg) 0 0 0
Total yield of Rabi Crop 3 (Kg) 0 0 0
Cost of Rabi Crop 1 (Rs./Kg) 7 7 7
Cost of Rabi Crop 2 (Rs./Kg) 0 0 0
Cost of Rabi Crop 3 (Rs./Kg) 0 0 0
Yield of other Crop 1 (Kg/Katha) 0 0 0
Yield of other Crop 2 (Kg/Katha) 0 0 0
Total yield of other Crop 1 (Kg) 0 0 0
Total yield of other Crop 2 (Kg) 0 0 0
Cost of other Crop 1 (Rs/Kg) 0 0 0
Cost of other Crop 2 (Rs/Kg) 0 0 0
Cost of total produce (Rs/Katha) 559 559 419.25
Total cost of produce (Rs) 39120 39120 29340
Required yield increase (Kg/Katha) to compensate the cost of Irrigation 7.59 6.21
Comparison of costs considering own tubewell
Tubewell irrigation is economical - Go ahead
Comparison of costs considering rented tubewell (Rs/hr)
Tubewell irrigation is economical - Go ahead
Exit Save Back

Fig. 1

about wide adoption and acceptance of conjunctive use practices among water users due to many constraints and mainly due to tremendous difference in cost of canal water and groundwater. Since canal water is available at much cheaper rates, water users don't prefer groundwater utilization unless they feel that their production will reduce drastically in the absence of water.

A user interactive decision support tool was developed both in Hindi and English versions using Visual Basic platform and demonstrated among water users in order to create awareness and explore the possibility of conjunctive use in the canal command. Farmers found this tool quite helpful in understanding and analyzing the concept of conjunctive use and taking appropriate decision for its adoption under the prevailing constraints.

Acknowledgements

This publication is an output from a project funded by the UK Department for International Development (DFID) for the benefits of developing countries. Authors acknowledge with thanks the financial support provided by DFID, NRSP, U.K. in carrying out this study under R-7830 Project.

References

Prasad, R.K. 1993. Conjunctive use of surface water and groundwater. *Proc. of National workshop on action for optimum utilization of water resources,* WAPCOS, New Delhi, India: 33–49.

Upadhyaya, A., Singh, A.K., Bhatnagar, P.R. & Sikka, A.K. 2004. Is conjunctive use a feasible option? A simple tool for participatory evaluation of the economic benefits of groundwater exploitation. *Poster presented in the Workshop on Realizing Potential: Livelihood, Poverty and Governance* at NASC Complex, Pusa Campus, New Delhi, August 3-4, 2004.

Bioaccumulation and Aquatic Food Chain-emerging Issues in Heavy Metal Toxicity

45

C. Ghosh[1], Ridhima Sud[2], K. Anjana[1] and A.J. Urfi[2]

[1]Centre for Environmental Management of Degraded Ecosystems, School of Environmental Studies, University of Delhi, Delhi–110007, India

[2]Department of Environmental Biology, School of Environmental Studies, University of Delhi, Delhi–110007, India

ABSTRACT

Environmental contamination with heavy metals poses a serious risk to the ecosystem and human health. The determination of metal ions concentration in natural water systems has received increased attention for monitoring environmental pollution because metals are non-biodegradable and find their way into food chain through a number of pathways and accumulate in different organs of human beings or other organisms. It has been reported that Zn, Cu and Cr are synergistically many times more toxic to fish as compared to their individual toxicity. The aim of the present study was to evaluate the occurrence of heavy metal toxicity in aquatic food chain. Water samples were collected from three locations along the Delhi segment of Yamuna river: i) site I (Wazirabad, in western Delhi); ii) site II (near ITO in Central Delhi) and iii) site III (Okhla in south-east Delhi). The selection of these three sites also indicates the changes in water quality as it enters and leaves the city. Quantification of heavy metal concentration in water and soil samples was conducted and compared with concentrations reported for the period from 1999-2003 by Central Pollution Control Board (CPCB). *In situ* analysis was undertaken for few physico-chemical parameters whereas composite samples were taken for other analysis. Sediment samples were taken across the river using a dredge. Physico-chemical parameters of water samples were analyzed according to the laboratory standard methods. The heavy metal concentrations were determined using Atomic Absorption spectrophotometer. The reported and generated value of heavy metal concentrations in

Yamuna water samples indicated that the concentrations of iron were dominantly higher in three sites. Interestingly, the concentration at site III was comparatively lower (4.5-6.1 ppm) than site I and site II (4-12 ppm) zone. Low concentrations of lead, copper, chromium and zinc in water sample were also reported at three places of Yamuna indicating their source of contamination. A similar trend was also observed in the soil samples with higher concentration values. Metals like lead (0.029 ppm in water and 0.014 mg/g in sediments) and zinc (0.012 ppm in water and 0.565 mg/g in sediments) were exceeding the environmentally safe concentrations as per standard guidelines of CPCB, which leads to presence of higher concentration of zinc (0.024-0.43 mg/g) in the fish tissue samples. This can be utilized as an effective indicator of environmental conditions. Presence of such persistent pollutants in the watercourse not only creates unfavorable environment for fish but also paucity of fish food organisms and fish eating birds.

1 INTRODUCTION

Heavy metals are known toxicants, which inflict acute disorders in aquatic organisms. The presence of heavy metals in the aquatic ecosystem and its impact on flora and fauna has been reported by many investigators (Hemelraad *et al.*, 1987; Radhakrishnaiah, 1988; Kargin & Cogun, 1999 and AY *et al.*, 1999). The tendency of metals to accumulate in organisms is one of the important properties of metals and the bioavailability of trace metals is the key factor determining tissue concentrations. The bioaccumulation of metals depends on availability and persistence of the contaminants in water/food and physicochemical properties of the toxicants. Fishes, which occupy higher trophic levels of food pyramid in the aquatic ecosystem are important targets of bio-magnification of trace metals and also act as possible bio-transfer routes to human beings. Generally, health related adverse effects are observed after chronic exposure to low doses of heavy metals.

Focus on the bioaccumulation of contaminants arose in the 1960s when incidents such as toxicity from methyl mercury residues in the shellfish and avian reproductive failure due to chlorinated pesticide residues in the aquatic species started surfacing. It is known that water quality parameters influence metal accumulation by affecting metal speciation (Bradley and Sprague, 1985). Metals may exist as free ions and inorganic complexes, which affect bioavailability. The bioconcentration of metals varies greatly with the species of organism, exposure concentration, environmental and physicochemical factors and the specific metal characteristics (Memmert, 1987). Evidence for biomagnification in aquatic based food webs has generally been restricted to contaminants that are persistent, halogenated and very hydrophobic. According to Thomann (1989), the extent of biomagnification is dependent on the species composition of the food web and source of contaminant. Various metals show different affinity to various organs and are accumulated within them in different amounts. Total quantity and distribution of metal is also related to the way of uptake, fish species, age and size, feeding habits and individual susceptibility. Bioaccumulation patterns of metals in fish tissue can be utilized as an effective indicator of environmental metal contamination (Larsson *et al.*, 1985, OECD, 1991, Jorgensen & Pederson, 1994). The most important disaster with heavy metals was reported in 1932 due to the mercury pollution known as *Minamata disease.* In the year 1998, The Coto de Donana reserve in Spain was contaminated when toxic chemicals leaked into water from a dam burst. Mud contaminated with sulphur, lead, copper, zinc and cadmium was released into the Rio Guadimar causing permanent damage to Europe's largest bird sanctuary as well as Spain's agriculture and fisheries (Puura *et al.*, 2002).

In India, although the industrial development has not reached the level attained in the developed countries, more toxic compounds, hitherto unknown are being detected in increasing numbers in our watercourses. It is largely because the production of chemicals resulting in the generation of toxic and hazardous substances has been continuously on the rise during the last three decades. Several studies (Misra *et al.*, 1991 & 2002; Muralidharan, 1995 & Jayakumar, 2001) have been reported incidences of heavy metal accumulation in living matter from water bodies near coastal Bombay, Baroda, Chandigarh, Lucknow, Chennai, Raipur and Kanpur. The study carried out by Nag and Jaffery (1994) observed accumulation of metals such as zinc, manganese, copper, nickel, mercury and lead in water, sediments and fish samples.

Regularly about 40% sewage and industrial waste water is disposed of on land in India. This important human activity that contaminates soil and aqueous streams with large quantities of toxic metals (Seaward and Richardson, 1990) and the remaining is drawn into different water bodies with or without little treatment (Mohan Rao, 1973). With rapid industrialization and urbanization, there is production of huge quantities of wastewater in Delhi also, which is regularly discharged into the Yamuna River through 19 storm water drains (either open or closed) laid in different parts of Delhi (Raina *et al.*, 1984). The sewage drain which carries industrial and community wastes do not actually enters the Yamuna, but are discharged into the trunk and outfall sewers which go to the sewage treatment plants. After treatment the water is used for irrigation as a fertilizer since it is rich in organic matter (Lunt 1963).

In the late 1970s, many years before the inception of Yamuna action plan (1993), the Central Pollution Control Board (CPCB) has estimated that there were 359 industrial units discharging effluents in river water. According to the 1996 survey, 98,000 industrial units were non-conforming ones, operating outside the 28 industrial areas of Delhi and immediately after a court order, 1,228 units were closed down (CPCB, 1999). In the year 2000, CPCB observed that there were 42 industrial units in Delhi directly responsible for water polluting the river Yamuna. From April 2005, the city has 10 common effluent treatment plants with a capacity of 133 MLD, set up by the Delhi Small-scale Industries Development Corporation but ironically only 53 MLD of industrial waste reaches these plants.

The major objectives of the present study were to assess the magnitude of metal contamination at the entry and exit points of the river system in Delhi and to discuss the bioaccumulation pattern of available metals among various tissues of fish, which is a bio-indicator of aquatic pollution. Thus, interpreting the level of heavy metals in fishes will not only reflect the exposure but also give an indication of temporal and spatial extent of metal accumulation as well as an assessment of its potential impact on human health (when consumed) and other organisms (if they have been exposed to elevated levels of metals or if consumed by predators).

2 MATERIAL AND METHODS

2.1 Meteorological Data

The climate regime of Delhi falls under the "semi arid type" – an influence by the city's inland position and the prevalence of continental air during the major part of the year. It is characterized by extreme dryness associated with hot summer and cold winters. Meteorological data were collected from the Indian Meteorological Department (IMD). The average minimum and maximum ambient temperatures were measured at 10 cm above ground level.

2.2 Study Area

The river Yamuna, a major tributary of river Ganga originates from the Yamunotri glacier at an elevation of 6387 meters above sea level. The Yamuna cannot be designated as a continuous river particularly in dry seasons (almost 9 months), but can be segmented in five distinguished independent segments due to characteristic hydrological and ecological conditions. On the basis of area, the catchment basin of Yamuna amounts to 40.2% of the Ganga Basin and 10.7% of the total landmass of the country. The low perennial flow in Yamuna and the huge quantity of waste it receives have given it the dubious distinction of being one of the most polluted rivers of the country.

Delhi, the capital city of India is facing the challenges of sanitation and environmental degradation due to increasing population and urbanization. The river enters Delhi after traversing a route of about 224 km. The total number of industries in Delhi was around 18500 in the year 1961 which steeply increased to about 93000 by 1993 (CPCB, 1999). CPCB has also identified 45 (20 large and 25 medium) significant water-polluting industries in the NCT (National Capital Territory) of Delhi which together discharge 120 MLD of the estimated 320 MLD discharge from industrial sources through various drains.

Three study areas were selected representing of the water quality of the river as it enters and leaves the city.

Site I (Wazirabad): The entry point of Yamuna river, the water was visibly clean and water treatment plant is operating there.

Site II (ITO): This highly polluted segment of the river receives discharge from various sewer drains, industrial effluents especially thermal wastewater and detergent inflow from laundry service providers.

Site III (Okhla): The exit point of the river, lies between Site II (ITO) and Site III (Okhla). The major sources of heavy metals pollution are Okhla Industrial Area and Indraprastha Thermal Power Plant. At the sampling site, a large number of migratory birds were also present.

The stretch between Site I and Site III in Delhi is only 2% of its catchment area but it contributes about 80% of the river's pollution load.

2.3 Sampling Procedure

For water analysis, *in situ* analysis was undertaken for few physico-chemical parameters whereas composite samples were taken for other analysis. Sediment samples were taken across the river using a dredge. The water samples were collected in thoroughly cleaned polyethylene bottles ensuring proper preservation as per the standard methods (APHA, 1989). For sludge samples polythene bags (zipper pouches) were used.

2.4 Sample Digestion and Analysis

Various physicochemical characteristics (pH, temperature, DO, COD, TDS and conductivity) were analyzed according to the laboratory standard methods (APHA, 1989). The heavy metal concentration was determined by digesting the water samples in concentrated HNO_3 and then analyzed by Atomic Absorption Spectrophotometer.

2.5 Reported Data Collection

The Central Pollution Control Board (CPCB), a Government organization has been monitoring water quality in the river Yamuna since 1974. We have collected the published heavy metal concentration's data of Yamuna for the comparative assessment of last five years (1999-2003).

3 RESULTS AND DISCUSSION

3.1 Meteorological Data

In Delhi, the ambient temperature ranges between 18.7 degree Celsius (mean min.) and 40.3 degree celsius (mean max.) in the summer months. Both day and night temperatures dip sharply in the winter season. During the greater part of the year, dry air prevails all over the city. Humidity increases in the monsoon months reaching up to 77 per cent.

During our experiment, the air temperature varied between 19.6°C and 25.2°C. The humidity recorded at the time of sampling was varied between 37% and 61% (Table 1).

Table 1 On site Meteorological data during experimental period units: air temperature (°C), relative humidity (%).

Parameters analyzed	*Site I*	*Site II*	*Site III*
Air temperature	19.6	25.2	23
Relative humidity	61	42	37

3.2 Comparative Assessment of Reported and Generated Values of Water Quality Analysis for the Month of January 2004 and 2005

The higher soil temperature at Site III (approximately 2°C higher) as compared to Site II and Site I can be attributed to the interferences of the industrial effluents especially from the Thermal power industry and agricultural discharges from irrigation water returned from fields. Interestingly, transparency was highest at site I as it is the entry point of Yamuna river in Delhi state (Table 2). The presence of water treatment plant at the upstream of the site of sampling is also a factor. At Site II it was nil due to its proximity to the Rajghat power plant and high pollution load from other industrial sources.

Table 2 *In situ water* analysis for three sites, January 2005 (Generated data) Units: water temperature (°C), soil temperature (°C), transparency (cm).

Parameters analyzed	*Site I*	*Site II*	*Site III*
Water temperature	17.2	17.1	16.7
Transparency	clear	nil	5.5
Soil temperature	14.4	13	16.5

On comparing the trend between the reported data and generated data, an increment of pH and dissolved oxygen (DO) values in Site I and Site III from 2004 to 2005 is noticeable. As Site II is a highly polluted segment of the river, the pH level is always low in comparison to other two sites (Site I and Site III) (Tables 3, 4 and 5).

Table 3 Reported water monitoring data of January, 2004 (Source CPCB, 2004) Units: DO (mg/L), COD (mg/L), TDS (mg/L), conductivity (μs/cm).

Parameters analyzed	*Site I*	*Site II*	*Site III*
pH	7.60	7.32	7.34
DO	8	Nil	Nil
COD	7	6.5	99
TDS	322	668	724
Conductivity	413	1360	1540

Table 4 Reported water monitoring data of January, 2005 (Source CPCB, 2005) (unpublished) Units: DO (mg/L), COD (mg/L), TDS (mg/L), conductivity (μs/cm).

Parameters analyzed	*Site I*	*Site II*	*Site III*
pH	8.08	7.21	7.42
DO	8.2	Nil	Nil
COD	4	101	55
TDS	254	960	580
Conductivity	475	1500	832

Table 5 Generated water monitoring data of January 2005. Units: DO (mg/L), COD (mg/L), TDS (mg/L), TSS (mg/L), TS (mg/L).

Parameters analyzed	*Site I*	*Site II*	*Site III*
Water pH	8.88	8.66	8.69
Soil pH	7.28	7.21	7.21
DO	8.2	Nil	3.58
COD	56	182.4	126.4
TDS	400	600	400
TSS	0	200	200
TS	400	800	600

The increase in Chemical Oxygen Demand (COD) at Site II in 2005 from 2004 is an indication of increased pollution load (Tables 3 and 4). Due to proximity of the sampling site to the bank, the generated values of COD (Table 5) might be higher than those reported by CPCB (Table 4).

The water chemistry of the system controls the rate of absorption and desorption of metals to and from sediments. Metals may be desorbed from the sediments if the water experiences decrease in pH and redox-potential and an increase in conductivity and salinity.

3.3 Metal Behaviour

In solution, heavy metals can be present either as free ions or complexed with inorganic and organic forms. Potential toxicity of metals and their compounds depends on its bioavailability. Again bioavailability depends on biological parameters and on the physicochemical properties of metals, their ions and their compounds. In environmental samples, bio-available metals are generally soluble in nature and are adsorbed to colloids or soil surface. The chemical nature can be changed through oxidation or reduction. However, the elemental nature remains the same because metals are neither thermally decomposed nor microbiologically degradable. Consequently, metals are difficult to remove from the environment. In addition, total metal concentrations in the environment do not necessarily reflect the degree of biological metal availability making it difficult to assess accurately the extent of risks posed by metals.

3.4 Metal Toxicity to Aquatic Organisms

Aquatic organisms may be adversely affected by heavy metals in the environment. Slightly elevated metal levels in natural waters may cause the following sub lethal effects in aquatic organisms: 1) histological or morphological change in tissues; 2) changes in physiology, such as suppression of growth and development, poor swimming performance, changes in circulation; 3) change in biochemistry, such as enzyme activity and blood chemistry; 4) change in behavior; 5) and changes in reproduction.

Living organisms require trace amounts of some heavy metals, including cobalt, copper, iron, manganese, molybdenum, vanadium, strontium and zinc. Excessive levels of essential metals, however, can be detrimental to the organism. Non-essential heavy metals of particular concern to surface water systems are cadmium, chromium, mercury, lead, arsenic and antimony. Many organisms are able to regulate the metal concentrations in their tissues. Fish and crustaceans can excrete non-essential metals, such as mercury and cadmium, although this is usually met with less success.

The major problem associated with heavy metals is their ability to bio-accumulate in the tissues of organisms and tendency to enter the food chain either directly or indirectly.World Health Organization (WHO) has developed guideline values to define the quality of water that can be safely consumed by everyone throughout their life times. These are given in Table 6.

Table 6 Guideline values of World Health Organization (WHO).

	Arsenic	*Cadmium*	*Chromium*	*Lead*	*Mercury*	*Iron*	*Manganese*
Permissible limit (mg/l)	0.05	0.005	0.05	0.05	0.01	0.3	0.1

The standards for metal contaminants in fish muscle have been as 100 mg/kg, 250 mg/kg, 1 mg/kg and 10 mg/kg dry weight for Cu, Zn, Cd and Pb respectively (USEPA, 1978).

3.5 Reported Heavy Metal Characteristics of River Yamuna

Heavy metals concentrations of the Yamuna river water (22 km stretch along Delhi region) were measured during past 5 years period from 1999-2003 by CPCB (CPCB report, 2003). The profile of heavy metal concentration was selected at three different stations, Site I (Fig. 1), Site II (Fig. 2) and Site III (Fig. 3) respectively.

If we compare all the results, we could clearly observe that the concentration of iron is dominantly higher in all three places. Interestingly, at site III concentration is comparatively lower (4.5-6.1 ppm) than Site I and Site II (4-12 ppm) area (Fig. 1-3). Iron oxides are the most important soil constitutes helping the metal adsorption and speciation in the soil.

Very low concentration of copper, chromium and zinc were also reported at three places of Yamuna indicating their source of contamination (Fig. 1-3).

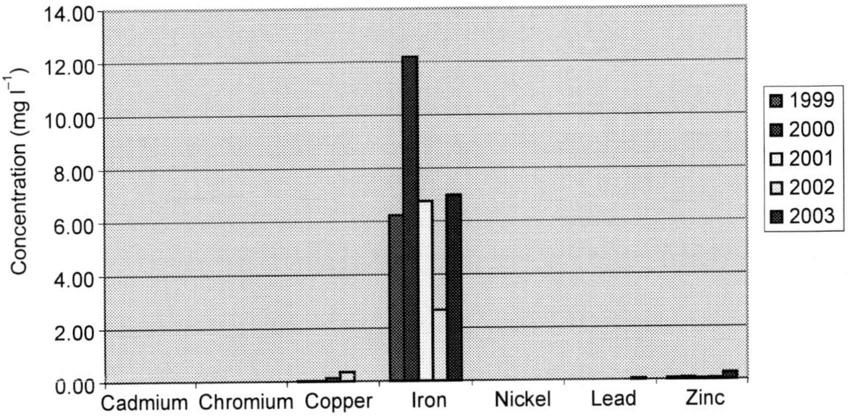

Fig. 1 Profile of concentrations of heavy metals in water samples at Site I.

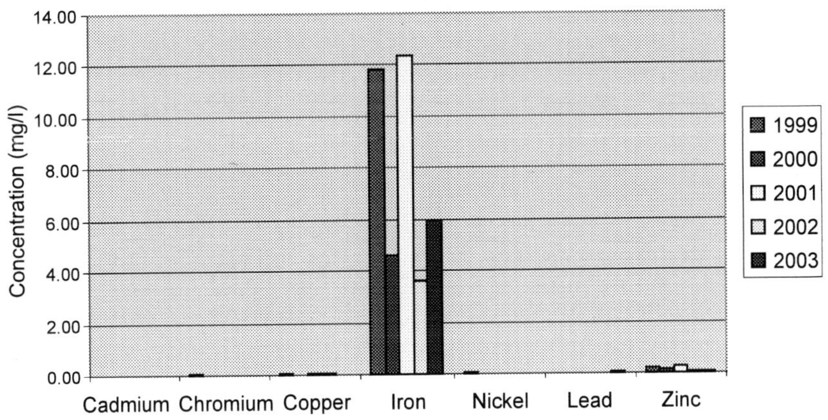

Fig. 2 Profile of concentrations of heavy metals in water samples at Site II.

3.6 Assessment of Heavy Metal Concentration in the Generated Data

Table 7, demonstrated that Iron in water and soil samples was quite high and varied between 0.086-0.331-mg/l in water and 15.2 to 23.6 mg/g in soil. Among them Site II shows, higher

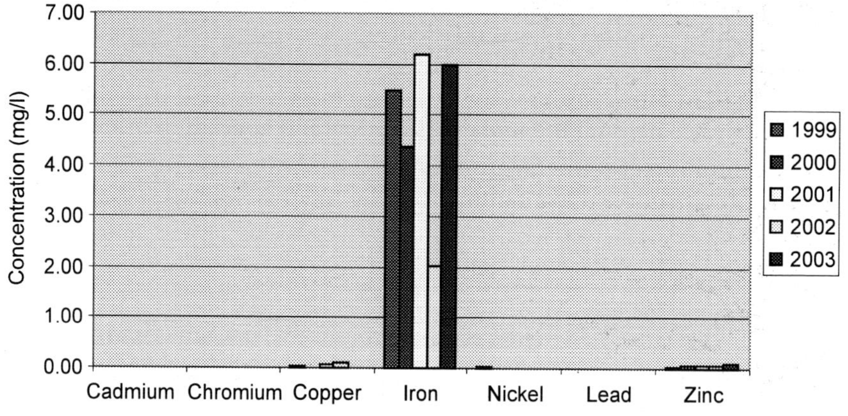

Fig. 3 Profile of concentrations of heavy metals in water samples at Site III.

Table 7 Heavy metal concentration from three sites based on the present study. Units: water (mg/l), soil and sludge (mg/g dry wt. soil)

S. no.	*Site Visited*	*Sample Type*	*Cadmium (Cd)*	*Chromium (Cr)*	*Copper (Cu)*	*Iron (Fe)*	*Nickel (Ni)*	*Lead (Pb)*	*Zinc (Zn)*
1.	Site I	Water	NT	NT	NT	0.112	NT	0.029	0.122
2.	Site I	Soil	NT	NT	NT	15.2	NT	NT	0.41
3.	Site I	Sludge	NT	NT	0.004	25.7	NT	NT	0.73
4.	Site II	Water	NT	NT	NT	0.331	NT	0.029	0.14
5.	Site II	Soil	NT	NT	0.11	23.6	NT	NT	0.72
6.	Site II	Sludge	NT	NT	0.033	21.7	NT	NT	0.92
7.	Site III	Water	NT	NT	NT	0.086	NT	NT	0.01
8.	Site III	Soil	NT	NT	NT	21.1	NT	0.014	0.54
9.	Site III	Sludge	NT	NT	NT	5.7	NT	NT	0.21

*NT-Not Traceable

concentration (0.331 ppm) for water and 23.6 mg/g for soil sample. But quite interestingly, sludge sample shows a slightly different trend, where Site I shows higher values (25.7 mg/g) than Site II (21.7 mg/g) and Site III (5.7 mg/g) area respectively.

Zinc concentration was high in water samples at Site I (0.12 mg/l) and at Site II (0.14 mg/L), whereas comparatively low concentration of Zn was seen at Site III (0.01 mg/L) (Table 7). Site II shows highest concentration of zinc in soil and sludge samples (0.72 mg/g & 0.92 mg/g) compared to Site III (0.54 mg/g & 0.21mg/g) and Site I (0.41 mg/g & 0.73 mg/g).

Table 7, also shows that highest concentration of lead (0.029 mg/L) was present in the water samples at Site I and Site II whereas, in the soil samples the maximum concentration of Lead (0.0 14 mg/g) was found to be at Site III. Higher concentration of copper was noticed at Site II in both soil (0.11 mg/g) and sludge samples (0.033 mg/g). This indicates that the trend of specific pollution load at different sampling sites of river Yamuna.

The generated and reported heavy metal concentrations in water and sediment samples show that the quality of heavy metals in Yamuna water was exceeding the environmentally safe concentrations (as shown in the following Table 8).

Unsafe concentrations of Zinc in every sampling site was noticed when the critical value for Zn is 0.0016-mg/L in water and 0.12 mg/g in sediments. Similar higher concentration was also reported on Lead value (in water samples) and copper concentration (in sediments) (Table 8). Accumulation of high Zinc levels in gonads has been found to be detrimental to fish health, affecting its reproductive potential in the long run (Muralidharan, 1995). Fish food organisms such as Cyclops and Daphnia are more sensitive to metals like Zinc (Sultana and Rao, 1998).

Retarded growth, anaemia and delayed reproduction of fish have been noticed at an exposure of 2 ppm of zinc, copper and chromium in combination for 120 days. This indicates that the combination of metals is many times more toxic than individual toxicity (Jarvinen and Ankley, 1999).

Table 8 Comparison of standard safe concentrations of heavy metals in water and sediment samples and the generated values of the same. Units: Water- (mg/l); sediments- (mg/g dry weight).

Name of Heavy metal	*Safe concentration*		*Observed concentration*	
	Water	*Sediments*	*Water*	*Sediments*
Cadmium	0.00016	0.014	nil	nil
Zinc	0.0016	0.12	0.01-0.14	0.21-0.92
Nickel	0.0014	0.0074	nil	nil
Lead	0.002	0.86	0-0.029	nil
Chromium	0.002	0.27	nil	nil
Copper	0.007	0.0006	nil	0-0.33

3.7 Bioaccumulation of Critical Heavy Metals in Fish Fauna of River Yamuna

The ecological implications of such residues are manifold. Apart from posing a public hazard for the fish eater, heavy metal accumulation in the tissues of fish and other organisms causes physiological disorders such as necrosis of liver, damage of nephrons in the kidney, hematological aberrations, decline in growth rate, fecundity and enlargement of gall bladder.

Metals analyzed in fish tissue included Cd, Cr, Ni, Pb and Zn. Linkages between tissue residues and effects have been made for liver, kidney, gill, spleen and whole body data. CPCB has been clearly reported that the higher concentration of zinc accumulated (0.022-0.43 mg/g) in the fish tissue samples throughout the season. Interestingly, the concentration was minimum during summer months (0.022-0.024 mg/g) and maximum during winter months (0.197- 0.43 mg/g) at both the side. According to Jarvinen and Ankley (1999), muscle, liver and gill residues of zinc appear to be similar in several studies of fish. Nickel appears to accumulate primarily in gill tissue and in kidney, although liver, muscle and body residues are also significant. Lead appears to accumulate preferentially in kidney and spleen, although liver and gill tissues also can contain significant amounts. The magnitude of accumulation is generally low in tissues such as muscle where the metabolic activity is relatively less. Earlier works have

Table 9 Reported Bioaccumulatory data of fish fauna by Central Pollution Control Board. Units: (mg/g).

Heavy Metals		*Wazirabad Barrage (Site I)*			*Okhla Barrage(Site II)*		
		Summer (Mar-June)	*Monsoon (July-Sept)*	*Winter (Oct-Feb)*	*Summer (March-June)*	*Monsoon (Jul-Sept)*	*Winter (Oct-Feb)*
Cadmium	Min.	NT	NT	NT	NT	NT	NT
	Max.	NT	NT	NT	NT	NT	NT
	Avg.	NT	NT	NT	NT	NT	NT
	N.	6	12	12	16	12	18
Chromium	Min	NT	NT	NT	NT	NT	NT
	Max.	NT	NT	NT	NT	NT	NT
	Avg.	NT	NT	NT	NT	NT	NT
	N.	6	12	12	16	12	18
Nickel	Min.	NT	NT	NT	NT	NT	NT
	Max.	NT	0.006	0.005	NT	0.009	NT
	Avg.	NT	0.0015	0.003	NT	0.0035	0.009
	N.	6	12	12	16	12	18
Lead	Min.	NT	NT	NT	NT	NT	NT
	Max.	NT	NT	NT	NT	NT	0.39
	Avg.	NT	NT	NT	NT	NT	0.064
	N.	6	12	12	16	12	18
Zinc	Min.	0.004	0.030	0.032	0.008	0.035	0.013
	Max.	0.040	0.970	0.997	0.046	0.131	3.99
	Avg.	0.024	0.128	0.197	0.022	0.063	0.431
	N.	6	12	12	16	12	18

N—Number of observation; Min—Minimum; Max—Maximum; Avg.—Average.

also ascertained a fact that accumulation of metals namely, cadmium and lead in kidney is more due to the induction in the synthesis of metallothionin (Sastry and Shukla, 1993; Lionett *et al.*, 2001). Further the variation in the levels of accumulation in different organs in the fishes could be attributed to the differences in the physiological role of each organ. Regulatory ability, behavior and feeding habits are other factors that could influence the accumulation pattern among organs (Kotze *et al.*, 1999).

4 CONCLUSION

If the last five years data of three study sites are compared, in accordance with our first objective, there is almost no change in the trend of heavy metal fluctuations. This clearly indicates that there is no further precaution taken up by the industry or government agencies to reduce the higher concentrations of Fe, Zn and Cu. Different heavy metal concentrations are exceeding the safe level as evidenced in the reported as well as the experimental values. Specially, due to widespread use of zinc in the industries and absence of cost effective treatment technologies for the effluents, it is released especially by small and medium scale industries to the environment in excess quantities, which being toxic to the biological system. This is alarming for the health of target consumer groups. The levels of zinc reported in the present study are likely to be harmful to piscivorous birds, which feed in very large numbers in the surrounding water bodies of Okhla and Wazirabad area.

As far as Delhi is concerned, the biomagnification processes in the aquatic food chain have not been mapped out clearly. Erratic releases of unsafe concentrations of heavy metals, together with high summer temperature and the anoxic condition of the Yamuna river, pose serious threats to the biotic community. Bio-monitoring processes to quantify bioaccumulation rate have become an inevitable necessity for water quality management studies now.

Acknowledgements

The authors are grateful to Central Pollution Control Board (CPCB) for providing data and the experimental facilities for conducting this study.

References

APHA. 1989. In: standards methods for the examination of water and wastewater. Publ. American Public health association, Washington. 17th Edition.

Ay, O., Kalay M., Tamer, L. & Canli, M. 1999. Copper and lead contamination in tissues of a freshwater fish tilapia zillii and its effects on the Bronchial Na, K-ATPase activity. Bull. Environ. Contam. Toxicol. 62: 160–168.

Bradley, R.W. & Sprague, J. B. 1985. Accumulation of zinc by rainbow trout as influenced by pH, water hardness and fish size. Environ. Toxicol. Chem. 4: 685.

Central Pollution Control Board. 1999. Parivesh' Highlights, CPCB publishers.

Central Pollution Control Board. 2003. Parivesh' Highlights, CPCB publishers.

Hemelraad, J., Kleinveld, H.A., DeRoss, A.M., Holwerda, D.A. & dZandee, DI. 1987. Cadmium kinetics in freshwater clams. III. Effects of zinc on uptake and distribution of cadmium in Anadonta cygnea. Arch. Environ. Contam. Toxicol. 16: 95–101.

Jarvinen, A.W and Ankley, G.T. 1999. Linkage of ejects to tissue residues-Development of a comprehensive database for Aquatic organisms exposed to inorganic and organic chemicals. Penscola, FL: Society of Environmental Toxicology and Chemistry Press

Jayakumar (2001). Levels of Heavy metals in commercial fresh wayter fishes of Coimbatore. M.Phil. dissertation thesis submitted to Bharathiar University.

Jorgensen, L.A. and Pedersen, B. 1994. Trace metals in fish used for time trend analysis and as environmental indicators. Mar. pollut. Bull., 28 (4): 235–243.

Kargin, F. & Cogun, H.Y. 1999. Metal interactions during accumulation and elimination of zinc and cadmium in tissues of the freshwater fish tilapia nilotica. Bull. Environ. Contam. Toxicol. 63: 511–519.

Kotze, P.,Hdu Preez, H. & van Vuren, J. H. J.1999. Bioaccumulation of Cu and Zn in Oreochromis mo ssambicus and Clarias gariepinus from the Olifants River Mpumalanya, South Africa. Water SA, 25 (1): 99–110.

Larsson, A., Haux, C. & Sjöbeck, M. 1985. Fish physiology and metal pollution: results and experience from laboratory and field studies. Ecotoxicol. Environ. Saf. 9: 250–281.

Lionetto, M.G.. Giordano M. E., Caricato, R., pascariello, M. F., Marinsci, L. & Schettino, T. 2001. Biomonitering of heavy metal contamination along the Salento coast (Italy) by metallothionein evaluation in Mytilus galloprovincialis and Mulls barbatus. Aquatic. Conserv.: Mar. Fresh W. Ecosyst., 11: 305–310.

Lunt, H.A. 1963. The case for sludge as a soil in power waste and sewage works 100, 294–301.

Memmert U. 1987. Bioaccumulation of Zinc in two fresh water organisms (Daphnia magna, crustacea and Brachydanio rerio, pisce), water Res., 21: 99.

Misra, V., Jaffery, F. N. & Viswanathan, P. N. 1991. Risk analysis in hazardous industries. Regul. Toxicol. Pharmacol. 13: 62– 69.

Misra S.M., Borana, K., Pani, S., Bajpai, A. and Bajpai, A.K. (2002). Assessment of heavy metal concentration in Grass carp (*Ctenopharyngodon idella*). Poll .Res. 21 (1): 69–71.

Muralidharan, S. (1995). Heavy metal contamination in and around the aquatic environs of Keoladeo National Park, Bharatpur. Ph.D. thesis, university of Rajasthan.

Nag, D. & Jaffery, F. N. 1994. Adverse health effects of environmental chemicals: Indian scenario. In.: Chemical safety: International reference Manual (ed), M. Richardson, VCH Publishers,: 111–125.

Organisation for Economic Corporation and Development (OECD) (1991). The state of the environments, Paris.

Puura, E., Marmo, L. & Alessandro, M.D. 2002. Proceedings of workshop on mine and quarry waste- the Burden from the past. 27-28th, 2002, Orta, Italy.

Rao, G.J.M. 1973. Waste collection treatment and disposal in India. Ind. J. Environ. Hlth. 15: 222–235.

OnLine Journal of Biological Sciences.2001, 1(5): 414-416. © Asian Network for Scientific Information

Radhakrishnaiah, K. 1988. Accumulations of copper in the organs of freshwater fish Labeo rohita (Hamilton) on exposure to lethal and sub-lethal concentration of copper. J. Environ. Biol. 319–326.

Raina, V., Shah, A.R. & Ahmad, S.R. 1984. Pollution studies of river Jhelam-I, An assessment of water quality, Ind. J Environ. Hlth. 26(3): 187–201.

Sastry, K.V. & Shukla, V. 1993. Uptake and distribution of cadmium in tissues of Channa punctatus. J. Environ. Boil., 14 (2): 137–142.

Seaward, M.R.D. & Richardson, D.H.S. 1990 Atmospheric sources of metal pollution and effect on vegetation. In heavy metal tolerance in plants: Evolutionary Aspects: In Shaw A.J. Beca Raton(ed), CRC Press: 75-92, PCB publication NAAQM/8/1996-97.

Sultana, R. and Rao, D.P. (1998). Bioaccumulation Patterns of Zinc, Copper, Lead and Cadmium in Grey mullet, Mugil Cephalus from harbour waters of Vishakapatanam, India. Bull. Environ. Contam. Toxicol., 60: 949–955.

Thomann, R.V. 1989. Bioaccumulation model of organic chemical distribution in aquatic food chains, Environ. Sci. Tecnol. 23, 699.

US Environmental Protection Agency (EPA) (1978). Metal bioaccumulation in fishes and aquatic invertibrates. G.R Philips and R.C. Russo, EPA - 600/3/-78-103. Washington DC 116.

Groundwater Quality Monitoring in the Vicinity of Pirana Landfill Site, Ahmedabad, India: Adequacy of Conventional Hydrochemical Approach

46

Umesh Kumar Singh and Manish Kumar*
School of Environmental Sciences, Jawaharlal Nehru University, New Delhi-110 067, India

ABSTRACT

The severity of leachate that contains myriad chemicals has been recognized for many years. The Municipal solid waste (MSW) in contact with moisture/water produces a leachate that can be hazardous or otherwise deleterious to the potential use of water resources due to quality deterioration. For effective disposal of wastes, there is need to know its quality, quantity and nature of toxicity. The present article deals with the adequacy assessment of hydrochemical data obtained through analysis of leachates and groundwater samples, for the impact assessment of landfill on groundwater quality. The empirical relationships between parameters showing abnormal value along with comparative chemical nature of MSW, leachates and groundwater were emphasized for the evaluation of the impact of landfill. The results obtained in this study suggest that even if there is no specific indicator parameter evident in groundwater samples, the trend of some variables of anthropogenic origin viz. heavy metals, nitrate, can be effectively used to establish an empirical relationship among the chemical nature of MSW, leachate and groundwater which subsequently can facilitate the visualization of impact on groundwater quality. Predominance of Fe and Zn in both water and solid waste samples is due to iron and tin based industry as well as Zinc based waste in the area. The comprehensive statistical analysis like factor and cluster analysis of chemical data and computational modeling work using MODFLOW, NETPATH will further enhance the understanding of leachate plume movement and its impact on groundwater quality in the city.

*Corresponding author: manish.env@gmail.com

1 INTRODUCTION

An escalating attention and recognition is being given world-wide to the pollution of groundwaters by municipal and industrial "non-hazardous" solid waste landfills. In India, the wastes are largely dumped in landfill areas without proper segregation and treatment. Moreover, majority of landfill areas are not well designed with modern practices of sanitary land filling, which leads to ground and surface water contamination. These low-lying dumping areas due to their poor site selection, are also prone to flooding and polluting adjoining areas, which further increase the pollution threat. Due to rapid change in the standard of living in India, there is substantial increase in generation of solid wastes in both absolute and per capita terms. If the waste is not collected and disposed off efficiently and effectively, it create environmental problem. For effective disposal of wastes, there is need to know its quality, quantity and nature of toxicity. Waste disposal by landfill site is very popular and the ever increasing demand for larger space for domestic and industrial wastes from urban areas makes them a necessary part of the human life cycle. Landfill sites commonly use the space available in disused quarries or special purpose built structures. Our interest in landfill site lies in assessing the pollution threat to groundwater since the former contain harmful substances. The pollution threat assumes more importance in regions where there are fractures or weak zones just beneath the landfill or surrounding it.

The impact of landfill sites on groundwater has been attempted by different researchers in different perspectives (Baedecker & Apgar, 1984; Carpenter *et al.*, 1990, 1991; Flyhammar 1995; Gaily & Gorelick 1993; Klefsted *et al.*, 1975; Kjeldsen and others 1993; Laine *et al.*, 1982; Loizidon & Kapetanios 1993; Meju, 1993; Newell *et al.* 1990; Robinson & Maris, 1982; Tejero *et al.* 1993; and Vendrame & Pinho, 1997). Most of the works cited above are focused on assessing the pollution risk to groundwater due to landfill by using invasive as well as non-invasive methods. In view of lack of proper documentation on the characteristics of the waste, investigations on landfill become a very difficult task. Abu-Rukah & El-Aloosy (1997) studied the various metal ions migration in the El-Akader landfill site and concluded that there is a migration of metal ions to the deep layers. In their study Gharaibeh *et al.* (1989) indicated the possibility of surface water, ground-water and soil pollution in the El-Akader region. Jawad *et al.* (1998) concluded in their study that the increase in the value of nitrate in the El-Akader region is due to waste-water dumped at the disposal site.

To assess the impact of leachates on the underground water, three general procedures can be followed: (1) Experimental determination and estimation through mathematical modeling (2) Isotopic characterization of MSW and groundwater and 3) By comparing chemical nature of MSW & leachates with groundwater chemistry. In the present study we have used third procedure to evaluate the impact of leachates on groundwater. The concentrations of different constituents in the leachate are determined by the nature of the waste, by the amount of recharge water that passes through the landfill and by factors that control groundwater velocity. The waste material contains both organic as well as inorganic substances. There are heavy metals present in the domestic wastes. The presence of different kinds of substances in the waste and their combined effect on the groundwater quality do help in establishing an empirical relationship between the nature of waste, leachate and groundwater. Some plots were made on the basis of leachate hydrochemical nature and an attempt has been made to evaluate not only the impact but the extent of it on groundwater quality. Study reveals the adequacy of comparative use of hydrochemical data for the preliminary and cost-effective assessment of landfill impact on groundwater quality.

2 STUDY AREA

Ahmedabad, the biggest City of Gujarat, lies on 23^0 1' North Latitude and 72^0 37' East Longitude on the bank of River of Sabarmati. In general, the climate of Ahmedabad remains hot and moist. Winters are cool and pleasant (November-February), but summers (April-June) are extremely dry, hot and the temperature can be as high as 46°C at times. The city experiences good rainfall during the monsoon season (July-September). It experiences a harsh summer season, and is lashed by the south-west monsoon.

Ahmedabad is one of the most important industrial cities in India. It has maximum number of chemical and textile industries and is seventh largest generator of total municipal waste in India, while on the basis of per capita production, it holds fifth position in the country (Table-1). The city has only one landfill site at Pirana with 84 acres of land, which is receiving municipal solid waste since last 24 years. The total solid waste generated from the city is around 1100-1200 tonnes/day (Source: Supervisor, Ahmedabad Municipal Corporation). Nearly 95% of the solid waste is dumped in the Pirana landfill site i.e. around 1000 tonnes/day goes to landfill area at Pirana and nearly 50 tonnes/day consumed by the excel factory for composting. There is no any particular solid waste disposal method used in Pirana landfill site at Ahmedabad.

Table 1 Solid waste generation of various cities in India (Yedla and Parikh, 2001)

City	*Mumbai*	*Delhi*	*Calcutta*	*Chennai*	*Hyderabad*	*Bangalore*	*Ahmadabad*
Tonnes/day	5000	4600	3692	3124	2800	2700	1600
Per Capita (Kg/Day)	0.44	0.48	0.38	0.657	0.38	0.48	0.58

2.1 Physical Characterisation of the Solid Waste in the Area

Organic Matter -72%, Paper-5.69%, Plastic-6.74%, Metals -0.79%, Glass - 1.38% and Earth Materials -13.40%.

3 METHODOLOGY

The Sampling was done for solid waste, leachate and groundwater in June, 2003. The solid waste samples were collected 0-25 cm from the ground level at Piarana landfill area in Ahmedabad with help of Quadrate area 1m x 1m. Four samples were collected from the each corner and one from the center and homogenized well. Municipal bulldozers were used for digging. Stainless steel trowel was used for collecting solid waste samples. These samples were collected in air tight plastic bags to avoid any chemical contaminations. The groundwater samples were collected from sources scattered around the landfill sites. There were eleven groundwater samples, five solid wastes and one leachate sample taken in total. The samples were collected from points which were at a lower elevation as compared to the landfill site. EC and pH were measured *in-situ* with help of Orion pen electrode. All sampling bottles were soaked with 1:1 HNO_3 and washed using detergents. These bottles were then rinsed with

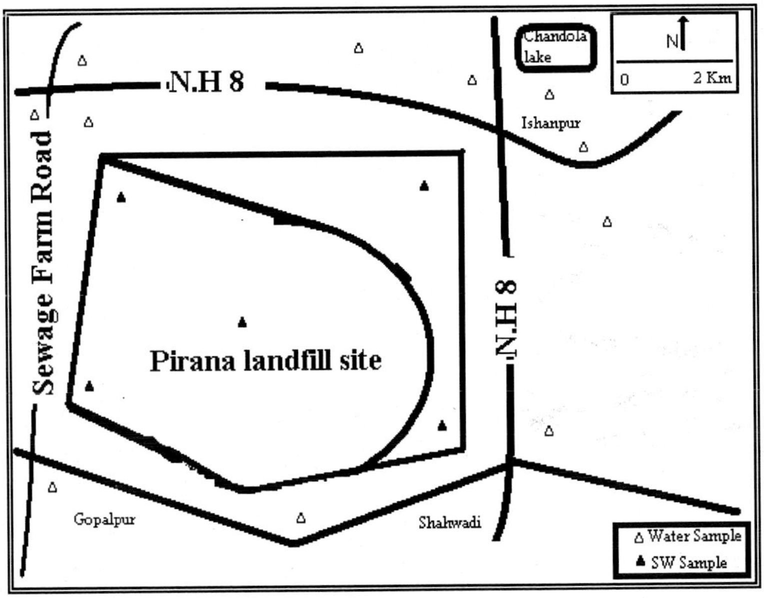

Fig. 1 Schemedic Diagram of Priana Landfill sites Of Ahmadabad with Sampling location

double-distilled water. At the time of sampling, sampling bottles were thoroughly rinsed 2-3 times using groundwater to be sampled. Water samples were filtered using 0.45mm Millipore filter paper and acidified with nitric acid (Ultrapure Merck) for cation analysis and with HBO_3 was used as preservative for nitrate analysis (Kumar et al., 2006). For anion analysis, these samples were stored below 4°C. Major cations like Ca^{2+}, Na^+, K^+ and Li^+ were analysed on flame photometer and heavy metals were analysed using atomic absorption spectrophotometer (Schimadzu AAS 6800). The chemical analyses were carried out as per the standard procedure given in APHA (1995). Apart from the collection and analysis of water samples, other data were collected from the relevant departments/organizations.

4 RESULT AND DISCUSSION

The results obtained from physical and chemical analysis of groundwater, leachate and Municipal Solid Waste (MSW) from Pirana landfill area of Ahmedabad is given in Table-2 and 3. The average value of pH of MSW is 8.76 with very narrow range of 8.55 to 9.01 as it is evident from the low standard deviation. This very alkaline nature of solid waste also leads to the high pH of leachate which is 8.35 while groundwater samples being showing alkali nature showing pH below 7.5 with two exceptional samples. This is perhaps due to lower elevation of this place as compared to the landfill site. It has been observed that pH values exceed from the standard prescribed for compost in Municipal Solid Waste (Management and Control) Rules, 2000.

EC has its importance in the analysis as this serve various purposes, often it effects the degree of mineralization and assess the effects of total concentration of ions on chemical equilibrium, physiological effects on plants and animals. In contrast to pH there is high variation in the range of EC of MSW showing the complex nature and origin of solid waste that

Table 1 Chemical analysis data of Water Samples (All units are in mg/L except pH and EC (μS/cm).

	pH	EC	TDS	Ca^{2+}	Mg^{2+}	Na^{+}	K^{+}	Fe	Mn	Cl^{-}	SO_4^{2-}	HCO_3^{-}	NO_3^{-}	PO_4^{3-}	Zn	Cd	Cr	Pb	Ni
AW1	7.05	1160	769	383	98	22.2	1.53	0.012	0.002	112	76.8	380	44.5	0.178	0.02	0.004	0.002	BD	0.003
AW2	7.03	1485	919	346	83	22.5	27	0.002	0.012	140	164.5	400	73.5	0.214	0.03	BD	BD	0.042	0.003
AW3	7.03	1345	893	336	68	24.3	2.19	0.025	0.007	165	152.6	400	40	0.178	0.04	BD	0.003	0.041	0.0025
AW4	7	1430	949	367	88	25.33	25	0.01	BD	160	170.5	400	66	0.196	0.034	0.003	0.003	BD	BD
AW5	7.04	1390	915	370	85	20.4	1.96	0.008	BD	165	148.6	390	65	0.339	0.008	0.003	0.004	BD	0.006
AW6	7.04	1390	918	383	90	21.1	1.86	0.007	0.005	192	139.6	370	64	0.339	0.032	0.003	0.007	BD	BD
AW7	7.19	1325	880	380	86	23	2.25	0.005	0.017	190	96.7	400	41	0.643	0.01	0.001	BD	0.011	0.007
AW8	7.04	1250	824	366	72	23	2.5	0.005	0.013	117	132.4	400	32	0.196	0.05	0.005	BD	0.017	0.003
AW9	7.04	1170	772	351	85.5	26	2.5	0.013	0.005	110	94.5	380	42	0.214	0.06	0.001	0.003	BD	0.0035
AW10	7.19	1200	795	328	85	26.5	2.35	0.01	0.006	117	105	390	40	0.357	0.09	0.003	BD	0.003	BD
AW11	7.04	1210	803	350	90	27	2.2	0.015	0.012	115	96.5	400	41	0.375	0.031	0.004	0.03	0.003	BD
Leachate	8.35	26500	1758	89	306	470	135	7.1	0.26	4360	130	110	178	13.56	2.01	0.01	0.1	0.12	0.06

Table 2 Chemical analysis data of Muncipal Solid Waste (MSW) Samples.
(All units are in mg/kg except pH)

Station No	*pH*	*EC(mS/cm)*	*Cu*	*Zn*	*Ni*	*Cr*	*Fe*	*Mn*	*Pb*	*Cd*	*Hg*
AS 1	8.89	0.12	15.92	154.72	3.87	9.46	16800	400	14.17	1.22	0.07
AS 2	8.72	1.31	26.50	138.36	0.01	BD	15300	400	14.73	BD	0.08
AS 3	8.63	1.97	32.63	126.82	BD	BD	1680	400	9.13	1.58	0.05
AS 4	9.01	0.15	32.63	148.95	BD	26.15	12900	500	14.73	BD	BD
AS 5	8.55	1.73	17.59	157.60	BD	26.15	9800	300	10.25	0.06	0.02
Max	9.01	1.97	32.63	157.60	3.87	26.15	16800	500	14.73	1.58	0.08
Min	8.55	0.12	15.92	126.82	0.01	9.46	9800	300	9.13	0.06	0.02
Average	8.76	1.06	25.05	145.29	1.94	20.59	14320	400	12.60	0.96	0.06
St. dev	0.19	0.87	8.00	12.68	2.73	9.64	2987	70.71	2.70	0.79	0.03

are being dumped here. MSW samples are showing less EC than that of leachate, perhaps due to high dissolution of ionic species in water. Groundwater samples are showing consistent EC value which is evident from the very narrow range of 1160 to 1485µS/cm. This range is very close to the permissible limit of 1400 µS/cm for drinking water (WHO 1995). Although only two samples are showing higher values than that of permissible limit of EC but all other samples are also very near to this limit and likely to cross the limit in near future. This nature of groundwater does indicate the radial nature of impact of landfill with slight increase towards to hydraulic gradient. It may be because the long working history of the landfill.

Heavy metals can be used as the major indicator of anthropogenic impact whether from industrial effluents or landfill leachate. Chemical analysis of MSW shows the general trend of heavy metal in the order of abundance Fe>Mn>Zn>Cu>Pb>Cr>Ni>Cd>Hg (Fig-2). Almost similar trend is observed in leachate with only exception of Zn at the second place and Mn at third. Groundwaters also show the identical trend which indicates the high impact of landfill leachates on groundwater quality (Fig-3). The results of heavy metals analyses reveal high value of Iron, may be because of the iron and tin based waste in landfill site. It has been observed that Mn value nearly falls within the prescribed standard except few samples. The low concentration is perhaps due to its requirement for several microbial enzymatic activities. Manganese has direct toxicological significance as it regulates the concentration of other element, including toxic and heavy metals. The major anthropogenic sources for manganese are coal burning and incineration of municipal waste. The observed value of Copper is within the prescribed limit while zinc concentration is high may be due to the presence of Zn based waste viz zinc plated materials or fertilizers and cement in it. The value of Ni and Cr for most of the stations were within the limit of prescribed standard.

The observed value of the Pb, Cd and Hg was within the prescribed standard of Municipal Solid Waste (Management and Control) Rules, 2000 and of USEPA (Table 2).

It appears that there is no indicator parameter or group of parameters, other than few heavy metals whose presence or concentration range would permit the unambiguous identification of all the leachate-impacted groundwaters at this site. This is because there is no

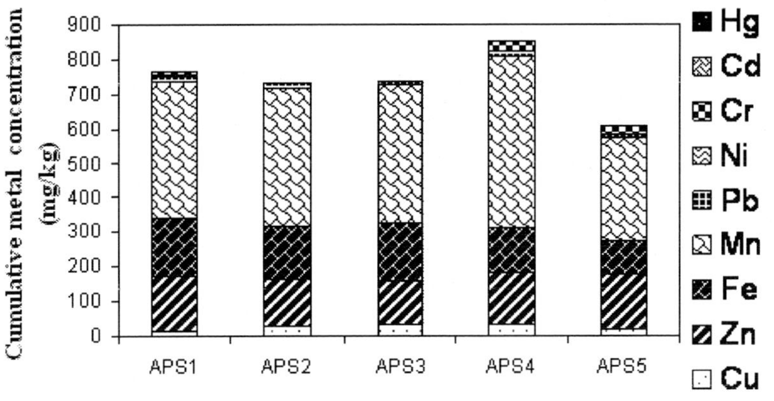

Fig. 2 Stacked Column to compare the contribution of each type of metal to a total value in MSW. (Iron concentration is divided by the factor of 100).

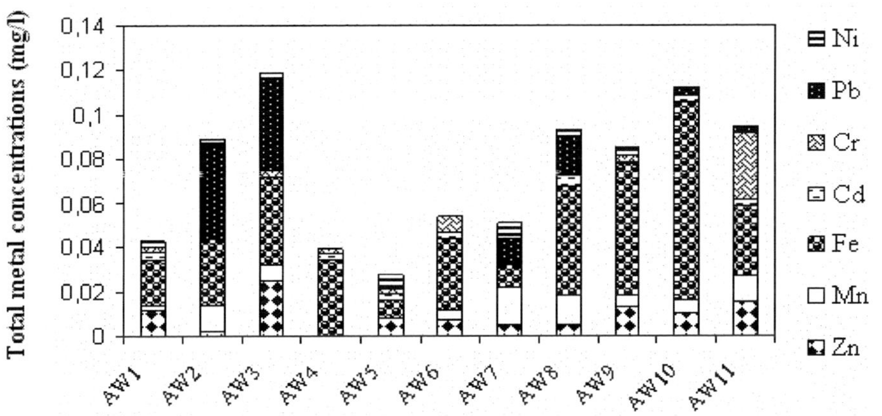

Fig. 3 Stacked Column to compare the contribution of each type of metal to a total value in the groundwater samples.

indicator compound(s) that is unique to landfill leachate (i.e. that was neither present in uncontaminated waters nor in the piezometer materials) and cost effective to include in the analytical protocol for the most of groundwaters. The extent of landfill-leachate in the groundwater can be evaluated by considering the chemistry of groundwaters and hydrogeological information about probable flow paths.

Initially, a series of arbitrary tests was established. This was based upon a comparison of "background" groundwater chemistries with leachate (Barker et al., 1988). The following observations were made:-

(1) leachate is intermediate in total inorganic load; (2) Mg dominates Ca in leachate, Ca dominates Mg in background waters and leachate has the highest K/Mg ratio (3) leachate has the highest concentrations of EC. Based on this observation, Mg vs. K cross plot were plotted which reveals three strongly impacted groundwater sample (Fig. 4).

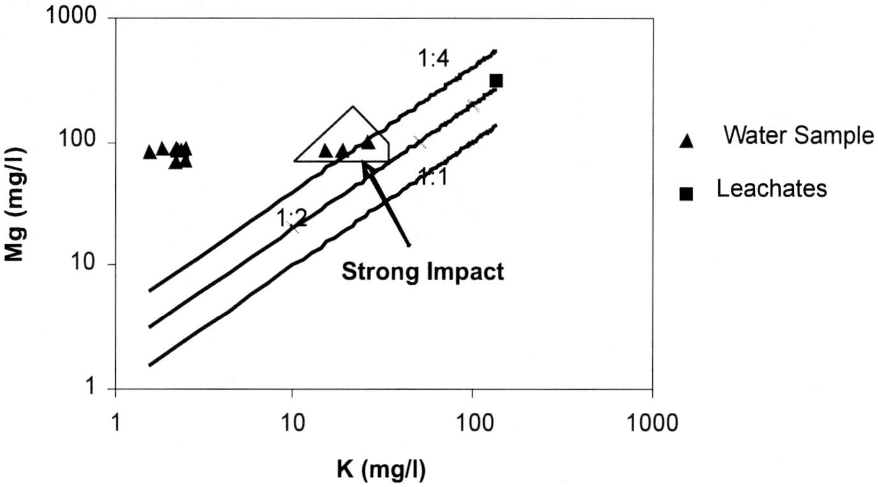

Fig. 4 A cross plot of Mg vs. K with leachate sample to observ the impact on Groundwater sample.

The common sources of nitrate in water are the leachates from the landfill sites whether in use or abandoned, local agricultural fields, domestic sewage. However, because of laminar flow condition and due to difference in specific gravity and viscosity between nitrate slugs and the natural freshwater, some diffusion bound to occur (Kumar et al., 2007). But no such natural sources are reported to exist in the area for NO_3^-. The leachate is having NO_3^- as high as 178 mg/L which directly shows its impact on the groundwater. Other than Fe, rest of the heavy metals are either in between desirable and permissible limit or below desirable limits.

5 CONCLUSIONS

Although, there is no specific indicator parameter, but almost similar trends of heavy metals in all three kind of samples do indicate an empirical relation between MSW, leachate and groundwater sample. K/Mg ratio also indicates that three water samples heavily impacted from leachate contamination. Predominance of Fe and Zn in both water and solid waste samples are due to iron and tin based industry as well as Zinc based waste in the area.

In order to palliate the pollution risks of the groundwater, it is necessary to:

- Stop further dumping in this urban landfill as contaminants found to be spread radially;
- Install the system for leachates collection, treatment and drainage ;;
- Set an impermeable surface at the site of landfill (with clay or with a high density polyethylene geomembrane) in order to limit the infiltration of leachates and to avoid the contamination of groundwater
- Control the groundwater quality and survey its evolution (especially the metallic elements) in time and in space, by choosing a representative network of wells.

Study reveals that general quality parameters including some heavy metals are adequate enough for the preliminary assessment of landfill impact on groundwater quality. However the principle component analysis and cluster analysis using these hydrochemical data can further provide the meaningful insight into the impact processes. Future work may focus on detailed modeling work in order to trace contaminant transport in the area with the help of MODFLOW or NETPATH for better understanding of leaching and its impact on groundwater quality in the city.

Acknowledgements

First and second author thanks University Grant Commission and Council of Scientific and Industrial Research (India) respectively for financial grant through research fellowship, which made this work possible.

References

Abu-Rukah, Y.H. & El-Aloosy, A.S. (1997). Various variables migration in landfill site using statistical explanation. A case study El-Akader landfill site, North Jordan. *Journal of Institute of Mathematics and Computer Sciences (Math. ser.)*, 10: 113–130.

APHA (1995), Standard methods for the examination of water and wastewater, 19th edn. American Public Health Association Washington, DC.

Baedecker, M. J. and Apgar, M. A. (1984) 'Hydrochemical Studies at a Landfill in Delaware', in: J.Bredehoeft (ed.), *Groundwater Contamination*, National Academy Press, Washington, DC, pp.127–138.

Barker, J. F., Barbash, J. E., & Labonte, M. (1988). Groundwater contamination at a landfill sited on fractured carbonate and shale. Contaminant Hydrology, 3,127–138.

Carpenter, P. J., Calkin, S. F. and Kuffmann, R. S., (1991) 'Assessing a fractured landfill cover using electrical resistivity and seismic refraction techniques', *Geophysics* 56, 1896–1904.

Carpenter, P. J., Kaufmann, R. S. and Price, B.: 1990, 'Use of resistivity soundings to determine landfill structure', *Groundwater* 28, 569–575.

Flyhammar, P. (1995). Leachate quality and environmental effects of active Swedish municipal landfill. In: Cossu, R., Christensen, H. T. & Stogmann, R. (Eds), *Proceeding Sardinia 95.Fifth Int. Landfill symposium* Vol. II. pp 551–557.

Gaily, R.M. & Gorelick, S.M. (1993). Design of optimal, reliableplunie capture schemes: application to the Gloucester landfill groundwater contamination problem. *Groundwater* 31:107–114.

Gharaibeh, S.H. & Masad, A. (1989). Die Problomatite der Ahallbeseitingung in Jordan. Fallstudie Fur Enstasicklungs lander. *Wasser und Boden*, 10: 620-622.

Jawad, A., Al-Shereideh, S.A., Abu-Rukah, Y. & Al Qadat, K. (1998). Aquifer groundwater quality and flow in the Yarmouk river of northern Jordan. *Journal of Environmental Systems*, 26:265-287

Kjeldsen, P., Bjarg, P., Rugger, K., Pedsersen, J.K., Skov, B., Forverskov, A., Wurtz, S. & Christensen T.H. (1993). Assessing the variability in leachate from an old municipal landfill In: *Presidents Sardinia 93, Fourth int. landfill Symposium*, Gosso R. Christensen, H.T., and Stegman, R.R. (Eds), pp 1514–1531.

Klefsted, G., Sendlein, L. V. and Palmquist, R. C. (1975) 'Limitations of electrical resistivity methods in landfill investigations', *Groundwater* 13, 418–437.

Kumar, M., Kumari, K., Ramanathan, AL., Saxena, R., (2007). A comparative evaluation of groundwater suitability for irrigation and drinking purposes in two agriculture dominated districts of Punjab, India. Environ. geology, 53,553-574.

Kumar, M., Ramanathan, AL., Rao, M.S., Kumar, B., (2006) Identification and evaluation of hydrogeochemical processes in the groundwater environment of Delhi, India Environ. geology, 50,1025-1039.

Laine, D. L., Parra, J. O. and Owen, T. E. (1982) 'Application of an Automatic Earth Resistivity System for Detecting Groundwater Migration under a Municipal Landfill', in: *Proceedings of NWWAConference on Surface and Borehole Geophysical Methods in Groundwater Investigations,* 12–14 February 1982, pp. 34–51.

Loizidon, M. & Kapetanios, E. (1993). Effect of leachate from landfills in under groundwater quality. *The Science of the Total Environment,* 128: 69–81.

Meju, M. A. (1993) 'Geophysical Mapping of Polluted Groundwater in a Closed Landfill Site', in:*Proceedings of the Third International Congress of Brazilian Geophysical Society,* Expanded.

Newell. G.J., Hopkins, L.P. & Bedient, P.B. (1990). A hydrogeologic data for groundwater modeling. *Journal of Groundwater,* 28: 703–714.

Robinson, H. C., Barber, C. and Maris, P. J.: 1982, 'Generation and treatment of leachate from domestic wastes in landfills', *Water Pollut. Control* 54, 465–478.

Tejero, M.I., Fantelli, L.M., Diaz, I.R. & Azanto, N.M. (1993). Characteristics and treatment of leachate in the Meruleo landfill (Spain). In: Cossu, R., Christensen, H.I. & Stegman, R. (Eds), *Proceeding Sardinia 93, farteInt. landfill symposium,* pp. 1033–1042.

Vendrame, I. and Pinho, M. F. (1997) 'Groundwater Quality in Taubate Landfill, Brazil', in: J. Chilton (ed.), *Groundwater in the Urban Environment,* A. A. Balkema, Rotterdam, pp. 559–564.

WHO (1995) Guidelines for drinking water quality-II pp. 333. Geneva, Environmental Health Criteria, 5.

Yedla. S.and Parikh. J., (2001) Solid Waste Management-Current Status and Strategies for the Future 12-14 December 2002, Bangalore, India.

Recent Developments in Analytical Techniques for Monitoring Inorganic Pollutants in Groundwater Samples

47

V. Balaram
National Geophysical Research Institute. Hyderabad–500007, India

ABSTRACT

Basic aspects of groundwater pollution and the role of different instrumental analytical techniques are highlighted with examples. Trace metal analysis of groundwater is being carried out by a variety of optical, nuclear and mass spectrometric techniques such as inductively coupled plasma-atomic emission spectrometry (ICP-AES), X-ray fluorescence spectrometry (XRF) and inductively coupled plasma-mass spectrometry (ICP-MS). Recent advances in some of these techniques have taken the detection limits of several elements to ppt (pg/g) and sub-ppt levels. Determination of anions such as Cl^-, NO_3^-, etc. in groundwater samples is extremely important and ion-chromatography (IC) has become an established technique for this application. Use of isotopic data for identifying pathways in the investigation of sewage pollution in groundwater is also discussed. Finally, the importance of information on different organic and inorganic species of certain metals in groundwaters, the analytical techniques used and quality requirements of data are also presented.

1 INTRODUCTION

Nearly 97.5% of the water on the earth's surface is salty. Of the remaining 2.5%, glaciers constitute 68.5% and about 30% is fresh groundwaters. Groundwater reserves act as standby resources for overcoming drought. With the tremendous increase in human population and its activities, there has been an unforeseen depletion of groundwater reserves. Increased use of water for agricultural purposes, urbanization and demand for supply of drinking water through public utilities, ever growing industrialization accompanied by increased power production, mining and other activites have greatly depleted the groundwater resources. Increase in population and climate changes will further raise the water demand by about 50% by 2025. This will increase the population under water-stress to about 3.4 billion and most of them will be in the developing countries. Demand for irrigation water will be maximum in Asia since the largest percentage of irrigated land lies in this continent. Continued withdrawal of groundwater has already depleted the water table. Besides, the use of fertilizers and pesticides, discharge of industrial effluents, domestic and industrial landfills have caused the infiltration of several potentially dangerous heavy metals and other pollutants into the already scarce groundwater reserves.

2 GROUNDWATER MONITORING STUDIES

Monitoring and protection of groundwater resources from pollution has become a high-priority area of research in recent years. Groundwater monitoring for various pollutants is a complex job requiring the application of time tested sampling and analytical procedures resulting in the collection of high quality data. Maximum permissible levels of different elements in drinking water have been legislated and accepted by the relevant national and international environmental protection agencies. The application of isotopic and speciation data in the monitoring of groundwater is also extremely important. Therefore, it is necessary to analyse several batches of groundwater samples on routine basis for inorganic pollutants (elemental, isotopic and molecular information) in the monitoring programmes.

3 DIFFERENT ANALYTICAL TECHNIQUES AND RECENT ADVANCES

Several modern analytical techniques have attained capability of measuring most of the elements in the Periodic Table at sub-ng/g levels in the groundwater samples. Some of the more popular analytical techniques that are being applied to the analysis of groundwater samples are spectrophotometry, flame photometry, flame atomic absorption spectrometry (F-AAS), graphite furnace atomic absorption spectrometry (GF-AAS), ion chromatography (IC), X-ray flouresence (XRF), inductively coupled plasma atomic emission spectrometry (ICP-AES), inductively coupled plasma-mass spectrometry (ICP-MS), high resolution inductively coupled plasma-mass spectrometry (HR-ICP-MS) and inductively coupled plasma time-of-flight-mass spectrometry (ICP-TOF-MS). Mass spectrometric techniques offer the capability of precise estimation of isotopic concentrations/ratios in addition to sensitive detection of different elements. XRF which was widely accepted as a technique for the estimation of metals in a variety of solid samples, is now being used for the analysis of groundwater samples using a micro-droplet method. The technique is good for the analysis of Pb, As, Cr, Se and Cd and especially effective for low Z elements such as Na, F and B. Ion chromatography is now a well established technique for the precise estimation of common anions like chloride, nitrate fluoride, etc., in groundwater samples (Jackson, 2001). Though, there have been some spectacular advancements in the field of ICP-MS by the development of HR-ICP-MS and ICP-TOF-MS, the original quadrupole based ICP-MS has revolutionized the inorganic analysis of groundwater samples. Currently, it has been widely accepted as a powerful tool for trace and ultra-trace (< pg/g) elemental as well as isotopic analyses (Balaram, 2002).

4 SPECIATION STUDIES

In light of the difference in the toxicity and physiological behaviour among the chemical forms, the need to determine individual chemical species instead of total element contents is now widely recognized. Once the toxic elements have entered into the environment, their physical and chemical properties, toxicity, mobility and biotransformation are controlled to a large extent by their physicochemical form. Particularly for those in contact with living organisms, it is necessary to determine not only the total content of the element but also to gain an indication of its individual chemical and physical form. The knowledge of speciation is necessary for estimation of biological availability, reactivity and toxicity of the element.

One of the most common approaches adopted by many laboratories for obtaining information on different species of elements has been by coupling the separating powers of chromatographic techniques with the element-specific detection capability offered by atomic and mass spectrometric techniques. The specific advantages of ICP-MS and HR-ICP-MS techniques, such as extremely high sensitivity, limited interference effects, wide element coverage and speed would make them powerful analytical tools in future for speciation analysis of groundwater samples (Menzel *et al.*, 2001).

5 ISOTOPIC STUDIES

Perhaps the most characteristics application of isotopes in environmental monitoring is their use as tracers. Boron isotopes can be used to trace and measure pollution and contamination pathways in the investigation of sewage pollution in groundwater. In addition, numerous radio-isotopes have been added to the earth's ecosystem mainly through nuclear industry. Tracking their transport through different compartments of the global ecosystem provides new insights into the dynamics of the hydrological cycle. The $^{234}U/^{238}U$ activity ratio is extensively used as a geochemical tool to investigate movement and flow relationships in major hydrothermal units, information that is particularly important when considering nuclear waste disposal. Cizdziel *et al.* (2005) have demonstrated that high resolution ICP-MS (HR-ICP-MS) is a fast, sensitive and reliable technique for measuring uranium isotopes in groundwater. Compared to conventional techniques such as thermal ionization mass spectrometry (TIMS), HR-ICP-MS reduces sample size requirements, minimizes the likelihood of contamination and increases productivity making it an attractive alternative for laboratories that analyze large number of groundwater samples for isotopic studies.

6 QUALITY REQUIREMENTS

In groundwater monitoring efforts, accurate and precise results only enable valid conclusions to be drawn about the different types of groundwater and mark the risks related to human consumption and other specific applications such as agriculture, aquaculture, etc. This can be best assured by the implementation of quality assurance (QA) and quality control (QC) measures.

7 CONCLUSIONS

Speed of analysis, sensitivity, the range of determinable elements and the accuracy of the determination are the significant factors for acceptance of an analytical technique for the analysis of groundwater samples in monitoring studies. ICP-MS and HR-ICP-MS have revolutionized the area of inorganic chemical analysis in recent times with extremely important analytical characteristics such as multi-element capability, very low detection limits, small sample quantities required for analysis, element versatility, and speciation and isotopic capabilities. In view of these advantages, these techniques are receiving a great deal of attention at present for elemental, isotopic and speciation analyses of groundwater samples for water resource monitoring programmes around the world.

References

Balaram, V. 2002. Recent trends in the inorganic analysis of environmental materials. *Research Journal of Chemistry and Environment* 6 (2): 69–79.

Cizdziel, J., Farmer, D., Hodge, V., Lindley, K. & Stetzenbach, K. 2005. $^{234}U/^{238}U$ isotope ratios in groundwater from Southern Nevada: a comparison of alpha counting and magenetic sector ICP-MS. *Science of the Total Environment* 350: 248–260.

Jackson, P.E. 2001. Determination of inorganic ions in drinking water by ion chromatography. *Trends in Analytical Chemistry* 20 (6-7): 320–329.

Menzal C.M., Berner Z and Stuben 2001. Coupling size exclusive chromatography and ICP-MS investigate the speciation of platinum group elements in environmental samples. *Geostandards. Newsletter*, 25 (2-3) 239–251.

Study of Medium Strength Wastewater Using Two-Stage Biological Treatment (Biotower Followed by Activated Sludge Process)

48

H.S. Jeswani
Sardar Patel College of Engg., Andheri (W), Mumbai University, Mumbai, India

ABSTRACT

The characteristics of wastewater from the industrial estate of Ankleshwar (South Gujarat) were studied in order to provide an optimal solution for the treatment of wastewater. The wastewater from the industrial estate of Ankleshwar could be classified as medium strength consisting mainly of pharmaceutical and chemical industrial effluents from over 50 units. A relationship between the BOD loading and the treatment efficiency provided in the lab scale biotower was studied. The entire treatment was given keeping the COD (Chemical Oxygen Demand) and the BOD (Biological Oxygen Demand) as the main criteria and wherever it was necessary TSS (Total Suspended Solids) and pH were also analyzed. The trend in the efficiency was observed. The remaining micro suspended solids were settled in the intermediate clarifier after the biotower reactor. An attempt is made in study to reduce the cost of the entire project of common effluent treatment plant using two-stage biological treatment (Biotower followed by activated sludge process) instead of using conventional activated sludge process.

1 INTRODUCTION

Trickling filters were first developed in 1863 AD but became so popular that in 1962 AD there were about 3500 sewage treatment plants in U.S.A. employing trickling filters for biological treatment. The main reason for their popularity was the ease of operation and ability to take shock loads. But with the advent of varities of industrial wastes there was a need to modify the normal trickling filter process. The answer to the requirement was biofiltration towers utilizing plastic media instead of stone media. The main advantages of utilizing biotowers are as follows:

(i) They utilize less space as height could be increased to 40 feet.

(ii) The plastic media reduce the structural requirement.

(iii) The specific surface area of plastic media is very high hence better air transfer is possible and less plugging.

(iv) They take shock loads better as compared to conventional activated sludge systems.

Biotowers have been employed in single-stage, multi-stage or in series with the activated sludge process depending on the requirement of the system to remove organics and dissolved solids. They have been used to treat the wastes of various industries like distilleries, breweries, refineries, pharmaceutical, fertilizer industries with great advantage as partial or entire treatments.

2 BACKGROUND

For the development of any country, industrialization is very essential in addition to agriculture. Industries provide the development and employment to local population. The State Governments are encouraging the industries mainly for further development and providing employment to local population.

The State Governments, according to the need of employment and looking for the availability of raw materials, transport, etc. have established several industrial estates and sectors and in Gujarat they are known as Gujarat Industrial Development Corporation (G.I.D.C.). According to the development of parent city the government decides the type of industrial processes to be carried out in such industrial estates.

Chemical industries (organic or inorganic) have a very high pollution potential and are not permitted near populated areas. Generally, they are part of such industrial estates. In South Gujarat chemical industries are located in Vapi, Ankleshwar and Valsad areas. The chemical industries of Ankleshwar, Panoli and Jhaghadiya G.I.D.C. have majority of chemical industries and each industry discharges its waste into nearby Amlakhadi receiving stream.

Nearly six hundred small and big industries manufacturing dye and dye intermediates, petrochemicals, pharmaceuticals, etc. are present which dispose off their effluents in Amlakhadi.

The wastewater of Amlakhadi has a TDS of 8000 to 10000 mg/L, COD in the range of 1000 to 4100 mg/L and BOD in range of 400 to 1200 mg/L. A study was conducted to see the treatability of the wastewater and to provide economical alternative for the treatment of wastewater.

3 COLLECTION AND ANALYSIS OF THE SAMPLES

Six sets of samples were collected as grab samples at different dates and they were analyzed. The results obtained were as follows:

Table 1 Analysis of effluent collected on different dates from Amlakhadi common.

Sr No.	*Parameter*	*Date of Collection*					
		24.7.02	2.8.02	10.8.02	17.8.02	29.8.02	7.9.02
1.	pH	6.9	7.0	7.1	7.3	6.8	7.3
2.	Colour	Blackish brown	Brownish	Brownish	Brownish	Brownish	Brownish
3.	COD (mg/L)	4722	3657	1363	1890	2463	2216
4.	BOD (mg/L)	1450	1050	360	675	880	740
5.	TSS (mg/L)	714	554	260	370	413	396
6.	TDS (mg/L)	7734	6200	7566	6874	7910	7694
7.	NH_3N (mg/L)	238	282	291	176	196	182
8.	Total P(mg/L)	4.5	5.5	6.0	4.5	5.5	4.5

4 TREATMENTS GIVEN AND RESULTS

All the six sets of collected samples were first given physiochemical treatment, followed by two-stage biological treatment and finally the tertiary treatment.

4.1 Physiochemical Treatment

The pH of the water was more or less neutral but for the working of biotower the best pH range is about 7.8 to 8.5 because at the pH of 7 the sulphur bacteria thrive, producing odours, and at higher pH the microorganisms would die due to alkalinity. Secondly, the sedimentation was done to reduce the solids. So the pH was increased using lime. After addition of lime, proper mixing followed by sedimentation for 30 minutes was done. The initial and final pH, BOD, COD were recorded and the results were as follows:

The reductions in TSS are more or less constant independent of the influent concentration of suspended solids.

Table 2 Properties of Amlakhadi effluent before and after treatment.

Sr No.	*Date*	*Lime Treatment*	*pH*	*COD (mg/L)*	*BOD (mg/L)*	*TSS (mg/L)*	*% TSS Reduction*
1.	24.7.02	Before	7.9	4722	1450	714	63.5
		After	8.5	4036	1275	260	
2.	2.8.02	Before	7.0	3657	1050	554	67.5
		After	7.93	3180	940	180	
3.	10.8.02	Before	7.1	1363	360	260	60.76
		After	7.7	1250	340	102	
4.	17.8.02	Before	7.3	1890	675	370	64.86
		After	7.6	1760	625	130	
5.	29.8.02	Before	6.8	2463	880	413	65.6
		After	7.4	2300	765	172	
6.	7.9.02	Before	7.3	2510	840	390	63.08
		After	7.6	2227	720	144	

4.2 Biological Treatment

The biological treatment given was aimed to reduce the entire cost of the project. As the wastewater was having medium strength and varying characteristics hence anaerobic treatment was non-reliable because the main disadvantage is the sensitiveness of the system towards pH, heavy metals, sulphides, alkalinity. Moreover, start up time is higher and operational control is required. Hence, a two-stage aerobic biological treatment, i.e. a biotower followed by activated sludge process was suggested.

First stage biotower is a better option because:

(i) It reduces shock loads on the activated sludge process.

(ii) Utilizes less space as compared to a trickling filter.

(iii) The use of biotower at first stage reduces the contact period of activated sludge process and hence economizes the entire project.

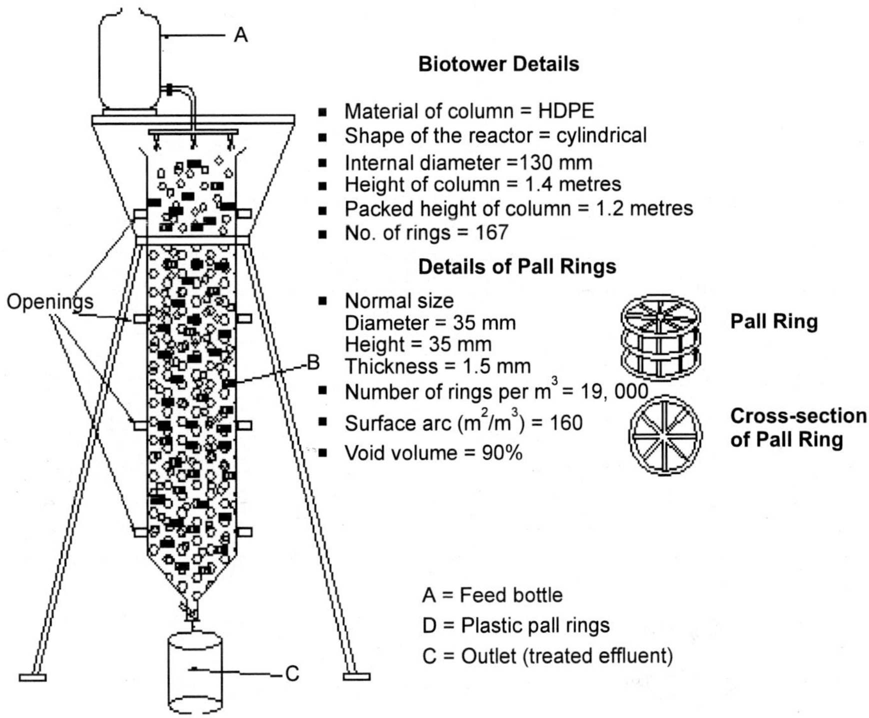

Lab scale biotower.

(i) 1st stage treatment (Biotower)

The influent and the effluent parameters were analyzed for six BOD loadings.

The main parameters analyzed were pH, BOD, COD and TSS. The TSS was analyzed after 30 minutes of effluent settlement. The results obtained for six different loadings, viz. 0.95, 1.15, 1.4, 1.8, 1.92 and 2.6 kg/m^3/d are shown in Table 3.

Table 3 Average reductions in BOD, COD and TSS.

Organic Loading kg/m³/d	*Average COD (mg/l)*		*Average BOD (mg/l)*		*Average TSS (mg/l)*		*Reduction %*		
	Inlet	*Outlet*	*Inlet*	*Outlet*	*Inlet*	*Outlet*	*COD*	*BOD*	*TSS*
0.95	1760	840	625	251	130	62	52.5	59.8	52.02
1.15	2300	1184	765	340	172	90	48.5	55.54	47.7
1.4	3180	2016	940	435	180	110	36.6	53.72	38.66
1.8	4036	2619	1275	644	260	165	34.66	49.51	36.2
1.92	1250	578	340	139.4	102	47	46.4	59.03	53.76
2.6	2227	1614	720	454	144	96	27.6	36.9	33.32

As seen from the table, with an increase in loading there is a decrease in percentage efficiency (Fig. 1), however, for a loading of 1.92 kg/m^3/d a good reduction is obtained which may be the breakeven point. But the influent wastewater during that loading was diluted with rainwater as a result it shows low influent BOD, COD and TSS.

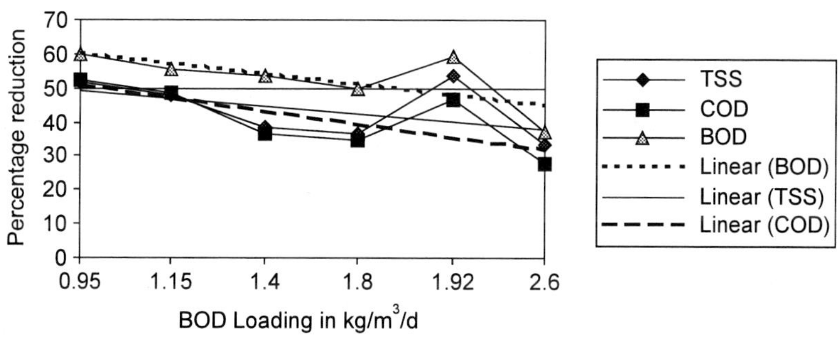

Fig. 1 Reductions at various loadings.

(ii) 2nd Stage Biological Treatment (Activated Sludge Treatment)

The treatment given by the biotower did not reduce the COD and BOD to required standards of disposal and hence a second stage treatment became necessary. Activated sludge process was chosen as the second stage treatment as it could give the required BOD reduction. Moreover, as the 1st stage attached growth treatment is given hence uniform organic loading is possible without shock loads. Initially activated sludge was brought from a full-scale running plant and acclimatized with the effluent of the biotower.

The above acclimatized biomass was used to carry out 2 batch tests on the effluent of biotower. The 2 batches were run for 72 hours by aerating with compressed air and samples were withdrawn after every 24 hours. The parameters analyzed were pH, BOD and COD.

The aim of running the batch was to see the biodegradability of the effluent from the biotower and secondly to analyze the further treatability of the effluent for tertiary treatment. The results of the batches are given in Table 4. The activated sludge process reduced BOD to about 90%, however, the COD was observed to be 390 to 408 mg/L which was not reduced to expected 250 mg/L as per GPCB norms for marine disposal.

In order to reduce COD to required level tertiary treatment using alum and lime was given.

Table 4 Reduction in COD and BOD at 48 hrs in activated sludge process.

Ba. No.	*COD (mg/l)*		*% Reduction*	*BOD (mg/l)*		*% Reduction*
	Inlet	*Outlet*		*Inlet*	*Outlet*	
1.	1075	397	63.06	340	30	91.2
2.	1052	430	59.12	330	40	87.8

*MLSS = 1800, F/M = 0.3

4.3 Tertiary Treatment

(i) Optimization of Alum Dose

The varying dose of alum was added to effluent from activated sludge process and optimize dose was found by jar test for coagulation followed by sedimentation (Table 5 and Fig. 2).

pH = 7.7 mg/L

Initial COD = 398 mg/L

Initial Turbidity = 10 NTU

Initial Colour = 140 Pt Co Scale

Table 5 Optimization of alum dose.

Dose mg/L	*COD mg/L*	*Turbidity NTU*
400	320.8	45
500	264.7	30
600	275.5	33
800	298.7	48
1000	324.6	64
1200	360.9	79

4.3.1 Colour Removal

At the optimum dose of alum the colour was found to be 87.5 on Pt Co scale. Thus, a reduction of 37.5%. From this, it can be concluded that most of the COD was due to colour.

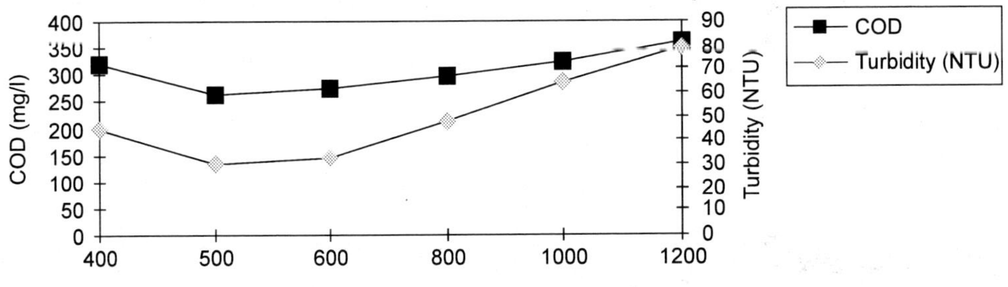

Fig. 2 Optimization of alum dose.

(ii) Optimization of Lime Dose

Once the alum dose was optimized, lime was added in varying amount to wastewater as the reduction by alum alone did not meet GPCB norms. The results obtained were as follows and it was found that effluent now met the GPCB standards of marine disposal (COD < 250).

Table 6 Optimization of lime dose.
Initial pH = 7.7 mg/L
Initial COD = 398 mg/L

Dose mg/L	*pH*	*COD mg/L*	*% Reduction in COD*
50	8.26	250.64	37.02
100	8.71	220.72	44.54
200	9.54	198.6	50.1
500	10.6	220.62	44.56
700	11.28	280.45	29.53

The Fig. 3 shows that optimum dose of lime to be added is 200 mg/L. Thus, the optimum alum and lime dose comes out to be 500 mg/L of alum + 200 mg/L of lime.

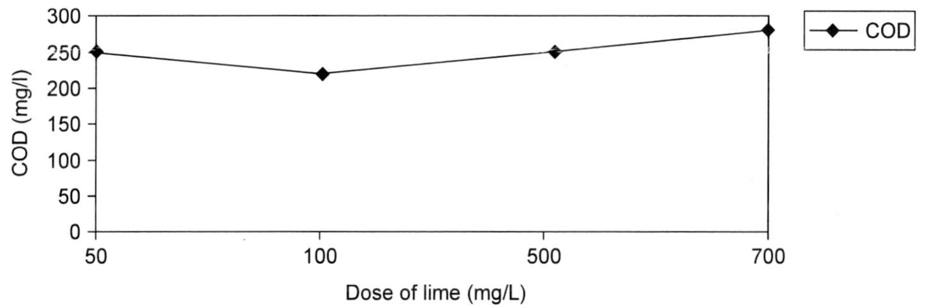

Fig. 3 Optimization of lime dose.

5 SUMMARY TABLE

Raw Wastewater

pH = 6.5 – 7.3
COD = 1360 – 4800
BOD = 360 – 1450 mg/L
Sulphate = 2100 – 3000 mg/L
Chloride = 2700 – 3500 mg/L TSS
= 260 – 750 mg/L
NH_3N = 180 – 290 mg/L
Total P = 4.5 – 6 mg/L
TDS = 6200 – 7910 mg/L

↓

Contd.

Contd.

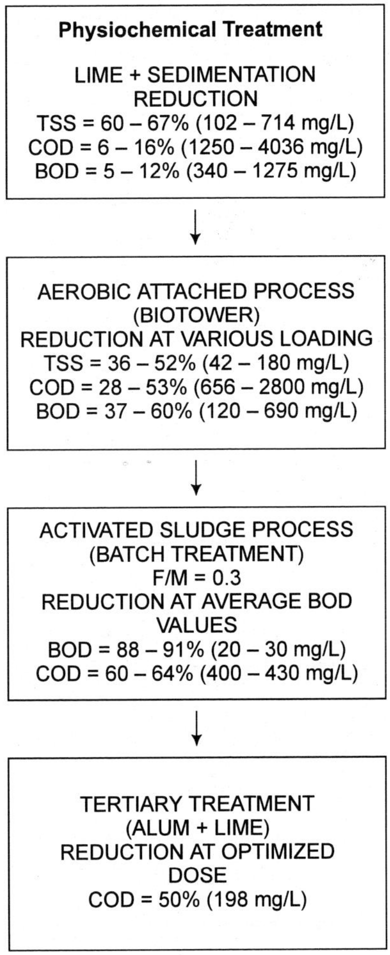

6 CONCLUSIONS

The conclusions drawn from the treatment given on the Amlakhadi Common Wastewater plant are as follows:

(1) Raw Wastewater

- Looking to the characteristics of the wastewater it was found that it was having medium strength which could be treated easily by aerobic treatment.

(2) Suggested Treatment

- The activated sludge treatment was one of the alternatives, but the operation cost of activated sludge is very high, so a biotower with plastic media was used as a 1st stage treatment.

(3) Treatment Given

- Sedimentation performed gave 65% reduction in suspended solids but only 6 – 16% of COD and 5 – 12% of BOD removal. This indicates that most of the compounds which impart COD were in soluble form.

- The biotower reactor was used as a first stage biological treatment to economize the whole treatment. It gave about 45% reduction in COD and about 55% reduction in BOD on an average on various organic loadings.
- The trend in the efficiency observed was that, with an increase in organic loading there was a decrease in COD% and BOD% reduction and this decrease was more or less linear. However, for a loading of 1.92 $kg/m^3/d$ the COD, BOD and TSS reductions obtained were higher than at lower loadings which may be because of higher acclimatization or due to lower strength of BOD during that period.
- The TSS reduction after settlement of the effluent of biotower were very encouraging and were on an average 45%, thus a very important result can be concluded that primary sedimentation can be omitted if proper screening facilities are provided which screen the incoming solids which may clog the nozzles of the distribution system. The remaining micro suspended solids can be settled in the intermediate clarifier provided after biotower reactor.
- Activated sludge process followed by biotower reactor gave about 90% of BOD reduction and about 62% of COD reduction in 48 hours. Thus, it can be concluded that the activated sludge can be used along with the biotower to achieve high removal of organics.
- The tertiary treatment using alum and lime became essential to meet the marine disposal standards for COD. The remaining COD was mainly due to the colour because a colour removal 37% resulted in COD removal of about 33% by addition of optimum dose of alum alone.
- Addition of optimum lime dose to alum dose resulted in 50% of COD removal.

References

Callely A.G, Foster C.F, and Stafford, "Treatment of Industrial Effluent", John Wiley and Sons, New York, Fourth Edition, 1982.

Sastry C.A., Hashim M.A., Agamuthu P, "Waste Treatment Plants", Narosa Publishing House, 1989.

Sarner Erik, "Removal of particulate and dissolved organics in aerobic fixed film biological processes", Journal of Water Poll. Control Federation, Vol. 58, 1986.

Eckenfelder W.W, Jr., "Industrial Water Pollution Control", McGraw-Hill International Editions, 1989.

Metcalf and Eddy, "Wastewater Engineering Treatment, Disposal and Reuse", Tata McGraw-Hill Publishing, New Delhi, 1995.

"Standards Methods for examination of water and wastewater", 17th Edition, APHA, AWWA, 1995.

Solenquist J, Parker Denny S, and Dosh Thomas J.," Carbon Oxidation – Nitrification in Synthetic Media Trickling Filters", Journal of Water Poll. Control, Vol. 46, Oct. 1974.

Shriver Larry E, Bowers Delores M, "Operational practices to upgrade trickling filter plant performance", Journal of Water Poll. Control Federation, November 1975.

Hao Oliver J., Davis Allen P, Phull Kotuku K , "Literature Review", Journal of Water Poll. Control, Vol. 62, June 1990.

Chipperfield P.N.J., Askew M.W and Benton J. H., "Multi Stage, Plastic Media Treatment Plants". Journal of Water Poll. Control Federation, Vol. 44 October 1972.

Richards Tyler, Reinhart Debra, "Evaluation of plastic media in trickling filter", Journal of Water Poll. Control Federation, Vol. 58, July 1986.

Hydrogeochemistry of Fluoride Contamination in Groundwater of Dungarpur District, Rajasthan

49

A.K. Keshari* and S.D. Dhiman**

*Department of Civil Engineering, I.I.T. Delhi, Hauz Khas, New Delhi 110016, India
Email: akeshari@hotmail.com

**Department of Civil Engineering, Birla Vishvakarma Mahavidyalaya (Engineering College), Vallabh Vidyanagar 388 120, India
Email: dhimansanjay18@hotmail.com

ABSTRACT

In semi-arid and arid regions, water scarcity is already a crucial problem. The growing groundwater contamination problem in these regions is making the problem further acute and complicated. In turn, availability of safe drinking water is becoming major challenging task at the state and national level. The present study reveals alarming level of fluoride in many villages of Dungarpaur district located in Rajasthan, India, where groundwater is the main source of the drinking water supply, and thus is at the stake of fluorosis risk. High fluoride water is of concern for drinking water supplies, as fluoride concentration has surpassed the permissible limit of 1.5 mg L^{-1} (Indian Standard) in many wells. Hydrogeochemical investigations are carried for the shallow as well as deep wells of Dungarpur district, Rajasthan. The hydrochemical data were analyzed and saturation states of various minerals are computed using PHREEQC geochemical code to investigate prevailing hydrogeochemistry in the study area. Fluoride concentration ranges from 1-6 mg L^{-1}. It is observed that twenty one percent of the shallow wells have fluoride concentration more than the prescribed limit of 1.5 mg L^{-1} and thirty four percent of the deep wells have fluoride concentration above 1.5 mg L^{-1}. The fluoride concentration is also correlated with the concentration of certain major dissolved constituents. It is observed that fluoride is inversely related to dissolved calcium in the groundwater, and solubility of calcium and fluoride containing minerals control the fluoride concentration in the groundwater. Results obtained in the present study will be useful to water resources managers and planners in chalking out strategic measures to control fluoride contamination in groundwater and to abate the fluorosis problem.

1 INTRODUCTION

Occurrence of fluorine in groundwater has drawn global attention due to its considerable adverse impact on human physiology. The assimilation of fluoride into the human body from potable water at the level of 1 mg L^{-1} enhances bone development and prevents dental caries. The maximum tolerance limit of fluoride in drinking water specified by the World Health Organization (WHO, 1984) is 1.5 mg L^{-1}. The high concentration of fluoride has been reported in many parts of Indian subcontinent and is becoming a serious concern for the drinking water supply (Jacks et al. 2000; Keshari and Dhiman, 2001, Dhiman and Keshari, 2003). Ingestion of water with fluoride concentration above 1.5 mg L^{-1} results in dental fluorosis characterized initially by opaque white patches, staining, mottling and pitting of teeth. Skeletal fluorosis may occur when fluoride concentration in drinking water exceeds 4-8 mg L^{-1}, which leads to increase in bone density, calcification of ligaments, rheumatic or arthritic pain in joints or muscles along with stiffness and rigidity of joints, bending of the vertebral column, etc. The disease may be present in an individual at subclinical, chronic, or acute levels of manifestation (Teotia and Teotia 1988). The health implications have been investigated by many researchers Kumaran et al. (1971), Dwarkanath and Subburam (1991) Gopalakrishanan et al. (1991), Chinoy et al. (1992), Ozha et al. (1993), Maithani et al. (1998), Susheela (1999) and Dhiman and Keshari (2003). Choubisa et al. (1996) carried out a survey in fifteen villages of Dungarpur district in Rajasthan for prevalence of dental and skeletal fluorosis in villagers and domestic animals. Fluoride concentration (mean) in drinking water of these villages varied between 1.7 to 6.1 mg L^{-1} An overall 73 and 82.9 percent prevalence of dental fluorosis was observed in children and adult, respectively. Cent percent prevalence of fluorosis in children and adults was observed at 5.2 and 3.8 mg L^{-1} fluoride respectively. The prevalence of skeletal fluorosis in adults was 32.5 %, and the highest prevalence (60%) was observed at 6.1 mg L^{-1}-fluoride concentration.

High fluoride concentration in drinking water has been reported from many parts of India. Three states viz., Gujarat, Rajasthan and Andhra Pradesh in India are considered to have very high fluoride concentrations in groundwater. The fluoride content in water can be derived from the weathering of fluoride bearing minerals like fluorite, fluoroapatite, cryolite and a number of other silicates, oxides, carbonates, sulfates and phosphates such as topaz, sellaite and villaumite (Fuge, 1988). Minerals such as micas, amphiboles and clays like kaolinite and bentonite contain inorganic fluorides (Reimann and de Caritat, 1998). The primary anthropogenic emission sources of inorganic fluorides are phosphate fertilizer, aluminium, lead and zinc smelting operations, coal burning facilities, oil refining, steel production, copper and nickel production, brick and ceramic manufacturing, fluoride containing pesticides, glass and enamel making, production of glues and adhesives. Over and above this, fluoride in water can be derived from precipitation in form of rain, snow, or dry fall out.

A number of hydrogeochemical studies have been carried out in context of fluoride contamination in the India. Trivedi et al. (1990); Chandrawanshi and Patel (1999); Gupta (1991); Datta et al. (1999); Jacks et al. (2000) and Chadha and Tamta (2001). Keshari and Dhiman

(2001) investigated the causes of fluoride contamination in the Western Indian Aquifer (WIA) systems, to quantify the level of fluoride contamination and to elucidate the hydrogeochemical processes responsible for mobilization of fluoride in the groundwater systems. Results reveal that presence of fluorite mineral and dissolution of fluorite are possibly the causes for fluoride contamination. In this particular study hydrogeochemical investigations are carried for the shallow and deep wells of Dungarpur district, Rajasthan. The saturation state of minerals is investigated using PHREEQC developed by Parkhurst and Appelo (1999).

2 STUDY AREA

Dungarpur district is situated in southern most part of Rajasthan. In East and North its borders on Banswara and Udaipur district respectively while it adjoins the State of Gujarat in South & West. Dungarpur district is the smallest district of the state covering 3770 sqkm. The most of the part of district is hilly. The climate of the district is dry and temperature rises in May with mean daily maximum of 41.50°C and absolute maximum 45°C. The coldest month is January with temperatures falling to around 5°C during night but with day temperatures still well above 20°C. Primary crops sown in the district are maize, rice, jawar, wheat & small grains, secondary crops are various lentils & oil seeds. There is lack of irrigation facilities in the district. The average rainfall of district is 710 mm. The main minerals of district are soapstone, fluorite, serpentine, marble slab, granite and masonry stone. Table 1 gives the geological and other details of the district.

Table 1 Geological and other information on the area.

Latitude	23° 20' N to 24° 01' N
Longitude	73° 22' E to 74° 23'
Total area	3770 sq. km
Mean Annual Rainfall	673.4 mm (1957-92)
Annual Potential Evapotranspiration	1380.6 mm
Total Population	1107037 (2001 census)
Major rivers	Mahi and Som
Soils	Red loam and hilly soils
Major Geology	Bhilwara Supergroup and Aravalli supergroup intruded by granites, Bhilwara Supergroup represented by gneiss, schist and migmatities; Aravalli Supergroup represented by schists, gneiss, quartzite, phyllite, dolomites and migmatities. The Aravalli group is intruded by rocks of Rakhabdev suite comprising of serpentinites and talc carbonate rock. Local shallow alluvium is found near river courses (Jethra, 1998)
Groundwater development as on 1.1.95	28.93%

3 FLUORIDE HYDROGEOCHEMISTRY

Hydrogeochemical investigations are carried for the shallow and deep wells of Dungarpur district, Rajasthan. The hydrochemical data for shallow and deep wells reported by Jethra (1998) are used for the study. Fluoride concentrations range from less than 1 to 6 mg L^{-1} in the groundwaters of Dungarpur district. Twenty one percent of the shallow wells have fluoride concentration more than the prescribed limit of 1.5 mg L^{-1} and thirty four percent of the deep wells have fluoride concentration above 1.5 mg L^{-1}. The fluoride concentration is correlated with the concentration of certain major dissolved constituents namely calcium, magnesium, sodium bicarbonate and alkalinity. The relationships are plotted (Figs. 1-4). Fluoride is generally inversely related to dissolved calcium in the groundwater. The association of fluoride with magnesium is similar. The hyperbolic curve suggests that solubility of calcium and fluoride containing minerals control the concentration (Fig. 1). Fluoride is generally directly related to dissolved sodium and bicarbonate concentration.

The saturation state of various minerals is investigated using PHREEQC geochemical code. The ionic strength and saturation indices of phases for the groundwater from the shallow and deep wells are shown in Tables 2-3, respectively. Results of aqueous speciation modeling reveal that the groundwater from shallow wells is undersaturated with anhydrite, aragonite, gypsum, carbon dioxide and halite and supersaturated with calcite and dolomite moreover the groundwaters are undersaturated to supersaturated with respect to fluorite. For the deep aquifers, the groundwaters are undersaturated with fluorite. They are undersaturated and supersaturated with calcites and dolomites at certain places.

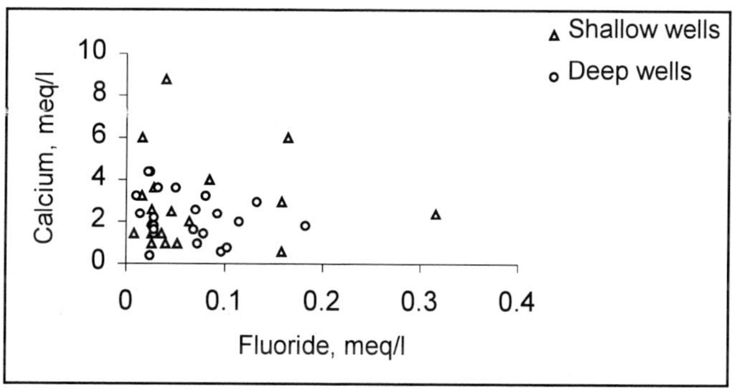

Fig. 1 Relationship between calcium and fluoride

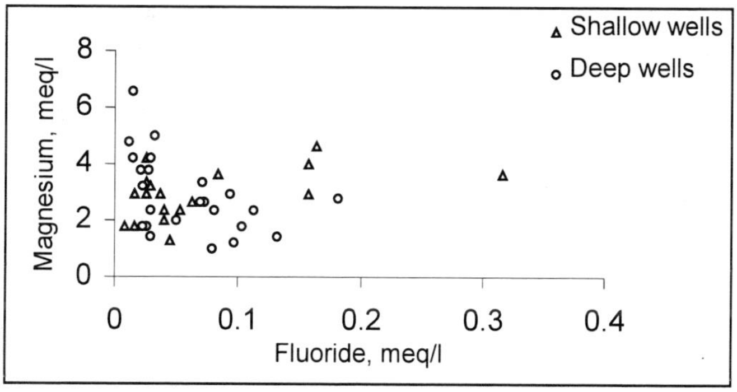

Fig. 2 Relationship between magnesium and fluoride

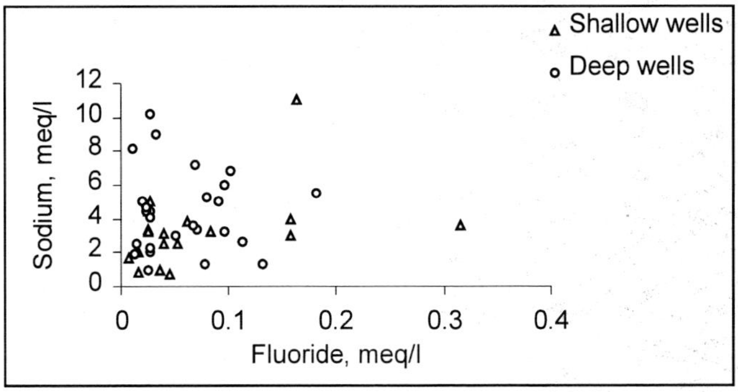

Fig. 3 Relationship between sodium and fluoride

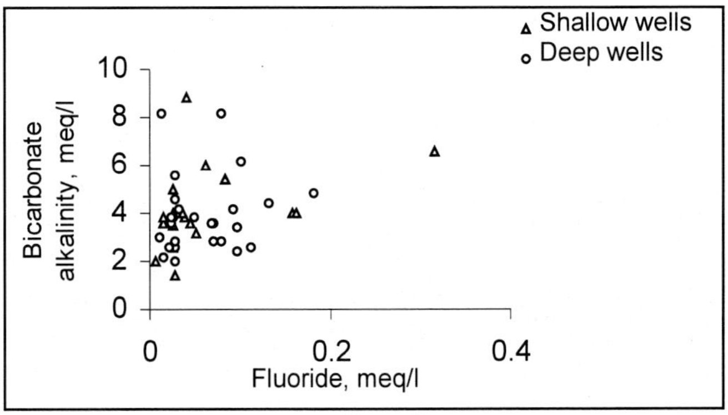

Fig. 4 Relationship between bicarbonate alkalinity and fluoride

The Bhilwara supergroup comprises of gneisses, schists, migmatities and impure marble and Aravalli Supergroup represented by schists, gneiss, quartzite, phyllite, dolomites and migmatities. The Aravalli group is intruded by rocks of Rakhabdev suite comprising of serpentinites and talc carbonate rock (Jethra, 1998). Enrichment of fluoride in groundwater and surface water is usually associated with dissolution of solid phases containing anomalous concentration of fluorite, fluoroapatite. Many micas can also be enriched in this element (e.g. biotite, muscovite, lepidolite) and their alteration during diagenesis may lead to release of fluoride in groundwater. In alkaline waters, the OH^- radical, because of its similar radius to that of F^- (OH^- = 1.32 A°; F^- = 1.36 A°) may exchange for F^- in minerals which allow structural OH^- - F^- substitutions. The source of additional fluoride at higher pH can be by exchange of hydroxyl ions for fluoride ions from weather micas, such as biotite that are present in the aquifer in the northern and eastern part of the district. The general increase in pH with the decrease in dissolved calcium is shown in Fig. 5 which suggests carbonate dissolution in the aquifer.

Table 2 Ionic strength and saturation indices of mineral phases for groundwater from shallow wells

Location	*Ionic strength*	*Anhy -drite*	*Arago -nite*	*Calcite*	*CO_2*	*Dolo -mite*	*Fluorite*	*Gypsum*	*Halite*
Antsari	$7.9x10^{-3}$	−2.98	0.52	0.66	−3.17	1.84	−1.59	−2.76	−7.06
Aspur	$1.69x10^{-2}$	−2.33	0.25	0.39	−2.26	1.05	−0.26	−2.11	−6.16
Badki	$7.9x10^{-3}$	−2.88	0.43	0.58	−2.70	1.61	−1.76	−2.66	−7.77
Baroda	$9.05x10^{-3}$	−3.2	0.1	0.25	−2.76	1.32	−0.87	−2.98	−6.89
Bhilura	$1.59x10^{-2}$	−2	0.74	0.88	−2.39	1.86	−0.69	−1.78	−6.65
Dungarpur	$1.72x10^{-3}$	−1.75	0.24	0.39	−3.13	0.85	−1.68	−1.53	−6.22
Gorada	$8.98x10^{-3}$	−2.24	−0.01	0.13	−1.93	0.14	−2.13	−2.02	−7.78
Hatai	$1.34x10^{-2}$	−2.93	0.67	0.87	−2.35	1.94	0.26	−2.11	−6.92
Jasala	$6.94x10^{-3}$	−2.93	0.7	0.85	−2.8	1.55	−1.28	−2.71	−8.11
Kanaba	$1.88x10^{-2}$	−1.82	0.3	0.44	−2.12	0.72	−1.97	−1.6	−6.78
Karawaba	$1.58x10^{-2}$	−2.15	0.57	0.71	−2.47	1.77	−1.89	−1.93	−6.34
Kua	$8.44x10^{-3}$	−2.98	0.02	0.17	−2.47	0.77	−1.83	−2.76	−6.97
Manpur	$6.97x10^{-3}$	−2.9	0.48	0.63	−3.37	1.51	−3.12	−2.68	−7.12
Nanthod	$1.95x10^{-2}$	−2.51	0.54	0.69	−2.78	1.74	−2.01	−2.29	−6.39
Nayadera	$1.12x10^{-2}$	−2.66	0.36	0.51	−3.05	1.47	−2.08	−2.44	−6.54
Nayagaon	$2.77x10^{-2}$	−1.78	0.95	1.1	−2.8	2.22	−0.02	−1.56	−5.5
Peeth	$2.1x10^{-2}$	−2.93	1.16	1.31	−2.09	2.19	−1.01	−2.71	−6.66
Ratanpur	$1.02x10^{-2}$	−2.96	0.26	0.4	−2.82	1.41	−2.22	−2.74	−6.75
Sabla	$1.17x10^{-2}$	−2.78	0.48	0.62	−2.28	1.51	−1.17	−2.56	−6.85

Table 3 Ionic strength and saturation indices of mineral phases in the groundwater from deep wells.

Location	*Ionic strength*	*Anhy -drite*	*Arago -nite*	*Calcite*	*CO_2*	*Dolo -mite*	*Fluo -rite*	*Gyp -sum*	*Halite*
Antri	$6.83x10^{-3}$	−3.52	0.2	0.35	−3.28	1.14	−1.23	−3.3	−6.74
Badamil	$7.46x10^{-3}$	−3.18	0.63	0.78	−2.84	1.7	−1.95	−2.96	−7.93
Bichwara	$1.77x10^{-2}$	−2.01	−0.27	−0.12	−1.95	−0.18	−1.18	−1.79	−6.04
Biluda	$9.78x10^{-3}$	−2.65	−0.79	−0.65	−1.92	−0.75	−1.33	−2.43	−6.67
Chikholi	$6.3x10^{-3}$	−2.17	−0.46	−0.31	−1.68	−0.17	−2.03	−1.95	−5.76
Chundwara	$2.17x10^{-2}$	−2.15	−0.57	−0.42	−1.69	−0.54	−2.63	−1.93	−5.76
Damri	$1.17x10^{-2}$	−2.86	−0.08	0.06	−2.08	0.61	−1.14	−2.64	−6.47
Devalkhas	$2x10^{-2}$	−2.51	−0.24	−0.1	−1.27	0.38	−2.55	−2.29	−7.05
Dungarpur-HB	$1.53x10^{-2}$	−2.51	0.27	0.41	−2.11	1.24	−1.88	−2.29	−6.46
Dungarpur-PHED	$1.7x10^{-2}$	−1.95	0.73	0.87	−2.7	1.75	−1.76	−1.73	−6.25
Dungarpur	$1.14x10^{-2}$	−2.65	−0.77	−0.63	−2.04	−0.95	−1.94	−2.43	−6.4
Galiakot	$1.4x10^{-2}$	−2.36	0.01	0.16	−2.62	0.69	−2.4	−2.14	−6.63
Gamra	$9.4x10^{-3}$	−3.11	−0.86	−0.72	−1.89	−0.99	−1.26	−2.89	−6.31
Jhantri	$1.28x10^{-2}$	−2.43	0.15	0.29	−2.08	0.46	−1.1	−2.21	−6.64
Karoda	$1.89x10^{-2}$	−2.16	0.02	0.17	−1.4	0.34	−0.83	−1.94	−6.29
Karwara	$8.7x10^{-3}$	−3.07	−1.17	−1.03	−1.66	−1.27	−2.69	−2.85	−6.65
Mandav	$1.37x10^{-2}$	−2.17	0.08	0.23	−2.14	0.78	−0.33	−1.95	−6.5

Table 3 contd.

Table 3 contd.

Munger	1.74×10^{-2}	–1.98	0.22	0.37	–2.41	0.98	–1.03	–1.76	–6.01
Naval Shyam	1.74×10^{-2}	–3.17	0.41	0.56	–3.04	1.25	–1.93	–2.95	–7.1
Peeth	2.25×10^{-2}	–2.12	0.06	0.2	–1.99	0.68	–1.61	–1.9	–5.69
Punali	9.17×10^{-3}	–3.05	0.39	0.53	–2.24	0.87	–0.3	–2.83	–7.67
Ramgarh	1.18×10^{-2}	–2.52	–0.04	0.1	–2.46	0.41	–0.65	–2.37	–6.88
Rasta	7.7×10^{-3}	–3.11	–0.55	–0.4	–1.92	–0.73	–1.88	–2.89	–7.05
Riche	5.84×10^{-3}	–3.06	0.46	0.6	–3	1.19	–1.03	–2.84	–7.77
Sarthuna	1.48×10^{-2}	–2.61	0.4	0.54	–2.45	1.31	–0.79	–2.39	–6.31
Thakarda	1.14×10^{-2}	–2.52	–0.36	–0.21	–1.95	–0.08	–1.19	–2.3	–6.67

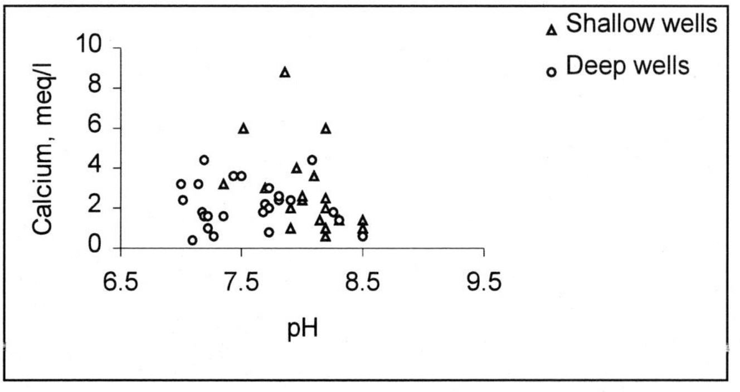

Fig. 5 Association of calcium concentration and pH

4 CONCLUSIONS

High fluoride concentrations are a cause of concern for drinking water supplies in the Dungarpur district. Twenty one percent of the shallow wells have fluoride concentration more than the prescribed limit of 1.5 mg L^{-1} and thirty four percent of the deep wells have fluoride concentration above 1.5 mg L^{-1}. Fluoride concentrations are correlated with calcium, magnesium, and sodium and bicarbonate alkalinity. Aqueous speciation modeling is carried out for the hydrochemical data for the shallow and deep wells in Dungarpur district using PHREEQC. The groundwaters are undersaturated to supersaturated with fluorite in shallow wells and undersaturated in the deep wells.

References

Chadha, D.K., and Tamta, S.R. (2001). Hydrogeological aspects of high fluoride content groundwater in India. *Proc. International Workshop on Fluoride in drinking water: Strategies, Management and Mitigation*, Bhopal, India, 184-195.

Chandrawanshi, C.K., and Patel, K.S. (1999). Fluoride deposition in Central India. *Environmental Monitoring and Assessment*, 55(2): 251-265.

Chinoy, N.J., Narayana, M.V., Sequeria, E., Joshi, S.M., Barot, J.M., Purohit, R.M., Parikh, D.J., and Ghodasara, N.B. (1992). Studies on effects of fluoride in 36 Villages of Mehsana District, North Gujarat Fluoride, 25(3): 101-110.

Choubisa, S.L., Sompura, K., Bhatt, S.K., Choubisa, D. K., Pandya, H., Joshi S.C., Choubisa, L. (1996). Prevalence of fluorosis in some villages of Dungarpur District of Rajasthan. *Indian J. Environmental Health*, 38(2): 119-126.

Datta, P.S., Tyagi, S.K., Mookerjee, P., Bhattacharya, S.K., Gupta, Navindu, and Bhatnagar, P.D. (1999). Groundwater NO_3 and F contamination processes in Puskhar valley, Rajasthan as reflected from ^{18}O isotopic signature and 3H recharge studies. *Environmental Monitoring and Assessment*, 56: 209-219.

Dhiman, S.D. and Keshari, A.K. (2003) Quantifying uncertainties using fuzzy logic for groundwater driven contaminant exposure assessment, In: Groundwater Quality Modeling and Management under Uncertainty (Ed. Srikanta Mishra) "Proceedings of World Water and Environmental Resources Congress 2003, June 23-26, 2003, Philadelphia, USA, EWRI-ASCE Publication 2003, pp 236-247

Dwarkanath, M., and Subburam, V. (1991) Incidence of fluorosis in a small village population, India. *Indian J. Environmental Health*, 33(2): 182-186.

Fuge, R. (1988). Sources of halogens in the environment, influences on human and animal health. *Environ. Geochem. Health*, 10(20): 51-61.

Gopalakrishanan, S., Narasimhan, S., and Seenipandian, C. (1991). Incidence of fluorosis in 10 villages near Tuticorin: Suggestions for prevention and control, *Indian J. of Environmental Health*, 11(2): 118-123.

Gupta, S.C. (1991). Chemical character of groundwater in Nagaur district, Rajasthan. *Indian J. Environmental Health*, 33(3): 341-349.

Jacks, G., Bhattacharya, P., and Singh, K.P. (2000). High fluoride groundwaters in India, In: *Proc. International Conf. on Groundwater Research*, Copenhagen, Denmark, 193-194.

Jethra, M.S. (1998) Hydrogeology and groundwater resources of Dungarpur district, Rajasthan, Report, CGWB, GOI.

Keshari, A.K. and Dhiman, S.D. (2001) Genesis of fluoride contamination in the western Indian aquifers. 3rd International Conf. on *Future Groundwater Resources at Risk*, Lisbon, Portugal, 25-27 June 2001. Theme 3.

Kumaran, P., Bhargava, G.N., and Bhakuni, T.S. (1971). Fluoride in groundwater and endemic fluorosis in Rajasthan, *Indian J. Environmental Health*, 13(4): 316-324.

Maithani, P.B., Gurjar, R., Banerjee, R., Balaji, B.K., Ramachandran, S., and Singh, R. (1998). Anomalous fluoride in groundwater from western part of Sirohi district, Rajasthan and its crippling effects on human health. *Current Science*, 74(9): 773-777.

Ozha, D.D, Bohra, J.L, Jain, P.C. (1993). Occurrence of excessive fluoride in ground waters of Western Rajasthan. *Current-Agriculture*, 17(1/2): 75-78.

Parkhurst, D.L., and Appelo, C.A.J. (1999). User's guide to PHREEQC (Version 2) - A computer program for speciation, batch reaction, one dimensional transport and inverse geochemical calculations: USGS/WRI-99-4259, U.S. Geol. Surv., Denver, Colo.

Riemann, C. and P.de Caritat (1998). Chemical Elements in the Environment. Fact sheets for the Geochemist and Environmental Scientist. Springer Verlag, New York, N.Y.

Susheela, A. K. (1999). Fluorosis management programme in India. *Current Science*, 77(10): 1250-1256.

Teotia, S.P.S., and Teotia, M (1988). Endemic skeletal fluorosis: clinical and radiological variants, *Fluoride*, 21: 39-44.

Trivedi, A.R., Ghevaria, Z.G., Parikh., and Jain, N.M. (1990). Incidence of high fluoride in groundwater in parts of Sabarkanta district, Gujarat, Western India, Hydrology Journal of IAH, 13(1): 35-43.

WHO (1984). Guidelines for drinking water quality, Vol 2, Health criteria and other supporting information, World Health Organization, Geneva.

Author Index

Subject Index